Lecture Notes in Computer Science

Lecture Notes in Artificial Intelligence   **15710**

Founding Editor

Jörg Siekmann

Series Editors

Randy Goebel, *University of Alberta, Edmonton, Canada*
Wolfgang Wahlster, *DFKI, Berlin, Germany*
Zhi-Hua Zhou, *Nanjing University, Nanjing, China*

The series Lecture Notes in Artificial Intelligence (LNAI) was established in 1988 as a topical subseries of LNCS devoted to artificial intelligence.

The series publishes state-of-the-art research results at a high level. As with the LNCS mother series, the mission of the series is to serve the international R & D community by providing an invaluable service, mainly focused on the publication of conference and workshop proceedings and postproceedings.

Qinghua Zhang · Christopher Henry ·
Richard Jensen · Xinbo Gao · Guoyin Wang ·
JingTao Yao · Chris Cornelis · Shuyin Xia
Editors

Rough Sets

International Joint Conference, IJCRS 2025
Chongqing, China, May 11–13, 2025
Proceedings, Part III

Springer

Editors
Qinghua Zhang (iD)
Chongqing University of Posts
and Telecommunications
Chongqing, China

Richard Jensen (iD)
Aberystwyth University
Aberystwyth, UK

Guoyin Wang (iD)
Chongqing Normal University
Chongqing, China

Chris Cornelis (iD)
Ghent University
Ghent, Belgium

Christopher Henry (iD)
University of Manitoba
Winnipeg, MB, Canada

Xinbo Gao (iD)
Chongqing University of Posts
and Telecommunications
Chongqing, China

JingTao Yao (iD)
University of Regina
Regina, SK, Canada

Shuyin Xia (iD)
Chongqing University of Posts
and Telecommunications
Chongqing, China

ISSN 0302-9743   ISSN 1611-3349 (electronic)
Lecture Notes in Artificial Intelligence
ISBN 978-3-031-92740-9   ISBN 978-3-031-92741-6 (eBook)
https://doi.org/10.1007/978-3-031-92741-6

LNCS Sublibrary: SL7 – Artificial Intelligence

Preface

This volume contains the papers presented at IJCRS 2025, the 11th International Joint Conference on Rough Sets. As a premier global forum for research on rough set theory and its diverse applications, IJCRS is dedicated to advancing the field by providing researchers and practitioners with a platform to exchange ideas, showcase cutting-edge research, and explore both theoretical developments and real-world implementations.

IJCRS 2025 was organized by the International Rough Set Society (IRSS) and co-hosted by Chongqing University of Posts and Telecommunications, Chongqing Normal University, and the Granular Computing and Knowledge Discovery Technical Committee of the Chinese Association for Artificial Intelligence (CAAI). The conference took place from May 11 to 13, 2025, in Chongqing, China. The origins of IJCRS trace back to the Joint Rough Set Symposium (JRS), which was first held in Toronto, Canada, in 2007. Since then, the conference has been hosted in various locations worldwide, including Chengdu, China (2012); Halifax, Canada (2013); Granada and Madrid, Spain (2014); Tianjin, China (2015, where the term "IJCRS" was first introduced); Santiago, Chile (2016); Olsztyn, Poland (2017); Quy Nhon, Vietnam (2018); Debrecen, Hungary (2019); Havana, Cuba (2020, held virtually); Bratislava, Slovakia (2021, hybrid); Suzhou, China (2022, hybrid); Kraków, Poland (2023, hybrid); and Halifax, Canada (2024, hosted by Saint Mary's University).

Continuing this tradition of excellence, IJCRS 2025 aimed to foster interdisciplinary collaboration and drive innovation in rough set research. We welcomed scholars from around the world to Chongqing to collectively advance the theory and applications of rough sets. The conference covered a wide range of topics, including fundamental research in rough set theory, its integration with other computational intelligence methods, and practical applications across various domains. The program featured keynote speeches by internationally renowned experts, technical paper presentations, workshops, and special sessions. This year, we received 187 full-paper submissions, all of which underwent a rigorous single-blind review process. Each paper was evaluated by at least three experts in the field, with some authors required to revise and resubmit their work for further assessment before a final decision was made. Ultimately, 90 high-quality papers were accepted as official conference proceedings. Additionally, 12 extended abstracts were included to highlight ongoing research and recently published findings. The success of the conference was made possible by the invaluable contributions of all authors, reviewers, and program committee members.

To further enrich the academic discussions, IJCRS 2025 held two workshops and three special sessions. The workshops focused on Three-Way Decision in Advancing Machine and Human Intelligence and Granular-Ball Computing, while the special sessions covered Soft Clustering: Applications and Approaches, Exploring the Intersection of Soft Computing and Generative Methods, and Mathematical Foundations of General Rough Sets and Their Application to AI and Machine Learning. In addition, the 2025 Data Mining Competition was held as part of the conference, inviting participants to design efficient feature selection and classification algorithms based on rough set theory. The competition aimed to explore novel approaches and techniques for large-scale data analysis. We extend our gratitude to the competition reviewers and all participants for their enthusiastic involvement.

The successful organization of IJCRS2025 would not have been possible without the support of numerous individuals and institutions. This conference and publication were supported in part by the National Natural Science Foundations of China under Grant 62221005 and 62276038. We extend our deepest gratitude to all authors, reviewers, session organizers, and program committee members for their dedication and hard work, which ensured the high academic quality of the conference. We also sincerely appreciate the generous support from our hosting and sponsoring institutions, which enabled the smooth execution of this event.

We were pleased to invite researchers from around the world to join us in Chongqing, a city rich in culture and innovation, to discuss the latest advances in rough set theory, share research insights, and explore opportunities for future collaboration. We hope you enjoyed IJCRS 2025!

May 2025

Qinghua Zhang
Christopher Henry
Richard Jensen
Xinbo Gao
Guoyin Wang
JingTao Yao
Chris Cornelis
Shuyin Xia

Organization

Honorary Chairs

Jiye Liang   Shanxi University, China
Andrzej Skowron   Systems Research Institute of the Polish Academy of Sciences, Poland
Yiyu Yao   University of Regina, Canada

Conference Chairs

Xinbo Gao   Chongqing University of Posts and Telecommunications, China
Guoyin Wang   Chongqing Normal University, China
JingTao Yao   University of Regina, Canada
Chris Cornelis   Ghent University, Belgium

Program Committee Chairs

Qinghua Zhang   Chongqing University of Posts and Telecommunications, China
Christopher Henry   University of Manitoba, Canada
Richard Jensen   Aberystwyth University, UK

Local Organizing Chairs

Weisheng Li   Chongqing University of Posts and Telecommunications, China
Shuyin Xia   Chongqing University of Posts and Telecommunications, China
Tao Jia   Chongqing Normal University, China
Weihua Xu   Southwest University, China
Yabin Shao   Chongqing University of Posts and Telecommunications, China

Local Organizing Committee

Sen Zhao	Chongqing University of Posts and Telecommunications, China
Qin Xie	Chongqing University of Posts and Telecommunications, China
Dawei Dai	Chongqing University of Posts and Telecommunications, China
Binbin Sang	Chongqing Normal University, China
Xu Zhang	Chongqing University of Posts and Telecommunications, China
Fan Zhao	Chongqing University of Posts and Telecommunications, China
Xiaoyu Lian	Chongqing University of Posts and Telecommunications, China

Steering Committee Chairs

Duoqian Miao	Tongji University, China
Witold Pedrycz	University of Alberta, Canada
Roman Słowiński	Poznań University of Technology, Poland

Data Mining Competition Chairs

Dominik Slezak	University of Warsaw, Poland
Sen Zhao	Chongqing University of Posts and Telecommunications, China
Xiaoyu Lian	Chongqing University of Posts and Telecommunications, China

Rough Set School and Tutorial Chairs

Piotr Artiemjew	University of Warmia and Mazury in Olsztyn, Poland
Chengyuan Chen	Chongqing University of Science and Technology, China
Qinghua Hu	Tianjin University, China
Tianrui Li	Southwest Jiaotong University, China

Neil Mac Parthaláin   Aberystwyth University, UK
Jiang Xie   Chongqing University of Posts and
                                   Telecommunications, China

Special Session Chairs

Nouman Azam   National University of Computer and Emerging
                                   Sciences, Pakistan
Soma Dutta   University of Warmia and Mazury in Olsztyn,
                                   Poland
Dun Liu   Southwest Jiaotong University, China
Weizhi Wu   Zhejiang Ocean University, China
Ling Wei   Northwest University, China
Hongxuan He   Chongqing University of Posts and
                                   Telecommunications, China

Workshop Chairs

Mengjun Hu   Saint Mary's University, Canada
Hong Yu   Chongqing University of Posts and
                                   Telecommunications, China
Jie Yang   Zunyi Normal University, China
Guan Wang   Chongqing University of Posts and
                                   Telecommunications, China

Publicity Chairs

Stefania Boffa   University of Milano-Bicocca, Italy
Andrzej Janusz   Queensland University of Technology, Australia
Jusheng Mi   Hebei Normal University, China
Yanhong She   Xi'an Shiyou University, China
Shu Zhao   Anhui University, China
Fan Yang   Shandong University, China

Publication Chairs

Yuhua Qian	Shanxi University, China
Marcin Szczuka	University of Warsaw, Poland
Xiaodong Yue	Shanghai University, China
Jianhang Yu	Chongqing University of Posts and Telecommunications, China

Web Chairs

Chengxin Hong	Chongqing University of Posts and Telecommunications, China
Yan Yang	Chongqing University of Posts and Telecommunications, China
Ying Yang	Chongqing University of Posts and Telecommunications, China

Treasurers

Xingzhen Wang	Chongqing University of Posts and Telecommunications, China
Junhong Zhao	Chongqing University of Posts and Telecommunications, China
Hui Huang	Chongqing University of Posts and Telecommunications, China
Degang Chen	Chongqing University of Posts and Telecommunications, China

Financial Committee

Jinhai Li	Kunming University of Science and Technology, China
Mingwen Shao	China University of PetroleumEast China), China
Xiaoxia Zhang	Chongqing University of Posts and Telecommunications, China
Xinran Zhou	Chongqing University of Posts and Telecommunications, China

Registration Chairs

Chunlin Li	Chongqing University of Posts and Telecommunications, China
Ke Xu	Chongqing University of Posts and Telecommunications, China

Registration Committee

Man Gao	Chongqing University of Posts and Telecommunications, China
Nanfang Luo	Chongqing University of Posts and Telecommunications, China
Yutai Wang	Chongqing University of Posts and Telecommunications, China
Yifan Wang	Chongqing University of Posts and Telecommunications, China
Fan Chen	Chongqing University of Posts and Telecommunications, China
Jinyuan Ni	Chongqing University of Posts and Telecommunications, China
Meng Yang	Chongqing University of Posts and Telecommunications, China
Jiancu Chen	Chongqing University of Posts and Telecommunications, China

Program Committee Members

A. Mani	Indian Statistical Institute, Kolkata, India
Lubomir Antoni	Jozef Šafárik University in Košice, Slovakia
Roberto G. Aragón	University of Cádiz, Spain
Piotr Artiemjew	University of Warmia and Mazury in Olsztyn, Poland
Nouman Azam	National University of Computer and Emerging Sciences, Pakistan
Jaume Baixeries	Universitat Politècnica de Catalunya, Spain
María José Benítez Caballero	University of Cádiz, Spain
Stefania Boffa	Università degli Studi di Milano-Bicocca, Italy
Henri Bollaert	Ghent University, Belgium
Nizar Bouguila	Concordia University, Canada
Andrea Campagner	Università degli Studi di Milano-Bicocca, Italy

Mihir Chakraborty	Jadavpur University, India
Shradha Chavan	Symbiosis International Deemed University, India
Yingxiao Chen	Beijing University of Posts and Telecommunications, China
Yumin Chen	Xiamen University of Technology, China
Zehua Chen	Taiyuan University of Technology, China
Mu-Chen Chen	National Yang Ming Chiao Tung University, Taiwan
Davide Ciucci	Università degli Studi di Milano-Bicocca, Italy
M. Eugenia Cornejo	University of Cádiz, Spain
Chris Cornelis	Ghent University, Belgium
Zoltán Ernő Csajbók	University of Debrecen, Hungary
Jianhua Dai	Hunan Normal University, China
Dayong Deng	Zhejiang Normal University, China
Tingquan Deng	Harbin Engineering University, China
Murat Diker	Hacettepe University, Turkey
Pawel Drozda	University of Warmia and Mazury in Olsztyn, Poland
Soma Dutta	University of Warmia and Mazury in Olsztyn, Poland
Zied Elouedi	Tunis Higher Institute of Management, Tunisia
Luis Farinas	Centre National de la Recherche Scientifique, France
Andrew Fisher	Saint Mary's University, Canada
Hamido Fujita	Iwate Prefectural University, Japan
Can Gao	Shenzhen University, China
Anna Gomolinska	University of Białystok, Poland
Rafal Gruszczynski	Nicolaus Copernicus University, Poland
Shenming Gu	Zhejiang Ocean University, China
Quang-Thuy Ha	VNU-University of Engineering and Technology, Vietnam
Christopher Hinde	Loughborough University, UK
Mengjun Hu	Saint Mary's University, Canada
Feng Hu	Chongqing University of Posts and Telecommunications, China
Qinghua Hu	Tianjin University, China
Ryszard Janicki	McMaster University, Canada
Jouni Jarvinen	University of Turku, Finland
Richard Jensen	Aberystwyth University, UK
Xiuyi Jia	Nanjing University of Science and Technology, China
Chunmao Jiang	Fujian University of Technology, China
Jinyuan Shi	Tianjin University, China

Raavee Kadam	Saint Mary's University, Canada
Ting Ke	Tianjin University of Science and Technology, China
Michal Kepski	University of Rzeszow, Poland
Md. Aquil Khan	Indian Institute of Technology Indore, India
Marzena Kryszkiewicz	Warsaw University of Technology, Poland
Yasuo Kudo	Muroran Institute of Technology, Japan
Sergei O. Kuznetsov	National Research University Higher School of Economics, Russia
Tamás Kádek	University of Debrecen, Hungary
Guangming Lang	Changsha University of Science and Technology, China
Dajiang Lei	Chongqing University, China
Oliver Urs Lenz	Leiden University, Netherlands
Tianrui Li	Southwest Jiaotong University, China
Min Li	Tianjin University of Science and Technology, China
Jinhai Li	Kunming University of Science and Technology, China
Huaxiong Li	Nanjing University, China
Jiye Liang	Shanxi University, China
Churn-Jung Liau	Academia Sinica, Taiwan
Pawan Lingras	Saint Mary's University, Canada
Caihui Liu	Gannan Normal University, China
Guilong Liu	Beijing Language and Culture University, China
Sujuan Liu	Tianjin University of Science and Technology, China
Jiangtao Liu	Tianjin University of Science and Technology, China
Junfang Luo	Southwestern University of Finance and Economics, China
Domingo López-Rodríguez	University of Málaga, Spain
Vijay Mago	York University, Canada
Jesús Medina	University of Cádiz, Spain
Ernestina Menasalvas	Universidad Politécnica de Madrid, Spain
Claudio Meneses	Universidad Católica del Norte, Chile
Duoqian Miao	Tongji University, China
Marcin Michalak	Silesian University of Technology, Poland
Tamás Mihálydeák	University of Debrecen, Hungary
Sonajharia Minz	Jawaharlal Nehru University, India
Mikhail Moshkov	King Abdullah University of Science and Technology, Saudi Arabia
Michinori Nakata	Josai International University, Japan

Bala Krushna Tripathy	Vellore Institute of Technology, India
Li-Shiang Tsay	North Carolina Agricultural and Technical State University, USA
Bay Vo	Ho Chi Minh City University of Technology, Vietnam
Zhen Wang	Weihai Institute of Beijing Jiaotong University, China
Ye Wang	Chongqing University of Posts and Telecommunications, China
Guoyin Wang	Chongqing University of Posts and Telecommunications, China
Wei Wei	Shanxi University, China
Jun Xie	Taiyuan University of Technology, China
Jianfeng Xu	Nanchang University, China
Xin Yang	Southwestern University of Finance and Economics, China
Hailong Yang	Shaanxi Normal University, China
Jilin Yang	Sichuan Normal University, China
Tian Yang	Hunan Normal University, China
Yiyu Yao	University of Regina, Canada
Ning Yao	Tongji University, China
Minda Yao	China University of Mining and Technology, China
JingTao Yao	University of Regina, Canada
Jianhang Yu	Chongqing University of Posts and Telecommunications, China
Hong Yu	Chongqing University of Posts and Telecommunications, China
Ying Yu	East China Jiaotong University, China
Xiaodong Yue	Shanghai University, China
Yanhui Zhai	Shanxi University, China
Jianming Zhan	Hubei Institute for Nationalities, China
Chuanlei Zhang	Tianjin University of Science and Technology, China
Yuanjian Zhang	Shanghai University, China
Yan Zhang	California State University San Bernardino, USA
Qinghua Zhang	Chongqing University of Posts and Telecommunications, China
Li Zhang	Soochow University, China
Hongyun Zhang	Tongji University, China
Shu Zhao	Anhui University, China
Huilai Zhi	Shanghai University, China
Bing Zhou	Sam Houston State University, USA

Jie Zhou   Shenzhen University, China
Wojciech Ziarko   University of Regina, Canada
Beata Zielosko   University of Silesia in Katowice, Poland

Contents – Part III

Three-Way Data Analytics and Decision

Three-Way Data Analytics and Decision

3WOS: Finding the Pillars of Strength in Three-Way Oversampling with Density Clustering for Imbalanced Data Synthesis

Yu Fang[1,2] , Tianrui Li[1(✉)] , and Fan Min[2]

[1] School of Computing and Artificial Intelligence, Southwest Jiaotong University,
Chengdu 611756, China
yufang@my.swjtu.edu.cn, trli@swjtu.edu.cn
[2] School of Computer Science and Software Engineering, Southwest Petroleum
University, Chengdu 610500, China
minfan@swpu.edu.cn

Abstract. In tackling the issues of overfitting and limited generalization in machine learning models, especially with imbalanced and limited data, oversampling methods have been widely used to enhance model discrimination by generating synthetic minority samples. However, these methods often struggle to capture the expressive nuances of the minority class due to noisy interference. In response to this challenge, we propose a three-way oversampling (3WOS) method that partitions data into three regions via density clustering, then synthesizes samples in purified regions to avoid noise. The 3WOS operates in two stages: first, a three-way density clustering (3WDC) process uses density ratios to divide data and identify the FringeRegion (Fr) as a source of benchmark representative samples, aiding in similar data recognition and noise mitigation. Second, a data synthesis strategy creates a purified region from the representative samples, where the FringeRegion provides meaningful, discriminative data for synthesis through linear interpolation. This approach avoids data redundancy and overlap, ensuring a more robust minority representation. Extensive experiments validate 3WOS effectiveness, demonstrating significant improvements in model performance, particularly in enhancing minority class representation.

Keywords: Imbalanced data · Three-way decision · Density-based clustering · Oversampling

1 Introduction

Imbalanced datasets are characterized by substantial differences in sample quantities across distinct classes. Typically, the majority class dominates, while the

This research was supported by the National Natural Science Foundation of China (Nos. 62176221, 61572407); Sichuan Science and Technology Program (Nos. 2024NSFTD0036, 2024ZHCG0166); The Nanchong Municipal Government Universities Scientific Cooperation Project of China (Nos. 23XNSYSX0062, 23XNSYSX0084).

Q. Zhang et al. (Eds.): IJCRS 2025, LNAI 15710, pp. 3–18, 2025.
https://doi.org/10.1007/978-3-031-92741-6_1

minority class represents a smaller fraction. This poses a ubiquitous challenge across diverse domains, spanning disease diagnosis [9], medical image interpretation [19], multi-class text classification [7], and financial forecasting [11]. The overarching objective in these contexts is to glean valuable insights from datasets characterized by significant class imbalances. Conventional machine learning methods may struggle to accurately capture the distributional features of imbalanced datasets, consequently leading to diminished classification performance. Particularly in real-world scenarios, where minority classes often signify critical contexts, inaccurate predictions for these classes can result in severe losses and potentially hazardous misinterpretations [17]. Consequently, the effective resolution of imbalanced data scenarios stands as a pivotal concern within the realm of data process.

In recent years, there has been a growing interest in investigating the fundamental origins of imbalanced data synthesis challenges within the research community. These studies have highlighted that the intricacies of synthesis extend beyond imbalance in data quantities among classes, encompassing various latent factors including: class overlap, noise inference, and limited expressiveness of minority class. To address those issues, three commonly used categories of methods exist: *conventional data sampling, cost-sensitive learning and ensemble learning.*

The *conventional data sampling* approaches aim to mitigate class distribution imbalances by employing various techniques such as oversampling, undersampling, or their combination. Currently, the Synthetic Minority Over-sampling Technique (SMOTE) stands as the prevailing oversampling approach. Despite SMOTE's demonstrated success in mitigating certain imbalanced learning challenges, it frequently carries the risk of producing noisy minority class data due to the influence of parameter k [13]. In contrast to conventional oversampling techniques, the Generative Adversarial Network (GAN) based approaches broaden the scope of synthetic data generation [3]. However, in practical scenarios, GANs often suffer from mode collapse and instability in low-data regimes, limiting their efficacy for minority-class synthesis.

The *cost-sensitive learning* methods involve leveraging the overall misclassification cost rather than overall accuracy to steer the learning process [5], where the expense of misclassifying minority classes outweighs that of majority classes, thereby incentivizing classifiers to prioritize minority classes to minimize the overall misclassification cost [16]. Despite its potential benefits, cost-sensitive learning methods are less prevalent than resampling techniques due to their inability to directly discern misclassification costs from data, posing challenges in setting such costs effectively.

The *ensemble learning* integrates and expands ensemble techniques including various resampling strategies with bagging or boosting [18]. Alternatively, enhancements are made to training mechanisms or prediction rules, including the modification of loss functions, incorporation of class cost dependencies, or adjustment of decision thresholds, aimed at refining standard classification methods to prioritize learning from minority class. However, they often fail to address the problem itself, specifically the data distribution issue.

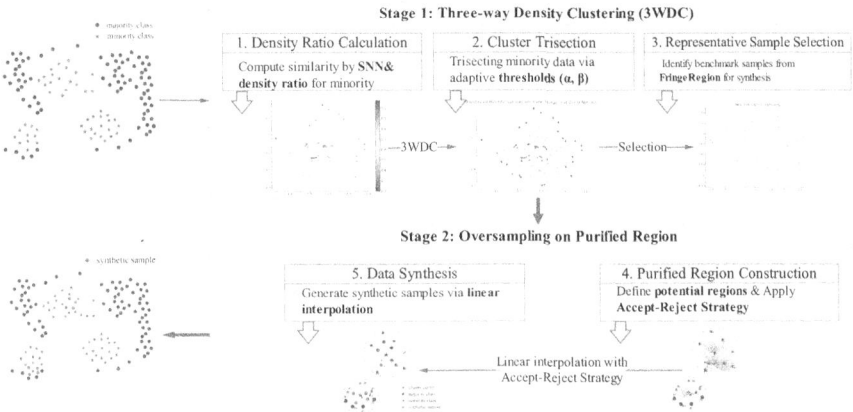

Fig. 1. The two-stage framework of 3WOS.

This paper proposes a Three-way Oversampling (3WOS) method to address imbalanced data synthesis, comprising two key stages (Fig. 1). In the first stage, for the sake of seeking the most representative data for synthesizing in stage two, a three-way density clustering (3WDC) plays a role of selector for identifying the benchmark representative candidates. Firstly, it calculates density ratio of each data. Then, all data undergoes a tripartition into three regions by measuring their density ratios, namely, CoreRegion, FringeRegion, and TrivialRegion. Thus, flexible class decision boundaries are described, wherein data within the same cluster exhibit higher similarity amongst themselves compared to those belonging to different clusters. The FringeRegion encapsulates data tightly linked with CoreRegion, thereby offering a transparent perspective on data of directly density-reachable from the CoreRegion, thereby constituting the most informative data. Finally, the FringeRegion serves as provisional benchmark dataset for minority classes in subsequent evaluation phases of stage two.

In the second stage, a purification region is defined for data synthesizing strategy. Initially, the undetermined benchmark data are subjected to purification region assessment, wherein the presence of other data within the region formed by each benchmark data and core data is evaluated. This procedure ensures precise identification of benchmark samples suitable for synthetic sample generation. Following this, the benchmark data are leveraged for data synthesis employing linear interpolation.

To summarize, the main contributions of the proposed 3WOS include:

1. The introduction of density ratio into three-way clustering marks a novel approach in similarity measurement;

2. A three-way density clustering is designed to partition a given set of data into distinct groups;
3. A purified region is set as data synthesizing strategy to promote the meaningful and discriminative data from FringeRegion are identified.

Numerical examples based on a range of imbalanced benchmark classification problems demonstrate that with the substantial enhancements, 3WOS achieves superior data synthetizing capabilities in comparison to the state-of-the-art oversampling methods with elevated classification accuracy.

The rest of this paper is organized as follows. Section 2 introduces the basic notations. Technical details of the proposed 3WOS are presented in Sect. 3. Numerical examples on imbalanced data problems are given in Sect. 4 for comparative performance demonstration. This paper is concluded in Sect. 5.

2   Notations of Three-Way Clustering

In contrast to traditional hard clustering methods, three-way clustering (3WC) offers a versatile framework for characterizing the uncertain associations between individuals and clusters [21]. This approach facilitates a three-way decision [20], interpreting rough clustering by categorizing an object as belonging to either a lower approximation or the intersection of two upper approximations. The foundational concepts of three-way clustering are defined as follows:

Given a universe $U = \{x_1, \ldots, x_n\}$ along with a set of clusters $C = \{C_1, \ldots, C_i, \ldots, C_k\}$, a crisp clustering imposes three fundamental properties on C_i: (i) $C_i \neq \emptyset$, $i = 1, \ldots, k$, (ii) $\bigcup_{i=1}^{k} C_i = U$, (iii) $C_i \cap C_j = \emptyset$, $i \neq j$.

In contrast to conventional cluster descriptions, a three-way cluster C_i can be characterized by employing a pair of sets, denoted as:

$$C_i = (\text{Co}(C_i), \text{Fr}(C_i)), \tag{1}$$

where $\text{Co}(C_i) \subset U$ and $\text{Fr}(C_i) \subset U$.

Let $\text{Tr}(C_i) = U - (\text{Co}(C_i) \cup \text{Fr}(C_i))$, the three sets, $\text{Co}(C_i)$, $\text{Fr}(C_i)$ and $\text{Tr}(C_i)$ form the CoreRegion, FringeRegion, and TrivialRegion, respectively, of a cluster. That can be defined as:

$$\begin{aligned} \text{CoreRegion}(C_i) &= \text{Co}(C_i), \\ \text{FringeRegion}(C_i) &= \text{Fr}(C_i), \\ \text{TrivialRegion}(C_i) &= U - (\text{Co}(C_i) \cup \text{Fr}(C_i)). \end{aligned} \tag{2}$$

$\text{CoreRegion}(C_i)$ represents the set of objects that unequivocally belong to the cluster C_i; $\text{TrivialRegion}(C_i)$ comprises objects that unequivocally do not belong to C_i; $\text{FringeRegion}(C_i)$ encompasses objects that exhibit uncertainty regarding their membership in C_i, as they may or may not belong to it.

In this paper, we encounter distinct requirements concerning $\text{Co}(C_i)$ and $\text{Fr}(C_i)$ and adhere to the following properties:

$$
\begin{align}
&\text{(I)} \quad \text{Co}(C_i) \neq \emptyset, \quad i = 1, \ldots, k, \\
&\text{(II)} \quad \bigcup_{i=1}^{k}(\text{Co}(C_i) \cup \text{Fr}(C_i)) = U, \tag{3} \\
&\text{(III)} \quad \text{Co}(C_i) \cap \text{Co}(C_j) = \emptyset, \quad i \neq j.
\end{align}
$$

Property (I) ensures that each cluster is non-empty. Property (II) stipulates that for any $x \in U$, it must belong to more than one cluster. Property (III) asserts that the CoreRegion of clusters are pairwise disjoint. The outcomes of three-way clustering can be represented as: $\mathbb{C} = \{(\text{Co}(C_1), \text{Fr}(C_1)), (\text{Co}(C_2), \text{Fr}(C_2)), \ldots, (\text{Co}(C_k), \text{Fr}(C_k))\}$.

3 Proposed 3WOS

3.1 Three-Way Density Clustering

Figure 2 outlines the 3WDC methodology in three stages: **1. Representative Sample Selection** (Fig. 2(a)), compute the k-nearest neighbors for minority class samples, evaluate density ratios, and select key samples (e.g., x_1, x_2, x_3); **2. Cluster Construction** (Fig. 2(b)), form clusters by identifying centroids (red stars) and directly density-reachable data (green stars). Outliers (blue stars) and transitional samples are partitioned into three subsets: $\text{Co}(C_i)$, $\text{Fr}(C_i)$, and $\text{Tr}(C_i)$; **3. Result Aggregation** (Fig. 2(c)), combine all clusters into the universe $U = \{\text{Co}(U) \cup \text{Fr}(U) \cup \text{Tr}(U)\}$, preserving the three-way structure.

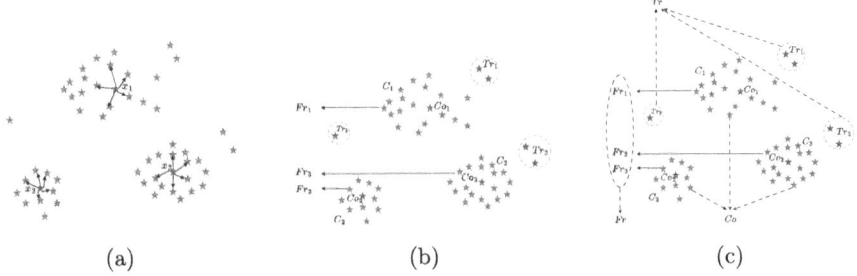

| (a) | (b) | (c) |

Fig. 2. Design methodology for three-way density clustering (3WDC). (Color figure online)

The specifics of the key process are formulated as follows:

Compute Similarity. In this study, we employ the Shared Nearest Neighbor (SNN) similarity measure, a widely utilized metric in density-based clustering algorithms [12]. Given two data points x_i and x_j, mathematically, SNN similarity can be expressed as:

$$SNN(x_i, x_j) = |SN_k(x_i) \cap SN_k(x_j)|, \tag{4}$$

where the $SN_k(x_i)$ and $SN_k(x_j)$ represent the sets of k nearest neighbors for x_i and x_j respectively.

Compute Density Ratio. According to the Eq. (4), the $SN_k(x)$ is set of k nearest neighbors from x, given a sample $x_i \in SN_k(x)$, its density and corresponding ratio are defined as:

$$ds(x_i) = \sum_{x \in SN_k(x_i)} (SNN(x_i, x) \cdot 1\{x_i \in SN_k(x)\}), \tag{5}$$

$$dr(x_i) = \frac{ds(x_i)}{\frac{1}{k} \sum_{x \in SN_k(x_i)} ds(x)}, \tag{6}$$

where $1\{\cdot\}$ is an indicator function that equals 1 if x_i is within the k nearest neighbors of x (i.e., $x_i \in SN_k(x)$), and 0 otherwise. According to the principle of three-way decision, the clustering centroid can be identified as a sample with a density ratio not lower than the threshold α.

Obtain Data of Directly Reachable. Given a sample x_i, the data set of directly reachable can be defined as:

$$R_k(x_i) = \{x_i\} \cup SN_k(x_i) \cup NSN_k(x_i), \tag{7}$$

where $SN_k(x_i)$ is the set of k nearest neighbors of x_i, and $NSN_k(x_i)$ is set of reverse k nearest neighbors of x_i.

The definition of $R_k(\cdot)$ ensures two properties: **1) Direct density reachability**: Samples near cluster centroids are directly density-reachable from those centroids; **2) Symmetry**: If sample x_j is density-reachable from centroid x_i, then x_i is reciprocally reachable from x_j.

The parameter k dynamically adjusts the size of density-reachable sets: Smaller k narrows the set, suppressing cluster formation; Larger k broadens the set, preventing cluster fragmentation. This tunability enables $R_k(\cdot)$ to adaptively balance cluster granularity and cohesion.

Construct Three-Way Density Clusters. Three-way clustering adopts the concept of CoreRegion and FringeRegion to represent clusters, rather than a single set. Constructing the CoreRegion and FringeRegion is one of the key tasks of 3WC, and the strategies to achieve this goal vary depending on the application. Inspired by sampling from distributions, we apply density ratio to evaluate the three regions. That provides more applicable schema for constructing clusters [10]. Following is the schema of three-way density clusters construction:

1. Identifying CoreRegion. It aims to find the main clustering area of the samples and label the relatively concentrated data points as centroids. Through the construction of centroids, the CoreRegion of three-way density clustering is formed.

2. Identifying FringeRegion. According to the Eq. (7), the set of directly density reachable forms the FringeRegion of the cluster, representing the transition and boundaries between other clusters. i.e., the FringeRegion of three-way density clustering is obtained by summing the data of directly density reachable from centroids.
3. Identifying TrivialRegion. There still exist discrete samples that are neither belong to FringeRegion nor CoreRegion, which are considered as noise samples. Those samples are identified to belong to TrivialRegion.

By trisecting the clusters into CoreRegion, FringeRegion and TrivialRegion before the second stage of oversampling, the core samples with the highest density ratio in the cluster and the noise samples in the cluster are filtered out through three density clustering operations, which avoids the influence of noise to a certain extent. The definition of a three-way density cluster is represented as:

Given a universe $U = \{x_1, \ldots, x_n\}$, a pair of thresholds (α, β), and k hard clustering results $C = \{C_1, \ldots, C_i, \ldots, C_k\}$, for any $x_i \in U$, its set of directly density reachable is denoted as $R_k(C_i)$. Using evaluation function $v(x)$, which represents the density ratio as defined in Eq. (6), a cluster C_i is trisected into three disjoint regions:

$$
\begin{aligned}
\mathrm{Co}(C_i) &= \{x_i \in U \mid v(x_i) > \alpha\}, \\
\mathrm{Fr}(C_i) &= \{x_i \in U \mid x_i \in R_k(C_i) \cap \beta \le v(x_i) \le \alpha\}, \\
\mathrm{Tr}(C_i) &= \{x_i \in U \mid v(x_i) < \beta\},
\end{aligned}
\tag{8}
$$

where $\mathrm{Co}(C_i)$ represents the core samples of the cluster C_i, characterized by having a density ratio greater than α; $\mathrm{Fr}(C_i)$ represents the fringe samples of the cluster C_i, which is the intersection of two sets: density-reachable samples and samples whose density ratios lie between β and α; $\mathrm{Tr}(C_i)$ includes the noise or outlier points, characterized by having a density ratio less than β. $R_k(C_i)$ can be derived from Eq. (7).

It is worth noting that the function $v(x)$ is a general mathematical expression that can serve as any evaluation factor depending on the requirements of different applications. In our study, we specifically define it as a density ratio function for two primary reasons: 1). Facilitating the determination of the CoreRegion; 2). Ensuring that the density ratio remains unaffected by changes in cluster density.

Outcomes of Three-Way Density Clustering. Based on the three-way density clusters, we can define the three-way density clustering on universe.

Algorithm 1: Stage one: three-way density clustering (3WDC)

---

1 **Input:** Universe U, threshold pair (α, β), and k hard clustering results
   $C = \{C_1, \ldots, C_i, \ldots, C_k\}$ obtained via SNN-based k-means;
2 **Output:** The three-way density clustering result $Co(U)$, $Fr(U)$, and $Tr(U)$;

  1: **for** each x_i in U **do**
  2:    $SN_k \leftarrow \mathcal{K}(x_i)$ // Compute the k nearest neighbors of x_i.
  3:    $SNN \leftarrow SN_k(x_i)$ by Eq. (4).
  4:    $ds \leftarrow \sum_{x \in SN_k} SNN(x_i, x)$ by Eq. (5).
  5:    $dr \leftarrow ds(x_i)$ by Eq. (6).
  6:    Calculate $Co(C_i)$, $Fr(C_i)$, $Tr(C_i)$ by Eq. (8).
  7:    Calculate $Co(U)$, $Fr(U)$, $Tr(U)$ by Eq. (9).
  8: **end for**
  9: **return** $\mathbb{C} = \{Co(U), Fr(U), Tr(U)\}$.

Given a universe $U = \{x_1, \ldots, x_n\}$ and a set of three-way density clusters $C = \{C_1, \ldots, C_i, \ldots, C_k\}$, the universe U can be partitioned into distinct regions:

$$Co(U) = \bigcup_{i=1}^{k} Co(C_i),$$

$$Fr(U) = \bigcup_{i=1}^{k} Fr(C_i), \qquad (9)$$

$$Tr(U) = \bigcup_{i=1}^{k} Tr(C_i),$$

where $Co(U)$ represents CoreRegion of the universe, consisting of core samples from all clusters; $Fr(U)$ denotes the FringeRegion, comprising fringe samples from all clusters; $Tr(U)$ signifies the TrivialRegion, including samples identified as noise or outliers from all clusters.

According to the main procedures of three-way density clustering, Algorithm 1 is designed. The implementation of the algorithm can be divided into following steps:

Step 1 (Line 2). Use the k nearest neighbor algorithm to obtain the neighbors of sample x_i.

Step 2 (Lines 3–5). Calculate the density ration of x_i according to the Eqs. (5) and (6).

Step 3 (Line 6). According to the Eq. (7), calculate the data of directly reachable. Applying the evaluation function $v(\cdot)$ and the threshold pair (α, β) to perform the three-way density clustering using Eq. (8). Obtain the CoreRegion, FringeRegion of C_i.

Step 4 (Lines 7–9). Output the three-way density clustering results $\mathbb{C}$ according to the Eq. (9).

3.2   Oversampling on Purified Region

As shown in Fig. 3, the oversampling process operates in three stages: **1. Ordered Neighbor Identification** (Fig. 3(a)), for a sample $x \in \mathrm{Fr}(C_i)$, compute its k-nearest neighbors (with k adaptively determined by $\mathrm{Fr}(C_i)$'s density) and obtain an ordered set $D'(x_i)$. The farthest neighbor (e.g., x_7) defines the boundary of potential regions; **2. Benchmark Sample Selection** (Fig. 3(b)), construct potential synthesis regions as circles centered between x and each neighbor, with radius scaled to half their pairwise distance. A neighbor (e.g., x_3) is accepted as a benchmark sample only if its corresponding circle is empty; regions containing majority class samples (e.g., $y_1, \ldots, y_7$) or overlapping samples (e.g., x_6) are rejected; **3. Synthetic Sample Generation** (Fig. 3(c)), valid benchmark samples drive the synthesis of new samples (e.g., s_1, s_2, s_3, s_4) within purified regions. If majority samples exist, iterative region refinement ensures non-overlapping synthetic data.

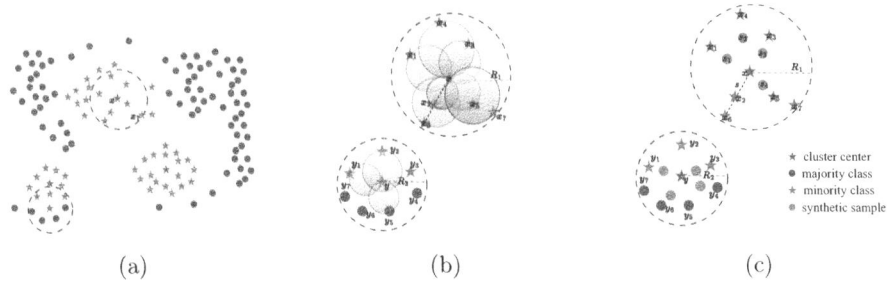

(a)   (b)   (c)

Fig. 3. Design methodology for oversampling on purified region.

To address overlap and redundancy in imbalanced classification, 3WOS employs a purified region-based synthesis strategy. This approach selectively expands minority class regions while preserving the underlying data distribution, enhancing the representativeness of synthetic samples and mitigating redundancy. Unlike methods reliant on idealistic assumptions or sensitive parameter tuning, 3WOS identifies sparse candidate regions within the $\mathrm{Fr}(C_i)$ to improve synthesis quality and classifier performance. The oversampling proposes three major processes:

Seek Potential Region. To refine the identification of potential regions within $\mathrm{Fr}(C_i)$ based on representative samples and cluster boundaries which are covered by $\mathrm{Fr}(U)$, a specific region is delineated by evaluating the distances from the centroid to all samples in $\mathrm{Fr}(C_i)$. This process involves centroid calculation and sorting all samples in $\mathrm{Fr}(C_i)$ by their distances from the centroid x_i of cluster C_i. The centroid x_i of the $\mathrm{Fr}(C_i)$ is determined as follows:

$$x_i = \begin{cases} \frac{1}{n} \sum_{j=1}^{n} x_j, & \text{if } \text{Fr}(C_i) \neq \emptyset, \ x_j \in \text{Fr}(C_i), \\ \text{centroid}(\text{Co}(C_i)), & \text{otherwise,} \end{cases} \tag{10}$$

where $n = |\text{Fr}(C_i)|$ denotes the number of samples in $\text{Fr}(C_i)$.

For a given $\text{Fr}(C_i)$ and its centroid x_i, the ordered set $D(x_i)$ is obtained using the distance function $d(\cdot)$:

$$D(x_i) = \{x_{i1}, \ldots, x_{ifinal}\}, \tag{11}$$

where x_{i1} represents the nearest sample to x_i and x_{ifinal} represents the farthest sample from x_i. The function $d(\cdot) : \text{Fr} \rightarrow D$ applies a distance metric to sort the samples within $\text{Fr}(C_i)$.

While various distance formulas exist, we opt for the Euclidean distance metric for simplicity and effectiveness in sorting the samples within $\text{Fr}(C_i)$.

Define the potential region $D'(x_i)$ as the set covered by the circle centered at x_i and $\bar{x}_{ij}$ representing the mean value between sample x_i and its j-th nearest neighbor x_{ij}. The potential region is defined as:

$$D'(x_i) = \{x_k | d(x_i, x_k) \leq d(x_i, \bar{x}_{ij}), x_k \in \{x_{i1}, \ldots, x_{ij}, \ldots, x_{ifinal}\}\}, \tag{12}$$

where $d(x_i, x_k)$ denotes the distance between x_i and x_k.

For a given universe U with its three-way density clustering result $\mathbb{C} = \{\text{Co}(U), \text{Fr}(U), \text{Tr}(U)\}$, the set of all potential regions can be expressed as:

$$D'(U) = \bigcup_{i=1}^{k} D'(x_i) \tag{13}$$

Identify Benchmark Samples. Let $D'(x_i)$ be the set representing the potential region covered by a circle centered at x_i, and $|D'(x_i)|$ denote the cardinality of this set. The **Accept-Reject** strategy for the purified region, denoted as $A(x_i)$, can be expressed as:

$$A(x_i) = \begin{cases} \{x_{ij}\}, & \text{if } |D'(x_i)| = 2, \\ \emptyset, & \text{if } |D'(x_i)| > 2, \end{cases} \tag{14}$$

where $\emptyset$ denotes an empty set. If $D'(x_i)$ contains exactly two elements (i.e., only x_i and x_{ij} in $D'(x_i)$), then x_{ij} is accepted as a benchmark sample for the set $A(x_i)$; otherwise, no benchmark sample is selected. This strategy ensures the avoidance of synthesizing overlapping samples within the potential region, providing a purified selection process based on the presence of a single additional sample within that region.

Synthesize Samples. Let x_s represent a random benchmark sample from potential region $A(x_i)$, for any $x_i \in X^+$, a synthetic sample *synt* can be defined as:

$$synt(x_i) = x_i + \xi \times (x_s - x_i), \tag{15}$$

Algorithm 2: Stage two: purified region oversampling

---

1 **Input:** Fringe regions $\mathrm{Fr}(U)$, minority set X^+, counting param k, synth size η
2 **Output:** Synthetic samples S
  1: Sort $\mathrm{Fr}(C_i)$ samples by distance to centroid $\rightarrow$ $D(x_i)$ by Eq. (11)
  2: **for** $x_i \in X^+$ **do**
  3:     Build potential region $D'(x_i)$ with k-nearest neighbors by Eq. (12)
  4:     **for** $x_{ij} \in D'(x_i)$ **do**
  5:         Compute midpoint $\bar{x}_{ij}$ and radius $d_j = \frac{1}{2}d(x_i, x_{ij})$
  6:         **if** $\nexists x_{il} \in D'(x_i)$ s.t. $d(x_{il}, \bar{x}_{ij}) \leq d_j$ **then**
  7:             Add x_{ij} to benchmark set $A(x_i)$ by Eq. (14)
  8:         **end if**
  9:     **end for**
 10: **end for**
 11: **while** $|S| < \eta$ **do**
 12:     Select $x_i \in X^+$ with $A(x_i) \neq \emptyset$
 13:     Generate $synt \leftarrow x_i + \xi(x_s - x_i)$ where $x_s \in A(x_i)$ by Eq. (15)
 14:     $S \leftarrow S \cup \{synt\}$ by Eq. (16)
 15: **end while**
 16: **return** S

---

where $\xi \in [0, 1]$ is a random number, X^+ denotes set of minority samples.

For all samples from potential region $A(x_i)$, the set of synthesized sample S can be define as:

$$S = \left\{ \mathrm{synt}_k \;\middle|\; \begin{array}{l} \mathrm{synt}_k = x_i + \xi_k \cdot (x_s - x_i), \\ x_i \in X^+, \; A(x_i) \neq \emptyset, \; x_s \in A(x_i), \; k = 1, 2, \ldots, \eta \end{array} \right\} \tag{16}$$

where η is the number of synthetic samples to be generated for the minority class.

According to the major operations of oversampling on purified region, the implementation of Algorithm 2 can be summarized as follows:

Step 1 (Line 1). Sort all samples in $\mathrm{Fr}(C_i)$ using the distance function $d(\cdot)$ to obtain $D(x_i)$ as defined in Eq. (11).

Step 2 (Line 3). Calculate the potential region $D'(x_i)$ for each sample x_i in $\mathrm{Fr}(C_i)$ using Eq. (12). The parameter ς is used to control the number $|final|$ of samples in $D(x_i)$.

Step 3 (Lines 4–9). Identify benchmark samples using the purified region **Accept-Reject** strategy described in Eq. (14).

Step 4 (Lines 11–14). Generate synthesized samples S using Eqs. (15) and (16). The parameter η controls the number of synthesized samples required, and ξ introduces a bias for sample generation.

Table 1. Description of experimental imbalanced datasets from the imbalanced-learn developers.

ID	Dataset	|Features|	Vol	CD	IR
1	Pima	8	768	268/500	1:1.87
2	Iris	4	150	50/100	1:2.00
3	Wine	13	178	48/130	1:2.71
4	Ecoli	7	336	52/284	1:8.60
5	Celiac-Disease	67	205	18/187	1:10.39
6	US-Crime	122	1994	150/1844	1:12.29
7	Yeast8	103	2417	178/2239	1:12.58
8	Libras-Move	90	360	24/336	1:14.00
9	Solar-Flare-M0	10	1389	66/1323	1:20.00
10	Car	6	1728	65/1663	1:25.58
11	Yeast6	8	1484	51/1433	1:28.00
12	Ozone-Level	72	2536	73/2463	1:33.74

4   Experimental Investigation

4.1   Experimental Setup

Datasets. Twelve imbalanced datasets are selected from the imbalanced-learn developers (ILD, https://imbalanced-learn.org/stable/datasets/index.html). Table 1 provides a summary of the imbalanced-learn datasets, where "Vol", "CD", and "IR" represent the dataset volume, class distribution, and imbalanced ratio, respectively.

Evaluation Metrics. To evaluate the effectiveness of the 3WOS method, we employed three commonly used base classifiers: Support Vector Machine (SVM), Random Forest (RF), and k Nearest Neighbors (KNN). The evaluation metrics selected for assessing classifier performance are F1-score and Cohen's κ-score, known for their robustness against class imbalance. Performance results for each dataset are presented as the average mean value over 10 runs using stratified 5-fold cross-validation, reported alongside Cohen's κ-score. The Meanrank provides a relative ranking comparison across all methods. To validate the significance of the Meanrank results and mitigate statistical anomalies, we conducted the Wilcoxon signed-rank test for further investigation.

Baselines. In our experiment, we compared the proposed 3WOS method against seven representative baseline oversampling methods: SMOTE [2], Borderline-SMOTE (BSMO) [6], ADASYN [8], MWMOTE [1], A-SUWO [14], k-means and SMOTE (KM-SMOTE) [4] and GB-SMOTE [15].

The parameter settings for the baseline methods are referred to the original papers and detailed as follows:

1. For SMOTE, BSMO, and ADASYN, the number of nearest neighbors k is set to 5.
2. For MWMOTE, the parameters are configured as follows: $k_1 = 5$, $k_2 = 3$, $k_3 = |S_{min}|/2$, $C_f = 5$, and $C_{MAX} = 2$.
3. For KM-SMOTE, the optimal number of clusters is selected from the set $\{2, 5, 7, 10\}$, and the *itr* threshold to identify secure sub-clusters is set to 1.
4. For A-SUWO, the parameter C_t is chosen from the set $\{1, 2\}$.
5. For GB-SMOTE, the Radial Basis Function (RBF) is selected as the kernel function. The ranges for the regularization parameter C and γ are set to $\{2^{-10}, 2^{-9}, \ldots, 2^{10}\}$ and $\{2^{-10}, 2^{-9}, \ldots, 2^3\}$, respectively.

4.2   Experimental Results and Analysis

Tables 2, 3, and 4 demonstrate that 3WOS has consistent superiority over baseline methods in F1-score and Cohen's κ-score across SVM, RF, and KNN classifiers. Notably, 3WOS achieves 3.09–4.19% higher F1-scores and 3.62–4.46% higher κ-score compared to clustering-based methods (A-SUWO, KM-SMOTE), attributed to its adaptive three-way density clustering, which automates cluster center selection, filters noisy samples, and strategically synthesizes minority samples in purified regions to minimize overlap.

Table 2. Average F1-score/κ-score by SVM classifier.

Dataset	SMOTE	BSMO	ADASYN	MWMOTE	A-SUWO	KM-SMOTE	GB-SMOTE	3WOS
1	0.727/0.716	0.754/0.735	0.738/0.714	0.745/0.738	**0.759/0.742**	0.747/0.730	0.750/0.740	0.746/0.738
2	0.816/0.805	0.832/0.819	0.840/0.835	0.887/0.879	0.938/0.926	0.923/0.919	0.935/0.925	**0.939/0.934**
3	0.906/0.889	0.920/0.809	0.922/0.915	0.923/0.917	0.915/0.9028	0.929/0.924	0.939/0.935	**0.943/0.940**
4	0.586/0.579	0.621/0.614	0.569/0.558	0.556/0.544	0.635/0.630	0.636/0.631	**0.676/0.666**	0.672/0.661
5	0.876/0.867	0.886/0.879	**0.915/0.907**	0.897/0.887	0.888/0.872	0.862/0.853	0.879/0.870	0.880/0.879
6	0.736/0.732	0.767/0.766	0.816/0.814	0.751/0.750	0.886/0.878	0.866/0.864	**0.891/0.885**	0.879/0.879
7	0.827/0.826	0.843/0.840	0.864/0.860	0.892/0.889	0.896/0.891	0.916/0.909	0.891/0.885	**0.941/0.938**
8	0.916/0.899	0.942/0.938	0.940/0.934	0.943/0.939	0.946/0.940	0.927/0.923	0.935/0.928	**0.961/0.959**
9	0.776/0.770	0.818/0.815	0.826/0.824	**0.840/0.837**	0.746/0.739	0.827/0.812	0.828/0.819	0.836/0.835
10	0.882/0.876	0.880/0.875	0.902/0.901	0.911/0.909	0.913/0.912	0.915/0.913	0.936/0.929	**0.955/0.952**
11	0.635/0.629	0.678/0.670	0.680/0.673	0.659/0.651	0.709/0.704	0.687/0.680	0.689/0.680	**0.714/0.712**
12	0.738/0.732	0.860/0.855	0.761/0.758	0.765/0.760	0.783/0.779	0.792/0.789	0.791/0.788	**0.878/0.876**
Average	0.785/0.777	0.817/0.801	0.814/0.808	0.814/0.808	0.835/0.826	0.836/0.829	0.840/0.836	**0.862/0.859**
Meanrank	7.92/7.4	4.75/5.6	4.33/5.3	4.0/4.6	3.25/3.8	3.25/4.4	2.67/3.3	2.33/1.6●

3WOS-RF excels on datasets 2, 4, 6, 9, and 12, aligning well with RF's ensemble learning. However, SVM outperforms on datasets 3, 7, and 8 due to its boundary precision, while KNN dominates on datasets 1, 5, 10, and 11, leveraging local data adaptability.

Wilcoxon signed-rank tests (Table 5) confirm statistical significance for 3WOS's improvements, rejecting the null hypothesis across all baselines. These results validate 3WOS's ability to preserve data distribution, enhance synthetic sample quality, and generalize across diverse imbalanced datasets.

Table 3. Average F1-score/κ-score by RF classifier.

Dataset	SMOTE	BSMO	ADASYN	MWMOTE	A-SUWO	KM-SMOTE	GB-SMOTE	3WOS
1	0.724/0.721	0.780/0.769	0.789/0.784	0.782/0.779	0.787/0.781	0.810/0.802	**0.818/0.809**	0.803/0.797
2	0.903/0.898	0.923/0.920	0.931/0.925	0.947/0.936	0.952/0.945	0.956/0.950	0.949/0.942	**0.968/0.957**
3	0.908/0.905	0.914/0.903	0.912/0.900	0.920/0.911	0.915/0.910	0.925/0.918	0.926/0.921	**0.935/0.931**
4	0.745/0.742	0.758/0.739	0.730/0.723	0.760/0.749	**0.807/0.790**	0.762/0.758	0.759/0.752	0.788/0.788
5	0.822/0.819	0.836/0.829	0.848/0.837	0.851/0.845	0.846/0.841	0.907/0.905	0.899/0.887	**0.917/0.914**
6	0.927/0.926	0.907/0.901	0.924/0.921	0.901/0.892	0.907/0.899	0.937/0.933	0.930/0.925	**0.938/0.927**
7	0.822/0.820	0.836/0.829	0.848/0.832	0.851/0.846	0.846/0.839	**0.917/0.912**	0.891/0.882	0.907/0.899
8	0.907/0.905	0.927/0.918	**0.948/0.937**	0.925/0.922	0.901/0.898	0.908/0.907	0.911/0.909	0.937/0.934
9	0.793/0.786	0.814/0.807	0.801/0.795	0.823/0.811	0.784/0.782	0.812/0.809	0.801/0.789	**0.845/0.843**
10	0.925/0.915	0.919/0.910	0.921/0.910	0.918/0.912	0.926/0.923	0.929/0.924	0.926/0.921	**0.931/0.927**
11	0.780/0.770	0.796/0.788	0.835/0.834	**0.859/0.856**	0.732/0.730	0.824/0.821	0.838/0.829	0.844/0.835
12	0.851/0.836	0.872/0.867	0.874/0.872	0.902/0.899	0.899/0.891	0.919/0.915	0.924/0.912	**0.931/0.928**
Average	0.842/0.837	0.857/0.848	0.863/0.856	0.870/0.863	0.859/0.852	0.884/0.880	0.881/0.873	**0.895/0.890**
Meanrank	7.1/6.7	6.1/6.4	5.3/5.4	4.5/4.3	5.4/5.3	2.9/2.7	3.3/3.6	1.5/1.6●

Table 4. Average F1-score/κ-score by KNN classifier.

Dataset	SMOTE	BSMO	ADASYN	MWMOTE	A-SUWO	KM-SMOTE	GB-SMOTE	3WOS
1	0.780/0.779	0.792/0.790	0.793/0.790	0.801/0.798	0.791/0.777	0.789/0.788	0.801/0.798	**0.823/0.816**
2	0.927/0.904	0.930/0.924	0.935/0.929	0.930/0.920	0.944/0.932	**0.950/0.941**	0.930/0.920	0.940/0.931
3	0.841/0.822	0.790/0.781	0.848/0.830	0.865/0.858	**0.905/0.889**	0.855/0.849	0.865/0.858	0.870/0.862
4	0.719/0.710	0.721/0.718	0.737/0.724	0.758/0.740	0.738/0.729	0.756/0.738	0.758/0.740	**0.769/0.768**
5	0.868/0.852	0.895/0.889	0.903/0.889	0.920/0.911	0.906/0.899	0.894/0.871	0.920/0.911	**0.939/0.934**
6	0.849/0.835	0.888/0.880	0.859/0.850	**0.931**/0.827	0.898/0.884	0.870/0.867	0.929/**0.915**	0.911/0.906
7	0.840/0.832	0.846/0.837	0.856/0.847	0.900/0.892	0.878/0.860	0.860/0.847	0.900/0.892	**0.923/0.919**
8	0.893/0.887	0.918/0.908	0.948/0.945	0.950/0.947	0.902/0.894	0.943/0.932	0.950/0.947	**0.960/0.957**
9	0.779/0.764	0.789/0.773	0.797/0.782	0.794/0.785	0.812/0.788	0.745/0.738	**0.830/0.825**	0.822/0.819
10	0.976/0.970	0.957/0.952	0.960/0.957	**0.975/0.969**	0.968/0.961	0.964/0.958	0.967/0.956	0.972/**0.969**
11	0.916/0.895	0.881/0.877	0.844/0.833	0.928/0.925	0.905/0.896	0.901/0.899	0.928/0.925	**0.940/0.936**
12	0.819/0.810	0.885/0.867	0.878/0.869	0.893/0.890	0.828/0.821	0.892/0.886	0.893/0.890	**0.927/0.919**
Average	0.851/0.838	0.858/0.850	0.863/0.854	0.887/0.879	0.873/0.861	0.868/0.860	0.888/0.881	**0.900/0.895**
Meanrank	6.8 /6.8	6.4/6.3	5.6/5.5	3.4/3.8	4.2/4.3	5.1/4.8	2.8/2.7	1.8/1.6●

As shown in Fig. 4, the parameter analysis reveals three distinct regimes: **Declining Phase** $(0.5, 0.5) \rightarrow (0.8, 0.2)$: Meanrank decreases monotonically with threshold adjustments; **Stable Phase** $(0.8, 0.2) \rightarrow (0.9, 0.1)$: Performance plateaus with minimal fluctuations (Δmeanrank < 0.03); **Degradation Phase** $(\alpha, \beta) > (0.9, 0.1)$: Overlap-induced performance deterioration ($\uparrow$ 18.7% meanrank increase). The empirical optimum $(\alpha, \beta) = (0.8, 0.2)$ achieves optimal balance between cluster density utilization and overlap prevention, validated through cross-dataset experiments (F1-score $\uparrow$ 4.2% vs. suboptimal thresholds). This parameter selection strategy enhances the 3WOS method's interpretability through clear density-ratio thresholds while maintaining $\geq$ 96.3% stability across different task domains.

Table 5. Wilcoxon Signed-rank Tests ($p < 0.05$) for 3WOS compared to baselines

	SMOTE	BSMO	ADASYN	MWMOTE	A-SUWO	KM-SMOTE	GB-SMOTE
SVM(F1-score)	4.88×10^{-4}	2.44×10^{-3}	6.84×10^{-3}	4.88×10^{-3}	3.42×10^{-3}	9.77×10^{-4}	6.40×10^{-3}
SVM(κ-score)	4.88×10^{-4}	3.33×10^{-3}	4.88×10^{-3}	7.65×10^{-3}	1.46×10^{-3}	4.88×10^{-4}	4.22×10^{-3}
RF(F1-score)	4.88×10^{-4}	4.88×10^{-4}	2.44×10^{-3}	2.44×10^{-3}	3.42×10^{-3}	3.22×10^{-4}	6.84×10^{-3}
RF(κ-score)	4.88×10^{-4}	4.88×10^{-4}	3.35×10^{-3}	9.77×10^{-4}	9.77×10^{-4}	2.44×10^{-3}	4.88×10^{-3}
KNN(F1-score)	9.77×10^{-4}	4.88×10^{-4}	4.88×10^{-4}	1.22×10^{-4}	2.10×10^{-3}	1.46×10^{-3}	5.22×10^{-3}
KNN(κ-score)	9.77×10^{-4}	4.88×10^{-4}	4.88×10^{-4}	5.82×10^{-3}	4.88×10^{-3}	9.77×10^{-4}	9.28×10^{-3}

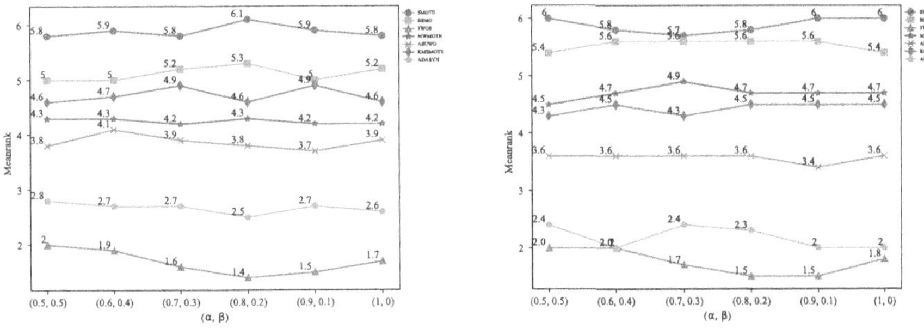

(a) F1-score as metric

(b) κ-score as metric

Fig. 4. Variation of meanrank with thresholds pair (α, β).

5   Conclusion

We propose 3WOS, a clustering-based oversampling model for imbalanced classification, leveraging adaptive three-way density clustering to partition minority class samples into CoreRegion, FringeRegion, and TrivialRegion. By using density ratios, 3WOS identifies latent data patterns through the FringeRegion, enriched with boundary proximal samples, to guide synthetic sample generation while preserving data distribution and minimizing overlap. A benchmark sample selection strategy ensures representative synthesis, enhancing classification performance. Experiments demonstrate 3WOS outperforms state-of-the-art methods on imbalanced datasets, though its efficacy diminishes in high-dimensional spaces due to the curse of dimensionality. Future work will integrate dimensionality reduction to address this limitation.

Conflict of Interest. The authors declare that there are no conflicts of interest regarding the publication of this paper.

References

1. Barua, S., Islam, M.M., Yao, X., Murase, K.: Mwmote-majority weighted minority oversampling technique for imbalanced data set learning. IEEE Trans. Knowl. Data Eng. **26**(2), 405–425 (2012)

2. Chawla, N.V., Bowyer, K.W., Hall, L.O., Kegelmeyer, W.P.: Smote: synthetic minority over-sampling technique. J. Artif. Intell. Res. **16**, 321–357 (2002)
3. Ding, H., Sun, Y., Huang, N., Cui, X.: VGAN-BL: imbalanced data classification based on generative adversarial network and biased loss. Neural Comput. Appl. **36**(6), 2883–2899 (2024)
4. Douzas, G., Bacao, F., Last, F.: Improving imbalanced learning through a heuristic oversampling method based on k-means and smote. Inf. Sci. **465**, 1–20 (2018)
5. Fang, Y., Gao, C., Yao, Y.: Granularity-driven sequential three-way decisions: a cost-sensitive approach to classification. Inf. Sci. **507**, 644–664 (2020)
6. Han, H., Wang, W., Mao, B.: Borderline-smote: a new over-sampling method in imbalanced data sets learning. In: International Conference on Intelligent Computing, pp. 878–887. Springer (2005)
7. Hasib, K.M., et al.: MCNN-LSTM: combining CNN and LSTM to classify multi-class text in imbalanced news data. IEEE Access **11**, 93048–93063 (2023)
8. He, H., Bai, Y., Garcia, E.A., Li, S.: Adasyn: adaptive synthetic sampling approach for imbalanced learning. In: 2008 IEEE International Joint Conference on Neural Networks, pp. 1322–1328. IEEE (2008)
9. Huang, C., et al.: Sample imbalance disease classification model based on association rule feature selection. Pattern Recogn. Lett. **133**, 280–286 (2020)
10. Koziarski, M.: Radial-based undersampling for imbalanced data classification. Pattern Recogn. **102**, 107262 (2020)
11. Li, T., Kou, G., Peng, Y., Philip, S.Y.: An integrated cluster detection, optimization, and interpretation approach for financial data. IEEE Trans. Cybern. **52**(12), 13848–13861 (2021)
12. Liu, R., Wang, H., Yu, X.: Shared-nearest-neighbor-based clustering by fast search and find of density peaks. Inf. Sci. **450**, 200–226 (2018)
13. Maldonado, S., Vairetti, C., Fernandez, A., Herrera, F.: FW-smote: a feature-weighted oversampling approach for imbalanced classification. Pattern Recogn. **124**, 108511 (2022)
14. Nekooeimehr, I., Lai-Yuen, S.K.: Adaptive semi-unsupervised weighted oversampling (A-SUWO) for imbalanced datasets. Expert Syst. Appl. **46**, 405–416 (2016)
15. Ren, J., Wang, Y., Cheung, Y., Gao, X., Guo, X.: Grouping-based oversampling in kernel space for imbalanced data classification. Pattern Recogn. **133**, 108992 (2023)
16. Tao, X., et al.: Self-adaptive cost weights-based support vector machine cost-sensitive ensemble for imbalanced data classification. Inf. Sci. **487**, 31–56 (2019)
17. Tsai, C., Weichao, L., Hu, Y., Yao, G.: Under-sampling class imbalanced datasets by combining clustering analysis and instance selection. Inf. Sci. **477**, 47–54 (2019)
18. Wang, Y., Gan, W., Yang, J., Wu, W., Yan, J.: Dynamic curriculum learning for imbalanced data classification. In: 2019 IEEE/CVF International Conference on Computer Vision (ICCV), pp. 5017–5026 (2019)
19. Yang, Y., Hu, Y., Zhang, X., Wang, S.: Two-stage selective ensemble of CNN via deep tree training for medical image classification. IEEE Trans. Cybern. **52**(9), 9194–9207 (2022)
20. Yao, Y.: The DAO of three-way decision and three-world thinking. Int. J. Approximate Reasoning **162**, 109032 (2023)
21. Yu, H., Wang, X., Wang, G., Zeng, X.: An active three-way clustering method via low-rank matrices for multi-view data. Inf. Sci. **507**, 823–839 (2020)

Three-Way Conflict Analysis Based on Fuzzy Preference Conflict Situation

Mingchuan Shang[1,2] and Guangming Lang[1,2(✉)]

[1] School of Mathematics and Statistics, Changsha University of Science and Technology, Changsha 410114, Hunan, People's Republic of China
[2] Hunan Provincial Key Laboratory of Mathematical Modeling and Analysis in Engineering, Changsha University of Science and Technology, Changsha 410114, Hunan, People's Republic of China
langguangming1984@126.com

Abstract. The rating of issues by agents plays a crucial role in the conflict situation table. However, existing studies predominantly focus on the direct rating of individual issue by individual agent, overlooking the variations in rating interpretations among different agents. This paper proposes a three-way conflict analysis model for fuzzy preference-based conflict situation. Initially, we construct a fuzzy preference matrix representing agents' evaluations of issues, from which fuzzy preference relations are derived to establish a fuzzy preference-based conflict situation. Then, we define the conflict degree and develop a fuzzy similarity matrix to classify agents into similarity equivalence classes. Based on Bayesian minimum risk theory, we calculate the threshold values to conduct three-way conflict analysis. Finally, an algorithm is designed to derive three-way conflict analysis rules, and a case study is presented to validate the feasibility and effectiveness of the proposed model.

Keywords: Three-way conflict analysis · Decision-theoretic rough set · Fuzzy preference relation · Conflict degree

1 Introduction

Conflict is undeniably one of the most fundamental characteristics of human behavior. In the book *The Power of Conflict: Speak Your Mind and Get the Results You Want* [1], Jon Taffer underscores the constructive role of conflict in fostering personal growth and enhancing problem-solving capabilities. Given its profound significance in daily life, the development of rigorous mathematical models for conflict analysis is essential. Pawlak [2] first introduced the concept of conflict situations. Building on this, Pawlak [3] designed an auxiliary function and defined three fundamental types of relationships between agents: alliance, conflict, and neutrality. To further formalize conflict analysis, Pawlak [4] employed a three-valued situation table to describe conflict problems, where the attitudes of agents toward all issues are selected from the set $\{+1, 0, -1\}$. Additionally, a distance function between two agents for the issues set was introduced, providing a more concrete framework for establishing relationships among agents.

© The Author(s), under exclusive license to Springer Nature Switzerland AG 2025
Q. Zhang et al. (Eds.): IJCRS 2025, LNAI 15710, pp. 19–32, 2025.
https://doi.org/10.1007/978-3-031-92741-6_2

To expand the research scope of conflict analysis, some scholars have drawn upon Yao's three-way decision theory [5–8], which focuses on problem-solving from a three-valued perspective and shares a close connection with Pawlak's conflict analysis model based on a three-valued situational table. Building on this foundation, Yao [9] proposes the three-way conflict analysis model, reformulating and extending Pawlak's conflict analysis model by refining the relationships among different agents with a rating of 0. In three-way conflict analysis, there are four types of tri-partitions: tri-partition of the agents set, tri-partition of agents pairs, tri-partition of the issues set, and tri-partition of issues pairs. Based on this, Lang et al. [10] proposed a general model for three-way conflict analysis that tri-partitions both the agents set and the issues set, and conducted in-depth research on conflict analysis models. Lang [11] further introduced a universal three-way conflict analysis model that tri-partitions agent pairs, demonstrating that other models are special cases of this model. Lang et al. [12] refined the conflict degree threshold for partition relationships from 0.5 to (α, β). By integrating Bayesian theory from decision-theoretic rough sets, (α, β) is determined through the minimization of decision costs. Additionally, an incremental algorithm was developed to handle dynamic three-valued situation table. In recent years, numerous scholars have integrated various types of fuzzy numbers, including triangular fuzzy numbers [13], Pythagorean fuzzy numbers [14–18], hesitant fuzzy numbers [19], and intuitionistic fuzzy numbers [20,21], into conflict situation table. Yang et al. [22] developed hybrid situation table based on three-valued, many-valued, and diverse fuzzy numbers, and defined feasible strategies within the hybrid situation table to address conflicts. Lu et al. [23] introduced multi-scale rating into situation table, constructing multi-scale situation table and proposing two approaches for tri-partitioning agent pairs.

Despite incorporating diverse rating types, existing studies are primarily based on direct evaluations of individual agents toward individual issues. This approach poses a potential limitation: consider two agents—one radical and the other conservative—both assigning a rating of 0.75 to the same issue. However, within their respective evaluation systems, a rating of 0.75 from the radical agent may reflect a high level of support for the issue, whereas the same rating from the conservative agent could imply a low level of support. To address this problem, Hu [24] introduced a preference-based three-way conflict analysis model. By utilizing the inherent consistency of agents in maintaining their own ratings, the model incorporates agents' preference attitudes toward pairs of issues within the issues set to construct conflict situation. Hu's approach effectively resolves issues stemming from discrepancies in the interpretation of ratings among agents. Building on Hu's insights, it is acknowledged that in real-world scenarios, agents' preference degree toward pairs of issues are unlikely to be entirely uniform. Therefore, the use of fuzzy preference relations is proposed to capture agents' preference degrees for issue pairs, thereby the concept of fuzzy preference relation is integrated into conflict analysis. The fuzzy preference relation proposed by Tanino [25,26] provides a methodological tool for decision-makers to assess and quantify preference intensities during pairwise comparisons of alternatives. Herrera-Viedma et al. [27] conducted a study on the consistency properties of fuzzy preference relation. This paper proposes a three-way conflict analysis model for fuzzy preference-based conflict situation by leveraging the properties of fuzzy preference relation and decision-theoretic rough set. The main contributions are summarized as follows:

(1) We explore the fuzzy preference relations of agents over issues, which not only enhance the Hu's model but also provide a more nuanced representation of agents' preference dagree. This approach improves the interpretability of data analysis in conflict analysis and offers new perspectives and insights for future research.

(2) We propose a novel model for conflict analysis, offering a scientifically grounded method for calculating conflict degree in fuzzy preference-based conflict situation. By incorporating fuzzy similarity matrices to partition agents into similarity equivalence classes, we further integrate decision-theoretic rough sets to conduct comprehensive conflict analysis.

(3) We design an algorithm to derive rules for three-way conflict analysis based on fuzzy preference conflict situation and demonstrate the application of the proposed model in decision-making through a illustrative example.

The remainder of this paper is organized as follows: Sect. 2 introduces the concept of preference-based conflict situations and the tri-partition of agent pairs within the Hu's model. Section 3 presents the main concepts of the proposed model, including fuzzy preference-based conflict situations, conflict degree, fuzzy similarity matrix, tri-partition of the agents set, and the associated algorithm. Section 4 illustrates the proposed model through an example and comparative analysis. Section 5 concludes the paper and provides an outlook on future research directions.

2 Preference-Based Conflict Situations

This section reviews and summarizes the fundamental concepts associated with preference-based conflict situations, including preference relations, conflict degree, and trisections of agent pairs.

Definition 1. *(Hu [24], 2025) Let $PS = (A, I, \{\succ_a^I | \ a \in A\})$ be a preference-based conflict situation, where A is the set of agents, I is the set of issues, $\succ_a^I \subseteq I \times I$ is the strict preference relation of an agent $a \in A$ over the issues. Correspondingly, $\prec_a^J \subseteq I \times I$ represents the converse preference relation, and $\sim_a^I \in I \times I$ is the indifference relation on the issues, which is formally defined as follows:*

$$\succ_a^I = \{(i,j) \in I \times I \mid a \ supports \ i \ more \ than \ j\};$$
$$\prec_a^I = \{(i,j) \in I \times I \mid a \ supports \ j \ more \ than \ i\};$$
$$\sim_a^I = \{(i,j) \in I \times I \mid \neg(i \succ_a^I j) \wedge \neg(j \succ_a^I i)\}.$$

For agent $a \in A$ and issues $i_1, i_2 \in I$, the relation $(i_1, i_2) \in \succ_a^I$ signifies that the agent a prefers i_1 over i_2; $(i_1, i_2) \in \prec_a^I$ denotes that the agent a prefers i_2

over i_1; $(i_1, i_2) \in \sim_a^I$ represents that the agent a has the same preference for i_1 and i_2.

The relationships between two agents can be classified into three distinct categories—agreement, partial agreement, and disagreement—based on their respective preference attitudes toward two identical issues. These relationships correspond to the conflict degrees $c_{ij}(a, b)$ between agents, denoted as $\delta^=$, $\delta^\approx$, and $\delta^\asymp$, where $\delta^=, \delta^\approx, \delta^\asymp \in \mathbb{R}$ ($\mathbb{R}$ representing the set of real numbers) and satisfying $\delta^= < \delta^\approx < \delta^\asymp$ and $\delta^= \leqslant 0 < \delta^\asymp$. The specific details are illustrated in Table 1 below.

Table 1. The case of conflict degrees between a and b w.r.t. i and j

Relationship	Preference by a and b	$c_{ij}(a, b)$
Agreement	$(i \succ_a^I j \wedge i \succ_b^I j) \vee (i \sim_a^I j \wedge i \sim_b^I j) \vee (j \succ_a^I i \wedge j \succ_b^I i)$	$\delta^=$
Partial agreement	$i \sim_a^I j \oplus i \sim_b^I j$	$\delta^\approx$
Disagreement	$(i \succ_a^I j \wedge j \succ_b^I i) \vee (j \succ_a^I i \wedge i \succ_b^I j)$	$\delta^\asymp$

Note: $\oplus$ denotes logical exclusive or.

Definition 2. *(Hu [24], 2025) In a $PS = (A, I, \{\succ_a^I \mid a \in A\})$, the conflict function $c_I : A \times A \longrightarrow \mathbb{R}$ on I is defined by: for $a, b \in A$,*

$$c_I(a, b) = \delta^= \left(\left| \succ_a^I \cap \succ_b^I \right| + \frac{\left| \sim_a^I \cap \sim_b^I \right| + |I|}{2} \right) + \delta^\approx \left(\left| \succ_a^I \cap \sim_b^I \right| + \left| \sim_a^I \cap \succ_b^I \right| \right) + \delta^\asymp \left(\left| \succ_a^I \cap \prec_b^I \right| \right),$$

where $| \bullet |$ denotes the cardinality of $\bullet$.

For agents $a, b \in A$ and issues set I, the value $c_I(a, b)$ stands for the conflict degree between a and b towards the issues set I. Since different decision-makers may provide different parameter values, the resulting conflict degree ranges will accordingly differ. Therefore, the max-min normalization method is employed to standardize the conflict degree within the range of $[0, 1]$.

Definition 3. *(Hu [24], 2025) In a $PS = (A, I, \{\succ_a^I \mid a \in A\})$, the normalized conflict function $\overline{c_I} : A \times A \longrightarrow [0, 1]$ on I is defined by: for $a, b \in A$,*

$$\overline{c_I}(a, b) = \frac{c_I(a, b) - min\ c_I}{max\ c_I - min\ c_I},$$

where $min\ c_I = \frac{\delta^= |I|(|I|+1)}{2}$, $max\ c_I = \frac{\delta^\asymp |I|(|I|+1)}{2}$.

For agents $a, b \in A$ and issues set I, the value $\overline{c_I}(a, b)$ stands for the conflict degree between a and b towards the issues set I. By Definition 3, we can perform a trisection of the set of all pairs of agents as follows.

Definition 4. *(Hu [24], 2025) In a $PS = (A, I, \{\succ_a^I \mid a \in A\})$, for $0 \leqslant \beta < \alpha \leqslant 1$, the alliance relation $R_I^=$, neutrality relation $R_I^\approx$, and conflict relation $R_I^\asymp$ on A w.r.t. issues set I are defined as:*

$$R_I^= = \{(a, b) \in A \times A \mid \overline{c_I}(a, b) < \beta\},$$
$$R_I^\approx = \{(a, b) \in A \times A \mid \beta \leq \overline{c_I}(a, b) \leq \alpha\},$$
$$R_I^\asymp = \{(a, b) \in A \times A \mid \overline{c_I}(a, b) > \alpha\}.$$

The triplet $R_{\bar{I}}^=$, $R_{\bar{I}}^\approx$ and $R_{\bar{I}}^\gtrless$ constitute a trisection of $A \times A$ towards I, satisfying the conditions: $R_{\bar{I}}^= \cap R_{\bar{I}}^\approx = \emptyset$, $R_{\bar{I}}^= \cap R_{\bar{I}}^\gtrless = \emptyset$, $R_{\bar{I}}^\approx \cap R_{\bar{I}}^\gtrless = \emptyset$, and $R_{\bar{I}}^= \cup R_{\bar{I}}^\approx \cup R_{\bar{I}}^\gtrless = A \times A$.

3 Three-Way Agent Relations Under Fuzzy Preference-Based Conflict Situations

The preference relations are insufficient for accurately characterizing agents' attitudes, this section introduces fuzzy preference-based conflict situation and discusses its properties. And a fuzzy similarity matrix is defined using the proposed conflict degree, which is then employed to partition agents into equivalence classes for trisections of agent pairs.

3.1 Fuzzy Preference-Based Conflict Situation

Definition 5. *Let I be a finite nonempty set of issues. $R \subseteq I \times I$ is a fuzzy preference relation on I, which is characterized by a membership function $r : I \times I \to [0,1]$. And the fuzzy preference relation is represented by a matrix $(r(i,j))_{\#(I) \times \#(I)}$, where $r(i,j)$ reflects the preference degree of i over j and $r(i,j) + r(j,i) = 1$ for $\forall i,j \in I$ and $\#(\bullet)$ denotes the cardinality of $\bullet$.*

Definition 6. *A conflict situation based on fuzzy preference is represented as a triplet $FS = (A, I, \{R_a^I \mid a \in A\})$ where:*

- *A is a finite nonempty set of agents;*
- *I is a finite nonempty set of issues;*
- *$R_a^I \subseteq I \times I$ is the fuzzy preference relation between issues with respect to an agent and can be equivalently represented by a fuzzy preference matrix $R_a^I = (r_a(i,j))_{\#(I) \times \#(I)}$.*

Remark 1. The following outlines the transformation between the two models and the consistency of the fuzzy preference relation.

(1) We can establish the connection between the conflict situation based on fuzzy preference and the preference-based conflict situation of Hu [24] when the membership function $r_a : U \times U \to \{0, 0.5, 1\}$. Specifically, $\succ_a^I = \{(i,j) \in I \times I \mid r_a(i,j) = 1\}$; $\sim_a^I = \{(i,j) \in I \times I \mid r_a(i,j) = 0.5\}$; $\prec_a^I = \{(j,i) \in I \times I \mid r_a(j,i) = 1\}$.

(2) In Definition 6, the problem of consistency in fuzzy preference needs to be considered. Specifically, if $r_a(i,j) = 0.7$ implies that the degree of preference an agent has for i over j is 0.7, and $r_a(j,k) = 0.8$ also implies that the preference degree of j for k is 0.8. According to the transitivity, the agent's preference for i over k cannot be less than 0.5, meaning a preference for i over k. Therefore, R_a^I must satisfy a certain condition [27] $r_a(i,j) \geqslant 0.5, r_a(j,k) \geqslant 0.5 \to r_a(i,k) \geqslant 0.5, \ \forall i,j,k \in I$ to ensure this consistency.

Theorem 1. *In a* $FS = (A, I, \{R_a^I \mid a \in A\})$, *for* $\forall i, j, k \in I$ *and an agent* $a \in A$. *The fuzzy preference degree* $r_a(i, j)$ *satisfies the following properties:*

(1) Reflexivity: $r_a(i, i) = 0.5$;
(2) Additive reciprocal: $r_a(i, j) + r_a(j, i) = 1$;
(3) Transitivity: $r_a(i, j) \geqslant 0.5, r_a(j, k) \geqslant 0.5 \rightarrow r_a(i, k) \geqslant 0.5$.

The properties in Theorem 1 specify that a rational conflict situation based on fuzzy preference should satisfy the following conditions:

(1) The preference degree of an agent for two identical issues is the median value of $[0, 1]$, which is 0.5, indicating that no preference relationship should exist between two identical issues, i.e., the agent should maintain a indifferent preference towards them.
(2) The agent's preference and converse preference between two issues exhibit an additive reciprocal, which also applies when the two issues are identical. Specifically, if i has a strong preference for j (i.e., $r_a(i, j)$ is large), the preference degree of j for i (i.e., $r_a(j, i)$) will be smaller, and vice versa.
(3) If an agent's preference for issue i over j is greater than or equal to 0.5, and the preference for issue j over k is also greater than or equal to 0.5, then the preference for issue i over k should also be greater than or equal to 0.5.

3.2   Three-Way Conflict Analysis Based on Decision-Theoretic Rough Sets

Definition 7. *Let* $FS = (A, I, \{R_a^I \mid a \in A\})$ *be a conflict situation based on fuzzy preference. For* $R_a^I = (r_a(i, j))_{\#(I) \times \#(I)}$, *the conflict function* $d_I : A \times A \longrightarrow [0, 1]$ *on* A *is defined by: for* $a, b \in A$,

$$d_I(a, b) = \frac{\sum_{i,j=1}^{\#(I)} |r_a(i, j) - r_b(i, j)|}{\#(I) \times \#(I)}.$$

Theorem 2. *In a* $FS = (A, I, \{R_a^I \mid a \in A\})$. *For* $\forall a, b, c \in A$, *the conflict degree* $d_I(a, b)$ *satisfies the following properties:*

(1) Non-negativity: $d_I(a, b) \geqslant 0$, *and specifically,* $d_I(a, a) = 0$;
(2) Symmetry: $d_I(a, b) = d_I(b, a)$;
(3) Triangle inequality: $d_I(a, b) + d_I(b, c) \geqslant d_I(a, c)$.

Proof. By Definition 7, the proof is straightforward.□

Definition 8. *In a* $FS = (A, I, \{R_a^I \mid a \in A\})$. *The fuzzy similarity matrix* M *of the issues set* I *is defined as follows:*

$$M = (s_I(a, b))_{\#(A) \times \#(A)},$$

where $s_I(a, b)$ *represents the similarity degree between* a *and* b *in the issues set* I, *i.e.* $s_I(a, b) = 1 - d_I(a, b)$.

Specifically, the fuzzy similarity matrix is a symmetric matrix with diagonal elements equal to 1. The elements are sorted in descending order of their values, denoted as $1 = t_1 > t_2 > \cdots > t_m$, where t_k represents the sorted fuzzy elements. We define $[a]^{t_k} = \{b \in A \mid s_I(a,b) \geqslant t_k\}$ as a t_k-level equivalence class of $a \in A$. The value of t_k can be interpreted as the "granularity" of partitioning equivalence classes. A higher value of t_k signifies a stronger indistinguishability relationship between agents, resulting in a finer partitioning, whereas a lower value of t_k indicates a weaker indistinguishability relationship, leading to a coarser partitioning. The choice of t_k is determined by the preferences of the decision-maker.

Definition 9. *In a $FS = (A, I, \{R_a^I \mid a \in A\})$. For $\forall X \subseteq A$ and $0 \leqslant \beta \leqslant \alpha \leqslant 1$, the evaluation function is $P(X|[a]^{t_k}) = \frac{|[a]^{t_k} \cap X|}{|[a]^{t_k}|}$. The allied, neutral, and conflict sets $AL_\beta^\alpha(X)$, $NE_\beta^\alpha(X)$, and $CO_\beta^\alpha(X)$ of X are defined as follows:*

$$AL_\beta^\alpha(X) = \{a \in A \mid P(X|[a]^{t_k}) > \alpha\};$$
$$NE_\beta^\alpha(X) = \{a \in A \mid \alpha \geqslant P(X|[a]^{t_k}) \geqslant \beta\};$$
$$CO_\beta^\alpha(X) = \{a \in A \mid P(X|[a]^{t_k}) < \beta\}.$$

The decision-theoretic rough set approach is applied to calculate the threshold values α and β in conflict analysis. Below, u_A, u_N, and u_C denote the actions assigned to $a \in A$ in categories $AL(X)$, $NE(X)$, and $CO(X)$, respectively. The losses incurred for taking actions u_A, u_N, and u_C are denoted by λ_{AA}, λ_{NA}, and λ_{CA}, respectively, when $a \in A$ belongs to $AL(X)$. Similarly, the losses associated with actions u_A, u_N, and u_C are represented by λ_{AC}, λ_{NC}, and λ_{CC}, respectively, when $a \in A$ belongs to $CO(X)$.

Theorem 3. *In a $FS = (A, I, \{R_a^I \mid a \in A\})$. The losses $\lambda_{AA}, \lambda_{NA}, \lambda_{CA}, \lambda_{CC}, \lambda_{NC}$, and λ_{AC}, where $0 \leq \lambda_{AA} \leq \lambda_{NA} \leq \lambda_{CA}$ and $0 \leq \lambda_{CC} \leq \lambda_{NC} \leq \lambda_{AC}$. Then*

(1) If $P(X|[a]^{t_k}) > \alpha$, then $a \in AL_\beta^\alpha(X)$;
(2) If $\alpha \geqslant P(X|[a]^{t_k}) \geqslant \beta$, then $a \in NE_\beta^\alpha(X)$;
(3) If $P(X|[a]^{t_k}) < \beta$, then $a \in CO_\beta^\alpha(X)$, where

$$\alpha = \frac{\lambda_{AC} - \lambda_{NC}}{\lambda_{AC} - \lambda_{NC} + \lambda_{NA} - \lambda_{AA}}, \beta = \frac{\lambda_{NC} - \lambda_{CC}}{\lambda_{NC} - \lambda_{CC} + \lambda_{CA} - \lambda_{NA}}, \gamma = \frac{\lambda_{AC} - \lambda_{CC}}{\lambda_{AC} - \lambda_{CC} + \lambda_{CA} - \lambda_{AA}}.$$

Proof. The expected losses $R(u_A|[a]^{t_k}), R(u_N|[a]^{t_k}), R(u_C|[a]^{t_k})$ associated with taking the individual action for an agent a can be expressed as:

$$R(u_A|[a]^{t_k}) = \lambda_{AA} P(X|[a]^{t_k}) + \lambda_{AC}(1 - P(X|[a]^{t_k}));$$
$$R(u_N|[a]^{t_k}) = \lambda_{NA} P(X|[a]^{t_k}) + \lambda_{NC}(1 - P(X|[a]^{t_k}));$$
$$R(u_C|[a]^{t_k}) = \lambda_{CA} P(X|[a]^{t_k}) + \lambda_{CC}(1 - P(X|[a]^{t_k})).$$

The Bayesian decision procedure suggests the following minimum-cost decision rules:

(A): If $R(u_A|[a]^{t_k}) \leqslant R(u_N|[a]^{t_k})$ and $R(u_A|[a]^{t_k}) \leqslant R(u_C|[a]^{t_k})$, then $a \in AL_\beta^\alpha(X)$;

(N): If $R(u_N|[a]^{t_k}) \leqslant R(u_A|[a]^{t_k})$ and $R(u_N|[a]^{t_k}) \leqslant R(u_C|[a]^{t_k})$, then $a \in NE_\beta^\alpha(X)$;

(C): If $R(u_C|[a]^{t_k}) \leqslant R(u_A|[a]^{t_k})$ and $R(u_C|[a]^{t_k}) \leqslant R(u_N|[a]^{t_k})$, then $a \in CO_\beta^\alpha(X)$.

Suppose $\lambda_{AA} \leq \lambda_{NA} \leq \lambda_{CA}$ and $\lambda_{CC} \leq \lambda_{NC} \leq \lambda_{AC}$, and the tie-breaking rule is applied when an object qualifies for multiple decision actions simultaneously, requiring the decision-maker to select a single action from the available alternatives. When $a \in AL_\beta^\alpha(X)$ and $a \in NE_\beta^\alpha(X)$ simultaneously, the decision is made in favor of $a \in AL_\beta^\alpha(X)$. Similarly, when $a \in CO_\beta^\alpha(X)$ and $a \in NE_\beta^\alpha(X)$ concurrently, the decision prioritizes $a \in CO_\beta^\alpha(X)$. Therefore the rules (A), (N), and (C) can be simply expressed as follows:

(A): If $P(X|[a]^{t_k}) > \alpha$ and $P(X|[a]^{t_k}) > \gamma$, then $a \in AL_\beta^\alpha(X)$;

(C): If $P(X|[a]^{t_k}) < \beta$ and $P(X|[a]^{t_k}) < \gamma$, then $a \in CO_\beta^\alpha(X)$;

(N): If $\beta \leqslant P(X|[a]^{t_k}) \leqslant \alpha$, then $a \in NE_\beta^\alpha(X)$, where

$$\alpha = \frac{\lambda_{AC} - \lambda_{NC}}{\lambda_{AC} - \lambda_{NC} + \lambda_{NA} - \lambda_{AA}}, \beta = \frac{\lambda_{NC} - \lambda_{CC}}{\lambda_{NC} - \lambda_{CC} + \lambda_{CA} - \lambda_{NA}}, \gamma = \frac{\lambda_{AC} - \lambda_{CC}}{\lambda_{AC} - \lambda_{CC} + \lambda_{CA} - \lambda_{AA}}.$$

Next, we provide Algorithm 1 for three-way conflict analysis based on fuzzy preference conflict situation as follows. And the time complexity of step 2 is $O(\#(A)^2 \cdot \#(I)^2)$, the time complexity of step 3, 4 and 5 is $O(\#(A)^2)$. Therefore, the time complexity of Algorithm 1 is $O(\#(A)^2 \cdot \#(I)^2)$.

Algorithm 1. The algorithm for three-way conflict analysis with fuzzy preference-based conflict situation.

Input: $FS = (A, I, \{R_a^I \mid a \in A\})$, X $(\emptyset \neq X \subseteq A)$, $\lambda_{AA}, \lambda_{NA}, \lambda_{CA}, \lambda_{CC}, \lambda_{NC}, \lambda_{AC}$;

Output: $AL_\beta^\alpha(X), NE_\beta^\alpha(X), CO_\beta^\alpha(X)$;

1: Input a conflict situation based on fuzzy preference $FS = (A, I, \{R_a^I \mid a \in A\})$;
2: Calculate conflict degree $d_I(a, b)$ for $a, b \in A$ according to Definition 7;
3: Calculate similarity degree $s_I(a, b)$ and get fuzzy similarity matrix M of the issues set I according to Definition 8;
4: Compute the t_k-level equivalence class $[a]^{t_k} = \{b \in A \mid s_I(a, b) \geqslant t_k\}$ of $a \in A$;
5: Compute the evaluation function $P(X|[a]^{t_k})$ and the threshold valuea α, β, γ according to Definitions 9 and Theorem 3;
6: Construct $AL_\beta^\alpha(X), NE_\beta^\alpha(X)$, and $CO_\beta^\alpha(X)$;
7: Output $AL_\beta^\alpha(X), NE_\beta^\alpha(X)$, and $CO_\beta^\alpha(X)$;

---

4 Comparison Analysis and Case Study

4.1 Comparison of Our Model and Other Existing Models

To better illustrate the advantages of the proposed model, we compare it with several existing models, as shown in Table 2. In this paper, the research environment of the conflict situation table is extended to fuzzy preference conflict situation, where the fuzzy preference relation of a single agent towards issue pairs is provided. This allows for a more precise description of an agent's preference degree for issue pairs, addressing the problem in Hu's model [24] of representing the preference degree of agents towards issue pairs in preference-based conflict situation. Clearly, traditional conflict analysis models [4,10,14] suffer from the problem where a single evaluation of agents on issues leads to different interpretations of the same rating in the conflict situation table. In conflict analysis, it is crucial to determine whether the relationship between agents is one of conflict, neutrality, or alliance. Therefore, selecting an appropriate threshold to distinguish relation is necessary. In the Pawlak [4] and Yao [10], an auxiliary function is used for classification, but the given thresholds $\{-1, 0, +1\}$ remain subjective. In this paper, we choose conditional probability to perform the classification, using loss functions and Bayesian minimum risk theory to characterize the threshold, which offers a more objective approach.

4.2   A Case Study

This section illustrates the application of three-way conflict analysis based on fuzzy preference by examining the conflict situation, which describes the attitudes of six countries in the Middle East conflict on five key issues. In this case, a conflict situation based on fuzzy preference $FS = (A, I, \{R_a^I \mid a \in A\})$, where

Table 2. Comparison of five models of conflict analysis.

Models	Threshold	Research environment	Preference relation	Fuzzy preference relation
Our paper	Loss function	Fuzzy preference based conflict situation	✓	✓
Hu [24]	Subjectivity	Preference based conflict situation	✓	×
Pawlak [4]	Subjectivity	Three valued situation table	×	×
Yao [10]	Subjectivity	Three valued, many valued situation table	×	×
Lang [14]	Loss function	Pythagorean fuzzy information system	×	×

$A = \{a_1, a_2, a_3, a_4, a_5, a_6\}$, $I = \{i_1, i_2, i_3, i_4, i_5\}$. The agents a_1, a_2, a_3, a_4, a_5, and a_6 represent Israel, Egypt, Palestine, Jordan, Syria, and Saudi Arabia, respectively; the issues i_1, i_2, i_3, i_4 and i_5 represent "Autonomous Palestinian state on the West Bank and Gaza", "Israeli military outpost along the Jordan River", "Israel retains East Jerusalem", "Israeli military outposts on the Golan Heights", "Arab countries grant citizenship to Palestinians who choose to remain within their borders", respectively. The following gives the fuzzy preference matrix $R_{a_1}^I$, $R_{a_2}^I$, $R_{a_3}^I$, $R_{a_4}^I$, $R_{a_5}^I$ and $R_{a_6}^I$ of agents under issues set I. Based on the Definition 6 presented above, each agent should have the following fuzzy preference matrix. For instance, $r_{a_1}(i_1, i_2) = 0.6$ denotes that the preference degree of agent a_1 for issue i_1 over i_2 is 0.6; $r_{a_1}(i_4, i_4) = 0.5$ denotes that a_1 maintains an indifferent preference between issues i_4 and i_4. The proposed model conducts data analysis in the situation table based on agents' fuzzy preferences over issue pairs. It overcomes the limitation of traditional conflict analysis, where agents provide single evaluations for individual issues, leading to identical ratings in the situation table but differing interpretations. Moreover, in this model, the fuzzy preference relation matrix of agents over issue pairs takes values in $[0, 1]$, enabling a precise representation of agents' preference degrees. This addresses the limitation in the Hu's model regarding the characterization of preference intensity.

$$R_{a_1}^I = \begin{bmatrix} 0.50 & 0.60 & 0.70 & 0.80 & 0.90 \\ 0.40 & 0.50 & 0.60 & 0.70 & 0.80 \\ 0.30 & 0.40 & 0.50 & 0.60 & 0.70 \\ 0.20 & 0.30 & 0.40 & 0.50 & 0.60 \\ 0.10 & 0.20 & 0.30 & 0.40 & 0.50 \end{bmatrix} , \quad R_{a_2}^I = \begin{bmatrix} 0.50 & 0.70 & 0.60 & 0.80 & 0.90 \\ 0.30 & 0.50 & 0.60 & 0.60 & 0.70 \\ 0.40 & 0.40 & 0.50 & 0.70 & 0.80 \\ 0.20 & 0.40 & 0.30 & 0.50 & 0.60 \\ 0.10 & 0.30 & 0.20 & 0.40 & 0.50 \end{bmatrix} ,$$

$$R_{a_3}^I = \begin{bmatrix} 0.50 & 0.40 & 0.65 & 0.80 & 0.95 \\ 0.60 & 0.50 & 0.75 & 0.65 & 0.80 \\ 0.35 & 0.25 & 0.50 & 0.65 & 0.70 \\ 0.20 & 0.35 & 0.35 & 0.50 & 0.65 \\ 0.05 & 0.20 & 0.30 & 0.35 & 0.50 \end{bmatrix} , \quad R_{a_4}^I = \begin{bmatrix} 0.50 & 0.70 & 0.90 & 0.60 & 0.80 \\ 0.30 & 0.50 & 0.70 & 0.40 & 0.60 \\ 0.10 & 0.30 & 0.50 & 0.20 & 0.40 \\ 0.40 & 0.60 & 0.80 & 0.50 & 0.70 \\ 0.20 & 0.40 & 0.60 & 0.30 & 0.50 \end{bmatrix} ,$$

$$R_{a_5}^I = \begin{bmatrix} 0.50 & 0.40 & 0.60 & 0.70 & 0.80 \\ 0.60 & 0.50 & 0.70 & 0.80 & 0.90 \\ 0.40 & 0.30 & 0.50 & 0.60 & 0.70 \\ 0.30 & 0.20 & 0.40 & 0.50 & 0.60 \\ 0.20 & 0.10 & 0.30 & 0.40 & 0.50 \end{bmatrix} , \quad R_{a_6}^I = \begin{bmatrix} 0.50 & 0.10 & 0.20 & 0.30 & 0.40 \\ 0.90 & 0.50 & 0.10 & 0.20 & 0.30 \\ 0.80 & 0.90 & 0.50 & 0.10 & 0.20 \\ 0.70 & 0.80 & 0.90 & 0.50 & 0.10 \\ 0.60 & 0.70 & 0.80 & 0.90 & 0.50 \end{bmatrix} .$$

The conflict degree $d_I(a_i, a_j)$ between two countries with respect to the set of issues I is calculated as presented in the Table 3. The fuzzy similarity matrix for the six countries is presented below, with the matrix elements ordered in descending order $1 > 0.9 > 0.862 > 0.852 > 0.816 > 0.808 > 0.804 > 0.792 > 0.784 > 0.756 > 0.748 > 0.7 > 0.648 > 0.632 > 0.62 > 0.6$. Then the t_k-level equivalence class can be partitioned by selecting $t_k = 0.808$, $[a_1]^{t_k} = \{a_1, a_2, a_5\}, [a_2]^{t_k} = \{a_1, a_2, a_3, a_5\}, [a_3]^{t_k} = \{a_2, a_3, a_5\}, [a_4]^{t_k} = \{a_4\}, [a_5]^{t_k} = \{a_1, a_2, a_3, a_5\}, [a_6]^{t_k} = \{a_6\}$.

$$M = \begin{bmatrix} 1 & 0.9 & 0.784 & 0.748 & 0.862 & 0.6 \\ 0.9 & 1 & 0.808 & 0.756 & 0.852 & 0.62 \\ 0.784 & 0.808 & 1 & 0.792 & 0.816 & 0.648 \\ 0.748 & 0.756 & 0.792 & 1 & 0.804 & 0.632 \\ 0.862 & 0.852 & 0.816 & 0.804 & 1 & 0.7 \\ 0.6 & 0.62 & 0.648 & 0.632 & 0.7 & 1 \end{bmatrix} .$$

Given a agents set $X = \{a_1, a_5\}$ (X means the probability of these countries forming an alliance is higher based on historical data). The conditional probability $P(X|[a]^{t_k})$ for every a can be calculated as:

$$P(X|[a_1]^{t_k}) = \frac{|[a_1]^{t_k} \cap X|}{|[a_1]^{t_k}|} = \frac{|\{a_1, a_2, a_5\} \cap \{a_1, a_5\}|}{|\{a_1, a_2, a_5\}|} = \frac{2}{3};$$

$$P(X|[a_2]^{t_k}) = \frac{|[a_2]^{t_k} \cap X|}{|[a_2]^{t_k}|} = \frac{|\{a_1, a_2, a_3, a_5\} \cap \{a_1, a_5\}|}{|\{a_1, a_2, a_3, a_5\}|} = \frac{1}{2};$$

$$P(X|[a_3]^{t_k}) = \frac{|\,[a_3]^{t_k} \cap X|}{|\,[a_3]^{t_k}\,|} = \frac{|\{a_2, a_3, a_5\} \cap \{a_1, a_5\}|}{|\{a_2, a_3, a_5\}|} = \frac{1}{3};$$

$$P(X|[a_4]^{t_k}) = \frac{|\,[a_4]^{t_k} \cap X|}{|\,[a_4]^{t_k}\,|} = \frac{|\{a_4\} \cap \{a_1, a_5\}|}{|\{a_4\}|} = 0;$$

$$P(X|[a_5]^{t_k}) = \frac{|\,[a_5]^{t_k} \cap X|}{|\,[a_5]^{t_k}\,|} = \frac{|\{a_1, a_2, a_3, a_5\} \cap \{a_1, a_5\}|}{|\{a_1, a_2, a_3, a_5\|} = \frac{1}{2};$$

$$P(X|[a_6]^{t_k}) = \frac{|\,[a_6]^{t_k} \cap X|}{|\,[a_6]^{t_k}\,|} = \frac{|\{a_6\} \cap \{a_1, a_5\}|}{|\{a_6\}|} = 0.$$

By Theorem 3, we have the thresholds α, β, and γ using Table 4 as follows:

$$\alpha = \frac{\lambda_{AC} - \lambda_{NC}}{\lambda_{AC} - \lambda_{NC} + \lambda_{NA} - \lambda_{AA}} = \frac{6-4}{6-4+4-0} = \frac{1}{3},$$

$$\beta = \frac{\lambda_{NC} - \lambda_{CC}}{\lambda_{NC} - \lambda_{CC} + \lambda_{CA} - \lambda_{NA}} = \frac{4-0}{4-0+10-2} = \frac{2}{7},$$

$$\gamma = \frac{\lambda_{AC} - \lambda_{CC}}{\lambda_{AC} - \lambda_{CC} + \lambda_{CA} - \lambda_{AA}} = \frac{6-0}{6-0+10-0} = \frac{3}{8}.$$

By Definition 9, we get the allied, neutral, and conflict sets $AL^{\alpha}_{\beta}(X)$, $NE^{\alpha}_{\beta}(X)$, and $CO^{\alpha}_{\beta}(X)$ of X as follows:

$$AL^{\alpha}_{\beta}(X) = \{a_1, a_2, a_5\}, \ NE^{\alpha}_{\beta}(X) = \{a_3\}, \ CO^{\alpha}_{\beta}(X) = \{a_4, a_6\}.$$

By comparing the results of conflict analysis with X, we observe the following: first, the analysis confirms that a_4, a_6 has never been part of the X alliance, as indicated by historical data. Second, while historical data also suggest that a_2 was not part of the X alliance, further analysis reveals that a_2 was, in fact, a member. Therefore, the conflict analysis method proposed in this study is crucial for identifying and rectifying inconsistencies in historical data, offering significant practical implications.

Table 3. Conflict degree for the Middle East conflict

$d_I(a_i, a_j)$	a_1	a_2	a_3	a_4	a_5	a_6
a_1	0	0.1	0.216	0.252	0.168	0.4
a_2	0.1	0	0.192	0.244	0.148	0.38
a_3	0.216	0.192	0	0.208	0.184	0.352
a_4	0.252	0.244	0.208	0	0.196	0.368
a_5	0.168	0.148	0.184	0.196	0	0.3
a_6	0.4	0.38	0.352	0.368	0.3	0

Table 4. Loss function related to $a, b \in A$

Action	$a \in AL(b)$	$a \in CO(b)$
u_A	$\lambda_{AA} = 0$	$\lambda_{AC} = 6$
u_N	$\lambda_{NA} = 4$	$\lambda_{NC} = 4$
u_C	$\lambda_{CA} = 10$	$\lambda_{CC} = 0$

5 Conclusion

Conflict analysis has developed rapidly and achieved significant progress. However, existing research predominantly relies on agents' direct ratings of issues, neglecting potential variations in their interpretations of these ratings. To address this limitation, this paper introduces a three-way conflict analysis model based on fuzzy preferences, utilizing agents' fuzzy preferences over issues to facilitate a more refined and comprehensive conflict analysis. The main conclusions of this paper are summarized as follows:

(1) We represent the preference relations of the agent over the issues set through the fuzzy preference matrice, thereby constructing a fuzzy preference conflict situation. And the transformation between our model and the Hu's model is achieved by constraining the degree of fuzzy preference.

(2) We determine the conflict degree between agents using the mean value and further construct a fuzzy similarity matrix based on agents similarity, thereby identifying the similarity equivalence classes for each agent. Based on the decision-theoretic rough set approach, we conduct a tri-partition of the agents set for conflict analysis.

(3) We designed an algorithm to obtain the rule for three-way conflict analysis based on fuzzy preference conflict situation and illustrated the detailed procedures of the model through a case study on the Middle East conflict.

The conflict situation based on fuzzy preferences provides a novel perspective for conflict analysis; however, certain limitations remain. In high-dimensional datasets, pairwise comparisons between issues significantly increase computational complexity. And while fuzzy preference relations effectively quantify preference levels, they do not provide decision-makers with clear and immediate conclusions, highlighting the need for further research and exploration. In future research, we may explore how to utilize fuzzy preferences to measure the fuzzy indifference relation. Additionally, the importance of agents in decision-making is a crucial issue worth investigating, and incorporating agent weights into the computation of conflict degrees could be considered. Finally, investigating fuzzy preference situations under incomplete information presents another promising direction for future study.

Acknowledgements. This work is supported by the National Natural Science Foundation of China (Nos. 62076040,12471431), the Scientific Research Fund of Hunan Provincial Education Department (No. 22A0233), and the Hunan Provincial Key Laboratory of Mathematical Modeling and Analysis in Engineering (No. 2018MMAEZD10).

References

1. Taffer, J.: The Power of Conflict, HarperCollins (2022)
2. Pawlak, Z.: About conflicts, ICS PAS Reports (1981). 451
3. Pawlak, Z.: On conflicts. Int. J. Hum Comput Stud. **21**, 127–34 (1984)
4. Pawlak, Z.: An inquiry into anatomy of conflicts. Inf. Sci. **109**, 65–78 (1998)
5. Yao, Y.Y.: Three-way granular computing, rough sets, and formal concept analysis. Int. J. Approximate Reasoning **116**, 106–125 (2020)
6. Yao, Y.Y.: Three-way decisions and cognitive computing, Cognitive. Computing **8**, 543–554 (2016)
7. Yao, Y.Y.: Interval sets and three-way concept analysis in incomplete contexts. Int. J. Mach. Learn. Cybern. **8**, 3–20 (2017)
8. Yao, Y.Y.: Three-way decision and granular computing. Int. J. Approximate Reasoning **103**, 107–123 (2018)
9. Yao, Y.Y.: Three-way conflict analysis: reformulations and extensions of the Pawlak model. Knowl.-Based Syst. **180**, 26–37 (2019)
10. Lang, G.M., Luo, J.F., Yao, Y.Y.: Three-way conflict analysis: a unification of models based on rough sets and formal concept analysis. Knowl.-Based Syst. **194**, 105556 (2020)
11. Lang, G.: A general conflict analysis model based on three-way decision. Int. J. Mach. Learn. Cybern. **11**(5), 1083–1094 (2020). https://doi.org/10.1007/s13042-020-01100-y
12. Lang, G.M., Mao, D.Q., Cai, M.G.: Three-way decision approaches to conflict analysis using decision-theoretic rough set theory. Inf. Sci. **406–407**, 185–207 (2017)
13. Li, X.N., Wang, X., Lang, G.M., Yi, H.J.: Conflict analysis based on three-way decision for triangular fuzzy information systems. Int. J. Approximate Reasoning **132**, 88–106 (2021)
14. Lang, G.M., Miao, D.Q., Fujita, H.: Three-way group conflict analysis based on Pythagorean fuzzy set theory. IEEE Trans. Fuzzy Syst. **28**, 447–61 (2020)
15. Lang, G.M., Ding, W.P., Miao, D.Q., Fujita, H., Yao, Y.Y.: Trisection-fusion and fusion-trisection methods of three-way conflict analysis with Pythagorean fuzzy information. Appl. Soft Comput. **164**, 111939 (2024)
16. Du, J.L., Liu, S.F., Liu, Y., Yi, J.H.: A novel approach to three-way conflict analysis and resolution with Pythagorean fuzzy information. Inf. Sci. **584**, 65–88 (2022)
17. Wang, T.X., Zhang, L.B., Huang, B., Zhou, X.Z.: Three-way conflict analysis based on interval-valued Pythagorean fuzzy sets and prospect theory. Artif. Intell. Rev. **56**, 6061–6099 (2023)
18. Zhao, J., Wan, R.X., Miao, D.Q.: Conflict analysis triggered by three-way decision and Pythagorean fuzzy rough set. Int. J. Comput. Intell. Syst. **17**, 17 (2024)
19. Yi, H.J., Zhang, H.M., Li, X.N., Yang, Y.P.: Three-way conflict analysis based on hesitant fuzzy information systems. Int. J. Approximate Reasoning **139**, 12–27 (2021)
20. Liu, Y., Lin, Y.: Intuitionistic fuzzy rough set model based on conflict distance and applications. Appl. Soft Comput. **31**, 266–273 (2015)

21. Wang, T.X., Huang, B.: Interval-valued intuitionistic fuzzy three-way conflict analysis based on cumulative prospect theory. J. Intel. Fuzzy Syst. **48**, 183–196 (2025)
22. Yang, H.L., Wang, Y., Guo, Z.L.: Three-way conflict analysis based on hybrid situation tables. Inf. Sci. **628**, 522–541 (2023)
23. Lu, C.Y., Yang, H.L., Guo, Z.L.: Three-way conflict analysis based on multi-scale situation tables. Appl. Intell. **55**, 293 (2025)
24. Hu, M.J.: Three-way conflict analysis with preference-based conflict situations. Inf. Sci. **693**, 121676 (2025)
25. Tanino, T.: Fuzzy preference orderings in group decision making. Fuzzy Sets Syst. **12**, 117–131 (1984)
26. Tanino, T.: Fuzzy preference relations in group decision making. Non-Conventional Prefer. Relat. Decision Making **301**, 54–71 (1988)
27. Herrera-Viedma, E., Herrera, F., Chiclana, F., Luque, M.: Some issues on consistency of fuzzy preference relations. Eur. J. Oper. Res. **154**, 98–109 (2004)

A Granular-Ball SVM Based
on Three-Way Decision

Rong Huang[1], Jie Yang[1,2(✉)], and Yanmin Liu[2]

[1] School of Computer, Jiangsu University of Science and Technology,
Zhenjiang 212100, Jiangsu, China
yj530966074@foxmail.com
[2] School of Physics and Electronic Science, Zunyi Normal University,
Zunyi 563002, Guizhou, China

Abstract. The granular-ball (GB)-based classifier exhibits adaptability in creating coarse-grained information granules as input, thereby enhancing its generality and flexibility. Nevertheless, current GB-based classifiers rigidly assign a specific class label to each data instance and lack the necessary strategies to address uncertain instances. Such certain classification approaches to uncertain instances may suffer considerable risks. To solve this problem, We introduced the three-way decision into granular-ball SVM (GBSVM) to construct a robust three-way granular-ball SVM (3WGBSVM) model for uncertain data, which categorizes data instances into certain classes and uncertain cases. Extensive comparative experiments are conducted with 4 GB-based classifiers on 6 public benchmark datasets. The results show that our model demonstrates robustness in managing uncertain data and effectively mitigates classification risks. Furthermore, our model almost outperforms the other comparative methods in both effectiveness and efficiency.

Keywords: granular-ball · SVM · three-way decision

1 Introduction

Granular computing (GrC) [1], as introduced by Zadeh, offers a powerful framework to improve the efficiency of knowledge discovery by processing information at varying levels of granularity. Currently, several classical models of GrC have emerged, such as rough set [2], fuzzy set [3], and quotient space [4], each offering unique advantages in handling uncertainty and complexity. In the classification field, GrC-based methods have shown remarkable improvements in classification accuracy and computational efficiency [5,6]. However, such classifiers suffer from poor robustness, low efficiency, and weak noise resistance due to their reliance on inputs at the finest granularity level.

Granular-ball computing (GBC) [7] is an effective, robust, and highly interpretable multi-granularity computation approach proposed by Xia, which aims to address the limitations of traditional GrC-based methods. GBC combines the

Q. Zhang et al. (Eds.): IJCRS 2025, LNAI 15710, pp. 33–43, 2025.
https://doi.org/10.1007/978-3-031-92741-6_3

idea of GrC with the concept of 'hyper-balls' to represent clusters of data points with similar characteristics. Over the years, research on GBC has made significant advancements. Xie [8] proposed both speedy and steady GB generation method based on the attention mechanism, which can significantly improve the efficiency and robustness while absolutely stabilizing. Cheng [9] introduced a novel manifold clustering algorithm based on GBC, which significantly reduces the time complexity of processing high-dimensional data. However, existing GB-based classification methods strictly classify data into determined category. The lack of a method for dealing with uncertain data in the situation of insufficient information leads to misclassification. It is well known that three-way decision (3WD) is a method proposed by Yao [10] to effectively deal with the uncertainty problem in the actual decision-making process, which is in line with human thinking and cognitive characteristics. Its core idea is to reduce the loss caused by wrong decisions and improve the interpretability of the model by introducing delayed decisions. Nowadays, the 3WD has been widely applied in many research fields. Three-way classification is an important research topic in recent years due to the combination of 3WD and machine learning. Yang [11] proposed a three-way classifier utilizing neighborhood rough sets, grounded in the principle of minimizing fuzziness loss. Combining the advantages of 3WD, to solve the constraints of the GB-based classification model under uncertainty, we introduce 3WD to build a three-way granular-ball support vector machine classifier, denoted by 3WGBSVM. In more detail, the shadowed set based on fuzzy set is constructed by fuzzy-rough transformation, and the 3WD model is established with the objective of minimizing the loss of fuzziness. The thresholds required for the 3WD rule are eventually obtained.

The rest of the paper is organized as follows: In Sect. 2, we review the preliminary definitions related to the content of the paper. The construction process of 3WGBSVM model is presented in Sect. 3. In Sect. 4, relevant experiments to validate and evaluate our proposed model are given. In Sect. 5, a detailed conclusion is presented.

2   Preliminaries

In this section, we will review some necessary definitions to facilitate a better understanding of this paper.

Definition 1. *(Center and Radius of Granular-ball) [7] Given a dataset $D = \{(x_i, y_i), i = 1, 2, \cdots, n\}$, where y_i denotes the label of x_i. Suppose GBs $= \{gb_1, gb_2, \cdots, gb_m\}$ is a GB space generated on D. The center C and radius r of a granular-ball gb are respectively defined as follows:*

$$C = \frac{1}{|gb|} \sum_{x \in gb} x \tag{1}$$

$$r = \frac{1}{|gb|} \sum_{x \in gb} |x - C| \tag{2}$$

where, C is the gravity of all objects in gb and r is the average distance from all objects in gb to C.

Definition 2. *(Label and Purity of Granular-ball) [7] Given a dataset $D = \{(x_i, y_i), i = 1, 2, \cdots, n\}$, where y_i denotes the label of x_i. Suppose $GBs = \{gb_1, gb_2, \cdots, gb_m\}$ is a GB space generated on D. The label l and purity p of a granular-ball gb are respectively defined as follows:*

$$l = \arg\max_{l_j \in L} |\{(x, y) \in gb | y = l_j\}| \tag{3}$$

$$p = \frac{|\{(x, y) \in gb | y = l_j\}|}{|gb|} \tag{4}$$

where L represents the set of label categories of samples contained in gb.

Definition 3. *(Three-way decision rules) [10] Suppose $X = \{x_1, x_2, \cdots, x_n\}$ is a non-empty finite set. $\mu_A(x)$ is the fuzzy membership of the object $x \in X$ on the fuzzy set A on X. For a given threshold pair (α, β), the three-way decision rules are formulated as follows:*

$$\begin{aligned} POS_{\alpha,\beta} &= \{x | \mu_A(x) \geq \alpha\} \\ BND_{\alpha,\beta} &= \{x | \alpha < \mu_A(x) < \beta\} \\ NEG_{\alpha,\beta} &= \{x | \mu_A(x) \leq \beta\} \end{aligned} \tag{5}$$

Due to the point-by-point input adopted by SVM [12], the computational load increases as the amount of data grows. To solve this problem, Xia [13] introduced the concept of granular-ball into SVM and proposed the granular-ball support vector machine (GBSVM). In GBSVM, data points are replaced with granular-balls, namely support granular-balls and non-support granular-balls [7]. Suppose the support planes are k_1 and k_2. Then these two support planes are constrained by two rules: (1) the support planes must be tangent to the support granular-balls; (2) the distance between the support planes and each granular ball should be no less than the corresponding radius. The objective function of the separable GBSVM is expressed as follows:

$$\begin{aligned} \min_{w,b} \quad & \tfrac{1}{2}||w||^2, \\ s.t. \quad & y_i(w \cdot c_i + b) - ||w||r_i \geq 1, i = 1, 2, \cdots, n. \end{aligned} \tag{6}$$

To address the inseparable classification problem, the slack variable ξ and penalty coefficient η are taken into Eq. 6. The inseparable GBSVM is obtained. Suppose $GBs = \{gb_1, gb_2, \cdots, gb_m\}$ is a GB space generated on D. Therefore, the inseparable GBSVM objective function is formulated as follows:

$$\begin{aligned} \min_{w,b,\xi_i} \quad & \tfrac{1}{2}||w||^2 + \eta \sum_{i=1}^{m} \xi_i, \\ s.t. \quad & y_i(wc_i + b) - ||w||r_i \geq 1 - \xi_i, \\ & \xi_i \geq 0, i = 1, 2, \cdots, m. \end{aligned} \tag{7}$$

where c_i and r_i are respectively the center and radius of gb_i.

3 Three-Way Granular-Ball SVM

The 3WD methodology has been combined with various kinds of machine learning methods to implement three-way learning for uncertain data analysis. This helps to reduce the risk of decision-making in many domains by postponing decision-making in uncertain situations. To further improve the classification performance and reduce the decision risk, we first introduce the GBs generated by Algorithm 1 into SVM to construct a GBSVM. Then, based on the fuzzy-rough transformation theory, we further propose a three-way granular-ball SVM, i.e., 3WGBSVM.

Algorithm 1. GB generation method [7]

Input: Datasets D;
Output: The set of GBs;
Initialize $GBs = \emptyset$;

1: Implement the 2-means clustering algorithm on D and generate two granular-balls gb_1^i and gb_2^i, where i is the iteration number, and it is initialized as 1;
2: **for** each gb_j^i **do**
3:    The center C of gb_j^i is calculated by Eq. 1, the radius R by Eq. 2, the purity P by Eq. 4;
4:    **if** $P < 1$ **then**
5:        split gb_j^i by applying the 2-means clustering algorithm on it;
6:    **end if**
7: **end for**
8: **if** the purity of each gb is higher than 1 **then**
9:    $GBs = GBs \cup gb$;
10: **else**
11:    $i = i + 1$;
12:    Go to step 2;
13: **end if**
14: **return** GBs;

The membership degree of training data mainly relies on the calculation of the distance from the data point to the class center. A larger distance implies a smaller membership degree, and vice versa. The definitions are as follows:

Definition 4. *(Fuzzy membership function) [14] Given a dataset $D = \{\{x_i, y_i\}, i = 1, 2, ..., n\}$, where y_i denotes the label of x_i. Let the average of "$y_i = +1$" denoted as x_+ and the average of "$y_i = -1$" denoted as x_-. Let the radius of "$y_i = +1$" be denoted as r_+ and "$y_i = -1$" be denoted as r_-, respectively:*

$$r_+ = \max_{\{x_i : y = 1\}} |x_+ - x_i| \tag{8}$$

$$r_- = \max_{\{x_i : y = -1\}} |x_- - x_i| \tag{9}$$

Then let the fuzzy membership be a function of the average and radius per class:

$$\mu(x_i) = \begin{cases} 1 - \frac{|x_i - x_+|}{r_+ + q}, & y_i = 1 \\ 1 - \frac{|x_i - x_-|}{r_- + q}, & y_i = -1 \end{cases} \quad (10)$$

where $q = 0.1$ is used to avoid the case $\mu(x_i) = 0$.

After the membership degree of training data is calculated, according to the fuzzy-rough transformation theory, the mapping is formed as follows:

Definition 5. *(Fuzzy-rough transformation) [15] Given a dataset $D = \{\{x_i, y_i\}, i = 1, 2, ..., n\}$, where y_i denotes the label of x_i. $\mu(x_i)$ is the fuzzy membership of the object $x_i \in D$. Then the mapping function $\mathbb{S} : D \to \{0, [0, 1], 1\}$ is defined as follows:*

$$\mathbb{S}(x_i) = \begin{cases} 0, & \mu(x_i) \leq \beta \\ \mu(x_i), & \beta < \mu(x_i) < \alpha \\ 1, & \mu(x_i) \geq \alpha \end{cases} \quad (11)$$

where α and β are two real numbers, $0 \leq \beta < \alpha \leq 1$.

The thresholds α and β are computed by optimizing the objective function for uncertainty variation as follows:

$$\underset{\alpha, \beta}{\arg \min} \, \mathbb{V}_{(\alpha, \beta)} \quad (12)$$

where $\mathbb{V}_{(\alpha, \beta)} = \sum\limits_{\mu(x_i) \leq \beta} S(x_i) + \sum\limits_{\mu(x_i) \geq \alpha} 1 - S(x_i) + \sum\limits_{\beta < \mu(x_i) < \alpha} |0.5 - S(x_i)|, \, 0 \leq \beta \leq 0.5, \, 0.5 \leq \alpha \leq 1$. The first two parts of $\mathbb{V}_{(\alpha, \beta)}$ denote the membership loss in a certain region and the last part denotes the membership loss in an uncertain region.

The performance of SVM are affected by many factors, including parameter selection and model configuration. Since each model under different training configurations may make different predictive labels for the same test point, considering this diversity, we quantify the membership of test data under multiple models, i.e., the commonality of the test point in the predictions of all models. Therefore, we define the membership of test data as follows.

Definition 6. *(Membership of test data) Given a non-empty finite set $X = \{x_1, x_2, \cdots, x_n\}$. Suppose a SVM training configuration parameter set is $configs = \{cg_1, cg_2, \cdots, cg_5\}$, and each SVM training configuration cg_j corresponds to a trained model $m_j \in models = \{m_1, m_2, \cdots, m_5\}$. For $\forall x_i \in X$, we obtain the set of predictive labels $pl_i = \{pl_1, pl_2, \cdots, pl_5\}$ for it by models. The membership degree of x_i is defined as follows:*

$$\mu_{test}(x_i) = \frac{|\{pl_j \in pl_i | pl_j = +1\}|}{|models|} \quad (13)$$

where pl_j is the class of labels predicted by model m_j, $pl_j \in \{+1, -1\}$.

The explicit construction process of the classification model is demonstrated in Fig. 1. First, a GBSVM model is trained by generating GBs based on Algorithm 1; The membership degree of each training point is calculated according to Eq. 10, and fuzzy-rough transformations are then performed according to Eq. 11. The optimal threshold pair (α, β) are obtained by Eq. 12. Finally, the labels of the test points are predicted based on the three-way decision rules.

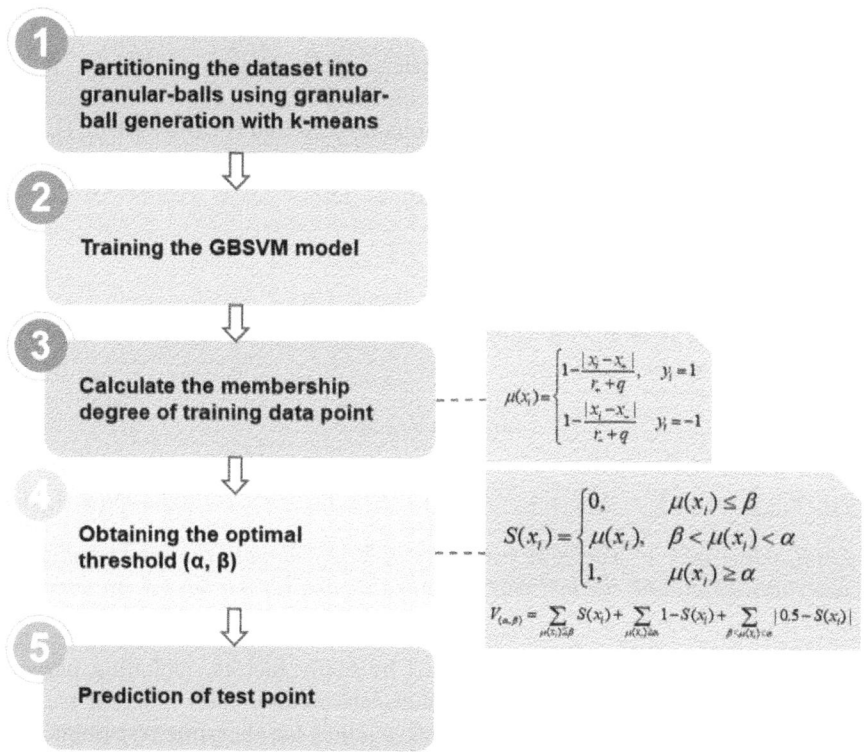

Fig. 1. The constructive process of 3WGBSVM

4   Experiment

In this section, we synthetically compare our proposed 3WGBSVM with each of the GB-based classifiers(GBSVM, GBKNN, GBKNN++) to provide a comprehensive overview of its effectiveness, efficiency, stability, and robustness. All the experiments and comparisons were implemented on 6 UCI benchmark datasets, which are described in detail in Table 1. To evaluate the generalization ability of the model across different datasets, 10-fold cross-validation was conducted on each dataset.

4.1   Effectiveness of 3WGBSVM VS Other Classifiers

Four classical machine learning evaluation metrics were used in the next comparison experiments to evaluate the effectiveness of the model: *Accuracy*, *Precision*,
Recall, and *F*1 *score*. The higher the value of these metrics, the better the classification performance. The comparison experiment contained three different classification algorithms with 3WGBSVM, i.e., GBSVM, GBKNN, and GBKNN++.
The average performance of four models on the above-mentioned four metrics
across 6 datasets is summarized in the *Average* row of Table 2. It presents that
3WGBSVM outperforms the other models on average across four metrics. Con-
cretely, 3WGBSVM achieves the highest scores in *Accuracy* (0.8740), *Precision*
(0.8910), *Recall* (0.8807), and *F*1 *score* (0.8833), showing its effectiveness in
classification tasks.

At the end of Table 2, a statistical analysis of the results is shown, which
includes *Win/Loss*, *Rank*, and *P − values*. *Win/Loss* reflects the number of
wins and losses from pairwise comparisons between 3WGBSVM and the other
models. Out of 72 pairwise comparisons, 3WGBSVM achieved 61 wins, demonstrating a clear advantage over its competitors. *Rank* indicates the model's overall performance, with a lower value indicating better performance. 3WGBSVM
achieved the highest ranking, followed by GBSVM.

Overall, 3WGBSVM is leading in effectiveness due to the introduction of
3WD, which allows the model to effectively handle uncertain data instances.
Unlike traditional classifiers, which can only perform hard classification, 3WGB-
SVM divides the data into certain and uncertain classes, thereby enhancing its
ability to manage ambiguous instances, reduce classification errors, and improve
accuracy.

Table 1. The information of experimental datasets

NO.	Datasets	Instances	Attributes
1	Breast Cancer	699	9
2	Raisin	901	6
3	Endgame	958	9
4	Ring	7400	20
5	High Diamond Ranked	9880	38
6	Elect	10000	13

4.2   Efficiency of 3WGBSVM VS Other Classifiers

We compared the prediction time of the above four models, and the results are
shown in Table 3. In terms of average efficiency, there is no significant difference
in the time taken by 3WGBSVM and GBSVM, with 3WGBSVM ranking first
(1.3333) and GBSVM ranking second (1.8333). And it is worth noting that
as the dataset size increases, 3WGBSVM consumes remarkably less time than

Table 2. The statistical analysis of various classifiers

NO.	Metrics	3WGBSVM	GBSVM	GBKNN	GBKNN++
1	Accuracy	0.9686	0.9600	0.9599	0.9556
	Precision	0.9952	0.9623	0.9613	0.9566
	Recall	0.957	0.9600	0.9599	0.9556
	F1 Score	0.9754	0.9603	0.9601	0.9553
2	Accuracy	0.8689	0.8600	0.8533	0.8522
	Precision	0.9005	0.8654	0.8579	0.8561
	Recall	0.8278	0.8600	0.8533	0.8522
	F1 Score	0.8611	0.8600	0.8534	0.8523
3	Accuracy	0.7703	0.7443	0.7359	0.7536
	Precision	0.7785	0.7571	0.7404	0.7816
	Recall	0.9140	0.7443	0.7359	0.7536
	F1 Score	0.8384	0.7320	0.7297	0.7585
4	Accuracy	0.9345	0.9382	0.7651	0.8295
	Precision	0.9924	0.9439	0.8400	0.8299
	Recall	0.8770	0.9382	0.7651	0.8295
	F1 Score	0.9311	0.9381	0.7506	0.8295
5	Accuracy	0.7259	0.7248	0.7087	0.6498
	Precision	0.7233	0.7249	0.7087	0.6502
	Recall	0.7304	0.7248	0.7087	0.6498
	F1 Score	0.7267	0.7248	0.7086	0.6498
6	Accuracy	0.9758	0.9658	0.9071	0.8596
	Precision	0.9562	0.9667	0.9069	0.8617
	Recall	0.9780	0.9658	0.9071	0.8596
	F1 Score	0.9669	0.9659	0.9069	0.8603
Average	Accuracy	**0.8740**	0.8655	0.8217	0.8167
	Precision	**0.8910**	0.8701	0.8359	0.8227
	Recall	**0.8807**	0.8655	0.8217	0.8167
	F1 Score	**0.8833**	0.8635	0.8182	0.8176
Statistics	win/loss	61/11	51/21	19/53	13/59
	rank	**1.4583**	1.8750	3.2083	3.4583
	p − values		0.312085	0.066564	0.011687

GBSVM. This means that the time required for 3WGBSVM is not significantly different from a classification model based on the same GB generation method. So this time cost is acceptable with the introduction of 3WD. Overall, 3WGBSVM provides a solution to reduce classification errors.

4.3   Robustness of 3WGBSVM VS Other Classifiers

To verify robustness of 3WGBSVM, we constructed 10%, 20%, 30%, and 40% noise ratios by varying the labels of the training set respectively. We represent the trend of *Accuracy* of each model on each dataset at different noise rates by line graphs in Fig. 2. As the noise level increases, the *Accuracy* of all models decreases to varying degrees across different datasets. However, the curve of

3WGBSVM stays at the highest position or the second highest position in the 6 datasets. This indicates that our model maintains a high accuracy as the noise rate increases. As in Fig. 3, the results are visualized using heat map to show the rank of *Accuracy* on each dataset at different noise rates. For example, it can be seen from Fig. 2f that on the Elect dataset with a noise rate of 10%, 3WGBSVM achieves the highest *Accuracy*, so marked as 1 in Fig. 3f.

Table 3. The prediction time required for various classifiers

NO.	3WGBSVM	GBSVM	GBKNN	GBKNN++
1	0.0075	**0.0008**	0.0075	0.0083
2	0.0093	**0.0027**	0.0095	0.0099
3	0.0102	**0.0051**	0.0106	0.0083
4	**0.0239**	0.0445	0.0450	0.2803
5	**0.0280**	0.1024	0.0884	1.1061
6	**0.0303**	0.0513	0.0725	0.7404
rank	**1.3333**	**1.8333**	2.8333	3.8333

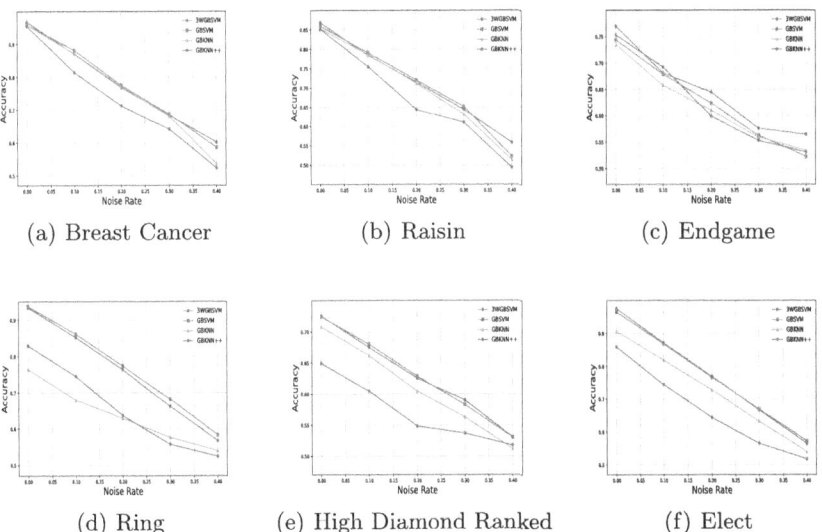

(a) Breast Cancer   (b) Raisin   (c) Endgame

(d) Ring   (e) High Diamond Ranked   (f) Elect

Fig. 2. The comparison of accuracy under different noise rates

The above advantages of 3WGBSVM are mainly attributed to the fact that in the classical GBSVM model, noise tends to affect the model's judgment of

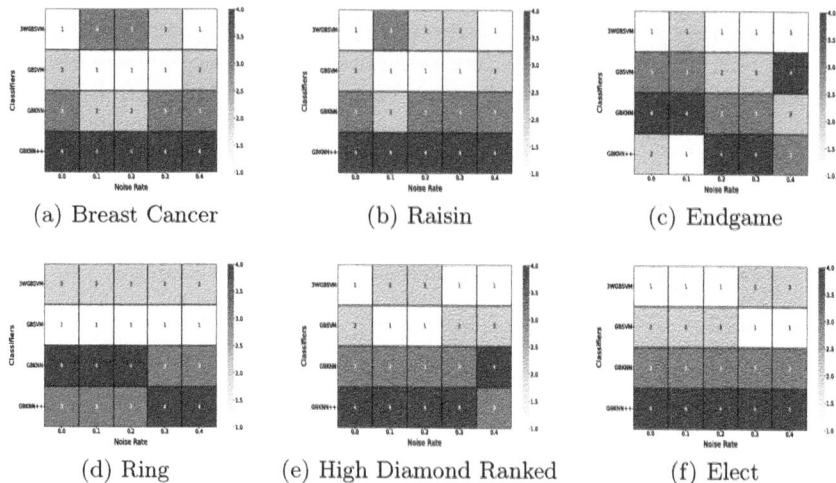

(a) Breast Cancer   (b) Raisin   (c) Endgame

(d) Ring   (e) High Diamond Ranked   (f) Elect

Fig. 3. Comparison on ranking of accuracy under different noise rates

the boundaries, leading to a degradation of classification performance. In contrast, 3WGBSVM offers more robust classification results when handling data uncertainty through the concept of three-way decision. This mechanism allows the model to make more accurate decisions when the data is noisy, preventing performance fluctuations caused by noise.

5   Conclusion

To enhance the performance of GBSVM and reduce decision risk, we proposed a three-way granular-ball support vector machine (3WGBSVM) based on three-way decision (3WD) theory. Through extensive experiments conducted on 6 UCI datasets, we demonstrated that the 3WGBSVM significantly outperforms other GB-based classifiers in terms of both effectiveness and robustness. Nevertheless, although our proposed model can effectively handle uncertain data, its processing mechanism relies on fixed rules, which may cause limitations in more complex application scenarios. Therefore, future research can focus on integrating incremental learning to design 3WGBSVM so that the model can dynamically handle changing data streams.

Acknowledgment. This work was supported by the National Science Foundation of China (Grant number 62466063), the Guizhou Provincial Department of Education Colleges and Universities Science and Technology Innovation Team (QJJ[2023]084), Science and Technology Project of Zunyi(ZSKRPT[2023] 3, ZSKH HZ [2023] 158), Key Laboratory of Evolutionary Artificial Intelligence in Guizhou (QJJ[2022] No.059).

References

1. Zadeh, L.A.: Fuzzy sets and information granularity. In: Fuzzy Sets, Fuzzy Logic, and Fuzzy Systems: Selected Papers, pp. 433–448. World Scientific (1996)
2. Yao, W., Han, S.-E.: A topological approach to rough sets from a granular computing perspective. Inf. Sci. **627**, 238–250 (2023)
3. Ali, I., Li, Y., Pedrycz, W.: Granular computing approach to evaluate spatio-temporal events in intuitionistic fuzzy sets data through formal concept analysis. Axioms **12**(5), 407 (2023)
4. Zhang, L., Zhang, B.: Multi-granular computing and quotient structure. In: Granular, Fuzzy, and Soft Computing, pp. 311–322. Springer, Cham (2023)
5. Guo, S., Zhao, H.: Hierarchical classification with multi-path selection based on granular computing. Artif. Intell. Rev. **54**(3), 2067–2089 (2021)
6. Niu, J., Chen, D., Li, J., Wang, H.: A dynamic rule-based classification model via granular computing. Inf. Sci. **584**, 325–341 (2022)
7. Xia, S., Liu, Y., Ding, X., Wang, G., Hong, Yu., Luo, Y.: Granular ball computing classifiers for efficient, scalable and robust learning. Inf. Sci. **483**, 136–152 (2019)
8. Xie, Q., et al.: GBG++: a fast and stable granular ball generation method for classification. IEEE Trans. Emerg. Top. Comput. Intell. **8**, 2022–2036 (2024)
9. Cheng, D., Liu, S., Xia, S., Wang, G.: Granular-ball computing-based manifold clustering algorithms for ultra-scalable data. Expert Syst. Appl. **247**, 123313 (2024)
10. Yao, Y.: The superiority of three-way decisions in probabilistic rough set models. Inf. Sci. **181**(6), 1080–1096 (2011)
11. Yang, J., Wang, X., Wang, G., Zhang, Q., Zheng, N., Di, W.: Fuzziness-based three-way decision with neighborhood rough sets under the framework of shadowed sets. IEEE Trans. Fuzzy Syst. **32**(9), 4976–4988 (2024)
12. Hearst, M.A., Dumais, S.T., Osuna, E., Platt, J., Scholkopf, B.: Support vector machines. IEEE Intell. Syst. Appl. **13**(4), 18–28 (1998)
13. Xia, S., Lian, X., Wang, G., Gao, X., Chen, J., Peng, X.: GBSVM: an efficient and robust support vector machine framework via granular-ball computing. IEEE Trans. Neural Netw. Learn. Syst. 1–15 (2024)
14. Lin, C.F., De Wang, S.: Fuzzy support vector machines. IEEE Trans. Neural Netw. **13**(2), 464–471 (2002)
15. Yue, X., Zhou, J., Yao, Y., Miao, D.: Shadowed neighborhoods based on fuzzy rough transformation for three-way classification. IEEE Trans. Fuzzy Syst. **28**(5), 978–991 (2020)

A Decision-Theoretic Formulation of Three-Way Conflict Analysis

Jialin Hou$^{(\boxtimes)}$ and Yiyu Yao

Department of Computer Science, University of Regina, Regina, SK S4S 0A2, Canada
{jhv842,Yiyu.Yao}@uregina.ca

Abstract. Classifying a pair of agents, based on their ratings on an issue, into one of alliance, conflict, or neutral relations is a basic issue in Pawlak conflict analysis. In this paper, we first review three models that are based on an auxiliary function, a distance function, and a pair of alliance and conflict measures, respectively. We show that the differences among the three models lie in their use of different partial orderings of the nine pairs of ratings given by two agents. Based on such an observation, in this paper, we give a decision-theoretic analysis by introducing a loss function. The loss function specifies the costs of assigning a pair of ratings to alliance, conflict, and neutrality. For any pair of ratings, we take an action with the minimum cost. The decision-theoretic analysis enables us to unify existing models. In other words, by using different loss functions, we can derive the existing models.

Keywords: Decision-theoretic analysis · Three-way decision · Three-way conflict analysis

1 Introduction

Pawlak [1,2] proposed a rough-sets-based conflict analysis model with a three-valued situation table, in which a set of agents rate a set of issues by using a three-valued scale of + (support), − (opposition), and 0 (neutral). A main task is to divide the relationships between agents into alliance, conflict, and neutrality relations, based on their ratings of a set of issues. Pawlak conflict analysis uses a three-valued situation table and a trisection of agent relations based on the triad of alliance-conflict-neutrality. Guided by the principles of three-way decision theory for thinking, problem-solving, and computing in threes [3–8], Yao [9] interpreted and reformulated Pawlak conflict analysis into three-way conflict analysis. He introduced additional triadic structures, including, for example, the triad of (alliance, weak alliance, non-alliance), the triad of (conflict, weak conflict, non-conflict), and others. In recent years, three-way conflict analysis has received much attention [10–20].

Y. Yao—This work was partially supported by a Discovery Grant from NSERC, Canada. The authors thank the reviewers for their constructive comments.

Q. Zhang et al. (Eds.): IJCRS 2025, LNAI 15710, pp. 44–57, 2025.
https://doi.org/10.1007/978-3-031-92741-6_4

For any issue, two agents may have one of the nine possible pairs of ratings, namely, $(-,-)$, $(-,0)$, $(0,-)$, $(0,0)$, $(0,+)$, $(+,0)$, $(-,+)$, and $(+,-)$. How to classify or measure these nine pairs in terms of alliance, conflict, and neutrality is of fundamental importance in three-way conflict analysis. By introducing an auxiliary function on the nine pairs, Pawlak [1] classified the two pairs $(-,-)$ and $(+,+)$ into alliance, the two pairs $(+,-)$ and $(-,+)$ into conflict, and four pairs $(-,0)$, $(0,-)$, $(0,+)$, and $(+,0)$ into neutrality. For the pair $(0,0)$, Pawlak classified it into alliance for the same agent, and into neutrality for two different agents.

Yao [9] introduced a distance function on the nine pairs to produce the Pawlak trisection, except that the pair $(0,0)$ is classified uniformly into alliance. Lang and Yao [18] suggested that an auxiliary function or a distance function may not fully reflect the meaning of the nine pairs with respect to alliance, conflict, and neutrality. Except for the extreme cases, any pair of ratings may reveal different degrees of alliance, conflict, and neutrality. They introduced an alliance-conflict ordering on the nine pairs and a pair of alliance and conflict measures for producing a trisection. The alliance-conflict ordering explicitly gives the constraints that alliance and conflict measures must satisfy. Yang et al. [21] extended this line of research by considering a bilattice with two orderings on the nine pairs, namely, a strength ordering and a rating ordering. They introduced measures of support-alliance, opposition-alliance, alliance strength, and opposition strength for classifying the nine pairs.

The orderings of the nine pairs with respect to alliance, conflict, and neutrality can provide more meaningful and qualitative information. In this paper, we provide a new ordering-based decision-theoretic formulation. Instead of defining various measures according to the orderings of the nine pairs, we introduce a loss function for classifying the nine pairs into alliance, conflict, and neutrality, respectively. The loss function is constrained by the ordering. We will choose a classification decision that has the minimum loss or cost. By using different loss functions, we will obtain different trisecting models.

The main objective of this paper is to establish a sound decision-theoretic basis for three-way conflict analysis. The rest of the paper is organized as follows. In Sect. 2, we review three conflict analysis models. In Sect. 3, we present a decision-theoretic formulation and use an example to illustrate our formulation. In Sect. 4, we provide a comparative examination of the decision-theoretic model in relation to the Pawlak [1] model and the Yao [9] model. Finally, in Sect. 5, we summarize the main results and point out future research directions.

2   Models of Three-Way Conflict Analysis

In this section, we recall some basic concepts of Pawlak conflict analysis and review models of three-way conflict. For the clarity of discussion, we focus mainly on a single issue. In a future paper, we will extend the results to multiple issues.

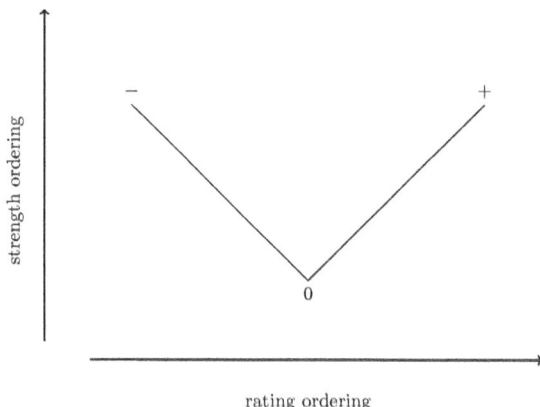

Fig. 1. Strength and rating orderings [21]

2.1 Three-Valued Situation Table and Two Orderings

An input of the Pawlak [1] conflict analysis is a three-valued situation table that represents the opinions of a set of agents on a set of issues. In a tabular form, the rows represent agents, the columns represent issues, and the cells represent ratings of an agent on an issue.

Definition 1. *A three-valued situation table is a triplet $S = (A, I, r)$, where A is a finite nonempty set of agents, I is a finite nonempty set of issues, and $r : A \times I \longrightarrow \{+, -, 0\}$ is a rating function. The rating of an agent $x \in A$ regarding to an issue $i \in I$ can be denoted as $r(x, i)$ with following interpretations: an agent x is positive on, or supports, an issue i if $r(x, i) = +$; is negative on, or opposes, an issue i if $r(x, i) = -$; is neutral (i.e., neither supports nor opposes) if $r(x, i) = 0$.*

By presenting the three values as $+$, $-$, and 0, we emphasize the qualitative nature of agent ratings. For easy computation, one may represent the three ratings numerically as $+1$, -1, and 0, which allows arithmetic operations. However, it may be commented that this numerical representation is not the only representation, and computations based on it may not always produce meaningful results. There are two important features of the three-valued ratings. First, the ratings $+$ and $-$ represent the two extreme opinions, i.e., support and opposition, and the rating 0 in the middle, represents a neutral state. Second, it is reasonable to assume that an agent with an extreme rating may have a strong commitment or hold an opinion firmer than another agent with a neutral rating. To reflect these two features, by adopting the idea from three-valued logic [22,23], Yao [9] introduced two orderings on $\{+, 0, -\}$. As shown in Fig. 1, horizontally, the rating order, $- \prec_r 0 \prec_r +$, represents the ratings from negative to neutral to positive; vertically, the strength order, $0 \prec_s -$ and $0 \prec_s +$, shows that the ratings of $+$ and $-$ are stronger than the rating 0. These two orderings form the basic context for understanding and analyzing a conflict situation.

2.2   Three Models of Three-Way Conflict Analysis

Pawlak conflict analysis deals with relationships between agents (dually, relation-ships between issues) based on ratings in a situation table. Pawlak [1] suggests to divide the relationships into the alliance, conflict, and neutrality three binary relations, based on an auxiliary function.

Definition 2. *In a situation table, for an issue* $i \in I$, *an auxiliary function* $\Phi_i : A \times A \longrightarrow \{+1, 0, -1\}$ *is defined by [1]: for* $(x, y) \in A \times A$,

$$\Phi_i(x, y) = \begin{cases} +1, & r(x, i) \cdot r(y, i) = +1 \vee x = y, \\ 0, & r(x, i) \cdot r(y, i) = 0 \wedge x \neq y, \\ -1, & r(x, i) \cdot r(y, i) = -1, \end{cases} \quad (1)$$

According to the values of the auxiliary function, the set of all pairs of agents $A \times A$ *is trisected into three pairwise disjoint binary relations:*

$$\begin{aligned} R_i^= &= \{(x, y) \in A \times A \mid \Phi_i(x, y) = +1\}, \\ R_i^\times &= \{(x, y) \in A \times A \mid \Phi_i(x, y) = -1\}, \\ R_i^\approx &= \{(x, y) \in A \times A \mid \Phi_i(x, y) = 0\}, \end{aligned} \quad (2)$$

where $R_i^=$, $R_i^\times$, *and* $R_i^\approx$ *denote, respectively, the alliance, conflict, and neutrality relations.*

There are nine possible pairs of ratings for two agents. The order of two ratings in a pair does not affect the alliance, conflict, and neutrality of the two agents. The two pairs $(0, -)$ and $(-, 0)$ are equivalent and are grouped together. The two pairs $(0, +)$ and $(+, 0)$ form another group, and the two pairs $(-, +)$ and $(+, -)$ form a third group. Each of the two $(+, +)$ and $(-, -)$ forms a group by itself. Pawlak treats the $(0, 0)$ differently. For two agents $x, y \in A$, if $x = y$, the alliance relation holds, and we label it as $(0, 0)^*$; If $x \neq y$, the neutrality relation holds, and we label it as $(0, 0)$. Therefore, there is a total of seven groups. The results of the Pawlak model may be explained by orderings on the seven groups, as shown in Fig. 2. More importantly, we arrange the seven groups according to two types of partial ordering. Vertically, we have three orderings denoted by solid lines (note that lines that can be derived from transitivity are not drawn). The top-down ordering shows an increase in conflict, the bottom-up ordering shows an increase in alliance, and the top-middle and bottom-middle orderings show an increase in neutrality. The dotted line shows the change in the support and opposition strength (again, lines that can be derived from transitivity are not drawn). The left-right ordering shows an increase in support, and right-left ordering shows an increase in opposition.

From the figure, we can conclude that the Pawlak trisection of $A \times A$ is determined by vertical orderings, with $(0, 0)$, $(-, 0)/(0, -)$, and $(0, +)/(+, 0)$ as the middle or neutral points. The horizontal orderings are not used. These orderings are useful when constructing support alliance coalitions and opposition alliance coalitions, respectively [21].

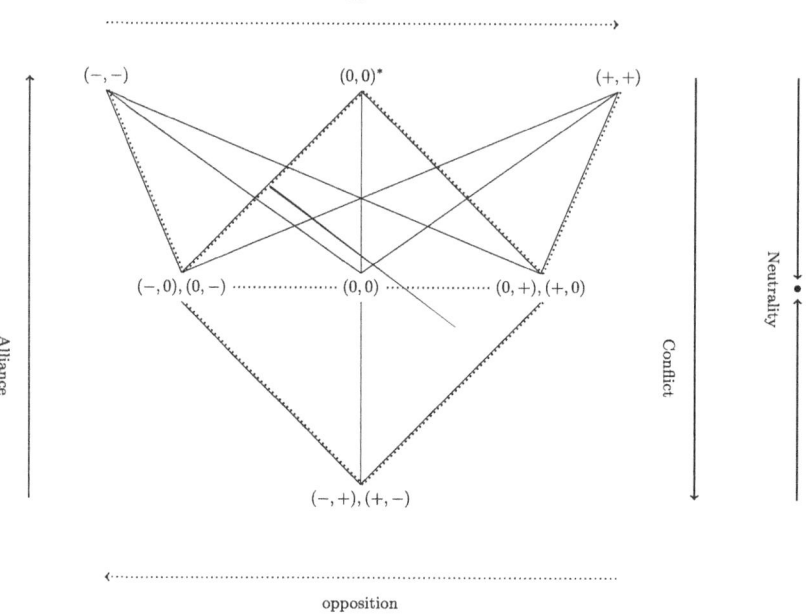

Fig. 2. Orderings of nine pairs of ratings: Pawlak model

Yao [9] argues that the pair of $(0,0)$ should be treated uniformly, independent of whether the pair of agents is formed by the same agent or two different agents. He suggests to use distance function to revise the Pawlak model.

Definition 3. *In a situation table, for an issue* $i \in I$, *distance function* $D_i :$ $A \times A \longrightarrow \{0, 0.5, 1\}$ *is defined by [9]: for* $(x, y) \in A \times A$,

$$D_i(x, y) = \frac{|r(x, i) - r(y, i)|}{2} = \begin{cases} 0, & r(x, i) = r(y, i), \\ 0.5, & r(x, i) \cdot r(y, i) = 0 \wedge r(x, i) \neq r(y, i), \\ 1, & r(x, y) \cdot r(y, i) = -1, \end{cases}$$

(3)

where $|r(x, i) - r(y, i)|$ *is the absolute value of the difference between the two ratings* $r(x, i)$ *and* $r(y, i)$. *According to the distance function, the alliance, conflict and neutrality relations are given by:*

$$\begin{aligned} R_i^= &= \{(x, y) \in A \times A \mid D_i(x, y) = 0\}, \\ R_i^\asymp &= \{(x, y) \in A \times A \mid D_i(x, y) = 1\}, \\ R_i^\approx &= \{(x, y) \in A \times A \mid D_i(x, y) = 0.5\}. \end{aligned}$$

(4)

The three relations are pairwise disjoint.

Figure 3 shows the orderings of the six groups of pairs for explaining the results of the Yao model. In this case, the pairs $(-, 0)/(0, -)$ and $(0, +)/(+, 0)$

present the middle or neutral points. In contrast to the Pawlak model, the pair $(0,0)$ is uniformly placed at the top. Alternatively, we may uniformly put $(0,0)$ in the middle, which will be characterized by Fig. 2 with $(0,0)^*$ removed.

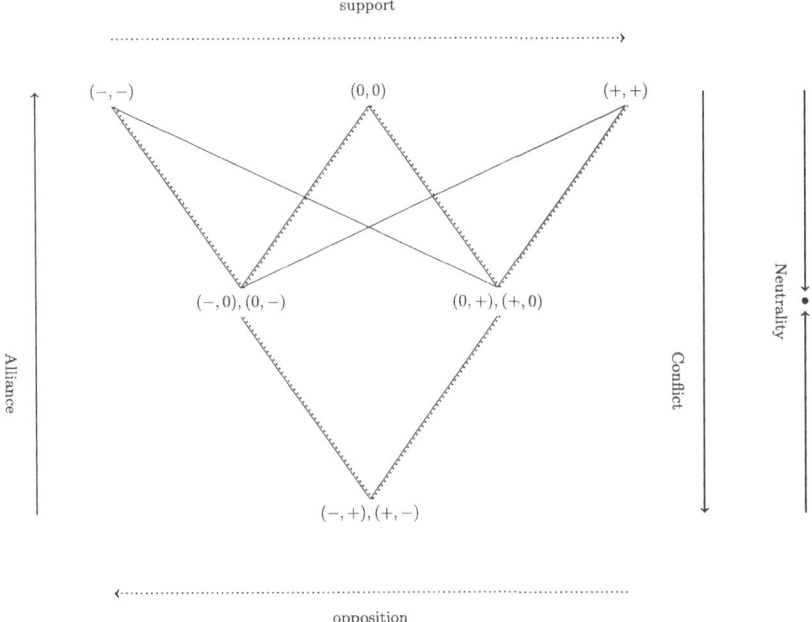

Fig. 3. Orderings of nine pairs of ratings: Yao model

Pawlak model and Yao model agree on orderings of all pairs, except for the pair $(0,0)$. There are two choices, that is, assigning $(0,0)$ to either alliance or neutrality. There may be ground for each of the choices. The assignment of the pairs $(-,0)/(0,-)$ and $(0,+)/(+,0)$ to neutrality may be questionable in some situations, as each of the pair also shows some degrees of alliance and conflict. To address these problems, Lang and Yao [18] consider other orderings of the six groups of pairs, as shown in Fig. 4. In this case, only the pair $(0,0)$ in the middle presents a neutral point, a weak alliance point, and a weak conflict point. This is consistent with the strength and rating orderings in Fig. 1. For example, it is reasonable to assume that the degree of alliance of two agents with rating pair $(0,0)$ should be smaller than the degree of alliance of two agents with rating pair $(-,-)$ or $(+,+)$, because the strength of 0 is weaker than the strength of both $-$ and $+$. It is also reasonable to assume that the degree of conflict of two agents with rating pair $(0,0)$ should be smaller than the degree of conflict of two agents with rating pairs $(-,0)/(0,-)$ or $(0,+)/(+,0)$, because the latter are either more negative or more positive than $(0,0)$.

From Fig. 4, we can see that Lang and Yao model consider four levels of alliance, conflict, and neutrality, instead of three levels of the Pawlak model and

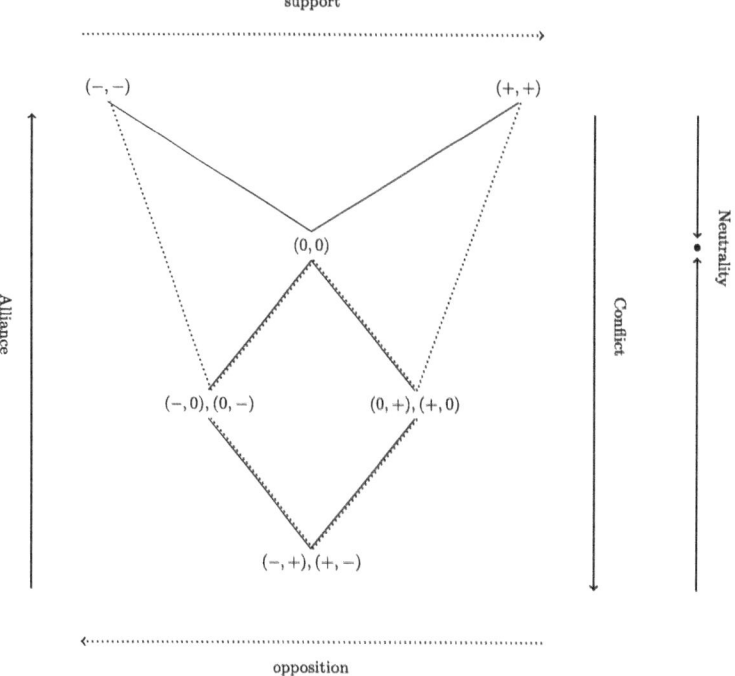

Fig. 4. Orderings of nine pairs of ratings: Lang and Yao model

Yao model. This perhaps better reflects all features of the three-value rating under the rating ordering and the strength ordering. Based on the orderings in Fig. 4, Lang and Yao [18] introduce alliance and conflict measures to trisect agent relationships.

3   A Decision-Theoretic Formulation

In this section, we propose a decision-theoretic approach from a cost-sensitive view to categorize the nine pairs of ratings into conflict, alliance, and neutrality three groups.

3.1   Loss Function

Given a pair of ratings, we may classify it into one of the alliance, conflict, and neutrality relations. Different classifications may have different losses or costs. For example, classifying $(+, +)$ into conflict should incur a higher loss or risk than classifying it into alliance. Formally, we can use a loss function $\lambda :$ $(\{+, 0, -\} \times \{+, 0, -\}, S) \longrightarrow \Re$, where $\Re$ is the set of real numbers and S is one of three relations, to map the cost of assigning a rating pair to each of the three relations. For simple analysis, Table 1 gives notations for representing various losses. Since the order of two ratings in a pair does not affect the relationship of the two agents, we assume that the losses of these pairs are equal. For example, $\lambda(0+, A) = \lambda(+0, A)$. We may use $\lambda(0+, A)$ and $\lambda(+0, A)$ interchangeably.

Table 1. Loss function

Ratings	Alliance	Conflict	Neutrality
$(-,-)$	$\lambda(--, A)$	$\lambda(--, C)$	$\lambda(--, N)$
$(+,+)$	$\lambda(++, A)$	$\lambda(++, C)$	$\lambda(++, N)$
$(0,0)$	$\lambda(00, A)$	$\lambda(00, C)$	$\lambda(00, N)$
$(-,0),(0,-)$	$\lambda(-0, A) = \lambda(0-, A)$	$\lambda(-0, C) = \lambda(0-, C)$	$\lambda(-0, N) = \lambda(0-, N)$
$(0,+),(+,0)$	$\lambda(0+, A) = \lambda(+0, A)$	$\lambda(0+, C) = \lambda(+0, C)$	$\lambda(0+, N) = \lambda(+0, N)$
$(-,+),(+,-)$	$\lambda(+-, A) = \lambda(-+, A)$	$\lambda(+-, C) = \lambda(-+, C)$	$\lambda(+-, N) = \lambda(-+, N)$

When classifying a pair of agents into alliance, conflict, or neutrality with respect to an issue, the corresponding losses obey the three orderings in Fig. 4. For example, if a pair of agents has the ratings $(-, -)$, it would have the lowest loss for classifying the pair of agents into alliance, the medium loss for classifying it into neutrality, and the highest loss for classifying it into conflict. We can similarly see the constraints on some other groups. Formally, we have the following constraints on the loss function below:

$$\begin{aligned}
(-, -): \quad & \lambda(--, C) > \lambda(--, N) > \lambda(--, A), \\
(+, +): \quad & \lambda(++, C) > \lambda(++, N) > \lambda(++, A), \\
(0, 0): \quad & \frac{\lambda(00, A)}{\lambda(00, C)} \geq \lambda(00, N), \\
(-, +), (+, -): \quad & \lambda(-+, A) > \lambda(-+, N) > \lambda(-+, C).
\end{aligned} \quad (5)$$

It should be noted that the orderings of rating pairs may not impose any constraints on the losses of certain pairs. For example, $\lambda(00, A)$ may be either greater than, equal, or less than $\lambda(00, C)$.

Based only on orderings in Fig. 4, it is a bit difficult to set and explain the constraints on the groups of $(-, 0)/(0, -)$ and $(+, 0)/(0, +)$. In this case, we also need to consider the strength ordering of the three ratings in Fig. 1 from a rating change point of view. Assume that the typical neutral point is $(0, 0)$, the typical alliance points are $(-, -)$ and $(+, +)$, and the typical conflict point is $(-, +)$. From $(-, 0)$ to $(0, 0)$, we need to change a rating from an extremely opposing attitude about an issue to neutral. From $(-, 0)$ to $(-, -)$, we need to change a rating from a neutral attitude about an issue to extremely opposing. Changing a rating from an extreme attitude to a neutral one is harder than changing from a neutral attitude to an extreme one. With such considerations, the losses of the two groups of ratings $(-, 0)/(0, -)$ and $(0, +)/(+, 0)$ should satisfy the following constraints:

$$(-,0),(0,-): \ \lambda(-0,N) \geq \frac{\lambda(-0,A)}{\lambda(-0,C)},$$

$$(0,+),(+,0): \ \lambda(0+,N) \geq \frac{\lambda(0+,A)}{\lambda(0+,C)}. \tag{6}$$

The losses of the pair $(0,-)/(-,0)$ for alliance and conflict can be assigned without any constraints. The same is true for the pair $(0,+)/(+,0)$.

In Fig. 4, alliance increased from the bottom-up, conflict increased from the top-down, and neutrality increased from the top-middle and bottom-middle. These hints that the loss function must be constrained by the orderings. Formally, the losses of assigning the nine pairs of ratings to alliance must satisfy the following conditions:

$$\lambda(-+,A) > \frac{\lambda(-0,A)}{\lambda(0+,A)} > \lambda(00,A) > \frac{\lambda(--,A)}{\lambda(++,A)}. \tag{7}$$

The above conditions are a reversal of the alliance ordering in Fig. 4. The loss of assigning the pair $(+,-)$ to alliance is the greatest since it has the highest conflict degree based on Fig. 4. The loss of assigning the pair $(-,-)/(+,+)$ to alliance is the smallest since it has the lowest conflict degree. For a similar reason, the losses of assigning the pairs $(0,0)$ and $(-,0)/(0,-)$ to alliance are in the middle. Since the degree of conflict incurred by $(0,0)$ is lower than $(-,0)/(0,-)$, the loss of assigning $(0,0)$ to alliance is smaller than $(-,0)/(0,-)$.

Similarly, the losses of assigning the nine pairs of ratings to conflict must satisfy the following conditions:

$$\frac{\lambda(--,C)}{\lambda(++,C)} > \lambda(00,C) > \frac{\lambda(-0,C)}{\lambda(0+,C)} > \lambda(-+,C). \tag{8}$$

For losses assigning pairs to the neutrality, from the top-middle direction, the losses of $(+,+)$ and $(-,-)$ are greater than $(0,0)$. From the bottom-middle direction, the losses of $(+,-)$ is greater than $(+,0)$ and $(-,0)$, and the latter two is large than $(0,0)$. Thus, the losses of assigning pairs to neutrality must satisfy the following conditions:

$$\begin{matrix} \lambda(--,N) \\ \lambda(++,N) \\ \lambda(-+,N) > \dfrac{\lambda(-0,N)}{\lambda(0+,N)} \end{matrix} > \lambda(00,N). \tag{9}$$

The pair $(0,0)$ has the minimum loss, the upper portion of the inequality represents top-middle part of the partial ordering of neutrality, and the lower portion represents the bottom-middle part.

3.2   Trisection of Agent Relationships

For a pair of agents $(x,y) \in A \times A$, the loss for assigning (x,y) to alliance with respect to an issue $i \in I$ is given by:

$$L_a(x,y|i) = \lambda(r(x,i)r(y,i),A), \tag{10}$$

the loss for assigning it to neutral is:

$$L_n(x, y|i) = \lambda(r(x, i)r(y, i), N), \tag{11}$$

and the loss for assigning it to conflict is:

$$L_c(x, y|i) = \lambda(r(x, i)r(y, i), C). \tag{12}$$

Based on the losses of the three actions, we have the following decision rules for action:

(A)   if $L_a(x, y|i) \leq L_n(x, y|i) \wedge L_a(x, y|i) \leq L_c(x, y|i)$,   **decide** $(x, y) \in R_i^=$.
(C)   if $L_c(x, y|i) \leq L_n(x, y|i) \wedge L_c(x, y|i) \leq L_a(x, y|i)$,   **decide** $(x, y) \in R_i^{\lessgtr}$.
(N)   if $L_n(x, y|i) \leq L_a(x, y|i) \wedge L_n(x, y|i) \leq L_c(x, y|i)$,   **decide** $(x, y) \in R_i^{\approx}$.

In order to produce three pairwise disjoint relations, we need to apply tie-breaking rules to choose only one action when two or three actions have the same cost. For example, a tie-breaking rule can be given by an ordering of alliance-conflict-neutrality, namely, we choose alliance before conflict and conflict before neutrality when ties occur. Other tie-breaking orderings include conflict-alliance-neutrality, neutrality-conflict-alliance, and so on. The selection of the tie-breaking rules depends on particular situations and applications.

3.3   An Example

We use an example to illustrate the main ideas of the decision-theoretic model. Consider a three-valued situation table, Table 2, where $A = \{a_1, a_2, a_3, a_4, a_5\}$ is a set of five agents and $I = \{i_1, i_2, i_3, i_4, i_5, i_6\}$ is a set of six issues. The rating $r(a, i) \in \{+, -, 0\}$ represents the opinion of agent a about issue i. For example, a_1 supports issues i_2 and i_5, opposes issues i_1 and i_6, and is neutral towards issues i_3 and i_4.

Table 2. A situation table

	i_1	i_2	i_3	i_4	i_5	i_6
a_1	$-$	$+$	0	0	$+$	$-$
a_2	$-$	$+$	$+$	$-$	$+$	$-$
a_3	$-$	0	$-$	$+$	0	0
a_4	$-$	0	0	0	$-$	$+$
a_5	0	$-$	0	$+$	0	$-$
a_6	0	$-$	0	$+$	0	$-$

The loss function with respect to these nine rating pairs are given in Table 3. For an agent pair (a_1, a_2) and an issue i_4, the losses for assigning the agent pair to alliance, neutrality, and conflict relations are given, respectively, by:

$$L_a(a_1, a_2|i_4) = \lambda(r(a_1, i_4)r(a_2, i_4), A) = \lambda(-0, A) = 0.30,$$
$$L_n(a_1, a_2|i_4) = \lambda(r(a_1, i_4)r(a_2, i_4), N) = \lambda(-0, N) = 0.40,$$
$$L_c(a_1, a_2|i_4) = \lambda(r(a_1, i_4)r(a_2, i_4), C) = \lambda(-0, C) = 0.35.$$

Since $L_a(a_1, a_2|i_4) < L_n(a_1, a_2|i_4) \wedge L_a(a_1, a_2|i_4) < L_c(a_1, a_2|i_4)$, we use the decision rule (A) to classify the agent pair (a_1, a_2) to the alliance relation.

If we use issue i_4, the losses for assigning the pair to alliance, neutrality, and conflict relations are given by:

$$L_a(a_1, a_2|i_4) = \lambda(r(a_1, i_4)r(a_2, i_4), A) = \lambda(0+, A) = 0.25,$$
$$L_n(a_1, a_2|i_4) = \lambda(r(a_1, i_4)r(a_2, i_4), N) = \lambda(0+, N) = 0.27,$$
$$L_c(a_1, a_2|i_4) = \lambda(r(a_1, i_4)r(a_2, i_4), C) = \lambda(0+, C) = 0.25.$$

In this case, we need to introduce tie-breaking rules. If we use the ordering alliance-conflict-neutrality for tie-breaking, we will assign (a_1, a_2) to alliance. If we use the ordering conflict-alliance-neutrality for tie-breaking, we will assign (a_1, a_2) to conflict.

Table 3. A loss function

Ratings	Alliance	Conflict	Neutrality
$(-,-)$	$\lambda(--, A) = 0$	$\lambda(--, C) = 0.85$	$\lambda(--, N) = 0.60$
$(+,+)$	$\lambda(++, A) = 0$	$\lambda(++, C) = 0.90$	$\lambda(++, N) = 0.70$
$(0,0)$	$\lambda(00, A) = 0.10$	$\lambda(00, C) = 0.40$	$\lambda(00, N) = 0.05$
$(-,0),(0,-)$	$\lambda(-0, A) = \lambda(0-, A) = 0.30$	$\lambda(-0, C) = \lambda(0-, C) = 0.35$	$\lambda(-0, N) = \lambda(0-, N) = 0.40$
$(0,+),(+,0)$	$\lambda(0+, A) = \lambda(+0, A) = 0.25$	$\lambda(0+, C) = \lambda(+0, C) = 0.25$	$\lambda(0+, N) = \lambda(+0, N) = 0.27$
$(-,+),(+,-)$	$\lambda(+-, A) = \lambda(-+, A) = 0.90$	$\lambda(+-, C) = \lambda(-+, C) = 0$	$\lambda(+-, N) = \lambda(-+, N) = 0.80$

4   A Comparative Examination of the Decision-Theoretic Model

To gain further insights, in this section, we compare our model with other models. Table 4 shows the results of the Pawlak model, Yao model, and our model. The first column is the nine pairs of ratings. The second column Φ_i is the Pawlak auxiliary function and the third column shows the result of Pawlak's assignment/classification. The fourth column D_i is the distance in the Yao model and the fifth column shows the result of Yao's assignment. The last column shows the assignment of our model and the extra conditions, in addition to those in Eq. (5) and Eq. (6), on the loss function.

The main difference between the Pawlak model and the Yao model is the treatment of rating pair $(0,0)$. Given a pair of agents (x, y), in the Pawlak

Table 4. Classification of alliance, conflict, and neutrality

$(r(x,i), r(y,i))$	Φ_i	Pawlak model	D_i	Yao model	Our model
$(-,-)$	+1	alliance	0	alliance	alliance
$(+,+)$	+1	alliance	0	alliance	alliance
$(0,0)$	+1	alliance$(x = y)$	0	alliance	alliance $(\lambda(00, C) > \lambda(00, A))$
	0	neutrality$(x \neq y)$			conflict $(\lambda(00, A) > \lambda(00, C))$
					neutrality $(\lambda(00, A) = \lambda(00, C))$
$(-,0), (0,-)$	0	neutrality	0.5	neutrality	alliance $(\lambda(-0, C) > \lambda(-0, A))$
					conflict $(\lambda(-0, A) > \lambda(-0, C))$
					neutrality $(\lambda(-0, A) = \lambda(-0, C))$
$(0,+), (+,0)$	0	neutrality	0.5	neutrality	alliance $(\lambda(0+, C) > \lambda(0+, A))$
					conflict $(\lambda(0+, A) > \lambda(0+, C))$
					neutrality $(\lambda(0+, A) = \lambda(0+, C))$
$(-,+), (+,-)$	-1	conflict	1	conflict	conflict

model, the rating pair $(0,0)$ represents alliance if $x = y$, and neutrality if $x \neq y$. In Yao's model, the rating pair $(0,0)$ represents alliance regardless of whether the two agents are the same or not. Both the Pawlak model and the Yao model treat the two groups of rating pairs $(-,0)/(0,-)$ and $(+,0)/(0,+)$ as neutral.

Our approach considers the assignment from a cost-sensitive view, i.e., we assign a pair to alliance, conflict, or neutrality by using the assignment having the minimum loss. According to Eq. (5), we assign pairs $(-,-)$ and $(+,+)$ to alliance, and assign the pair $(+,-)$ to conflict. These assignments are the same as that of the Pawlak model and the Yao model. For the other groups of rating pairs, the relation of each rating pair depends on the extra conditions satisfied by a loss function. In addition, tie-breaking rules should be applied if two or three assignments have the same loss. Consider the rating pair $(0,0)$. To obtain the Pawlak assignment, we need to have the constraint $\lambda(00, C) > \lambda(00, A) = \lambda(00, N)$ and use a tie-breaking rule to choose alliance when the rating pair is from the same agent and to use a tie-breaking rule to choose neutrality if the pair is from two different agents. To obtain the Yao assignment, under the condition, $\lambda(00, C) > \lambda(00, A) = \lambda(00, N)$, we use a tie-breaking rule to choose alliance. There are two other possible assignments. Under the condition $\lambda(00, A) > \lambda(00, C)$, we have $\lambda(00, A) > \lambda(00, C) = \lambda(00, N)$. We may use a tie-breaking to choose conflict. Under the conditions $\lambda(00, A) > \lambda(00, C) \wedge \lambda(00, C) > \lambda(00, N)$, we will choose neutrality. Similarly, we can impose extra conditions and use tie-breaking rules for the rest of pairs.

It should be noted that different combinations of conditions and tie-breaking rules may produce the same assignment. The extra conditions in Table 4 are just some examples. This also shows that the decision-theoretic model is flexible.

5   Conclusion

We introduce a decision-theoretic model to interpret Pawlak conflict analysis with respect to the task of dividing the relationships between agents into alliance, conflict, and neutrality binary relations. A basic assumption is that there are different losses or costs for assigning a pair of ratings of two agents to alliance, conflict, and neutrality. Furthermore, a loss function needs to satisfy the constraints of orderings on the nine possible pairs of ratings produced by two agents under the three-valued rating scale $\{+, 0, -\}$. The alliance, conflict, and neutrality orderings reflect the agreement and conflict of two agents. The support and opposition orderings reflect the opinions of two agents. These orderings serve as a basis for constructing a loss function. Given a loss function, we take an action that has the minimus loss of assigning a pair to alliance, conflict, or neutrality. Finally, we show that the existing models can be derived from the proposed decision-theoretic formulation by using a loss function satisfying some conditions.

The decision-theoretic provides a sound basis for three-way conflict analysis. In this paper, we only consider the case of a single issue. In a future paper, based on the results presented, we will systematically investigate of three-way conflict analysis with respect to a set of issues by aggregating results from individual issues. Another research direction is to examine a decision-theoretic formulation by extending the three-valued rating scale into a scale having more values.

References

1. Pawlak, Z.: An inquiry into anatomy of conflicts. Inf. Sci. **109**, 65–78 (1998)
2. Pawlak, Z.: Some remarks on conflict analysis. Eur. J. Oper. Res. **166**, 649–654 (2005)
3. Yao, Y.Y.: An outline of a theory of three-way decisions. In: International Conference on Rough Sets and Current Trends in Computing, pp. 1–17 (2012)
4. Yao, Y.Y.: Tri-level thinking: models of three-way decision. Int. J. Mach. Learn. Cybern. **11**, 947–959 (2020)
5. Yao, Y.: The geometry of three-way decision. Appl. Intell. **51**(9), 6298–6325 (2021). https://doi.org/10.1007/s10489-020-02142-z
6. Yao, Y.Y.: Human-machine co-intelligence through symbiosis in the SMV space. Appl. Intell. **53**, 2777–2797 (2023)
7. Yao, Y.Y.: The Dao of three-way decision and three-world thinking. Int. J. Approximate Reasoning **162**, 109032 (2023)
8. Suo, L., Yang, H., Li, Q.Y., Yang, H.L., Yao, Y.Y.: A review of three-way decision: triadic understanding, organization, and perspectives. Int. J. Approximate Reasoning **173**, 109268 (2024)
9. Yao, Y.Y.: Three-way conflict analysis: reformulations and extensions of the Pawlak model. Knowl.-Based Syst. **180**, 26–37 (2019)
10. Fan, Y., Qi, J.J., Wei, L.: A conflict analysis model based on three-way decisions. In: International Joint Conference on Rough Sets, pp. 522–532 (2018)

11. Hu, M.J., Lang, G.M.: A probabilistic approach to analyzing agent relations in three-way conflict analysis based on Bayesian confirmation. In: International Joint Conference on Rough Sets, pp. 319–333 (2022)
12. Zhi, H.L., Qi, J.J., Qian, T., Ren, R.S.: Conflict analysis under one-vote veto based on approximate three-way concept lattice. Inf. Sci. **516**, 316–330 (2020)
13. Sun, B.Z., Chen, X.T., Zhang, L.Y., Ma, W.M.: Three-way decision making approach to conflict analysis and resolution using probabilistic rough set over two universes. Inf. Sci. **507**, 809–822 (2020)
14. Lang, G.M., Luo, J.F., Yao, Y.Y.: Three-way conflict analysis: a unification of models based on rough sets and formal concept analysis. Knowl.-Based Syst. **194**, 105556 (2020)
15. Lang, G.: A general conflict analysis model based on three-way decision. Int. J. Mach. Learn. Cybern. **11**(5), 1083–1094 (2020). https://doi.org/10.1007/s13042-020-01100-y
16. Hu, M.J.: Modeling relationships in three-way conflict analysis with subsethood measures. Knowl.-Based Syst. **260**, 110131 (2023)
17. Luo, J.F., Hu, M.J., Lang, G.M., Yang, X., Qin, K.Y.: Three-way conflict analysis based on alliance and conflict functions. Inf. Sci. **594**, 322–359 (2022)
18. Lang, G.M., Yao, Y.Y.: New measures of alliance and conflict for three-way conflict analysis. Int. J. Approximate Reasoning **132**, 49–69 (2021)
19. Lang, G.M., Ding, W.P., Miao, D.Q., Fujita, H., Yao, Y.Y.: Trisection-fusion and fusion-trisection methods of three-way conflict analysis with Pythagorean fuzzy information. Appl. Soft Comput. **164**, 111939 (2024)
20. Hu, M.J.: Three-way conflict analysis with preference-based conflict situations. Inf. Sci. **693**, 121676 (2025)
21. Yang, H., Yao, Y.Y., Qin, K.Y.: A lattice-theoretic model of three-way conflict analysis. Knowl.-Based Syst. **288**, 111470 (2024)
22. Ciucci, D., Dubois, D.: A map of dependencies among three-valued logics. Inf. Sci. **250**, 162–177 (2013)
23. Yao, Y.Y., Wang, S., Deng, X.F.: Constructing shadowed sets and three-way approximations of fuzzy sets. Inf. Sci. **412**, 132–153 (2017)

Three-Way Conflict Analysis with Nonlinear Conflict Functions in Fuzzy Situation Tables

Jing Liu[1,2] and Guangming Lang[1,2(✉)]

[1] School of Mathematics and Statistics, Changsha University of Science and Technology, Changsha 410114, Hunan, People's Republic of China
langguangming1984@126.com
[2] Hunan Provincial Key Laboratory of Mathematical Modeling and Analysis in Engineering, Changsha University of Science and Technology, Changsha 410114, Hunan, People's Republic of China

Abstract. Three-way conflict analysis typically investigates conflict situations by analyzing the trisection of agent pairs, agents, and issues from the perspectives of either conflict or alliance. However, in fuzzy conflict situations, most existing studies rely on linear conflict functions to measure conflict degrees, which fail to accurately depict the intricate and dynamic trends of conflict intensity. In this paper, we aim to conduct three-way conflict analysis by utilizing a nonlinear conflict function within a fuzzy situation table. First, we propose a nonlinear conflict function to address the limitations of traditional linear conflict functions, providing a more precise and flexible representation of conflict intensity. Second, using the nonlinear conflict function, we analyze the relations between agent pairs, categorizing them into alliance, neutrality, and conflict relations. An algorithm is developed to identify the optimal trisection. Finally, based on a case study of the Middle East conflict, we implement case and comparison analyses to demonstrate the practical effectiveness and superiority of our proposed model in real-world fuzzy scenarios.

Keywords: Fuzzy situation Table · Nonlinear conflict function · Optimal trisection · Three-way conflict analysis

1 Introduction

As an inherent characteristic of human social interaction, conflict extensively permeates various domains, including international relations, organizational management, interpersonal communication, and resource allocation. In this regard, conflict analysis [1–3] serves as a vital instrument for understanding the underlying mechanisms of conflict and formulating effective management strategies. Initially, scholars explored conflict analysis from the perspective of rough sets. Pawlak [1,2] presented a conflict analysis model and discussed the anatomical structure of conflict, establishing a foundational framework for the subsequent research. Deja [3] further expanded the theoretical framework of conflict analysis

© The Author(s), under exclusive license to Springer Nature Switzerland AG 2025
Q. Zhang et al. (Eds.): IJCRS 2025, LNAI 15710, pp. 58–73, 2025.
https://doi.org/10.1007/978-3-031-92741-6_5

and systematically explored the basic conflict-related issues, along with developing algorithms to address them effectively.

As a decision model rooted in rough set theory, three-way decision partitions a universe into three distinct regions: acceptance, rejection, and deferment [4–6], offering a flexible and efficient methodology for managing uncertainty and fuzzy information. In recent years, three-way decision models have achieved remarkable success across diverse applications, particularly in medical diagnosis [7,8], risk assessment [9,10], spam filtering [11,12], and related domains. Notably, its unique classification strategy aligns closely with the requirements for trisecting agent pairs in conflict analysis, thereby significantly enhancing both the practical applicability and interpretability of conflict analysis approaches. Unsurprisingly, three-way conflict analysis, incorporating three-way decision and conflict analysis, has been widely studied by scholars [13–17]. In the realm of three-way conflict analysis, trisection and threshold determination emerge as two fundamental and central challenges.

Trisection: Generally speaking, trisection can be broadly categorized into approaches based on either one or two evaluation functions. (1) Trisection based on a single evaluation function: Yao [13] employed a distance function to trisect the set of all pairs of agents with respect to a specific issue. Xu and Jia [18] achieved the trisection of all pairs of agents on multiple issues by introducing a similarity function. Xu et al. [19] further defined a conflict function in fuzzy situation tables and utilized it to trisect all agents. By introducing a preference-based conflict function, Hu [20] trisected all pairs of agents as alliance, neutrality, and conflict relations. (2) Trisection based on two evaluation functions: Lang and Yao [21] proposed a dual-function framework, leveraging a pair of alliance and conflict measures to classify all agent pairs. Luo et al. [22] trisected the set of agent pairs and the set of issues with alliance and conflict functions.

Threshold Determination: Threshold determination can directly influence trisection results. Existing approaches to threshold determination can be broadly categorized into subjective and objective methods, each with its own advantages and limitations. (1) Subjective threshold determination: In some studies [19,23], trisection thresholds are determined subjectively by decision-makers based on their expertise or preferences, which lacks objectivity, potentially leading to less accurate trisection results. (2) Objective threshold determination: Lang et al. [14] integrated decision-theoretic rough sets into conflict analysis, employing loss functions given by the decision-maker and Bayesian decision theory to calculate the thresholds of trisection. Gao et al. [24] proposed a novel approach for computing relative loss functions in interval set conflict systems, significantly reducing the dependency on subjective inputs and improving the objectivity of threshold determination. In contrast to the aforementioned methods, Xu et al. [18] developed an innovative algorithm to identify the optimal threshold pair, which selected the most suitable pair from a finite number of threshold pairs by leveraging a comprehensive evaluation function of three regions.

Through the above discussion and analysis, this study aims to explore three-way conflict analysis using nonlinear conflict functions in fuzzy situation tables. Our motivations are outlined as follows:

(1) Traditional linear conflict functions assume that conflict degrees increase uniformly with the difference of ratings, which fails to capture its nuanced dynamics accurately. In reality, neutral relations are more sensitive to ratings, while conflict and alliance relations exhibit greater stability. To address this limitation, we propose a nonlinear conflict function and utilize it to analyze three types of binary relations between agents.

(2) Existing studies predominantly rely on Bayesian risk theory and loss functions to determine thresholds. However, these approaches require the introduction of six parameters in loss functions, which significantly increase the subjectivity and complexity of the models. To mitigate this limitation, we employ an evaluation function of threshold pairs to derive the optimal threshold pair and its corresponding trisection.

(3) While prior research, such as Xu et al. [19], aggregates conflict degrees across multiple issues using simple averaging, this approach overlooks the varying importance of different issues. To handle this, we calculate the relative weight of each issue with regard to a given subset of issues and apply a weighted average method to aggregate conflict degrees.

The remainder of this paper is structured as follows. Section 2 recalls the conflict functions in fuzzy situation tables. Section 3 constructs a three-way analysis model based on nonlinear conflict functions. Section 4 performs case and comparison analyses in the fuzzy Middle East conflict scenario to verify the effectiveness and superiority of the presented model. Section 5 concludes the paper and highlights potential directions for future research.

2   A Review of Conflict Functions in Fuzzy Situation Tables

First of all, we overview the definition of a fuzzy situation table [19].

Definition 1. *(Xu [19], 2022) A fuzzy situation table is defined as a quadruple* $T = (A, I, V, r)$, *where*

- $A = \{a_1, a_2, \cdots, a_m\}$ $(m \in \mathbb{Z}^+)$ *is a nonempty finite set of agents;*
- $I = \{i_1, i_2, \cdots, i_n\}$ $(n \in \mathbb{Z}^+)$ *is a nonempty finite set of issues;*
- $V = \bigcup_{i \in I} V_i$, *where* $V_i = [0, 1]$ *is the value domain of the issue* i;
- $r : A \times I \to V$ *is a rating function. For an agent* $a \in A$ *and an issue* $i \in I$, $r(a, i) \in V_i$ *represents the support degree of* a *regarding* i, *and* $1 - r(a, i)$ *denotes the opposition degree of* a *regarding* i.

The rating $r(a, i) \to 1$ indicates that an agent a has a high support degree on the issue i, that is, we regard that a is supportive to i. The rating $r(a, i) \to 0.5$ signifies that a has a high uncertain degree on i, namely, we reckon that a is neutral to i. The rating $r(a, i) \to 0$ manifests that a has a high opposition degree on i, that is, we consider that a is opposed to i.

Based on the fuzzy situation table, Xu et al. [19] defined a conflict function.

Definition 2. *(Xu [19], 2022) In a fuzzy situation table $T = (A, I, V, r)$, $X \subseteq A$ is a nonempty subset of A. A conflict function $\mathcal{C}_i^{Xu} : X \times X \rightarrow [0, 1]$ on an issue $i \in I$ is defined by: for two agents $a, b \in X$,*

$$\mathcal{C}_i^{Xu}(a, b) = |r(a, i) - r(b, i)|, \tag{1}$$

where $|\cdot|$ represents the absolute value of a number.

By taking the average, Xu et al. [19] gave the notion of a conflict function on multiple issues.

Definition 3. *(Xu [19], 2022) In a fuzzy situation table $T = (A, I, V, r)$, $X \subseteq A$ is a nonempty subset of A. A conflict function $\mathcal{C}_J^{Xu} : X \times X \rightarrow [0, 1]$ on a nonempty issue subset $J \subseteq I$ is defined by: for two agents $a, b \in X$,*

$$\mathcal{C}_J^{Xu}(a, b) = \frac{\sum\limits_{i \in J} \mathcal{C}_i(a, b)}{\#(J)}, \tag{2}$$

where $\#(\cdot)$ denotes the cardinality of a set.

3 Three-Way Conflict Analysis Based on Nonlinear Conflict Functions

In this section, we introduce a nonlinear conflict function and utilize it to trisect all pairs of agents.

With $d_i(a, b) = |r(a, i) - r(b, i)|$, the conflict function proposed by Xu et al. [19] can be re-represented as: $\mathcal{C}_i^{Xu}(a, b) = d_i(a, b)$, which essentially a straight line with respect to $d_i(a, b)$. In general, if the value of $\mathcal{C}_i(a, b)$ is large, then we regard that agents a and b on the issue i are in conflict; if the value of $\mathcal{C}_i(a, b)$ is small, then we argue that a and b on the issue i are allied; if the value of $\mathcal{C}_i(a, b)$ is middle, then we think that a and b on the issue i are neutral. In fact, the alliance and conflict relations between two agents are more difficult to change, while the neutral relation between two agents is relatively easy to change. In other words, the neutral relation is more sensitive to the change of ratings. Obviously, a linear conflict function fails to describe the above phenomenon. Therefore, we propose a nonlinear conflict function as follows.

Definition 4. *In a fuzzy situation table $T = (A, I, V, r)$, $X \subseteq A$ is a nonempty subset of A. A nonlinear conflict function $\mathcal{C}_i : X \times X \rightarrow [0, 1]$ on an issue $i \in I$ is defined by: for two agents $a, b \in X$,*

$$\mathcal{C}_i(a, b) = \begin{cases} 1 - e^{-4|r(a,i)-r(b,i)|^2 \ln 2}, & 0 \leq |r(a, i) - r(b, i)| \leq 0.5, \\ e^{-4(1-|r(a,i)-r(b,i)|)^2 \ln 2}, & 0.5 < |r(a, i) - r(b, i)| \leq 1, \end{cases} \tag{3}$$

which can be re-expressed as a function about $d_i(a, b)$:

$$\mathcal{C}_i(a, b) = \begin{cases} 1 - e^{-4(d_i(a,b))^2 \ln 2}, & 0 \leq d_i(a, b) \leq 0.5, \\ e^{-4(1-d_i(a,b))^2 \ln 2}, & 0.5 < d_i(a, b) \leq 1. \end{cases} \tag{4}$$

We view functions $\mathcal{C}_i^{Xu}(a,b)$ and $\mathcal{C}_i(a,b)$ as two functions of $d_i(a,b)$. The functions $\mathcal{C}_i^{Xu}(a,b)$ and $\mathcal{C}_i(a,b)$ with respect to $d_i(a,b)$ are shown in Fig. 1. By inspecting Fig. 1, we find:

(1) The overall change trend of the functions $\mathcal{C}_i^{Xu}(a,b)$ and $\mathcal{C}_i(a,b)$ is consistent. Both two functions increase as $d_i(a,b)$ increases.
(2) The function $\mathcal{C}_i^{Xu}(a,b)$ is a straight line, and the function $\mathcal{C}_i(a,b)$ is an S-shaped curve. The function $\mathcal{C}_i^{Xu}(a,b)$ varies uniformly with $d_i(a,b)$. When the value of $d_i(a,b)$ is large or small, the function $\mathcal{C}_i(a,b)$ changes slower; When the value of $d_i(a,b)$ is middle, the function $\mathcal{C}_i(a,b)$ changes faster.
(3) When $0 < d_i(a,b) < 0.5$, $\mathcal{C}_i^{Xu}(a,b) > \mathcal{C}_i(a,b)$; When $0.5 < d_i(a,b) < 1$, $\mathcal{C}_i^{Xu}(a,b) < \mathcal{C}_i(a,b)$; When $d_i(a,b) = 0, 0.5, 1$, $\mathcal{C}_i^{Xu}(a,b) = \mathcal{C}_i(a,b)$.

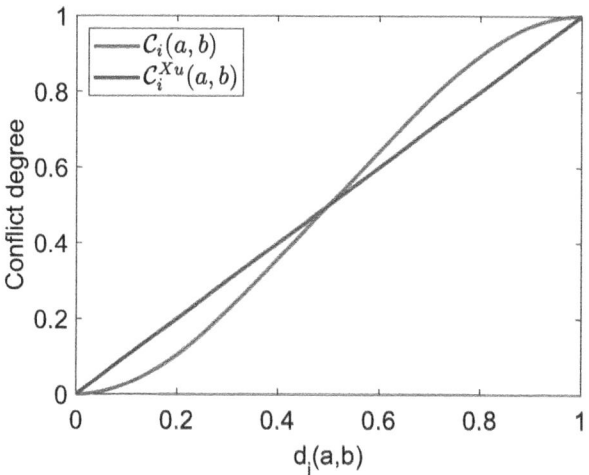

Fig. 1. Conflict functions $\mathcal{C}_i(a,b)$ and $\mathcal{C}_i^{Xu}(a,b)$.

In the following, we investigate some properties of the nonlinear conflict function regarding a single issue. The proof of Proposition 2 follows directly from Definition 4 and is therefore omitted.

Proposition 1. *In a fuzzy situation table $T = (A, I, V, r)$, $X \subseteq A$ is a nonempty subset of A. The function $\mathcal{C}_i : X \times X \to [0,1]$ holds the following properties: for two agents $a, b \in X$,*

(1) $\mathcal{C}_i(a,b)$ is continuous with respect to $d_i(a,b)$ in $[0,1]$;
(2) $\mathcal{C}_i(a,b)$ increases monotonically with the value of $d_i(a,b)$ in $[0,1]$;
(3) $\mathcal{C}_i(a,b)$ with respect to $d_i(a,b)$ is centrosymmetric about $(0.5, 0.5)$.

Proof. By viewing $\mathcal{C}_i(a,b)$ as a function of $d_i(a,b)$, we will prove Proposition 1 (1)–(3) one by one.

(1) Obviously, $C_i(a,b)$ is continuous when $d_i(a,b) \in [0,0.5) \cup (0.5,1]$. In the following, we only need to prove that $C_i(a,b)$ is continuous at $d_i(a,b) = 0.5$. Since

$$\lim_{d_i(a,b)\to 0.5^-} C_i(a,b) = \lim_{d_i(a,b)\to 0.5^+} C_i(a,b) = 0.5,$$

$C_i(a,b)$ is continuous at $d_i(a,b) = 0.5$. Thus, (1) is proved.

(2) The derivative of $C_i(a,b)$ with respect to $d_i(a,b)$ is calculated by:

$$\frac{d\, C_i(a,b)}{d\, d_i(a,b)} = \begin{cases} 8\ln 2 d_i(a,b)e^{-4(d_i(a,b))^2 \ln 2}, & 0 \le d_i(a,b) \le 0.5, \\ 8\ln 2(1 - d_i(a,b))e^{-4(1-(d_i(a,b)))^2 \ln 2}, & 0.5 < d_i(a,b) \le 1. \end{cases}$$

Since $\frac{d\, C_i(a,b)}{d\, d_i(a,b)} \ge 0$, (2) is proved.

(3) Suppose that $Q = (x,y)$ is any point on the function $C_i(a,b)$. Then the symmetry point of $Q = (x,y)$ about $(0.5,0.5)$ is $Q' = (1-x, 1-y)$. Since $Q = (x,y)$ is any point on the function $C_i(a,b)$, we have

$$y = \begin{cases} 1 - e^{-4x^2 \ln 2}, & 0 \le x \le 0.5, \\ e^{-4(1-x)^2 \ln 2}, & 0.5 < x \le 1. \end{cases}$$

In the following, we need to prove that $Q'(1-x, 1-y)$ is also a point on the function $C_i(a,b)$.

① When $0 \le x < 0.5$, we have $y = 1 - e^{-4x^2 \ln 2}$ and $0.5 < 1 - x \le 1$. Thus, $C_i(a,b) = e^{-4[1-(1-x)]^2 \ln 2} = e^{-4x^2 \ln 2} = 1 - y$.

② When $0.5 < x \le 1$, we have $y = e^{-4(1-x)^2 \ln 2}$ and $0 \le 1 - x < 0.5$. Thus, $C_i(a,b) = 1 - e^{-4(1-x)^2 \ln 2} = 1 - y$.

③ When $x = 0.5$, we have $y = 0.5$ and $1 - x = 0.5$. Thus, $C_i(a,b) = 0.5 = 1 - y$.

Therefore, (3) is proved.

Proposition 2. *In a fuzzy situation table $T = (A, I, V, r)$, $X \subseteq A$ is a nonempty subset of A. The function $C_i : X \times X \to [0,1]$ holds the following properties: for any $a, b \in X$,*

(1) $C_i(a,b) \in [0,1]$;
(2) $C_i(a,b) = 0 \Leftrightarrow r(a,i) = r(b,i)$;
(3) $C_i(a,b) = 1 \Leftrightarrow (r(a,i) = 1 \wedge r(b,i) = 0) \vee (r(a,i) = 0 \wedge r(b,i) = 1)$;
(4) $C_i(a,b) = 0.5 \Leftrightarrow |r(a,i) - r(b,i)| = 0.5$;
(5) $C_i(a,b) = C_i(b,a)$.

By weighted aggregation, we obtain the nonlinear conflict function regarding multiple issues. Below, we first discuss the relative weight.

Definition 5. *In a fuzzy situation table $T = (A, I, V, r)$, suppose that $W = (\omega(i_1), \omega(i_2), \cdots, \omega(i_n))$ is the weight vector of I, where $\omega(i) = \omega(i|I)$ denotes the weight of an issue i with respect to I, satisfying $\omega(i) \in [0,1]$ for any $i \in I$*

and $\sum_{i \in I} \omega(i) = 1$. *The relative weight of i with respect to a nonempty issue set $J \subseteq I$ is defined by:*

$$\omega(i|J) = \frac{\omega(i|I)}{\omega(J|I)} = \frac{\omega(i)}{\omega(J)} = \frac{\omega(i)}{\sum\limits_{j \in J} \omega(j)}, \tag{5}$$

where $\omega(J) = \omega(J|I) = \sum\limits_{j \in J} \omega(j)$ represents the weight of J with respect to I.

Obviously, the relative weight satisfies $\omega(i|J) \in [0,1]$ for any $i \in J$ and $\sum_{i \in J} \omega(i|J) = 1$.

Definition 6. *In a fuzzy situation table $T = (A, I, V, r)$, $X \subseteq A$ is a nonempty subset of A. A nonlinear conflict function $\mathcal{C}_J : X \times X \rightarrow [0,1]$ on a nonempty issue subset $J \subseteq I$ is defined by: for two agents $a, b \in X$,*

$$\mathcal{C}_J(a, b) = \sum_{i \in J} \big(\omega(i|J) \times \mathcal{C}_i(a, b)\big). \tag{6}$$

Next, we study some properties of the nonlinear conflict function regarding multiple issues. Their proofs follow directly from Definition 6 and are therefore omitted.

Proposition 3. *In a fuzzy situation table $T = (A, I, V, r)$, $X \subseteq A$ is a nonempty subset of A. The function $\mathcal{C}_J : X \times X \rightarrow [0,1]$ holds the following properties: for any $a, b \in X$:*

(1) $\mathcal{C}_J(a, b) \in [0, 1]$;
(2) $\mathcal{C}_J(a, b) = 0 \Leftrightarrow \forall i \in J, \mathcal{C}_i(a, b) = 0$;
(3) $\mathcal{C}_J(a, b) = 1 \Leftrightarrow \forall i \in J, \mathcal{C}_i(a, b) = 1$;
(4) $\forall i \in J, \mathcal{C}_i(a, b) = 0.5 \Rightarrow \mathcal{C}_J(a, b) = 0.5$;
(5) $\mathcal{C}_J(a, b) = \mathcal{C}_J(b, a)$.

Based on nonlinear conflict degrees, we perform the trisection of the set of all agent pairs with regard to multiple issues.

Definition 7. *In a fuzzy situation table $T = (A, I, V, r)$, $X \subseteq A$ is a nonempty subset of A, and β and α are two given thresholds satisfying $0 \le \beta \le \alpha \le 1$. The alliance relation $\mathbb{AR}_J(X^2)$, neutrality relation $\mathbb{NR}_J(X^2)$, and conflict relation $\mathbb{CR}_J(X^2)$ of X^2 on a nonempty issue subset $J \subseteq I$ are defined by:*

$$\mathbb{AR}_J(X^2) = \{(a, b) \in X \times X \mid 0 \le \mathcal{C}_J(a, b) \le \beta\}, \tag{7}$$

$$\mathbb{NR}_J(X^2) = \{(a, b) \in X \times X \mid \beta < \mathcal{C}_J(a, b) < \alpha\}, \tag{8}$$

$$\mathbb{CR}_J(X^2) = \{(a, b) \in X \times X \mid \alpha \le \mathcal{C}_J(a, b) \le 1\}. \tag{9}$$

The triplet $\langle\langle \mathbb{AR}_J(X^2), \mathbb{NR}_J(X^2), \mathbb{CR}_J(X^2) \rangle\rangle$ constitutes a trisection of $X \times X$ with respect to J, satisfying the conditions: $\mathbb{AR}_J(X^2) \cap \mathbb{NR}_J(X^2) = \emptyset, \mathbb{AR}_J(X^2) \cap \mathbb{CR}_J(X^2) = \emptyset, \mathbb{NR}_J(X^2) \cap \mathbb{CR}_J(X^2) = \emptyset$, and $\mathbb{AR}_J(X^2) \cup \mathbb{NR}_J(X^2) \cup \mathbb{CR}_J(X^2) = X \times X$.

Remark 1. We discuss some special cases of the trisection of all pairs of agents $\langle\langle \mathbb{AR}_J(X^2), \mathbb{NR}_J(X^2), \mathbb{CR}_J(X^2)\rangle\rangle$.

- When $X = A$, the triplet $\langle\langle \mathbb{AR}_J(A^2), \mathbb{NR}_J(A^2), \mathbb{CR}_J(A^2)\rangle\rangle$ constitutes a trisection of $A \times A$ with respect to an issue subset J.
- When $J = i$, the triplet $\langle\langle \mathbb{AR}_i(X^2), \mathbb{NR}_i(X^2), \mathbb{CR}_i(X^2)\rangle\rangle$ constitutes a trisection of $X \times X$ with respect to a single issue i.
- When $J = I$, the triplet $\langle\langle \mathbb{AR}_I(X^2), \mathbb{NR}_I(X^2), \mathbb{CR}_I(X^2)\rangle\rangle$ constitutes a trisection of $X \times X$ with respect to the complete issue set I.
- When $X = A$ and $J = i$, the triplet $\langle\langle \mathbb{AR}_i(A^2), \mathbb{NR}_i(A^2), \mathbb{CR}_i(A^2)\rangle\rangle$ constitutes a trisection of $A \times A$ with respect to a single issue i.
- When $X = A$ and $J = I$, the triplet $\langle\langle \mathbb{AR}_I(A^2), \mathbb{NR}_I(A^2), \mathbb{CR}_I(A^2)\rangle\rangle$ constitutes a trisection of $A \times A$ with respect to the complete issue set I.

In the following, we develop an approach for determining the optimal threshold pair of the above trisection by means of evaluation functions.

Definition 8. *In a fuzzy situation table* $T = (A, I, V, r)$, $X \subseteq A$ *is a nonempty subset of* A, $J \subseteq I$ *is a nonempty subset of* I, *and* (β, α) *is the threshold pair of* $\langle\langle \mathbb{AR}_J(X^2), \mathbb{NR}_J(X^2), \mathbb{CR}_J(X^2)\rangle\rangle$. *An evaluation function* $\mathcal{E} : [0,1] \to \mathbb{R}^+$ *of the threshold* β *is defined by:*

$$\mathcal{E}(\beta) = \sum_{(a,b)\in\mathbb{CR}_J(X^2)\cup\mathbb{NR}_J(X^2)} \big|\beta - \mathcal{C}_J(a,b)\big|, \tag{10}$$

and an evaluation function $\mathcal{E} : [\beta, 1] \to \mathbb{R}^+$ *of the threshold* α *is defined by:*

$$\mathcal{E}(\alpha) = \sum_{(a,b)\in\mathbb{AR}_J(X^2)\cup\mathbb{NR}_J(X^2)} \big|\alpha - \mathcal{C}_J(a,b)\big|. \tag{11}$$

Definition 9. *In a fuzzy situation table* $T = (A, I, V, r)$, $X \subseteq A$ *is a nonempty subset of* A, $J \subseteq I$ *is a nonempty subset of* I, *and* (β, α) *is the threshold pair of* $\langle\langle \mathbb{AR}_J(X^2), \mathbb{NR}_J(X^2), \mathbb{CR}_J(X^2)\rangle\rangle$. *An evaluation function* $\mathcal{E} : [0,1] \times [\beta, 1] \to \mathbb{R}^+$ *of the threshold pair* (β, α) *is defined by:*

$$\mathcal{E}(\beta, \alpha) = \mathcal{E}(\beta) + \mathcal{E}(\alpha). \tag{12}$$

Apparently, there is an infinite number of threshold pairs that satisfy the conditions $\beta \in [0,1]$ and $\alpha \in [\beta, 1]$. If we assume that both thresholds retain k ($k \in \mathbb{Z}^+$) decimal places, then we obtain a finite number of threshold pairs. According to the above evaluation function $\mathcal{E}(\beta, \alpha)$, we regard the threshold pair corresponding to the minimum evaluation value as the optimal threshold pair.

Definition 10. *In a fuzzy situation table* $T = (A, I, V, r)$, *suppose that the thresholds* β *and* α *all retain* k ($k \in \mathbb{Z}^+$) *decimal places. The trisection* $\langle\langle \mathbb{AR}_J^*(X^2), \mathbb{NR}_J^*(X^2), \mathbb{CR}_J^*(X^2)\rangle\rangle$ *with respect to* (β^*, α^*) *is optimal if*

$$\mathcal{E}(\beta^*, \alpha^*) = \min_{\beta\in[0,1],\alpha\in[\beta,1]} \mathcal{E}(\beta, \alpha). \tag{13}$$

where (β^*, α^*) *is called the optimal threshold pair.*

We summarize Algorithm 1 for identifying the optimal trisection of the set of all agent pairs and the corresponding optimal threshold pair. In Algorithm 1, the time complexity of Lines 2–8 is $O\big((\#(X))^2 \times \#(J)\big)$, the time complexity of Lines 9–27 is $O\big((5k+1) \times (10k+1) \times (\#(X))^2\big)$, the time complexity of Lines 28–30 is $O\big((5k+1) \times (10k+1)\big)$. Thus, the total time complexity of Algorithm 1 is $O\big((5k+1) \times (10k+1) \times (\#(X))^2\big)$.

4   Case and Comparison Analyses

In this section, we perform case and comparison analyses in the fuzzy Middle East conflict scenario to verify the effectiveness and superiority of the presented model.

4.1   Case Analysis

We utilize the example of the fuzzy Middle East conflict in the literature [19] to reproduce the identification of optimal threshold pair and corresponding trisection.

The Middle East conflict $T = (A, I, V, r)$ is given by Table 1, where $A = \{a_1, a_2, a_3, a_4, a_5, a_6\}$ and $I = \{i_1, i_2, i_3, i_4, i_5\}$. The agents a_1, a_2, a_3, a_4, a_5, and a_6 represent Israel, Egypt, Palestine, Jordan, Syria, and Saudi Arabia, respectively. The issues i_1, i_2, i_3, i_4, and i_5 denote "Autonomous Palestinian state on the West Bank and Gaza", "Israeli military outpost along the Jordan River", "Israel retains East Jerusalem", "Israeli military outposts on the Golan Height", and "Arab countries grant citizenship to Palestinians who choose to remain within their borders", respectively.

Taking $X = \{a_1, a_2, a_3, a_4, a_6\}$ and $J = \{i_1, i_2, i_3, i_4\}$, the detail processes for computing optimal threshold pair and corresponding trisection are shown as follows:

(1) By Eq. (3), the conflict degrees between two agents in X on the issues i_1, i_2, i_3, and i_4 are shown in Tables 2, 3, 4 and 5, respectively.
(2) Experts give the relative importance of individual issues in the Middle East conflict based on their extensive experience, namely, the weight vector of the issue set is $W = (0.1, 0.25, 0.3, 0.2, 0.15)$. By Eq. (5), the relative weight of the issue i_1 with respect to J is computed as:

$$w(i_1|J) = \frac{0.1}{0.1 + 0.25 + 0.3 + 0.2} = \frac{2}{17}.$$

Similarly, the relative weights of other issues with respect to J are listed as follows:

$$w(i_2|J) = \frac{5}{17}, \quad w(i_3|J) = \frac{6}{17}, \quad w(i_4|J) = \frac{4}{17}.$$

Algorithm 1: The algorithm of identifying optimal threshold pair and trisection.

Input : A fuzzy situation table $T = (A, I, V, r)$, $X \subseteq A$, and $J \subseteq I$;
The weight vector of issues $W = (\omega(i_1), \omega(i_2)..., \omega(i_n))$;
Thresholds β and α all retain k $(k \in \mathbb{Z}^+)$ decimal places.
Output: Optimal trisection $\langle\langle \mathbb{AR}_J^*(X^2), \mathbb{NR}_J^*(X^2), \mathbb{CR}_J^*(X^2)\rangle\rangle$;
Optimal threshold pair (β^*, α^*).

---

1 **begin**
2   **for** *two agents* $a, b \in X$ **do**
3   **for** *each issue* $i \in J$ **do**
4   **Compute:** $\mathcal{C}_i(a, b)$ by Equation (3).
5   **Compute:** $\omega(i|J)$ by Equation (5).
6   **end**
7   **Compute:** $\mathcal{C}_J(a, b)$ by Equation (6).
8   **end**
9   **for** *the threshold* $\beta = 0 : \frac{k}{10} : 1$ **do**
10   **for** *the threshold* $\alpha = \beta : \frac{k}{10} : 1$ **do**
11   **Initialize:** $\mathbb{AR}_J(X^2) = \emptyset$, $\mathbb{NR}_J(X^2) = \emptyset$, and $\mathbb{CR}_J(X^2) = \emptyset$.
12   **for** *two agents* $a, b \in X$ **do**
13   **if** $\mathcal{C}_J(a, b) \leq \beta$ **then**
14           $\mathbb{AR}_J(X^2) = \mathbb{AR}_J(X^2) \cup (a, b)$ by Equation (7).
15   **else if** $\mathcal{C}_J(a, b) < \alpha$ **then**
16           $\mathbb{NR}_J(X^2) = \mathbb{NR}_J(X^2) \cup (a, b)$ by Equation (8).
17   **else**
18           $\mathbb{CR}_J(X^2) = \mathbb{CR}_J(X^2) \cup (a, b)$ by Equation (9).
19   **end**
20   **end**
21   **end**
22   **end**
23   **Compute:** $\mathcal{E}(\beta) = \sum\limits_{(a,b) \in \mathbb{CR}_J(X^2) \cup \mathbb{NR}_J(X^2)} |\beta - \mathcal{C}_J(a, b)|$ by Equation (10).
24   **Compute:** $\mathcal{E}(\alpha) = \sum\limits_{(a,b) \in \mathbb{AR}_J(X^2) \cup \mathbb{NR}_J(X^2)} |\alpha - \mathcal{C}_J(a, b)|$ by Equation (11).
25   **Compute:** $\mathcal{E}(\beta, \alpha) = \mathcal{E}(\beta) + \mathcal{E}(\alpha)$ by Equation (12).
26   **end**
27   **end**
28   **if** $\mathcal{E}(\beta^*, \alpha^*) = \min\limits_{\beta \in [0,1], \alpha \in [\beta, 1]} \mathcal{E}(\beta, \alpha)$ **then**
29     $\langle\langle \mathbb{AR}_J^*(X^2), \mathbb{NR}_J^*(X^2), \mathbb{CR}_J^*(X^2)\rangle\rangle$ and (β^*, α^*) are optimal by Equation (13).
30   **end**
31 **end**
32 **Return:** $\langle\langle \mathbb{AR}_J^*(X^2), \mathbb{NR}_J^*(X^2), \mathbb{CR}_J^*(X^2)\rangle\rangle$, (β^*, α^*).

Table 1. A fuzzy situation table of Middle East conflict.

	i_1	i_2	i_3	i_4	i_5
a_1	0.1	0.9	0.8	1	0.7
a_2	0.9	0.4	0	0.2	0.1
a_3	1	0.1	0.3	0.2	0.5
a_4	0.5	0.1	0.3	0.4	0.2
a_5	0.8	0.2	0.2	0.1	0.3
a_6	0.6	0.8	0.1	0.5	0.9

Table 2. The conflict degrees between two agents in X^* on the issue i_1.

	a_1	a_2	a_3	a_4	a_6
a_1	0	0.8950	0.9727	0.3583	0.5000
a_2	0.8950	0	0.0273	0.3583	0.2208
a_3	0.9727	0.0273	0	0.5000	0.3583
a_4	0.3583	0.3583	0.5000	0	0.0273
a_6	0.5000	0.2208	0.3583	0.0273	0

$^*X = \{a_1, a_2, a_3, a_4, a_6\}$.

Table 3. The conflict degrees between two agents in X^* on the issue i_2.

	a_1	a_2	a_3	a_4	a_6
a_1	0	0.5000	0.8950	0.8950	0.0273
a_2	0.5000	0	0.2208	0.2208	0.3583
a_3	0.8950	0.2208	0	0	0.7792
a_4	0.8950	0.2208	0	0	0.7792
a_6	0.0273	0.3583	0.7792	0.7792	0

Table 4. The conflict degrees between two agents in X^* on the issue i_3.

	a_1	a_2	a_3	a_4	a_6
a_1	0	0.8950	0.5000	0.5000	0.7792
a_2	0.8950	0	0.2208	0.2208	0.0273
a_3	0.5000	0.2208	0	0	0.1050
a_4	0.5000	0.2208	0	0	0.1050
a_6	0.7792	0.0273	0.1050	0.1050	0

$^*X = \{a_1, a_2, a_3, a_4, a_6\}$.

Table 5. The conflict degrees between two agents in X^* on the issue i_4.

	a_1	a_2	a_3	a_4	a_6
a_1	0	0.8950	0.8950	0.6417	0.5000
a_2	0.8950	0	0	0.1050	0.2208
a_3	0.8950	0	0	0.1050	0.2208
a_4	0.6417	0.1050	0.1050	0	0.0273
a_6	0.5000	0.2208	0.2208	0.0273	0

$^*X = \{a_1, a_2, a_3, a_4, a_6\}$.

(3) By Eq. (6), the conflict degree of agents a_1 and a_2 regarding J is calculated as:

$$C_J(a_1, a_2) = 0.8950 \times \frac{2}{17} + 0.5 \times \frac{5}{17} + 0.8950 \times \frac{6}{17} + 0.8950 \times \frac{4}{17} = 0.7788.$$

Similarly, the conflict degrees between two agents in X on J are shown in Table 6.

(4) Taking $k = 1$, that is, the thresholds α and β all retain one decimal place. By Eqs. (11)–(12), we calculate the evaluation values of all threshold pairs (β, α), which are shown in Table 7. From Table 7, the optimal threshold β^* is 0.1 and the optimal threshold α^* is 0.2. The optimal trisection with respect

Table 6. The conflict degrees between two agents in X on J^*.

	a_1	a_2	a_3	a_4	a_6
a_1	0	0.7788	0.7647	0.6329	0.4595
a_2	0.7788	0	0.1461	0.2097	0.1930
a_3	0.7647	0.1461	0	0.0835	0.3603
a_4	0.6329	0.2097	0.0835	0	0.2759
a_6	0.4595	0.1930	0.3603	0.2759	0

$^*X = \{a_1, a_2, a_3, a_4, a_6\}$ and $J = \{i_1, i_2, i_3, i_4\}$.

Table 7. The evaluation values of all threshold pairs (β, α).

β	α										
	0	0.1	0.2	0.3	0.4	0.5	0.6	0.7	0.8	0.9	1
0	7.81	6.44	6.16	7.19	8.47	10.31	11.41	13.98	18.64	20.64	22.64
0.1		6.81	6.13	7.76	7.84	9.48	10.58	12.95	17.20	19.20	21.20
0.2			6.67	6.90	7.78	9.22	10.32	12.49	16.35	18.35	20.35
0.3				8.06	8.74	9.98	11.08	13.05	16.50	18.50	20.50
0.4					10.20	11.24	12.34	14.11	17.16	19.16	21.16
0.5						11.69	12.29	13.36	15.51	17.01	18.51
0.6							14.19	15.06	16.81	18.31	19.81
0.7								14.73	15.58	16.58	17.58
0.8									12.19	12.19	12.19
0.9										14.69	14.69
1											17.19

to $(0.1, 0.2)$ is given by:

$$\mathbb{AR}_J^*(X^2) = \{(a, b) \in X \times X \mid 0 \leq \mathcal{C}_J(a, b) \leq 0.1\}$$
$$= \{(a_1, a_1), (a_2, a_2), (a_3, a_3)(a_4, a_4), (a_6, a_6), (a_3, a_4), (a_4, a_3)\},$$
$$\mathbb{NR}_J^*(X^2) = \{a \in X \times X \mid 0.1 < \mathcal{C}_J(a, b) < 0.2\}$$
$$= \{(a_2, a_3), (a_3, a_2), (a_2, a_6), (a_6, a_2)\},$$
$$\mathbb{CR}_J^*(X^2) = \{a \in X \times X \mid 0.2 \leq \mathcal{C}_J(a, b) \leq 1\}$$
$$= \{(a_1, a_2), (a_2, a_1), (a_1, a_3), (a_3, a_1), (a_1, a_4), (a_4, a_1), (a_1, a_6),$$
$$(a_6, a_1), (a_2, a_4), (a_4, a_2), (a_3, a_6), (a_6, a_3), (a_4, a_6), (a_6, a_4)\}.$$

If $k = 2$, then the thresholds α and β all retain two decimal places. By Definition 10, the optimal threshold β^* is 0 and the optimal threshold α^* is 0.19. The optimal trisection with respect to $(0, 0.19)$ is given by:

$$\mathbb{AR}_J^*(X^2) = \{(a, b) \in X \times X \mid \mathcal{C}_J(a, b) = 0\}$$
$$= \{(a_1, a_1), (a_2, a_2), (a_3, a_3)(a_4, a_4), (a_6, a_6)\},$$

$$\mathbb{NR}_J^*(X^2) = \{a \in X \times X \mid 0 < C_J(a,b) < 0.19\}$$
$$= \{(a_2, a_3), (a_3, a_2), (a_3, a_4), (a_4, a_3)\},$$
$$\mathbb{CR}_J^*(X^2) = \{a \in X \times X \mid 0.19 \le C_J(a,b) \le 1\}$$
$$= \{(a_1, a_2), (a_2, a_1), (a_1, a_3), (a_3, a_1), (a_1, a_4), (a_4, a_1), (a_1, a_6),$$
$$(a_6, a_1), (a_2, a_4), (a_4, a_2), (a_2, a_6), (a_6, a_2), (a_3, a_6), (a_6, a_3),$$
$$(a_4, a_6), (a_6, a_4)\}.$$

If $k = 3$, then the optimal threshold β^* is 0 and the optimal threshold α^* is 0.192. In this case, the optimal trisection corresponding to $(0, 0.192)$ is the same as those corresponding to $(0, 0.19)$. Moreover, we also compute the optimal trisection for $k = 4$ and $k = 5$, which are the same as those for $k = 2$. We do not list the optimal trisection for $k = 3, 4$ and 5 one by one in our paper. In view of the time and space complexity of the algorithm, we do not continue to compute the optimal trisection for $k > 5$. By looking at the optimal trisection for $k = 1, 2, \ldots, 5$, we obtain the same optimal trisection for $k \ge 2$. In other words, when $k \ge 2$, the optimal trisection results are stable and robust in the case of the Middle East conflict.

4.2 Comparison Analysis

Based on the case of the Middle East conflict, we compare the presented non-linear conflict function with Xu et al.'s linear conflict function [19] and analyze their influence on the trisection.

Taking $X = \{a_1, a_2, a_3, a_4, a_6\}$ and $J = \{i_1, i_2, i_3, i_4\}$, in linear and nonlinear conflict functions, our approach for determining relative weights (i.e. Definition 5) are used to aggregate the conflict degrees about individual issues in J. The conflict degrees obtained by Xu et al.'s linear conflict function are shown in Table 8. The conflict degrees obtained by our nonlinear conflict function are shown in Table 6. With $\alpha = 0.75$ and $\beta = 0.25$, the trisection by Xu et al.'s linear conflict function and our nonlinear conflict function is shown in Table 9.

Table 8. The conflict degrees between two agents in X on J^* by Xu et al.'s linear conflict function.

	a_1	a_2	a_3	a_4	a_6
a_1	0	0.7118	0.7059	0.6000	0.4529
a_2	0.7118	0	0.2059	0.2882	0.2588
a_3	0.7059	0.2059	0	0.1059	0.3941
a_4	0.6000	0.2882	0.1059	0	0.3118
a_6	0.4529	0.2588	0.3941	0.3118	0

$^*X = \{a_1, a_2, a_3, a_4, a_6\}$ and $J = \{i_1, i_2, i_3, i_4\}$.

Table 9. The trisection by Xu et al.'s linear conflict function and our nonlinear conflict function when $\alpha = 0.75$ and $\beta = 0.25$.

Relations	Trisection by Xu et al.'s linear conflict function	Trisection by our nonlinear conflict function
$\mathbb{AR}_J(X^2)$	$\{(a_1,a_1),(a_2,a_2),(a_2,a_3),$ $(a_3,a_2),(a_3,a_3),(a_3,a_4),$ $(a_4,a_3),(a_4,a_4),(a_6,a_6)\}$	$\{(a_1,a_1),(a_2,a_2),(a_2,a_3),$ $(a_2,a_4),(a_2,a_6),(a_3,a_2),$ $(a_3,a_3),(a_3,a_4),(a_4,a_2),$ $(a_4,a_3),(a_4,a_4),(a_6,a_2),$ $(a_6,a_6)\}$
$\mathbb{NR}_J(X^2)$	$\{(a_1,a_2),(a_1,a_3),(a_1,a_4),$ $(a_1,a_6),(a_2,a_1),(a_2,a_4),$ $(a_2,a_6),(a_3,a_1),(a_3,a_6),$ $(a_4,a_1),(a_4,a_2),(a_4,a_6),$ $(a_6,a_1),(a_6,a_2),(a_6,a_3),$ $(a_6,a_4)\}$	$\{(a_1,a_4),(a_1,a_6),(a_3,a_6),$ $(a_4,a_1),(a_4,a_6),(a_6,a_1),$ $(a_6,a_3),(a_6,a_4)\}$
$\mathbb{CR}_J(X^2)$	$\emptyset$	$\{(a_1,a_2),(a_1,a_3),(a_2,a_1),$ $(a_3,a_1)\}$

From Table 6 and 8, when the conflict degrees are less than 0.5, conflict degrees by Xu et al.'s linear conflict function are greater than those by our nonlinear conflict function; when the conflict degrees are greater than 0.5, conflict degrees by Xu et al.'s linear conflict function are less than those by our nonlinear conflict function. The above phenomenon conforms to the characteristic of Fig. 1. According to Table 9, by replacing Xu et al.'s linear conflict function with our nonlinear conflict function to trisect all agent pairs, we find that there are more agent pairs in alliance and conflict relations and fewer agent pairs in the neutral relation. This is due to the fact that the nonlinear conflict function makes some medium conflict degrees larger to divide corresponding agent pairs into the conflict relation, and some conflict degrees smaller to divide corresponding agent pairs into the alliance relation, which is the concrete manifestation of the features of the nonlinear conflict function in Fig. 1.

5   Conclusion

In this paper, we have introduced a nonlinear conflict function within fuzzy situation tables to investigate the binary relations of agent pairs. Furthermore, we have developed a methodology for identifying optimal thresholds and their corresponding trisection outcomes. The principal contributions of this research are delineated as follows:

(1) We have rigorously defined a nonlinear conflict function, systematically investigated its connection with conventional linear conflict functions, and thoroughly examined its boundary conditions and symmetry properties.

(2) Leveraging the nonlinear conflict function, we have successfully classified all agent pairs into three distinct categories: alliance, neutrality, and conflict relations, and developed a computational approach to identify the optimal trisection thresholds.
(3) We have demonstrated how to calculate the optimal threshold pair and trisection and implemented comparison analysis under the scenario of fuzzy Middle East conflict, confirming the effectiveness and superiority of our proposed model in real-world conflict scenarios.

Future research directions will focus on three key aspects: (1) Extending the application of nonlinear conflict functions to analyze the trisection of an agent set and an issue set; (2) Exploring the construction of nonlinear conflict functions across diverse environmental contexts; (3) Developing advanced computational techniques for more efficient and precise trisection threshold determination.

Acknowledgements. We would like to thank the anonymous reviewers for their critical and constructive comments. This work is supported by the National Natural Science Foundation of China (Nos. 62076040, 12471431), the Scientific Research Fund of Hunan Provincial Education Department (No. 22A0233), the Hunan Provincial Key Laboratory of Intelligent Computing and Language Information Processing (No. 2018TP1018), the Hunan Provincial Key Laboratory of Mathematical Modeling and Analysis in Engineering (No. 2018MMAEZD10), and a Discovery Grant from the Natural Sciences and Engineering Research Council of Canada.

References

1. Pawlak, Z.: On conflicts. Int. J. Man Mach. Stud. **21**, 127–134 (1984)
2. Pawlak, Z.: An inquiry into anatomy of conflicts. Inf. Sci. **109**, 65–78 (1998)
3. Deja, R.: Conflict analysis. Int. J. Intell. Syst. **17**, 235–253 (2002)
4. Yao, Y.Y.: Three-way decisions with probabilistic rough sets. Inf. Sci. **180**(3), 341–353 (2010)
5. Yao, Y.: An outline of a theory of three-way decisions. In: Yao, J.T., et al. (eds.) RSCTC 2012. LNCS (LNAI), vol. 7413, pp. 1–17. Springer, Heidelberg (2012). https://doi.org/10.1007/978-3-642-32115-3_1
6. Yao, Y.Y.: Three-way decision and granular computing. Int. J. Approximate Reasoning **103**, 107–123 (2018)
7. Huang, X.F., Zhan, J.M.: TWD-R: a three-way decision approach based on regret theory in multi-scale decision information systems. Inf. Sci. **581**, 711–739 (2021)
8. Wang, J.J., Ma, X.L., Xu, Z.S., Zhan, J.M.: Three-way multi-attribute decision making under hesitant fuzzy environments. Inf. Sci. **552**, 328–351 (2021)
9. Wang, T.X., Li, H.X., Qian, Y.H., Huang, B., Zhou, X.Z.: A regret-based three-way decision model under interval type-2 fuzzy environment. IEEE Trans. Fuzzy Syst. **30**(1), 175–189 (2022)
10. Zhan, J.M., Ye, J., Ding, W.P., Liu, P.D.: A novel three-way decision model based on utility theory in incomplete fuzzy decision systems. IEEE Trans. Fuzzy Syst. **30**(7), 2210–2226 (2022)

11. Li, J., Deng, X., Yao, Y.: Multistage email spam filtering based on three-way decisions. In: Lingras, P., Wolski, M., Cornelis, C., Mitra, S., Wasilewski, P. (eds.) RSKT 2013. LNCS (LNAI), vol. 8171, pp. 313–324. Springer, Heidelberg (2013). https://doi.org/10.1007/978-3-642-41299-8_30
12. Zhou, B., Yao, Y.Y., Luo, J.G.: Cost-sensitive three-way email spam filtering. J. Intell. Inf. Syst. **42**, 19–45 (2014)
13. Yao, Y.Y.: Three-way conflict analysis: reformulations and extensions of the Pawlak model. Knowl.-Based Syst. **180**, 26–37 (2019)
14. Lang, G.M., Miao, D.Q., Cai, M.J.: Three-way decision approaches to conflict analysis using decision-theoretic rough set theory. Inf. Sci. **406–407**, 185–207 (2017)
15. Lang, G.M., Miao, D.Q., Fujita, H.: Three-way group conflict analysis based on Pythagorean fuzzy set theory. IEEE Trans. Fuzzy Syst. **28**(3), 447–461 (2020)
16. Zhi, H.L., Li, J.H., Li, Y.N.: Multilevel conflict analysis based on fuzzy formal contexts. IEEE Trans. Fuzzy Syst. **20**(12), 5128–5142 (2022)
17. Jiang, Q., Liu, Y., Yi, J.H., Forrest, J.Y.: A three-way conflict analysis model with decision makers' varying preferences. Appl. Soft Comput. **151**, 111171 (2024)
18. Xu, W.Y., Jia, B.: Three-way conflict analysis with similarity degree on an issue set. Int. J. Mach. Learn. Cybern. **15**, 405–427 (2024)
19. Xu, F., Cai, F., Song, H.L., Dai, J.H.: The selection of feasible strategies based on consistency measurement of cliques. Inf. Sci. **583**, 33–55 (2022)
20. Hu, M.J.: Three-way conflict analysis with preference-based conflict situations. Inf. Sci. **693**, 121676 (2025)
21. Lang, G.M., Yao, Y.Y.: New measures of alliance and conflict for three-way conflict analysis. Int. J. Approximate Reasoning **132**, 49–69 (2021)
22. Luo, J.F., Hu, M.J., Lang, G.M., Yang, X., Qin, K.Y.: Three-way conflict analysis based on alliance and conflict functions. Inf. Sci. **594**, 322–359 (2022)
23. Lang, G.M., Ding, W.P., Miao, D.Q., Fujita, H., Yao, Y.Y.: Trisection-fusion and fusion-trisection methods of three-way conflict analysis with Pythagorean fuzzy information. Appl. Soft Comput. **164**, 111939 (2024)
24. Gao, S., Yang, H.L., Guo, Z.L.: Three-way conflict analysis and resolution based on interval set information. Inf. Sci. **703**, 121938 (2025)

Three-Way Approximate Representation of Rough Fuzzy Multi-granularity Knowledge Spaces via Information Measure

Deyou Xia[1][(✉)], Ye Tang[2][(✉)], Man Gao[3], Gengyu Ge[2], and Daoping Yang[2]

[1] School of Computer Science and Engineering, Chongqing University of Science and Technology, Chongqing 401331, China
1025968052@qq.com
[2] School of Information Engineering, Zunyi Normal University, Zunyi 563002, China
595940640@qq.com
[3] School of Computer Science and Technology, Chongqing University of Posts and Telecommunications, Chongqing 400065, China

Abstract. Three-way approximations of fuzzy sets offer a computationally efficient and conceptually straightforward method for characterizing uncertainty in fuzzy information systems, which has increasingly garnered academic attention. At the core of three-way approximation theory lies the development of objective optimization functions to systematically partition the sample space into three mutually exclusive regions: positive, boundary, and negative domains. While existing methodologies have made progress in traditional approximation analysis, they exhibit limited exploration of how to effectively divide rough fuzzy multi-granularity knowledge spaces into these regions. This paper addresses this research gap by introducing a novel framework for three-way approximate representation of rough fuzzy multi-granularity knowledge spaces. We first propose a fuzzy knowledge granularity measure to quantify the inherent uncertainty of fuzzy concepts. A critical insight is that this uncertainty measure demonstrates a monotonic decay property within hierarchical rough fuzzy approximation spaces. Subsequently, based on the fuzzy knowledge granularity measure, the similarity between different rough fuzzy approximation spaces is defined. Theoretical analysis confirms that the proposed similarity function can be reduced to the information measure under specific conditions. Ultimately, the proposed framework employs a three-way approximation strategy to precisely segment rough fuzzy multi-granularity knowledge spaces into three disjoint regions. This work not only extends the theoretical boundaries of three-way approximation sets by incorporating multi-granularity perspectives but also establishes a systematic methodology for both theoretical refinement and practical applications in uncertain knowledge representation.

Keywords: Three-way approximations · Fuzzy sets ·
Multi-granularity · Information measure

Q. Zhang et al. (Eds.): IJCRS 2025, LNAI 15710, pp. 74–84, 2025.
https://doi.org/10.1007/978-3-031-92741-6_6

1   Introduction

Granular computing (GrC) [7,11,18], emerging as a foundational paradigm in artificial intelligence, revolves around the core constructs of granularity, hierarchical layering, and structured granular frameworks. This computational paradigm streamlines problem-solving complexity by strategically selecting rationally selected granularity levels to abstract system behavior. The theoretical foundations of GrC have synergistically evolved through integration with multiple mathematical formalisms: Fuzzy set theory [18], which enables continuous uncertainty modeling; Rough set theory [7], offering boundary-region based knowledge representation; Shadowed set theory [8], addressing incomplete information characterization; Three-way approximation [16], advancing region partitioning precision. By synthesizing these heterogeneous frameworks, contemporary research has developed hybrid uncertainty quantification mechanisms. These advancements facilitate granular-based analysis of epistemic and aleatory uncertainties in complex information systems, providing both theoretical foundations and practical methodologies for real-world applications.

Three-way approximation sets, originating from decision-theoretic rough set theory [15], serve to characterize the shadowed regions of shadowed sets. These methods have gained significant attention due to their simplicity and computational efficiency in representing fuzzy uncertainties, making them a focal point of interdisciplinary research. The fundamental objective of three-way approximation theory is to partition the sample space into three non-overlapping regions(positive, boundary, and negative) through strategically designed optimization functions. Early studies focused on defining boundary thresholds for three-way approximations. Representative approaches include interval-based methods $[0, 1]$ [8], fixed thresholds (e.g., 0.5 [1]), semantic averages [2], and adaptive parameters m [19]. Subsequent research integrated Bayesian decision theory [20] and game theory to derive interpretable thresholds (β, α) by optimizing cost functions, fuzzy entropy, or distance loss. Notably, Gao et al. [3] combined uncertainty and decision costs in game analysis, while Zhang et al. [20] developed a computationally efficient game-theoretic model using fuzzy entropy loss. Hu et al. [4] proposed structured rough set expressions for incomplete systems, while Stepaniuk [9] introduced decision granularity to improve approximate reasoning. Yang et al. [14] further integrated three-way approximation into rough fuzzy sets, developing a data-driven model for uncertainty minimization. Zhou et al. [21] even proposed an unsupervised learning framework for evaluating three-way approximation principles empirically. Building on this foundation, recent research has expanded three-way approximation theory by merging it with advanced data modeling techniques. For instance, rough set extensions [5], granular ball computing [12], and incremental rough set models [6] have enabled multi-granularity three-way approximations [17]; Neighborhood-based K-nearest neighbor classifiers [13] and three-way clustering have enhanced adaptability; These developments highlight the versatility of three-way approximations across diverse applications.

While conventional approximation frameworks have achieved advancements in traditional analysis, there remains a paucity of systematic approaches to precisely segment hierarchical rough fuzzy multi-granularity knowledge spaces into their constituent regions. To address these gaps, this paper presents a novel framework for three-way approximation of rough fuzzy multi-granularity knowledge spaces, which is shown in Fig. 1.

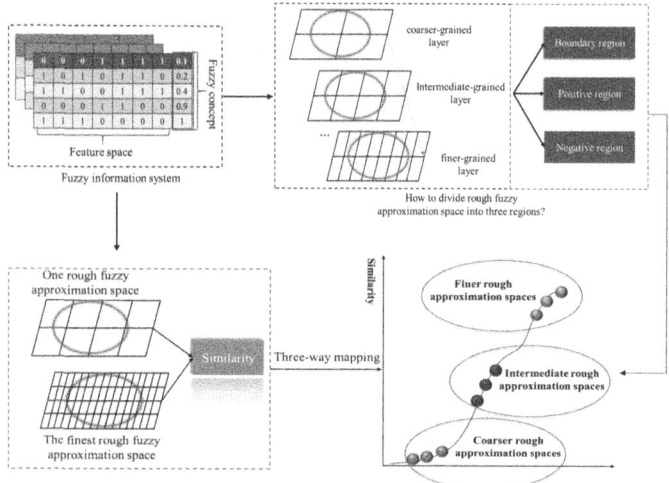

Fig. 1. A novel framework for three-way approximation of rough fuzzy multi-granularity knowledge spaces.

Our contributions are threefold: (1) A novel fuzzy knowledge granularity is proposed to quantify the uncertainty of fuzzy concepts, and the regulation of monotonically decreasing uncertainty with granularity is revealed; (2) A novel similarity function is constructed based on fuzzy knowledge granularity, and it is proven that this measurement method is an effective information measure; (3) By combining three-way approximation with granularity measure, a pair of interpretable thresholds (β, α) is successfully obtained, thus achieving a three-way representation of the rough fuzzy multi-granularity knowledge spaces.

The remainder of this paper is structured as follows. Section 2 provides foundational concepts, including rough sets, rough fuzzy sets, shadowed sets, and three-way approximation theory. Section 3 presents a novel framework for three-way representations of rough fuzzy multi-granularity spaces. Finally, Sect. 4 summarizes the key findings and discusses future research directions.

2   Preliminaries

To explain the content of this paper, some basic concepts about rough sets, rough fuzzy sets, shadowed sets, three-way approximations are reviewed in this section. A decision system is denoted by $\mathbb{S} = (\mathbb{U}, \mathbb{C} \cup \mathbb{D})$, in which $\mathbb{U}$ is a finite non-empty object set, $\mathbb{C}$ is a finite non-empty condition attribute set, $\mathbb{D}$ is the decision attribute set. If the decision value ranges from 0 to 1, then the decision system is referred to as a fuzzy decision system.

Definition 1. *[7] Let $\mathbb{S} = (\mathbb{U}, \mathbb{C} \cup \mathbb{D})$ be a decision system, an indistinguishable relation for a subset $\mathbb{R} \subseteq \mathbb{C}$ is defined as follows:*

$$\mathbb{IND}(\mathbb{R}) = \{(x, y) \in \mathbb{U} \times \mathbb{U} | \forall_{f \in \mathbb{R}} (f(x) = f(y))\}. \tag{1}$$

The $\mathbb{U}/\mathbb{IND}(\mathbb{R})$ on $\mathbb{U}$ is denoted by $\mathbb{U}/\mathbb{IND}(\mathbb{R}) = \{[x]_{\mathbb{IND}(\mathbb{R})} | x \in \mathbb{U}\}$, in which $[x]_{\mathbb{IND}(\mathbb{R})}$ denotes an equivalence class. For simplicity, let $\mathbb{U}/\mathbb{IND}(\mathbb{R}) \equiv \mathbb{U}/\mathbb{R}$ and $[x]_{\mathbb{IND}(\mathbb{R})} \equiv [x]_{\mathbb{R}}$, and $| \cdot |$ denotes the cardinality of a set.

Definition 2. *(Rough Sets [7]) Let $\mathbb{S} = (\mathbb{U}, \mathbb{C} \cup \mathbb{D})$ be a decision system, $\mathbb{X} \subseteq \mathbb{U}$, and $\mathbb{R} \subseteq \mathbb{C}$. Then, the lower and upper approximation sets of $\mathbb{X}$ are defined as follows:*

$$\underline{\mathbb{R}}(\mathbb{X}) = \{x \in \mathbb{U} | [x]_{\mathbb{R}} \subseteq \mathbb{X}\}, \quad \overline{\mathbb{R}}(\mathbb{X}) = \{x \in \mathbb{U} | [x]_{\mathbb{R}} \cap \mathbb{X} \neq \emptyset\}. \tag{2}$$

Definition 3. *(Fuzzy Sets [18]) Given a mapping on a non-empty finite set $\mathbb{U}$, i.e., $\widetilde{\mathbb{A}} : \mathbb{U} \to [0,1]$, $x| \to \widetilde{\mathbb{A}}(x)$, $\mathbb{A}$ is a fuzzy set on $\mathbb{U}$, in which $\forall x \in \mathbb{U}$, $\widetilde{\mathbb{A}}(x)$ is the membership function of $\widetilde{\mathbb{A}}$.*

Definition 4. *(Rough Fuzzy Sets [22]) Let $\mathbb{S} = (\mathbb{U}, \mathbb{C} \cup \mathbb{D})$ be a fuzzy decision system, $\widetilde{\mathbb{A}}$ be a fuzzy set on $\mathbb{U}$, and $\mathbb{R} \subseteq \mathbb{C}$. Then, the lower and upper approximation sets of $\widetilde{\mathbb{A}}$ are defined as a pair of fuzzy sets, and its membership degree are defined as follows:*

$$\underline{\widetilde{\mathbb{A}}}_{\mathbb{R}}(x) = inf\{\widetilde{\mathbb{A}}(y) | y \in [x]_{\mathbb{R}}\}, \quad \overline{\widetilde{\mathbb{A}}}_{\mathbb{R}}(x) = sup\{\widetilde{\mathbb{A}}(y) | y \in [x]_{\mathbb{R}}\}. \tag{3}$$

If $\underline{\widetilde{\mathbb{A}}}_{\mathbb{R}}(x) = \overline{\widetilde{\mathbb{A}}}_{\mathbb{R}}(x)$, then $\widetilde{\mathbb{A}}$ is a definable fuzzy sets; Otherwise, $\widetilde{\mathbb{A}}$ is a rough fuzzy set.

Definition 5. *(Shadowed Sets [8]) Given a pair of thresholds (α, β) with $0 \leq \beta < \alpha \leq 1$, assume that a fuzzy set $\widetilde{\mathbb{A}}$ on a non-empty finite domain $\mathbb{U}$ is mapping as a three-valued set $\{0, [0,1], 1\}$ by a function $\mathbb{S}^{(\alpha,\beta)}(\widetilde{\mathbb{A}})(x)$, which is called as a shadowed set and expressed as*

$$\mathbb{S}^{(\alpha,\beta)}(\widetilde{\mathbb{A}})(x) = \begin{cases} 1, & \alpha \leq \widetilde{\mathbb{A}}(x) \leq 1; \\ [0,1], & \beta < \widetilde{\mathbb{A}}(x) < \alpha; \\ 0, & 0 \leq \widetilde{\mathbb{A}}(x) \leq \beta. \end{cases} \tag{4}$$

As shown in Fig. 2, shadowed sets can be explained by two types of operations, i.e., elevation operation and reduction operation. Through elevation operation and reduction operation, it is clear that the shadowed sets $\mathbb{S}^{(\alpha,\beta)}(\widetilde{\mathbb{A}})$ consists of three areas: Elevated Area, Shadowed Area, and Reduced Area. In the realm

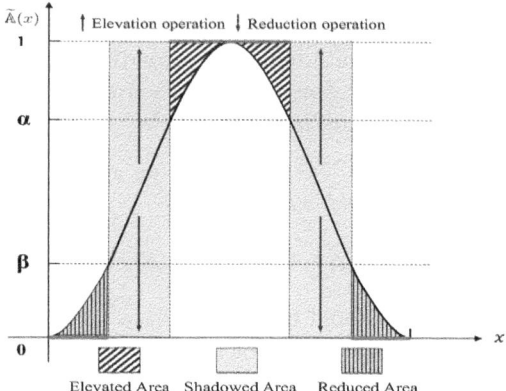

Fig. 2. Shadowed sets.

of three-way decisions, a shadowed set can be intuitively understood as comprising three distinct regions: the positive region, the boundary region, and the negative region, which are represented as

$$\mathbb{POS}(\mathbb{S}^{(\alpha,\beta)}(\widetilde{\mathbb{A}})) = \{x \in \mathbb{U}|\mathbb{S}^{(\alpha,\beta)}(\widetilde{\mathbb{A}})(x) = 1\},$$
$$\mathbb{BND}(\mathbb{S}^{(\alpha,\beta)}(\widetilde{\mathbb{A}})) = \{x \in \mathbb{U}|\mathbb{S}^{(\alpha,\beta)}(\widetilde{\mathbb{A}})(x) = [0,1]\}, \qquad (5)$$
$$\mathbb{NEG}(\mathbb{S}^{(\alpha,\beta)}(\widetilde{\mathbb{A}})) = \{x \in \mathbb{U}|\mathbb{S}^{(\alpha,\beta)}(\widetilde{\mathbb{A}})(x) = 0\}.$$

The main challenge with shadowed sets lies in constructing an optimization objective function that possesses semantic coherence to pinpoint the optimal thresholds (α, β). To solve this problem, Professor Yao proposed to replaced shadowed region $[0, 1]$ with 0.5, then a decision-theoretic three-way approximations of fuzzy sets was given to explain the thresholds (α, β) [16] from the principle of least cost.

Definition 6. *(Three-way approximations [16]) Given a pair of thresholds (α, β) with $0 \leq \beta < \alpha \leq 1$, assume that a fuzzy set $\widetilde{\mathbb{A}}$ on a non-empty finite domain $\mathbb{U}$ is mapping as a three-valued set $\{1, 0.5, 0\}$ by function $\mathbb{T}^{(\alpha,\beta)}(\widetilde{\mathbb{A}})(x)$, which is called as a three-way approximations and expressed as*

$$\mathbb{T}^{(\alpha,\beta)}(\widetilde{\mathbb{A}})(x) = \begin{cases} 1, & \alpha \leq \widetilde{\mathbb{A}}(x) \leq 1; \\ 0.5, & \beta < \widetilde{\mathbb{A}}(x) < \alpha; \\ 0, & 0 \leq \widetilde{\mathbb{A}}(x) \leq \beta. \end{cases} \qquad (6)$$

A three-way approximations of a fuzzy set can be explained by Fig. 3. In the realm of three-way decisions, a three-way approximations of fuzzy sets is readily conceptualized into three distinct regions, i.e., the positive, boundary, and negative regions, represented respectively as

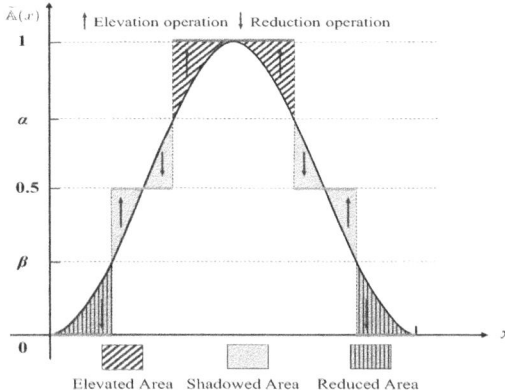

Fig. 3. A three-way approximations of a fuzzy set.

$$\mathbb{POS}(\mathbb{T}^{(\alpha,\beta)}(\widetilde{\mathbb{A}})) = \{x \in \mathbb{U} | \mathbb{T}^{(\alpha,\beta)}(\widetilde{\mathbb{A}})(x) = 1\},$$
$$\mathbb{BND}(\mathbb{T}^{(\alpha,\beta)}(\widetilde{\mathbb{A}})) = \{x \in \mathbb{U} | \mathbb{T}^{(\alpha,\beta)}(\widetilde{\mathbb{A}})(x) = 0.5\}, \tag{7}$$
$$\mathbb{NEG}(\mathbb{T}^{(\alpha,\beta)}(\widetilde{\mathbb{A}})) = \{x \in \mathbb{U} | \mathbb{T}^{(\alpha,\beta)}(\widetilde{\mathbb{A}})(x) = 0\}.$$

It is apparent that the transformation to three-way approximations through the mechanisms of elevation and reduction inevitably incurs a certain margin of error and decision cost. Subsequently, Theorem 1 explain how to compute the thresholds (α, β) from the principle of least cost.

Theorem 1. *([16]) Consider $\widetilde{\mathbb{A}}$ be a fuzzy set on $\mathbb{U}$, and $\mathbb{T}^{(\alpha,\beta)}(\widetilde{\mathbb{A}}) = \{0, 0.5, 1\}$. In the process of converting $\widetilde{\mathbb{A}}$ into $\mathbb{T}^{(\alpha,\beta)}(\widetilde{\mathbb{A}}) = \{0, 0.5, 1\}$, assume that $\mathbb{A} = \{a_1, a_0, a_{\downarrow 0.5}, a_{\uparrow 0.5}\}$ be a group of operations, and be a group of cost actions $\mathbb{C} = \{c_1, c_0, c_{\downarrow 0.5}, c_{\uparrow 0.5}\}$. Thereinto, c_1 represents the unit cost of viewing $\widetilde{\mathbb{A}}(x) = 1$ by a_1, c_0 represents the unit cost of viewing $\widetilde{\mathbb{A}}(x) = 0$ by a_0, $c_{\downarrow 0.5}$ represents the unit cost of reducing $\widetilde{\mathbb{A}}(x)$ to 0.5 by $a_{\downarrow 0.5}$, and $c_{\uparrow 0.5}$ represents the unit cost of elevating $\widetilde{\mathbb{A}}(x)$ to 0.5 by $a_{\uparrow 0.5}$. Assume that $0 \leq c_{\downarrow 0.5} \leq c_0$ and $0 \leq c_{\uparrow 0.5} \leq c_1$, if $\alpha = \frac{c_1 + 0.5 c_{\downarrow 0.5}}{c_1 + c_{\downarrow 0.5}}$ and $\beta = \frac{0.5 c_{\uparrow 0.5}}{c_0 + c_{\uparrow 0.5}}$, then the domain $\mathbb{U}$ will be divided into three disjoint regions with the minimum cost loss.*

3   Three-Way Representation of Rough Fuzzy Approximation Spaces

This section is dedicated to exploring the application of the three-way approximation concept to meticulously divide the hierarchical rough fuzzy approximation space into three distinct regions(the positive region, the boundary region, and the negative region). Initially, we introduce Definition 7, which outlines the structure of the hierarchical rough fuzzy approximation space.

Definition 7. *Let* $\mathbb{S} = (\mathbb{U}, \mathbb{C} \cup \mathbb{D})$ *be a fuzzy decision system,* $\widetilde{\mathbb{A}}$ *is a fuzzy set on* $\mathbb{U}$, *and* $\mathbb{R}_1 \subseteq \mathbb{R}_2, \ldots, \subseteq \mathbb{R}_k \subseteq \mathbb{C}$. *Then, a hierarchical rough fuzzy multi-granularity space structure is given as follows:*

$$\mathbb{HS} = \{\mathbb{GL}_1, \mathbb{GL}_2, \ldots, \mathbb{GL}_k\}. \tag{8}$$

Where $\mathbb{GL}_i = (\mathbb{U}/\mathbb{R}_i, \widetilde{\mathbb{A}})$ *denotes fuzzy granular layer.*

Following this, we present Definition 8, which elaborates on the nature of uncertainty inherent in each layer of the rough fuzzy granulation. This measure takes into account that the uncertainty of rough fuzzy approximation space is not only related to the size of knowledge granularity, but also to the membership degree of fuzzy decision concept.

Definition 8. *Let* $\mathbb{S} = (\mathbb{U}, \mathbb{C} \cup \mathbb{D})$ *be a fuzzy decision system,* $\widetilde{\mathbb{A}}$ *is a fuzzy set on* $\mathbb{U}$, $\mathbb{R} \subseteq \mathbb{C}$, *and* $\mathbb{U}/\mathbb{R} = \{\mathbb{P}_1, \mathbb{P}_2, \ldots, \mathbb{P}_n\}$. *Then, the rough fuzzy multi-granularity space in* $\mathbb{U}/\mathbb{R}$ *is quantified as follows:*

$$\mathbb{F}(\mathbb{R}, \widetilde{\mathbb{A}}) = \frac{1}{|\mathbb{U}| \sum_{x \in \mathbb{U}} \widetilde{\mathbb{A}}(x)} \sum_{i=1}^{n} |\mathbb{P}_i| \sum_{x \in \mathbb{P}_i} \widetilde{\mathbb{A}}(x). \tag{9}$$

According to Definition 8, Theorem 2 is then employed to substantiate the observation that the level of uncertainty exhibits a monotonic decrease.

Theorem 2. *Let* $\mathbb{S} = (\mathbb{U}, \mathbb{C} \cup \mathbb{D})$ *be a fuzzy decision system, and* $\widetilde{\mathbb{A}}$ *is a fuzzy set on* $\mathbb{U}$. *If* $\mathbb{R} \subseteq \mathbb{Q} \subseteq \mathbb{C}$, *then* $\mathbb{F}(\mathbb{R}, \widetilde{\mathbb{A}}) \geq \mathbb{F}(\mathbb{Q}, \widetilde{\mathbb{A}})$ *holds.*

Proof. Because $\mathbb{R} \subseteq \mathbb{Q}$, then $\mathbb{U}/\mathbb{Q} \preceq \mathbb{U}/\mathbb{R}$. For convenience, it is assumed that only one equivalence class $[x_1]_{\mathbb{R}}([x_1]_{\mathbb{R}} \in \mathbb{U}/\mathbb{R})$ is subdivided into two finer equivalence classes $[y_1]_{\mathbb{Q}}, [y_2]_{\mathbb{Q}}$. Therefore, let $[x_1]_{\mathbb{R}} = [y_1]_{\mathbb{Q}} \cup [y_2]_{\mathbb{Q}}$, $[x_2]_{\mathbb{R}} = [y_3]_{\mathbb{Q}}, \cdots$, $[x_m]_{\mathbb{R}} = [y_{m+1}]_{\mathbb{Q}}$, and $|[y_1]_{\mathbb{Q}}| = a$, $|[y_2]_{\mathbb{Q}}| = b$. According to Definition 8, we get that $\mathbb{F}(\mathbb{R}, \widetilde{\mathbb{A}}) - \mathbb{F}(\mathbb{Q}, \widetilde{\mathbb{A}}) = \dfrac{1}{|\mathbb{U}| \sum_{x \in \mathbb{U}} \widetilde{\mathbb{A}}(x)} \{(a + b) \sum_{x \in [y_1]_{\mathbb{Q}} \cup [y_2]_{\mathbb{Q}}} \widetilde{\mathbb{A}}(x) - a \sum_{x \in [y_1]} \widetilde{\mathbb{A}}(x) - b \sum_{x \in [y_2]} \widetilde{\mathbb{A}}(x)\}$. Let $\sum_{x \in [y_1]} \widetilde{\mathbb{A}}(x) = m$ and $\sum_{x \in [y_2]} \widetilde{\mathbb{A}}(x) = n$, then $\mathbb{F}(\mathbb{R}, \widetilde{\mathbb{A}}) - \mathbb{F}(\mathbb{Q}, \widetilde{\mathbb{A}}) = \dfrac{1}{|\mathbb{U}| \sum_{x \in \mathbb{U}} \widetilde{\mathbb{A}}(x)} \{(a + b)(m + n) - am - bn\} = \dfrac{an + bm}{|\mathbb{U}| \sum_{x \in \mathbb{U}} \widetilde{\mathbb{A}}(x)}$. Obviously, $\mathbb{F}(\mathbb{R}, \widetilde{\mathbb{A}}) - \mathbb{F}(\mathbb{Q}, \widetilde{\mathbb{A}}) \geq 0$. Therefore, Theorem 2 holds.

Proposition 1. *Let* $\mathbb{S} = (\mathbb{U}, \mathbb{C} \cup \mathbb{D})$ *be a fuzzy decision system,* $\widetilde{\mathbb{A}}$ *is a fuzzy set on* $\mathbb{U}$, $\mathbb{R} \subseteq \mathbb{C}$. *Then,* $\frac{1}{|\mathbb{U}|} \leq \mathbb{F}(\mathbb{R}, \widetilde{\mathbb{A}}) \leq 1$ *holds.*

According to Definition 8, we will provide Definition 9 to describe the similarity between two rough fuzzy approximation spaces, which will lead to the discussion on how to trisect the rough fuzzy approximation space into three parts.

Definition 9. *Let* $\mathbb{S} = (\mathbb{U}, \mathbb{C} \cup \mathbb{D})$ *be a fuzzy decision system,* $\widetilde{\mathbb{A}}$ *is a fuzzy set on* $\mathbb{U}$, $\mathbb{R} \subseteq \mathbb{Q} \subseteq \mathbb{C}$, *then the similarity between* $\mathbb{U}/\mathbb{R}$ *and* $\mathbb{U}/\mathbb{Q}$ *is defined as follows:*

$$\mathbb{S}(\mathbb{R}, \mathbb{Q}) = 1 - \mathbb{F}(\mathbb{R}, \widetilde{\mathbb{A}}) + \mathbb{F}(\mathbb{Q}, \widetilde{\mathbb{A}}). \tag{10}$$

Theorem 3. *Let* $\mathbb{S} = (\mathbb{U}, \mathbb{C} \cup \mathbb{D})$ *be a fuzzy decision system,* $\widetilde{\mathbb{A}}$ *is a fuzzy set on* $\mathbb{U}$, $\mathbb{R}_1 \subseteq \mathbb{R}_2 \subseteq \mathbb{R}_3 \subseteq \mathbb{C}$, *then* $\mathbb{S}(\mathbb{R}_1, \mathbb{R}_3) \leq \mathbb{S}(\mathbb{R}_2, \mathbb{R}_3)$ *holds.*

Proof. This can easily be concluded from Theorem 2.

Proposition 2. *Let* $\mathbb{S} = (\mathbb{U}, \mathbb{C} \cup \mathbb{D})$ *be a fuzzy decision system,* $\widetilde{\mathbb{A}}$ *is a fuzzy set on* $\mathbb{U}$, $\mathbb{R}, \mathbb{Q} \subseteq \mathbb{C}$. *Then,* $\frac{1}{|\mathbb{U}|} \leq \mathbb{S}(\mathbb{R}, \mathbb{Q}) \leq 1$.

Proof. This can easily be concluded from Theorem 3.

If each equivalence class $|[x]_\mathbb{R}| = 1$ in domain $\mathbb{U}$, then $\mathbb{U}/\mathbb{R}$ is called as the finest granularity of rough fuzzy approximation space, which denoted as ω. That is, $\omega = \{\{x_1\}, \{x_2\}, \ldots, \{x_n\}\}$. Subsequently, according to the principle of information measure [23], we will deduce that the similarity between the current rough fuzzy approximation space and the finest granularity of rough fuzzy approximation space is an information measure, as shown in Theorem 4.

Theorem 4. *Let* $\mathbb{S} = (\mathbb{U}, \mathbb{C} \cup \mathbb{D})$ *be a fuzzy decision system,* $\widetilde{\mathbb{A}}$ *is a fuzzy set on* $\mathbb{U}$, $\mathbb{R} \subseteq \mathbb{C}$, ω *denotes the finest granularity of rough fuzzy space. Then* $\mathbb{S}(\mathbb{R}, \omega)$ *is an information measure.*

Proof. (1) Based on Proposition 2, we can get that $\mathbb{S}(\mathbb{R}, \mathbb{Q}) \geq 0$; (2) Based on Theorem 3, if $\mathbb{R}_1 \subseteq \mathbb{R}_2$, we can get that $\mathbb{S}(\mathbb{R}_1, \omega) \leq \mathbb{S}(\mathbb{R}_2, \omega)$; (3) If $\mathbb{U}/\mathbb{R}_1 = \mathbb{U}/\mathbb{R}_2$, obviously, $\mathbb{S}(\mathbb{R}_1, \omega) = \mathbb{S}(\mathbb{R}_2, \omega)$. Therefore, according to the principle of information measure, Theorem 4 holds.

According to definition 9, the rough fuzzy approximation space can be represented as three approximation sets, as shown in Definition 10. Subsequently, based on the thresholds (α, β), we can divide the hierarchical coarse fuzzy multi-granularity knowledge space into three regions, as shown in Definition 11.

Definition 10. *Let* $\mathbb{S} = (\mathbb{U}, \mathbb{C} \cup \mathbb{D})$ *be a fuzzy decision system,* $\widetilde{\mathbb{A}}$ *is a fuzzy set on* $\mathbb{U}$, $\mathbb{R} \subseteq \mathbb{C}$, ω *denotes the finest granularity of rough fuzzy space. The three-way approximations set of* $\mathbb{S}(\mathbb{R}, \omega)$ *is defined as follow.*

$$\mathbb{T}^{(\alpha, \beta)}(\mathbb{S}(\mathbb{R}, \omega)) = \begin{cases} 1, & \alpha \leq \mathbb{S}(\mathbb{R}, \omega) \leq 1; \\ 0.5, & \beta < \mathbb{S}(\mathbb{R}, \omega) < \alpha; \\ 0, & 0 \leq \mathbb{S}(\mathbb{R}, \omega) \leq \beta. \end{cases} \tag{11}$$

Definition 11. *Let* $\mathbb{S} = (\mathbb{U}, \mathbb{C} \cup \mathbb{D})$ *be a fuzzy decision system,* $\widetilde{\mathbb{A}}$ *is a fuzzy set on* $\mathbb{U}$, *and* $\mathbb{R}_1 \subseteq \mathbb{R}_2, \ldots, \subseteq \mathbb{R}_k$. *According to the granularity measure* $\mathbb{D}(\mathbb{R}, \omega)$, *the hierarchical rough fuzzy multi-granularity space structure* $\mathbb{HS} = \{\mathbb{GL}_1, \mathbb{GL}_2, \ldots, \mathbb{GL}_k\}$ *will be divide into three fuzzy knowledge spaces as follows:*

$$\mathbb{POS}(\mathbb{T}^{(\alpha, \beta)}(\mathbb{HS})) = \{\mathbb{GL}_i \in \mathbb{HS} | \mathbb{T}^{(\alpha, \beta)}(\mathbb{S}(\mathbb{R}, \omega)) = 1\},$$
$$\mathbb{BND}(\mathbb{T}^{(\alpha, \beta)}(\mathbb{HS})) = \{\mathbb{GL}_i \in \mathbb{HS} | \mathbb{T}^{(\alpha, \beta)}(\mathbb{S}(\mathbb{R}, \omega)) = 0.5\}, \tag{12}$$
$$\mathbb{NEG}(\mathbb{T}^{(\alpha, \beta)}(\mathbb{HS})) = \{\mathbb{GL}_i \in \mathbb{HS} | \mathbb{T}^{(\alpha, \beta)}(\mathbb{S}(\mathbb{R}, \omega)) = 0\}.$$

Based on Definitions 10 and 11, the following Corollary 1 will explain how to use a pair of thresholds (β, α) to divide the hierarchical rough fuzzy approximation space into three regions with minimum similarity loss.

Corollary 1. *Suppose that $c_1 = c_0 = c_{\downarrow 0.5} = c_{\uparrow 0.5} = 1$. If $\alpha = 0.75$ and $\beta = 0.25$, then the hierarchical rough fuzzy multi-granularity space structure $\mathbb{HS} = \{\mathbb{GL}_1, \mathbb{GL}_2, \ldots, \mathbb{GL}_k\}$ will be divide into three fuzzy knowledge spaces three disjointed regions with minimum similarity loss.*

Proof. This can easily be concluded from Theorem 1.

To assist readers in gaining a clearer understanding of the core ideas presented in this paper, we will provide a detailed explanation and illustration with the aid of Example 1.

Example 1. Let $\mathbb{S} = (\mathbb{U}, \mathbb{C} \cup \mathbb{D})$ be a fuzzy decision system, $\widetilde{\mathbb{A}}$ is a fuzzy set on $\mathbb{U} = \{x_1, x_2, x_3, x_4, x_5\}$, $\widetilde{\mathbb{A}} = \frac{0}{x_1} + \frac{0.1}{x_2} + \frac{0.2}{x_3} + \frac{0.3}{x_4} + \frac{0.9}{x_5}$, and $\mathbb{C} = \{f_1, f_2, f_3, f_4, f_5\}$. Suppose that $\mathbb{R}_1 = \{f_1\}$, $\mathbb{R}_2 = \{f_1, f_2\}$, $\mathbb{R}_3 = \{f_1, f_2, f_3\}$, $\mathbb{R}_4 = \{f_1, f_2, f_3, f_4\}$, $\mathbb{R}_5 = \{f_1, f_2, f_3, f_4, f_5\}$, obviously, $\mathbb{R}_1 \subseteq \mathbb{R}_2 \subseteq \mathbb{R}_3 \subseteq \mathbb{R}_4 \subseteq \mathbb{R}_5$. Let $\mathbb{U}/\mathbb{R}_1 = \{x_1, x_2, x_3, x_4, x_5\}$, $\mathbb{U}/\mathbb{R}_2 = \{\{x_1\}, \{x_2, x_3, x_4, x_5\}\}$, $\mathbb{U}/\mathbb{R}_3 = \{\{x_1\}, \{x_2\}, \{x_3, x_4, x_5\}\}$, $\mathbb{U}/\mathbb{R}_4 = \{\{x_1\}, \{x_2\}, \{x_3\}, \{x_4, x_5\}\}$, $\mathbb{U}/\mathbb{R}_5 = \{\{x_1\}, \{x_2\}, \{x_3\}, \{x_4\}, \{x_5\}\}$. Therefore, the hierarchical rough fuzzy multi-granularity space structure $\mathbb{HS} = \{\mathbb{GL}_1, \mathbb{GL}_2, \mathbb{GL}_3, \mathbb{GL}_4, \mathbb{GL}_5\}$ is shown in Fig. 4. According to Definition 8 and Definition 9, we can get that $\mathbb{S}(\mathbb{R}_1, \omega) = 1 - \frac{1}{5*1.5}\{(5*1.5) - (0*1 + 1*0.1 + 1*0.2 + 1*0.3 + 1*0.9)\} = 0.2$. Because $0.0 < \mathbb{S}(\mathbb{R}_1, \omega) = 0.4 < 0.25$, then $\mathbb{GL}_1 = (\mathbb{U}/\mathbb{R}_1, \widetilde{\mathbb{A}})$ will be divided into the negative region. The rest can be done in the same manner, we get that $\mathbb{S}(\mathbb{R}_2, \omega) = 0.4$, $\mathbb{S}(\mathbb{R}_3, \omega) = 0.49$, $\mathbb{S}(\mathbb{R}_4, \omega) = 0.84$, $\mathbb{S}(\mathbb{R}_5, \omega) = 1$. Therefore, the hierarchical rough fuzzy multi-granularity space structure $\mathbb{HS} = \{\mathbb{GL}_1, \mathbb{GL}_2, \mathbb{GL}_3, \mathbb{GL}_4, \mathbb{GL}_5\}$ will be divide into three fuzzy knowledge spaces as follows:

$$\mathbb{POS}(\mathbb{T}^{(\alpha, \beta)}(\mathbb{HS})) = \{\mathbb{GL}_4, \mathbb{GL}_5\},$$
$$\mathbb{BND}(\mathbb{T}^{(\alpha, \beta)}(\mathbb{HS})) = \{\mathbb{GL}_2, \mathbb{GL}_3\}, \tag{13}$$
$$\mathbb{NEG}(\mathbb{T}^{(\alpha, \beta)}(\mathbb{HS})) = \{\mathbb{GL}_1\}.$$

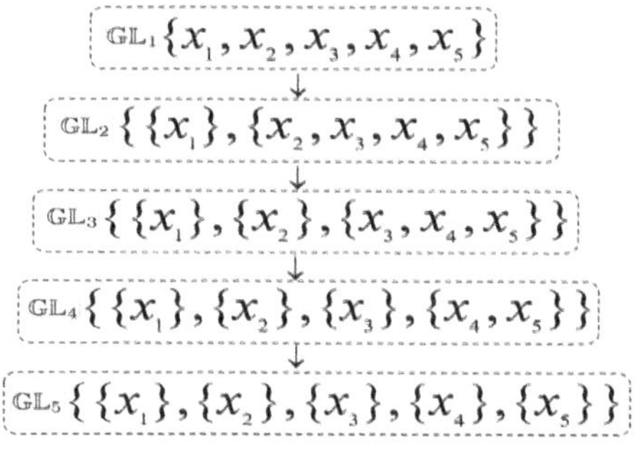

Fig. 4. The hierarchical rough fuzzy multi-granularity space structure.

4 Conclusion

This study is grounded in the three-way approximation theory framework, achieving the first extension from point elements to regional coverage through the development of a three-way approximation decomposition method specifically designed for rough-fuzzy multi-granularity knowledge spaces. We systematically establish relevant theorems and properties while acknowledging that current implementations remain limited in practical applications for knowledge acquisition. Nonetheless, the model's multidimensional feature representation capability demonstrates significant theoretical potential as a novel tool for complex data modeling. Future research will focus on two principal directions: (1) Exploring the applicative value of three-way approximation concepts in attribute reduction and clustering analysis; (2) Constructing an integrated learning mechanism tailored for multi-source heterogeneous data to enhance both practical applicability and generalization performance of the model.

Acknowledgements. This work was supported by the Guizhou Provincial Department of Education Colleges and Universities Science and Technology Innovation Team (QJJ[2023]084), and the National Natural Science Foundation of China (No. 62366033), Jiangxi Provincial Natural Science Foundation (No. 20232BAB202049).

References

1. Deng, X., Yao, Y.: Decision-theoretic three-way approximations of fuzzy sets. Inf. Sci. **279**, 702–715 (2014)
2. Gao, M., Zhang, Q., Zhao, F., Wang, G.: Mean-entropy-based shadowed sets: a novel three-way approximation of fuzzy sets. Int. J. Approximate Reasoning **120**, 102–124 (2020)
3. Gao, M., Zhang, Q., Zhao, F., Wu, C., Wang, G., Xia, D.: Constructing shadowed set based on game analysis of uncertainty and decision cost. Appl. Soft Comput. **147**, 110762 (2023)
4. Hu, M., Yao, Y.: Structured approximations as a basis for three-way decisions in rough set theory. Knowl.-Based Syst. **165**, 92–109 (2019)
5. Hu, Q., Yu, D., Liu, J., Wu, C.: Neighborhood rough set based heterogeneous feature subset selection. Inf. Sci. **178**(18), 3577–3594 (2008)
6. Li, D., Wang, T., Chen, J., Kawaguchi, K., Lian, C., Zeng, Z.: Multi-view class incremental learning. Inf. Fusion **102**, 102021 (2024)
7. Pawlak, Z.: Rough sets. IJICS. **11**(5), 341–356 (1982)
8. Pedrycz, W.: Shadowed sets: representing and processing fuzzy sets. IEEE Trans. Syst. Man Cybern. Part B **28**(1), 103–109 (1998)
9. Stepaniuk, J., Skowron, A.: Three-way approximation of decision granules based on the rough set approach. Int. J. Approximate Reasoning **155**, 1–16 (2023)
10. Wang, G.Y.: DGCC: data-driven granular cognitive computing. Granul Comput. **2**(4), 343–355 (2017)
11. Wang, G.Y.: Granular computing: from granularity optimization to multi-granularity joint problem solving. Granul Comput. **2**(3), 105–120 (2017)

12. Xia, S., Zhang, H., Li, W., Wang, G., Giem, E., Chen, Z.: GBNRS: a novel rough set algorithm for fast adaptive attribute reduction in classification. IEEE Trans. Knowl. Data Eng. **34**(3), 1231–1242 (2020)
13. Yang, J., et al.: Adaptive three-way KNN classifier using density-based granular balls. Inf. Sci. **678**, 120858 (2024)
14. Yang, J., Wang, X., Wang, G., Xia, D.: Constructing three-way decision of rough fuzzy sets from the perspective of uncertainties. Cogn. Comput. **16**(5), 2454–2470 (2024)
15. Yao, Y.: The superiority of three-way decisions in probabilistic rough set models. Inf. Sci. **181**(6), 1080–1096 (2011)
16. Yao, Y., Wang, S., Deng, X.: Constructing shadowed sets and three-way approximations of fuzzy sets. Inf. Sci. **412**, 132–153 (2017)
17. Yao, Y., Yang, J.: Granular rough sets and granular shadowed sets: three-way approximations in Pawlak approximation spaces. Int. J. Approximate Reasoning **142**, 231–247 (2022)
18. Zadeh, L.: Fuzzy sets. Inf. Control **8**(3), 338–353 (1965)
19. Zhang, Q., Xia, D., Liu, K., Wang, G.: A general model of decision-theoretic three-way approximations of fuzzy sets based on a heuristic algorithm. Inf. Sci. **507**, 522–539 (2020)
20. Zhang, Y., Yao, J.: Game theoretic approach to shadowed sets: a three-way tradeoff perspective. Inf. Sci. **507**, 540–552 (2020)
21. Zhou, J., Pedrycz, W., Gao, C., Lai, Z., Yue, X.: Principles for constructing three-way approximations of fuzzy sets: a comparative evaluation based on unsupervised learning. Fuzzy Sets Syst. **413**, 74–98 (2021)
22. Sun, B., Pedrycz, Ma, W., Zhao, H.: Decision-theoretic rough fuzzy set model and application. Inf. Sci. **283**, 180–196 (2014)
23. Beaubouef, T., Petry, F.E., Arora, G.: Information-theoretic measures of uncertainty for rough sets and rough relational databases. Inf. Sci. **109**, 185–195 (1998)

Three-Way Strategy Design for Conflict Analysis

Chengjun Shi[1,2], Mengjun Hu[1(✉)], Yilin Huang[1], and Yiyu Yao[2]

[1] Department of Mathematics and Computing Science, Saint Mary's University,
Halifax, NS B3H 3C3, Canada
Mengjun.Hu@smu.ca
[2] Department of Computer Science, University of Regina,
Regina, SK S4S 0A2, Canada

Abstract. Effective decision-making and strategy design remain challenging in conflict analysis, particularly when managing complex multi-agent interactions. This paper addresses this challenge by introducing a 3×3 strategy space with the three-valued rating scale. Following the Triading-Acting-Optimizing (TAO) model of three-way decision-making, we propose two dimensions of evaluation that form two triads, which are combined to support actionable strategy design. The first evaluation constructs a triad based on three levels of agreement among agents: strong-agreement, weak-agreement, and non-agreement. The second evaluation builds a triad based on agent-aggregated ratings: positivity, neutrality, and negativity. Combining these two dimensions, we define a 3×3 strategy space comprising nine distinct regions. Within this strategy space, we derive five decision rules to guide strategic choices: acceptance under strong agreement, acceptance under weak agreement, rejection under strong agreement, rejection under weak agreement, and deferment. An example demonstrates that the proposed model prioritizes high-confidence actions, supports phased implementations, and minimizes risky decisions, thereby reducing the likelihood of conflict.

Keywords: Three-way decision · Conflict analysis · Three-valued rating · Triadic thinking · Strategy design

1 Introduction

Conflict analysis examines the relationships and interactions among agents and issues in conflicting situations. Pawlak's foundational work [8,9] introduced conflict analysis using a three-valued situation table, classifying agent relations into three types: allied, neutral, and conflict. Yao [19] highlighted the conceptual alignment between conflict analysis and three-way decision (3WD) and proposed a fundamental model of three-way conflict analysis. Since then, many studies have extended the application of 3WD in conflict analysis [1–4,7,11,14,15].

This work was partially supported by Discovery Grants from NSERC, Canada (Mengjun Hu and Yiyu Yao) and the National Natural Science Foundation of China (Nos. 62276217, 62076040, Mengjun Hu). The authors thank reviewers for their constructive comments.

Q. Zhang et al. (Eds.): IJCRS 2025, LNAI 15710, pp. 85–97, 2025.
https://doi.org/10.1007/978-3-031-92741-6_7

While agent relations have been well studied in various contexts, the subsequent action of designing effective resolution strategies remains underexplored. Sun et al. [10, 11] introduced feasible consensus strategies using probabilistic rough sets. Xu et al. [13] proposed consistency measures to select feasible strategies. Yang, Wang, and Guo [15] defined optimal strategies as those maximizing conflict resolution while satisfying agent interests. Despite differences in interpreting and assessing feasibility and optimality, these works share a common view of strategy as a subset of issues, fundamentally distinct from the 3WD perspective, which perceives strategies as actions taken in decision-making. Real-world scenarios, such as legislative motions, peer-reviewed manuscript decisions, and committee voting, demand actionable decision rules. This gap motivates us to explore how to design actionable strategies to support decision-making in conflict analysis.

As an initial effort, this paper focuses on the three-valued rating scale and considers two dimensions: the agreement level among agents and their overall aggregated ratings, both with respect to a single issue. These two dimensions offer distinct yet complementary perspectives, both essential for understanding conflict and designing actionable strategies. A high agreement level does not necessarily imply a high overall rating, as agreement might reflect a unanimous oppositional or neutral attitude. Conversely, a low agreement level may arise when agents display clear but opposing attitudes, resulting in a near-neutral overall rating. These diverse scenarios complicate decision-making and strategy design in conflict analysis.

To address these challenges, we adopt the Triading-Acting-Optimizing (TAO) model of 3WD [18, 21] for conflict analysis. As the name suggests, the TAO model formulates 3WD in three steps: (1) Triading for decomposing the complex problem into three components, (2) Acting for developing targeted strategies for each component, and (3) Optimizing for refining the combination of the previous two steps to achieve desired outcomes. Following the TAO model, in the first Triading step with respect to conflict analysis, we consider the two dimensions mentioned above and formulate two triads. These dimensions are quantified by two evaluation functions: one assesses agent agreement levels, and the other aggregates agent ratings. Accordingly, the first triad consists of strong-agreement, weak-agreement, and non-agreement, while the second triad includes positivity, neutrality, and negativity. In the second Acting step, the combination of these two triads produces nine potential cases, forming a 3×3 strategy space that aligns with the concept of double triadic thinking [12]. The nine regions within this strategy space are further categorized into five decision-making cases, leading to five actionable strategies expressed as decision rules. While much of the literature focuses on the Triading step, we advance strategy design by emphasizing the Acting step, addressing how the diverse triadic results from existing studies can be leveraged in conflict resolution. The final Optimizing step will be explored in future work.

The remainder of this paper is organized as follows. Section 2 reviews the foundational concepts of 3WD. The next two sections present our approach for the Triading and Acting steps: Sect. 3 formulates the two triads from the two dimensions, and Sect. 4 introduces the strategy space and actionable strategy rules. Section 5 demonstrates these two steps through an example. Finally, Sect. 6 summarizes the contributions and outlines directions for future work.

2   An Overview of Three-Way Decision and Evaluation-Based Models

A recent study [21] on three-way decision refined the "Triading-Acting-Optimizing" model as illustrated in Fig. 1. Triading represents the procedure of constructing a triad, acting is a set of strategies to process or study the triad through processing or studying of the three elements, and optimizing describes the stage of seeking the most effective combination of triading and acting, aiming to achieve the desired outcome.

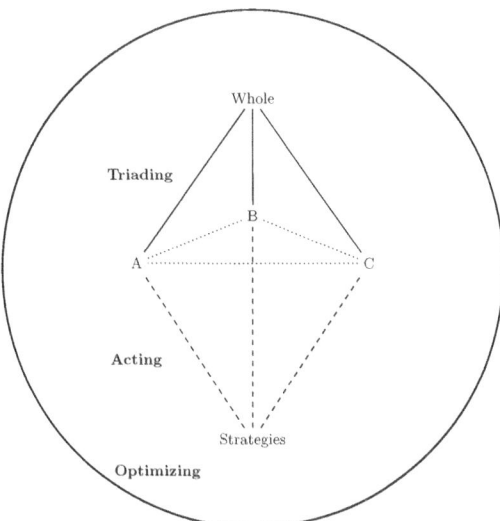

Fig. 1. TAO model of three-way decision [21]

The outline of three-way decision theory [17] formalizes decision-making by triading the universe into three regions: acceptance, rejection, and non-commitment. Yao [20, 21] generalized the three regions into H, L, and M, representing three regions with high, low, and medium evaluation values, respectively.

Definition 1. *Suppose U is the universe and $e : U \longrightarrow \mathbb{R}$ is an evaluation function that maps an element in U to a real value. Given a pair of thresholds*

(h, l) with $h > l$, the universe U is triaded into the following three regions:

$$H_{(h,l)}(e) = \{x \in U \mid e(x) \geqslant h\},$$
$$L_{(h,l)}(e) = \{x \in U \mid e(x) \leqslant l\},$$
$$M_{(h,l)}(e) = \{x \in U \mid l < e(x) < h\}. \tag{1}$$

In Definition 1, the pair of thresholds (h, l) plays a pivotal role in shaping the three regions. Existing studies addressed threshold determination through diverse methods: decision-theoretic rough sets [16], optimization by means of penalty mechanism [6], and automatic learning [5]. The interpretations of the three regions may vary depending on contexts. With the above evaluation-based model, the TAO model is interpreted as:

Triading: Building a triad $\langle H, L, M \rangle$ with a given evaluation function;
Acting: Devising three sets of strategies to process the elements in the triad;
Optimizing: Searching for the best combination of Triading and Acting.

The TAO model, structured with $\langle H, L, M \rangle$, offers flexibility and generality across diverse domains. The application scenarios and the choice of evaluation function can assign different meanings, semantics, and explanations to the triad. Considering the problem of financial investment, the triad can be explained as high, low, and medium risk. The corresponding strategies might be reject, invest and further analyze. Another example is peer reviewing, where the review scores can be categorized into high (publish), low (reject), and medium (revise).

The advantage of the TAO model enables user to make adaptive and targeted strategies. In this work, we focus on decision-making in conflict analysis by constructing two triads with two different evaluations in the Triading step. The agreement triad divides the set of issues according to the agreement level among agents, while another rating triad categorizes issues based on agent-aggregated ratings. The proposed approach exemplifies the feasibility and flexibility of the TAO model.

3   Triading of Issues from Two Dimensions

Conflict analysis usually starts with a situation table that represents the attitudes, opinions, or ratings of a set of agents towards a set of issues. There are four essential components in a situation table: (1) a set of agents, (2) a set of issues, (3) a domain of ratings, and (4) a rating function.

Definition 2. [1]   *A situation table is defined as a 4-tuple: $T = (A, I, V, r)$, where A is a finite nonempty set of agents, I is a finite nonempty set of issues, V is a nonempty domain set of values for ratings, and $r : A \times I \longrightarrow V$ is a rating function. The value $r(a, i) \in V$ is the rating of an agent $a \in A$ on an issue $i \in I$.*

Intuitively, the ratings may be either qualitative or quantitative, depending on specific contexts. In this paper, we consider a three-valued rating domain, which is also commonly studied in recent three-way conflict analysis research [1, 4,7,14]. Nevertheless, the ideas of our approach can be extended to other more complex rating scales. Specifically, a three-valued rating domain consists of three qualitative attitudes of positive, neutral, and negative, which are often encoded as a domain of three quantities $V = \{-1, 0, +1\}$. The situation table is then called a three-valued situation table.

3.1 Agreement-Based Triading of Issues

We present two intuitively reasonable ways to triad the issues. While the first approach reflects the most common intuition and aligns better with standard three-way classification models, we will argue that it is not ideal for designing actionable strategies in conflict analysis.

The prevalent interpretation of a triad takes two opposite extremes and one in-the-middle component, usually referred to as positive, boundary/neutral, and negative. Following this interpretation, one may consider three agreement levels among agents: agreement, disagreement, and hesitation/uncertainty. This idea could be quantified through a measurement of agreement degree among agents, which can be further decomposed into pair-wise agreement between agents. To facilitate the formulation, we introduce the following three sets of agent pairs in a set of agents $B \subseteq A$ and with respect to an issue $i \in I$:

$$\text{Agreed pairs}: \; \mathbb{A}_i^B = \{(a, b) \in B \times B \mid a \neq b, r(a, i) * r(b, i) = 1\},$$
$$\text{Disagreed pairs}: \; \mathbb{D}_i^B = \{(a, b) \in B \times B \mid a \neq b, r(a, i) * r(b, i) = -1\},$$
$$\text{Compatible pairs}: \; \mathbb{C}_i^B = \{(a, b) \in B \times B \mid a \neq b, r(a, i) * r(b, i) = 0\}.$$

A general formulation of the agreement degree should be a function regarding these three sets. Concrete specific measurements can vary across different contexts and assumptions.

Definition 3. *The agreement degree of an agent group with respect to an issue is given by a mapping $\Phi : 2^A \times I \longrightarrow [-1, +1]$, where $\Phi(B, i) = f(\mathbb{A}_i^B, \mathbb{D}_i^B, \mathbb{C}_i^B)$ and the function f should satisfy the following properties:*

(1) f increases with respect to $|\mathbb{A}_i^B|$,
(2) f decreases with respect to $|\mathbb{D}_i^B|$,
(3) f decreases with respect to $|\mathbb{C}_i^B|$,

where $|\cdot|$ represents the cardinality of a set.

The least value -1 denotes full disagreement, and the greatest value $+1$ indicates full agreement. A higher value of $\Phi(B, i)$ indicates a higher degree of agreement. Following Definition 1, the agreement degree can be used as the evaluation function to triad issues.

Definition 4. *Suppose $T = (A, I, V, r)$ is a three-valued situation table, $B \subseteq A$ a set of agents, and $\Phi : 2^A \times I \longrightarrow [-1, +1]$ a mapping for agreement degree.*

The set of issues I can be triaded, with respect to B, into an agreement region AGR, a disagreement region DIS, and an uncertainty region UNC as follows:

$$AGR = \{i \in I \mid \Phi(B,i) \geqslant \alpha\},$$
$$DIS = \{i \in I \mid \Phi(B,i) \leqslant \beta\},$$
$$UNC = \{i \in I \mid \beta < \Phi(B,i) < \alpha\}, \tag{2}$$

where α and β are thresholds satisfying $-1 \leqslant \beta < 0 < \alpha \leqslant +1$.

The agreement region AGR consists of issues where agents achieve consensus, that is, their agreement level is higher than α. On the opposite side, the disagreement region DIS consists of issues where agents have irreconcilable disagreement, that is, their agreement level is lower than β. The uncertainty region UNC consists of the remaining issues where agents exhibit uncertainty or partial agreement/disagreement.

Although intuitively reasonable, the triad $\langle AGR, DIS, UNC \rangle$ is not ideal for designing strategies in conflict analysis. Specifically, taking action on issues from the disagreement and uncertainty regions is challenging due to their inherent limitations. For issues in DIS, agents hold opposing attitudes and face certain conflicts. Taking a decisive action, whether through "acceptance" or "rejection", may provoke strong protest. For issues in UNC, agents' attitudes might be mixed, or the majority may remain neutral, making it difficult to justify definitive decisions on issues in this region. Even within the AGR region, the strength of agreement varies. A strong level of agreement supports immediate decisions, whereas a weaker agreement level calls for more cautious actions. To address these issues for reasonable strategy design, we put emphasis on the agreement perspective and define a triad of actionable regions.

Definition 5. *Suppose $T = (A, I, V, r)$ is a three-valued situation table, $B \subseteq A$ a set of agents, and $\Phi : 2^A \times I \longrightarrow [-1, +1]$ a mapping for agreement degree. The set of issues I can be triaded, with respect to B, into three actionable regions of strong-agreement SAGR, weak-agreement WAGR, and non-agreement NAGR as follows:*

$$SAGR = \{i \in I \mid \Phi(B,i) \geqslant \alpha\},$$
$$WAGR = \{i \in I \mid \beta \leqslant \Phi(B,i) < \alpha\},$$
$$NAGR = \{i \in I \mid \Phi(B,i) < \beta\}, \tag{3}$$

where α and β are a pair of thresholds satisfying $0 < \beta < \alpha \leqslant +1$.

We reuse α and β to denote different pairs of thresholds for simplicity. While the triad $\langle AGR, DIS, UNC \rangle$ looks at both agreement and disagreement, the triad $\langle SAGR, WAGR, NAGR \rangle$ focuses on the achievement of agreement. The strong-agreement region SAGR consists of issues upon which the group of agents holds a high level of agreement to proceed with an action. The weak-agreement region WAGR consists of issues upon which the group has a moderate level of agreement to proceed with an action. The non-agreement region NAGR consists of the remaining issues upon which the group is not confident with any action. The corresponding decisions for these three regions are, respectively, "agree-to-do", "weakly-agree-to-do", and "defer".

3.2   Rating-Based Triading of Issues

This subsection switches to a second dimension which is based on ratings from individual agents. With the context of three-valued ratings, a group of agents $B \subseteq A$ can be triaded, with respect to a single issue $i \in I$, as follows:

$$
\begin{aligned}
B_i^+ &= \{a \in B \mid r(a,i) = +1\}, \\
B_i^- &= \{a \in B \mid r(a,i) = -1\}, \\
B_i^0 &= \{a \in B \mid r(a,i) = 0\}.
\end{aligned}
\tag{4}
$$

The three subsets represent positive, negative, and neutral agents regarding the issue i. When it comes to the overall rating of the group, a common approach is to aggregate the ratings from individual agents. While many existing efforts have explored different aggregation methods, our focus is on triading based on the aggregation and the subsequent strategy design. Therefore, we define a general agent-aggregated rating function, which can be concretized by plugging in specific aggregation methods.

Definition 6. *The agent-aggregated ratings toward a single issue are given by a mapping $\Psi : 2^A \times I \longrightarrow [-1, +1]$, where $\Psi(B, i) = g(B_i^+, B_i^-, B_i^0)$ and the function g should satisfy the following properties:*

(1) g increases with respect to $|B_i^+|$,
(2) g decreases with respect to $|B_i^-|$,
(3) g approaches zero when $|B_i^0|$ approaches $|B|$.

An agent-aggregated rating -1 indicates that all agents in the group have a negative attitude toward the issue, while the value of $+1$ means that all agents have a positive attitude. A higher value of Ψ comes with more positivity, while a lower value means more negativity. By using Ψ as the evaluation function in Definition 1, we can construct the rating-based triad of the issue set I.

Definition 7. *Suppose $T = (A, I, V, r)$ is a three-valued situation table, $B \subseteq A$ a set of agents, and $\Psi : 2^A \times I \longrightarrow [-1, +1]$ a mapping for agent-aggregated ratings. The set of issues I can be triaded, with respect to B, into a positive region* POS, *a negative region* NEG, *and a neutral region* NEU *as follows:*

$$
\begin{aligned}
\text{POS} &= \{i \in I \mid \Psi(B, i) \geqslant h\}, \\
\text{NEG} &= \{i \in I \mid \Psi(B, i) \leqslant l\}, \\
\text{NEU} &= \{i \in I \mid l < \Psi(B, i) < h\},
\end{aligned}
\tag{5}
$$

where h and l are a pair of thresholds satisfying $-1 \leqslant l < 0 < h \leqslant +1$.

The triad $\langle \text{POS}, \text{NEG}, \text{NEU} \rangle$ reflects the overall attitude from a group of agents. The positive region POS consists of issues upon which most agents in the group have a positive rating. The negative region NEG consists of issues upon which most agents in the group have a negative rating. The neutral region NEU consists of the remaining issues upon which a majority of agents have a neutral rating or the group holds balanced positive/negative ratings. The three regions suggest the following three decisions: (1) acceptance for issues in POS, (2) rejection for issues in NEG, and (3) deferment for issues in NEU.

4   Acting Strategies Based on a 3 × 3 Strategy Space

The two dimensions of triading focus on two perspectives of agreement degrees and agent-aggregated ratings. By integrating the two dimensions, we proceed to the "Acting" step in the TAO model and construct a strategy space, which supports the subsequent design of targeted strategies. Table 1 lists the nine cases of combining the two triads.

Table 1. The possible combinations of Φ and Ψ

Possible cases	Suggested decisions
1. $(\Phi(B,i) \geqslant \alpha) \wedge (\Psi(B,i) \geqslant h)$	agree-to-do + acceptance
2. $(\beta \leqslant \Phi(B,i) < \alpha) \wedge (\Psi(B,i) \geqslant h)$	weakly-agree-to-do + acceptance
3. $(\Phi(B,i) < \beta) \wedge (\Psi(B,i) \geqslant h)$	deferral + acceptance
4. $(\Phi(B,i) \geqslant \alpha) \wedge (l < \Psi(B,i) < h)$	agree-to-do + deferral
5. $(\beta \leqslant \Phi(B,i) < \alpha) \wedge (l < \Psi(B,i) < h)$	weakly-agree-to-do + deferral
6. $(\Phi(B,i) < \beta) \wedge (l < \Psi(B,i) < h)$	deferral + deferral
7. $(\Phi(B,i) \geqslant \alpha) \wedge (\Psi(B,i) \leqslant l)$	agree-to-do + rejection
8. $(\beta \leqslant \Phi(B,i) < \alpha) \wedge (\Psi(B,i) \leqslant l)$	weakly-agree-to-do + rejection
9. $(\Phi(B,i) < \beta) \wedge (\Psi(B,i) \leqslant l)$	deferral + rejection

If either dimension calls for a deferral, the decision is deferred. Otherwise, we combine the suggested decisions to reach a final outcome. Figure 2 visualizes the nine possible outcomes within this strategy space, structured as a 3 × 3 matrix. The horizontal axis considers agreement degrees and splits the decisions into "agree" ($\Phi \geqslant \alpha$), "weakly-agree" ($\beta \leqslant \Phi < \alpha$), and "defer" ($\Phi < \beta$). The vertical axis considers agent-aggregated ratings and splits the decisions into "accept" ($\Psi \geqslant h$), "defer" ($l < \Psi < h$), and "reject" ($\Psi \leqslant l$).

From Table 1 and Fig. 2, we can derive five actionable strategies, expressed by the following decision rules:

(s1) If $i \in \text{SAGR} \cap \text{POS}$, then agents agree to accept i.
(s2) If $i \in \text{WAGR} \cap \text{POS}$, then agents weakly agree to accept i.
(s3) If $i \in \text{SAGR} \cap \text{NEG}$, then agents agree to reject i.

Agreement degree Φ

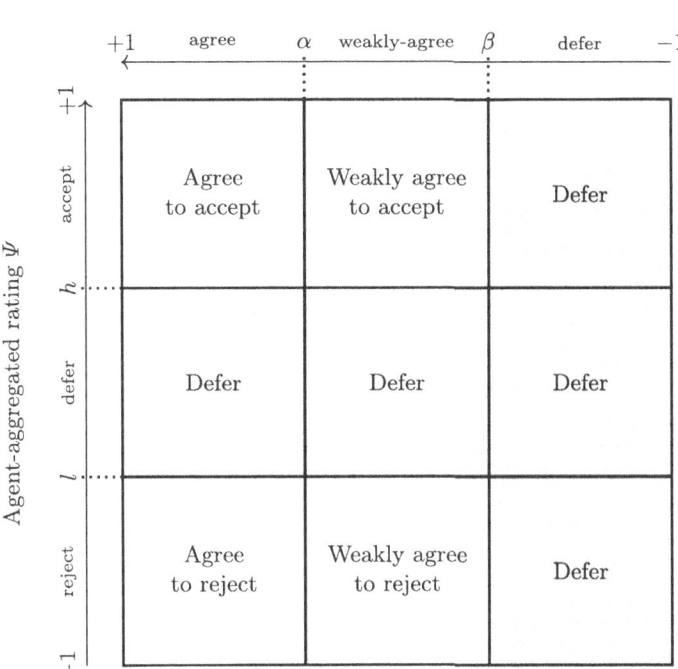

Fig. 2. The 3×3 strategy space

(s4) If $i \in \text{WAGR} \cap \text{NEG}$, then agents weakly agree to reject i.
(s5) If $i \in \text{NAGR} \cup \text{NEU}$, then agents defer the action on i.

These actionable strategies reflect a balance between agents' agreement and aggregated ratings. Strategies (s1) and (s3) suggest high-confidence actions, requiring both sufficient agreement and directional rating to minimize the risk. In these cases, immediate acceptance or rejection can be acted without hesitation. To address uncertainty in scenarios with weak agreement, strategies (s2) and (s4) allow cautious and phased implementation. Finally, strategy (s5) defers actions on issues with insufficient agreement or around-neutral ratings. The deferment prevents forced actions that could cause serious conflicts.

5   An Example

We demonstrate the proposed approach with the situation table given in Table 2. The table involves nine agents $A = \{a_1, a_2, \ldots, a_9\}$ and ten issues $I = \{i_1, i_2, \ldots, i_{10}\}$.

For illustrative purposes, we derive simple concrete formulations of the two functions Φ and Ψ from existing studies. In Pawlak's conflict model [8], the

Table 2. A three-valued situation table

$A \setminus I$	i_1	i_2	i_3	i_4	i_5	i_6	i_7	i_8	i_9	i_{10}
a_1	+1	0	−1	+1	0	−1	+1	0	+1	−1
a_2	+1	+1	0	−1	0	+1	+1	−1	+1	−1
a_3	−1	+1	0	0	+1	−1	−1	+1	+1	−1
a_4	−1	+1	+1	0	+1	−1	+1	−1	−1	−1
a_5	+1	+1	0	0	+1	−1	+1	0	0	−1
a_6	0	+1	+1	0	+1	−1	+1	0	+1	−1
a_7	−1	+1	0	0	−1	−1	+1	+1	+1	−1
a_8	−1	+1	−1	+1	−1	−1	+1	+1	+1	−1
a_9	−1	+1	+1	−1	−1	−1	+1	+1	+1	−1

relations between different agents are expressed through an auxiliary function. For two agents $a, b \in A$ and an issue $i \in I$, the auxiliary function is defined as:

$$\phi((a,b),i) = \begin{cases} +1, & \text{if } r(a,i) * r(b,i) = +1 \text{ or } a = b, \\ 0, & \text{if } r(a,i) * r(b,i) = 0 \text{ and } a \neq b, \\ -1, & \text{if } r(a,i) * r(b,i) = -1. \end{cases} \quad (6)$$

We simply consider the arithmetic mean of ϕ to measure the agreement degree $\Phi(B,i)$. To be meaningful, we exclude the agent pairs on B with identical agents. Let Δ_B denote the identity relation on B, that is, $\Delta_B = \{(a,a)|a \in B\}$. The agreement degree of $B \subseteq A$ on $i \in I$ is computed as:

$$\Phi(B,i) = \frac{1}{|B|(|B|-1)} \sum_{(a,b) \in B \times B \setminus \Delta_B} \phi((a,b),i). \quad (7)$$

We also consider the arithmetic mean to evaluate the agent-aggregated ratings as follows:

$$\Psi(B,i) = \frac{\sum_{a \in B} r(a,i)}{|B|}. \quad (8)$$

Table 3 gives the agreement degrees and agent-aggregated ratings with respect to all individual issues in our example. We take i_1 as an example to illustrate the computation. The agreement degree of all agents in A on i_1 is computed as:

$$\begin{aligned} \Phi(A,i_1) &= \frac{1}{|A|(|A|-1)} \sum_{(a,b) \in A \times A \setminus \Delta_A} \phi((a,b),i_1) \\ &= \frac{1}{|A|(|A|-1)} \sum_{(a,b) \in A \times A \setminus \Delta_A} r(a,i_1) * r(b,i_1) \\ &= \frac{1}{72}(r(a_1,i_1) * r(a_2,i_1) + r(a_1,i_1) * r(a_3,i_1) + \ldots \\ &\quad + r(a_9,i_1) * r(a_8,i_1)) \\ &= -\frac{1}{18} \approx -0.0556. \end{aligned}$$

And the agent-aggregated rating of A on i_1 is:

$$\Psi(A, i_1) = \frac{\sum_{a \in A} r(a, i_1)}{|A|} = \frac{+1 + 1 + (-1) + \cdots + (-1)}{9}$$

$$= -\frac{2}{9} \approx -0.2222.$$

Table 3. The agreement degrees and agent-aggregated ratings of A in the example

I	$\Phi(A, i)$	$\Psi(A, i)$
i_1	-0.0556	-0.2222
i_2	0.7778	0.8889
i_3	-0.0556	0.1111
i_4	-0.0556	0.0000
i_5	-0.0833	0.1111
i_6	0.5556	-0.7778
i_7	0.5556	0.7778
i_8	-0.0278	0.2222
i_9	0.3889	0.6667
i_{10}	1.0000	-1.0000

As illustrations, let us take the thresholds $\alpha = 0.75$ and $\beta = 0.25$. Then we have the following triad based on agreement degrees:

$$\text{SAGR} = \{i_2, i_{10}\},$$
$$\text{WAGR} = \{i_6, i_7, i_9\},$$
$$\text{NAGR} = \{i_1, i_3, i_4, i_5, i_8\}.$$

In another dimension, let us take the thresholds $h = 0.5$ and $l = -0.5$. Then we can construct the following triad based on ratings:

$$\text{POS} = \{i_2, i_7, i_9\},$$
$$\text{NEG} = \{i_6, i_{10}\},$$
$$\text{NEU} = \{i_1, i_3, i_4, i_5, i_8\}.$$

Accordingly, all the issues are categorized into five parts according to their Φ and Ψ values:

$$\text{SAGR} \cap \text{POS} = \{i_2\},$$
$$\text{WAGR} \cap \text{POS} = \{i_7, i_9\},$$
$$\text{SAGR} \cap \text{NEG} = \{i_{10}\},$$
$$\text{WAGR} \cap \text{NEG} = \{i_6\},$$
$$\text{NAGR} \cup \text{NEU} = \{i_1, i_3, i_4, i_5, i_8\}.$$

Then the actionable strategies are:

(1) agents agree to accept i_2,
(2) agents weakly agree to accept i_7 and i_9,
(3) agents agree to reject i_{10},
(4) agents weakly agree to reject i_6,
(5) the decisions for $\{i_1, i_3, i_4, i_5, i_8\}$ are deferred.

Our proposed approach makes the decision-making process transparent and derives five clear and easily-applied strategies in conflict analysis. This example demonstrates that the proposed model satisfies 3WD's core principle of simple-to-understand, easy-to-remember, and practical-to-use.

6   Conclusion

This paper integrates three-way decision theory with conflict analysis to develop a structured approach for decision-making and strategy design. By introducing two dimensions of evaluation, namely, the agreement degree Φ and agent-aggregated rating Ψ, we construct a strategy space that categorizes issues into nine regions. From these regions, we derive five strategies of taking actions: agree to accept/reject, weakly agree to accept/reject, and defer. An example demonstrates the framework's ability to prioritize high-confidence decisions, manage uncertainty through phased implementations, and avoid risks by deferring issues. This work bridges the stage of investigating agent and issue relations and the subsequent stage of strategy design in conflict analysis. Furthermore, it also enriches the theoretical principles of the TAO model of 3WD with practical strategy development to resolve conflict.

This work could be further extended in at least three directions. Firstly, one can formalize the functions for agreement degree and agent-aggregated rating. In the example, we use two simple functions to illustrate the strategy space. However, the interdependency of the two functions requires further investigation. Secondly, the determination of optimal thresholds remains a challenge, particularly balancing the trade-offs between two pairs of thresholds. The optimization of these thresholds may involve the cost-benefit analysis of the 3×3 strategy space, because they have a direct impact on the five decision rules. In addition, the deferred issues in the five regions could be further refined. Thirdly, the Optimizing step following the TAO model calls for studies to complete the whole process.

References

1. Hu, M.J.: Modeling relationships in three-way conflict analysis with subsethood measures. Knowl. Based Syst. **260**, 110131 (2023)
2. Hu, M.J.: Three-way conflict analysis with preference-based conflict situations. Inf. Sci. **693**, 121676 (2025)

3. Lang, G.M.: A general conflict analysis model based on three-way decision. Int. J. Mach. Learn. Cybern. **11**, 1083–1094 (2020)
4. Lang, G.M., Yao, Y.Y.: Formal concept analysis perspectives on three-way conflict analysis. Int. J. Approximate Reasoning **152**, 160–182 (2023)
5. Li, Y.S., Gao, F., Sha, M.Y., Shao, X.Y.: Sequential three-way decision with automatic threshold learning for credit risk prediction. Appl. Soft Comput. **165**, 112127 (2024)
6. Liu, J.B., Huang, S.T., Li, T.R., Liang, Q., Li, H.X., Hao, Z.F.: Models and algorithms for optimizing thresholds in fuzzy representation-based three-way decision. IEEE Trans. Fuzzy Syst. **32**, 4912–4926 (2024)
7. Luo, J.F., Hu, M.J., Lang, G.M., Yang, X., Qin, K.Y.: Three-way conflict analysis based on alliance and conflict functions. Inf. Sci. **594**, 322–359 (2022)
8. Pawlak, Z.: An inquiry into anatomy of conflicts. Inf. Sci. **109**, 65–78 (1998)
9. Pawlak, Z.: Some remarks on conflict analysis. Eur. J. Oper. Res. **166**, 649–654 (2005)
10. Sun, B.Z., Ma, W.M., Zhao, H.Y.: Rough set-based conflict analysis model and method over two universes. Inf. Sci. **372**, 111–125 (2016)
11. Sun, B.Z., Chen, X.T., Zhang, L.Y., Ma, W.M.: Three-way decision making approach to conflict analysis and resolution using probabilistic rough set over two universes. Inf. Sci. **507**, 809–822 (2020)
12. Suo, L., Yang, H.L., Yang, H., Yao, Y.Y.: Double triadic thinking and the 3×3 methods. J. Shaanxi Normal Univ. (Natural Science Edition) **52**(3), 1–10 (2024)
13. Xu, F., Cai, M.J., Song, H.L., Dai, J.H.: The selection of feasible strategies based on consistency measurement of cliques. Inf. Sci. **583**, 33–55 (2022)
14. Yang, H., Yao, Y.Y., Qin, K.Y.: A lattice-theoretic model of three-way conflict analysis. Knowl. Based Syst. **288**, 111470 (2024)
15. Yang, H.L., Wang, Y., Guo, Z.L.: Three-way conflict analysis based on hybrid situation tables. Inf. Sci. **628**, 522–541 (2023)
16. Yao, Y.Y.: Three-way decisions with probabilistic rough sets. Inf. Sci. **180**, 341–353 (2010)
17. Yao, Y.Y.: An outline of a theory of three-way decisions. In: Yao, J.T., et al. (eds.) Rough Sets and Current Trends in Computing. RSCTC 2012. LNCS (LNAI), vol. 7413, pp. 1–17. Springer, Berlin, Heidelberg (2012)
18. Yao, Y.Y.: Three-way decision and granular computing. Int. J. Approximate Reasoning **103**, 107–123 (2018)
19. Yao, Y.Y.: Three-way conflict analysis: reformulations and extensions of the Pawlak model. Knowl. Based Syst. **180**, 26–37 (2019)
20. Yao, Y.Y.: Set-theoretic models of three-way decision. Granular Comput. **6**, 133–148 (2021)
21. Yao, Y.Y.: The Dao of three-way decision and three-world thinking. Int. J. Approximate Reasoning **162**, 109032 (2023)

Three-Way Conflict Analysis:issue Reduction of Incomplete Three-Valued Situation Tables

Xuemei Ran[1,2] and Guangming Lang[1,2(✉)]

[1] School of Mathematics and Statistics, Changsha University of Science and Technology, Changsha 410114, Hunan, People's Republic of China
langguangming1984@126.com
[2] Hunan Provincial Key Laboratory of Mathematical Modeling and Analysis in Engineering, Changsha University of Science and Technology, Changsha 410114, Hunan, People's Republic of China

Abstract. In three-way conflict analysis, resolving conflicts based on fewer but more critical issues can greatly improve efficiency. However, most studies on issue reduction focus on complete three-valued situation tables, while research on incomplete ones remains limited. This paper examines the issue reducts of weak alliance, weak neutrality, and weak conflict relations in incomplete three-valued situation tables using discernibility functions. First, it introduces the concept of incomplete three-valued situation tables and defines weak alliance, weak neutrality, and weak conflict relations under missing values. Then, it constructs discernibility matrices and their corresponding functions for these weak relations. The definitions of weak alliance, weak neutrality, and weak conflict reducts are provided, and all reducts are derived by analyzing the properties of discernibility functions. Finally, the proposed method is applied to NBA labor negotiations, demonstrating the effectiveness of issue reduction in incomplete three-valued situation tables.

Keywords: Three-way conflict analysis · Discernibility matrix · Issue reduction · Incomplete three-valued situation table

1 Introduction

The concept of conflict was first introduced by Pawlak, with the aim of analyzing and resolving conflicts in various contexts such as government and economics [1,2]. Yao [3] later introduced the concept of three-way decisions, which divides the decision-making process into three categories: acceptance, rejection, and delay. In recent years, many researchers [4–11] have integrated three-way decisions with conflict analysis, leading to the development of several new mathematical models. For instance, Yao [4] redefined Pawlak model by incorporating a distance function, thus proposing a novel three-way conflict analysis model. Lang, Miao, and Fujita [5] applied Bayesian minimum risk theory to propose

© The Author(s), under exclusive license to Springer Nature Switzerland AG 2025
Q. Zhang et al. (Eds.): IJCRS 2025, LNAI 15710, pp. 98–111, 2025.
https://doi.org/10.1007/978-3-031-92741-6_8

a three-way conflict analysis model that classifies agent pairs into three categories. In the context of q-rung orthopair fuzzy information, Mandal et al. [6] introduced a three-way conflict analysis model based on regret theory, addressing more complex and uncertain conflict situations beyond current cognitive limits. Luo et al. [7] developed a weighted three-way conflict analysis model by incorporating attribute weights into existing distance functions. Hu [8] proposed a preference-based three-way conflict analysis model that prioritizes individual preferences. Tang, Chen, and Rao [9] explored three-way conflict analysis in fuzzy multiset information systems and introduced a novel method for threshold determination. Li et al. [11] developed a group weighted average operator and proposed an index to evaluate alliance stability.

The concept of the discernibility matrix was first introduced by Pawlak in rough set theory to represent the discernibility relations between objects with respect to attributes [12]. Skowron and Rauszer further investigated the properties of discernibility matrices and discernibility functions [13], demonstrating that all attribute reductions of an information system can be derived by computing the discernibility function. This discovery laid the theoretical foundation for subsequent attribute reduction algorithms and optimization methods [14–18]. For example, Yao and Zhao [14] proposed a reduction construction method based on a simplified discernibility matrix and introduced two heuristic algorithms. Sowkuntla and Prasad [15] developed an attribute reduction accelerator using a fuzzy discernibility matrix to expedite the reduction process. Wen, Xu, and Liang [16] proposed a parallel attribute reduction algorithm based on fuzzy discernibility matrices combined with a soft deletion mechanism. In incomplete information systems, Kryszkiewicz [17] assumed that the actual value of a missing attribute is selected from the attribute's domain and studied attribute reduction under this assumption using discernibility functions. Dai, Wang, and Huang [18] proposed an attribute reduction algorithm for incomplete interval-valued information systems.

In the study of issue reduction in three-valued situation tables, Lang [19] incorporated the discernibility matrix into three-way conflict analysis, enabling issue reduction by constructing discernibility matrices and their corresponding functions for alliance, conflict, and neutrality relations. Chen and Zhang [20] introduced a conflict analysis method based on three-way membership and similarity, developing a heuristic algorithm for issue reduction. Zhang and Chen [21] proposed a quantitative three-way conflict analysis model and realized issue reduction through regional preservation. While these studies focus on complete three-valued situation tables, real-world conflict scenarios may involve partially unknown agent attitudes. This highlights the need for issue reduction research in incomplete three-valued situation tables. The main contributions of this paper are as follows:

(1) We give an incomplete three-valued situation table, where missing values are assumed to belong to the domain {-1, 0, +1}. Additionally, we classify agent pairs into weak alliance, weak neutrality, and weak conflict relations.

(2) We define the concepts of weak alliance, weak neutral, and weak conflict reducts, Furthermore, we construct discernibility matrices and corresponding discernibility functions with respect to these weak relations.

(3) We apply the proposed method to NBA labor negotiations to analyze the weak alliance, weak neutrality, and weak conflict reducts in this real-world conflict, offering informed recommendations for decision-makers.

The paper is structured as follows: Sect. 2 reviews fundamental concepts of incomplete information systems in three-way conflict analysis. Section 3 examines the issue reduction process for weak alliance, weak neutrality, and weak conflict relations using discernibility functions. Section 4 applies the proposed method to NBA labor negotiations. Section 5 concludes the study and outlines future research directions.

2   Preliminaries

In this section, we briefly reviewed some concepts related to the discernibility matrix in incomplete information systems and the three-way conflict analysis model.

2.1   The Rough Set Model of an Incomplete Information System

In 1998, Kryszkiewicz introduced the concept of incomplete information systems, assuming that the true value of a missing attribute is selected from its domain.

Definition 1. *(Kryszkiewicz [17], 1998) An incomplete information system is a quadruple $IS^* = (U, C, V, f)$, where U is a finite non-empty set of objects, C is a finite non-empty set of attributes, $V = \bigcup\{V_c \mid c \in C\}$ is the set of possible values for the attributes across all objects, and $f : U \times C \to V$ is the information function that assigns a value from V to each object-attribute pair. Simultaneously, at least one attribute $c \in C$ includes a missing value in V.*

A missing value in an incomplete information system is denoted by $*$. An information system without missing values is called a complete information system.

Definition 2. *(Kryszkiewicz [17], 1998) In an incomplete information system $IS^* = (U, C, V, f)$, the similarity relation with respect to the attribute subset $B \subseteq C$ is defined as follows:*

$$SIM(B) = \{(x, y) \in U \times U \mid \forall c \in B, \, f(x, c) = f(y, c) \text{ or } f(x, c) = * \text{ or } f(y, c) = *\}.$$

The similarity relation can be viewed as a weak equivalence relation, meaning it is both reflexive and symmetric. It represents the largest set of object pairs that are indistinguishable based on a subset of attributes B.

Definition 3. *(Kryszkiewicz [17], 1998) In an incomplete information system $IS^* = (U, C, V, f)$, the discernibility matrix $\mathbb{M}^*$ is defined as:*

$$\mathbb{M}^* = [m^*(x,y)]_{\#(U) \times \#(U)},$$

where

$$m^*(x,y) = \{c \in C \mid (x,y) \notin SIM(c)\}.$$

The term $m(x,y)$ represents the set of attributes $c \in C$ such that $(x,y) \notin SIM(c)$, i.e., the attributes that can distinguish objects x and y. If $(x,y) \in SIM(c)$, then $m^*(x,y) = \emptyset$, meaning that x and y are indistinguishable with respect to attribute c. Since the discernibility matrix is symmetric, focusing on the triangular part can simplify the computation.

Definition 4. *(Kryszkiewicz [17], 1998) In an incomplete information system $IS^* = (U, C, V, f)$, given the discernibility matrix $\mathbb{M}^*$, the discernibility function $\triangle$ is defined as follows:*

$$\triangle = \prod_{(x,y) \in U \times U} \sum m^*(x,y),$$

where

$$\sum m^*(x,y) = \begin{cases} c_1 \vee c_2 \vee \cdots \vee c_k, & m^*(x,y) = \{c_1, c_2, \ldots, c_k\}; \\ 1, & m^*(x,y) = \emptyset. \end{cases}$$

and $\prod$ denotes the conjunction operation.

The term $\sum m^*(x,y)$ refers to the disjunctive expressions of all attributes in $m^*(x,y)$, while $\prod_{(x,y) \in U \times U} \sum m^*(x,y)$ represents the conjunctive normal form formed by the conjunction of these disjunctive expressions.

2.2 The Three-Way Conflict Analysis Model

In 1998, Pawlak developed a conflict analysis model using a three-valued situation table, which is based on three binary relations: alliance, neutrality, and conflict.

Definition 5. *(Pawlak [2], 1998) A three-valued situation table is a triplet $T = (A, I, r)$, where A is a non-empty finite set of agents, I is a non-empty finite set of issues, and $r : A \times I \rightarrow \{-1, 0, +1\}$ is a rating function. Specifically, $r(a,i) = +1$ indicates that agent a supports issue i, $r(a,i) = -1$ indicates opposition, and $r(a,i) = 0$ indicates neutrality towards i.*

Pawlak defined an auxiliary function that classifies all object pairs into three categories. Yao introduced a distance function to refine the Pawlak model.

Definition 6. *(Pawlak [2], 1998) In a three-valued situation table $T = (A, I, r)$, the auxiliary function $\phi_i : A \times A \to \{-1, 0, +1\}$ for an issue $i \in I$ is defined as follows: for any $a, b \in A$,*

$$\phi_i(a, b) = \begin{cases} +1, & r(a, i) \times r(b, i) = +1 \vee a = b, \\ 0, & r(a, i) \times r(b, i) = 0 \wedge a \neq b, \\ -1, & r(a, i) \times r(b, i) = -1, \end{cases}$$

where $\times$ denotes the multiplication operator.

If the auxiliary function $\phi_i(a, b) = +1$, it indicates that agents a and b have the same opinion on issue i. If $\phi_i(a, b) = 0$, it means that at least one of the agents' opinions on issue i is neutral. If $\phi_i(a, b) = -1$, it signifies that agents a and b hold opposite opinions on issue i. The following defines three binary relations: alliance, neutrality, and conflict.

Definition 7. *(Pawlak [2], 1998) In a three-valued situation table $T = (A, I, r)$, for each $i \in I$, the alliance relation AR_i^+, the neutral relation NR_i^0, and the conflict relation CR_i^- are defined as follows:*

$$AR_i^+ = \{(a, b) \mid \phi_i(a, b) = +1\};$$
$$NR_i^0 = \{(a, b) \mid \phi_i(a, b) = 0\};$$
$$CR_i^- = \{(a, b) \mid \phi_i(a, b) = -1\}.$$

The alliance relation AR_i^+ is reflexive, symmetric, and transitive, thus forming an equivalence relation. In contrast, both the neutral relation NR_i^0 and the conflict relation CR_i^- are symmetric but do not exhibit reflexivity or transitivity.

Definition 8. *(Yao [4], 2019) In a three-valued situation table $T = (A, I, r)$, the distance function $D_i : A \times A \to \{0, 0.5, 1\}$ with respect to an issue $i \in I$ is defined as follows: for two agents $a, b \in A$,*

$$D_i(a, b) = \frac{|r(a, i) - r(b, i)|}{2} = \begin{cases} 0, & r(a, i) = r(b, i); \\ 0.5, & r(a, i) \neq r(b, i) \wedge r(a, i) \times r(b, i) = 0; \\ 1, & r(a, i) \times r(b, i) = -1, \end{cases}$$

where $|\diamond|$ denotes the absolute value of $\diamond$.

If the distance function $D_i(a, b) = 0$, it indicates that agents a and b have identical opinions on issue i. If $D_i(a, b) = 0.5$, it implies that at least one of the agents holds a neutral opinion on issue i. If $D_i(a, b) = 1$, it denotes that agents a and b have opposing opinions on issue i.

Definition 9. *(Yao [4], 2019) In a three-valued situation table $T = (A, I, r)$, the alliance relation $AR_i^=$, neutrality relation $NR_i^\approx$, and conflict relation $CR_i^\asymp$ with respect to issue $i \in I$ are defined by:*

$$AR_i^= = \{(a, b) \in A \times A \mid D_i(a, b) = 0\},$$
$$NR_i^\approx = \{(a, b) \in A \times A \mid D_i(a, b) = 0.5\},$$
$$CR_i^\asymp = \{(a, b) \in A \times A \mid D_i(a, b) = 1\}.$$

The distance function defines three distinct relations: alliance, neutrality, and conflict, which partition the set of object pairs. The alliance relation AR_i^+ is reflexive, symmetric, and transitive, while the neutral NR_i^0 and conflict relations CR_i^- are symmetric but lack reflexivity and transitivity.

In three-way conflict analysis, Lang [19] introduced the concept of issue reducts within a complete three-valued situation table, examining reducts for alliance, conflict, and neutrality relations based on the discernibility function's properties. However, in real-world conflict scenarios, an agent's attitude toward a specific issue may be unknown, leading to missing attitude values. Therefore, it is crucial to extend this analysis to three-valued situation tables that incorporate missing values.

3   Issue Reduction in an Incomplete Three-Valued Situation Table

In this section, we discuss the issue reducts with respect to weak alliance, weak neutrality, and weak conflict relations in incomplete three-valued situation tables.

Definition 10. *An incomplete three-valued situation table is a triplet $T^* = (A, I, r)$, where A is a non-empty finite set of agents, I is a non-empty finite set of issues, and $r : A \times I \to \{-1, 0, +1, *\}$ is a rating function. There exists at least one issue $i \in I$ for which the attitude of some agent is undefined, denoted by $*$. Specifically, $r(a, i) = -1$ indicates that agent a opposes issue i; $r(a, i) = 0$ indicates that agent a is neutral toward issue i; $r(a, i) = +1$ indicates that agent a supports issue i; and $r(a, i) = *$ indicates that the attitude of agent a toward issue i is unknown or undefined.*

If the attitudes of all agents toward all issues are known, with no missing values, the situation table is a standard three-valued situation table.

Definition 11. *In an incomplete three-valued situation table $T^* = (A, I, r)$, the conflict measure $CN_i^* : A \times A \to \{0, 0.5, 1\}$ for an issue $i \in I$ is defined as follows: for two agents $a, b \in A$,*

$$CN_i^*(a, b) = \frac{|r(a, i) - r(b, i)|}{2} = \begin{cases} 0, & r(a, i) = r(b, i); \\ 0.5, & r(a, i) \neq r(b, i) \text{ and } r(a, i) \times r(b, i) = 0; \\ 1, & r(a, i) \times r(b, i) = -1. \end{cases}$$

where $r(a, i) \neq *$ and $r(b, i) \neq *$.

The function $CN_i^*(a, b)$ quantifies the degree of conflict between two objects a and b on the same issue i, excluding missing values.

Definition 12. *In an incomplete three-valued situation table $T^* = (A, I, r)$, let $J \subseteq I$ be a set of issues. The weak alliance relation $WR_i^=$, weak neutrality* relation $WR_i^{\approx}$, and weak conflict relation $WR_i^{\succ}$ with respect to issue $i \in J$ are defined as follows:

$$WR_{\bar{J}}^{=} = \{(a,b) \in A \times A \mid CN_i^*(a,b) = 0 \vee r(a,i) = * \vee r(b,i) = *\};$$
$$WR_{\bar{J}}^{\approx} = \{(a,b) \in A \times A \mid CN_i^*(a,b) = 0.5 \vee r(a,i) = * \vee r(b,i) = *\};$$
$$WR_{\bar{J}}^{\succ} = \{(a,b) \in A \times A \mid CN_i^*(a,b) = 1 \vee r(a,i) = * \vee r(b,i) = *\}.$$

A weak alliance relationship represents the most extensive set of object pairs that may be indistinguishable concerning the alliance relation. Likewise, a weak neutrality relationship encompasses the broadest set of object pairs that may be indistinguishable with respect to the neutrality relation. A weak conflict relationship denotes the largest set of object pairs that may be indistinguishable in terms of the conflict relation.

Property 1. In an incomplete three-valued situation table $T^* = (A, I, r)$, the following relations are tolerance relations:

(1) The weak alliance relation $WR_{\bar{J}}^{=}$ is a tolerance relation, i.e., $WR_{\bar{J}}^{=} = \bigcap_{i \in J} WR_i^{=}$;

(2) The weak neutrality relation $WR_{\bar{J}}^{\approx}$ is a tolerance relation, i.e., $WR_{\bar{J}}^{\approx} = \bigcap_{i \in J} WR_i^{\approx}$;

(3) The weak conflict relation $WR_{\bar{J}}^{\succ}$ is a tolerance relation, i.e., $WR_{\bar{J}}^{\succ} = \bigcap_{i \in J} WR_i^{\succ}$.

By Property 2, we can derive the sets of weak alliance, weak neutrality, and weak conflict relations over the entire issue set I.

Property 2. The three weak relations form a cover of A, i.e., $WR_{\bar{J}}^{=} \cup WR_{\bar{J}}^{\approx} \cup WR_{\bar{J}}^{\succ} = A \times A$.

Definition 13. *Consider an incomplete three-valued situation table $T^* = (A, I, r)$.*

(1) *A set $J^{=} \subseteq I$ is called a weak alliance reduct if it satisfies the following conditions:*
  (i) $WR_{J^{=}}^{=} = WR_I^{=}$;
  (ii) *for each $i \in J^{=}$, $WR_{J^{=}-\{i\}}^{=} \neq WR_I^{=}$.*
(2) *A set $J^{\approx} \subseteq I$ is called a weak neutrality reduct if it satisfies the following conditions:*
  (i) $WR_{J^{\approx}}^{\approx} = WR_I^{\approx}$;
  (ii) *for each $i \in J^{\approx}$, $WR_{J^{\approx}-\{i\}}^{\approx} \neq WR_I^{\approx}$.*
(3) *A set $J^{\succ} \subseteq I$ is called a weak conflict reduct if it satisfies the following conditions:*
  (i) $WR_{J^{\succ}}^{\succ} = WR_I^{\succ}$;
  (ii) *for each $i \in J^{\succ}$, $WR_{J^{\succ}-\{i\}}^{\succ} \neq WR_I^{\succ}$.*

A weak alliance reduct is the minimal subset of issues that preserves the weak alliance relation. Similarly, a weak neutrality reduct is the smallest set of issues that maintains the weak neutrality relation. A weak conflict reduct is the least extensive subset of issues that sustains the weak conflict relation.

Definition 14. *Let $T^* = (A, I, r)$ be an incomplete three-valued situation table. The discernibility matrices $\mathbb{WM}^\square$ ($\square \in \{=, \approx, \asymp\}$) corresponding to the weak alliance relation $WR_i^=$, weak neutrality relation $WR_i^\approx$, and weak conflict relation $WR_i^\asymp$ are defined as follows:*

(1) $\mathbb{WM}^= = [WM^=(a_j, a_k)]_{\#(A) \times \#(A)}$, *where*

$$WM^=(a_j, a_k) = \{i \in I \mid (a_j, a_k) \notin WR_i^=\};$$

(2) $\mathbb{WM}^\approx = [WM^\approx(a_j, a_k)]_{\#(A) \times \#(A)}$, *where*

$$WM^\approx(a_j, a_k) = \{i \in I \mid (a_j, a_k) \notin WR_i^\approx\};$$

(3) $\mathbb{WM}^\asymp = [WM^\asymp(a_j, a_k)]_{\#(A) \times \#(A)}$, *where*

$$WM^\asymp(a_j, a_k) = \{i \in I \mid (a_j, a_k) \notin WR_i^\asymp\}.$$

As stated in Definition 12, the weak alliance, neutrality, and conflict relations represent the largest sets of object pairs that remain indistinguishable within their respective categories. Consequently, object pairs that do not belong to these weak relations are considered distinguishable.

Definition 15. *Consider an incomplete three-valued situation table $T^* = (A, I, r)$. Given the discernibility matrices $\mathbb{WM}^\square$, the corresponding discernibility functions $\triangle^\square$ are defined as follows:*

(1) $\triangle^= = \bigwedge \{\bigvee WM^=(a_j, a_k) \mid WM^=(a_j, a_k) \neq \emptyset, \forall a_j, a_k \in A\}$;
(2) $\triangle^\approx = \bigwedge \{\bigvee WM^\approx(a_j, a_k) \mid WM^\approx(a_j, a_k) \neq \emptyset, \forall a_j, a_k \in A\}$;
(3) $\triangle^\asymp = \bigwedge \{\bigvee WM^\asymp(a_j, a_k) \mid WM^\asymp(a_j, a_k) \neq \emptyset, \forall a_j, a_k \in A\}$.

The expression $\bigvee WM^\square(a_j, a_k)$ denotes the disjunction of all the issues in $WM^\square(a_j, a_k)$. Meanwhile, the overall formula $\bigwedge \{\bigvee WM^\square(a_j, a_k)\}$ represents the conjunction of these disjunctions for every pair (a_j, a_k) where $WM^\square(a_j, a_k) \neq \emptyset$.

Property 3. All conjunctions in the minimal disjunctive normal form (DNF) of the function $\triangle^\square$ ($\square \in \{=, \approx, \asymp\}$) correspond to the issue reducts of the issue set I.

Through Property 3, we can obtain a method for calculating the complete set of issue reducts for weak alliance, weak neutrality, and weak conflict relations in incomplete three-valued situation tables. In other words, once we compute the discernibility functions for these relations, we can derive all the issue reducts corresponding to the weak alliance, weak neutrality, and weak conflict relations.

4   The Practical Application of Issue Reduction in Incomplete Three-Valued Situation Tables

This section applies the proposed method to NBA labor negotiations and analyzes the reduction results to provide insights for decision-makers. The NBA labor negotiations [22] involve ten agents, including the president, representatives of team owners, legal and financial experts, the president of the players' union, the executive director, and members of the negotiation committee. The negotiation committee consists of players from different income brackets and experience levels. These agents are denoted as a_1, a_2, a_3, a_4, a_5, a_6, a_7, a_8, a_9, and a_{10}. The negotiations focus on nine key issues, including the total salary cap for athletes, minimum wage standards, club owner commissions, the proportion of league revenue allocated to players, the maximum number of athletes a club can sign, age restrictions, working conditions, insurance coverage for players, and the recruitment ratio of foreign athletes. These issues are represented as i_1, i_2, i_3, i_4, i_5, i_6, i_7, i_8, and i_9. Table 1 shows the agents' attitudes on these issues, where -1 indicates opposition, 0 denotes neutrality, and $+1$ indicates support. Missing values, represented by $*$, indicate that the stance is unknown but falls within the set $\{-1, 0, +1\}$. In the following analysis, the proposed method will be used to compute the issue reductions under weak alliance, weak neutrality, and weak conflict relations, respectively.

(1) **Weak alliance reducts:** According to Definition 14, the discernibility matrix $\mathbb{WM}^{=}$ for the weak alliance relation is constructed.

Based on this matrix, the discernibility function $\triangle^{=}$ is then derived following Definition 15:

$$
\begin{aligned}
\triangle^{=} =\ & (i_2 \vee i_5) \wedge (i_2 \vee i_9) \wedge (i_4 \vee i_6) \wedge (i_4 \vee i_9) \wedge (i_7 \vee i_8) \wedge (i_1 \vee i_3 \vee i_7) \\
& \wedge (i_1 \vee i_5 \vee i_6) \wedge (i_3 \vee i_5 \vee i_8) \wedge (i_5 \vee i_6 \vee i_8) \wedge (i_5 \vee i_6 \vee i_9) \\
& \wedge (i_1 \vee i_3 \vee i_6 \vee i_8) \wedge (i_1 \vee i_6 \vee i_7 \vee i_9) \\
=\ & (i_5 \wedge i_6 \wedge i_7 \wedge i_9) \vee (i_1 \wedge i_2 \wedge i_4 \wedge i_5 \wedge i_7) \vee (i_1 \wedge i_2 \wedge i_4 \wedge i_5 \wedge i_8) \\
& \vee (i_1 \wedge i_2 \wedge i_4 \wedge i_6 \wedge i_8) \vee (i_1 \wedge i_2 \wedge i_4 \wedge i_8 \wedge i_9) \vee (i_1 \wedge i_2 \wedge i_6 \wedge i_8 \wedge i_9) \\
& \vee (i_1 \wedge i_4 \wedge i_5 \wedge i_7 \wedge i_9) \vee (i_1 \wedge i_4 \wedge i_5 \wedge i_8 \wedge i_9) \vee (i_1 \wedge i_5 \wedge i_6 \wedge i_8 \wedge i_9) \\
& \vee (i_2 \wedge i_3 \wedge i_4 \wedge i_5 \wedge i_7) \vee (i_2 \wedge i_3 \wedge i_4 \wedge i_6 \wedge i_7) \vee (i_2 \wedge i_3 \wedge i_4 \wedge i_6 \wedge i_8) \\
& \vee (i_2 \wedge i_3 \wedge i_6 \wedge i_7 \wedge i_9) \vee (i_2 \wedge i_3 \wedge i_6 \wedge i_8 \wedge i_9) \vee (i_2 \wedge i_4 \wedge i_5 \wedge i_6 \wedge i_7) \\
& \vee (i_2 \wedge i_4 \wedge i_5 \wedge i_7 \wedge i_8) \vee (i_2 \wedge i_4 \wedge i_6 \wedge i_7 \wedge i_8) \vee (i_2 \wedge i_6 \wedge i_7 \wedge i_8 \wedge i_9) \\
& \vee (i_3 \wedge i_4 \wedge i_5 \wedge i_7 \wedge i_9) \vee (i_3 \wedge i_4 \wedge i_5 \wedge i_8 \wedge i_9) \vee (i_3 \wedge i_5 \wedge i_6 \wedge i_8 \wedge i_9) \\
& \vee (i_4 \wedge i_5 \wedge i_7 \wedge i_8 \wedge i_9).
\end{aligned}
$$

Finally, the weak alliance reducts are obtained by applying Property 3. The results show that there are 22 reducts for the weak alliance relation, as follows:

Table 1. The attitudes of ten agents on nine issues [22].

A	I								
	i_1	i_2	i_3	i_4	i_5	i_6	i_7	i_8	i_9
a_1	+1	+1	+1	-1	0	-1	-1	-1	+1
a_2	+1	*	-1	+1	-1	0	*	+1	-1
a_3	+1	0	*	*	+1	-1	-1	*	0
a_4	+1	+1	+1	-1	-1	+1	+1	0	-1
a_5	-1	+1	+1	+1	*	-1	0	-1	-1
a_6	+1	+1	*	0	-1	*	-1	+1	0
a_7	*	+1	*	+1	+1	-1	+1	0	-1
a_8	+1	-1	+1	-1	0	+1	*	-1	-1
a_9	+1	+1	0	+1	+1	*	-1	-1	-1
a_{10}	-1	*	+1	-1	+1	0	0	-1	-1

$\{\{i_5, i_6, i_7, i_9\}, \{i_1, i_2, i_4, i_5, i_7\}, \{i_1, i_2, i_4, i_5, i_8\}, \{i_1, i_2, i_4, i_6, i_8\}, \{i_1, i_2, i_4, i_8, i_9\},$
$\{i_1, i_2, i_6, i_8, i_9\}, \{i_1, i_4, i_5, i_7, i_9\}, \{i_1, i_4, i_5, i_8, i_9\}, \{i_1, i_5, i_6, i_8, i_9\}, \{i_2, i_3, i_4, i_5, i_7\},$
$\{i_2, i_3, i_4, i_6, i_7\}, \{i_2, i_3, i_4, i_6, i_8\}, \{i_2, i_3, i_6, i_7, i_9\}, \{i_2, i_3, i_6, i_8, i_9\}, \{i_2, i_4, i_5, i_6, i_7\},$
$\{i_2, i_4, i_5, i_7, i_8\}, \{i_2, i_4, i_6, i_7, i_8\}, \{i_2, i_6, i_7, i_8, i_9\}, \{i_3, i_4, i_5, i_7, i_9\}, \{i_3, i_4, i_5, i_8, i_9\},$
$\{i_3, i_5, i_6, i_8, i_9\}, \{i_4, i_5, i_7, i_8, i_9\}\}.$

In the weak alliance reducts, issues i_2, i_4, and i_9 appear with the highest frequency. Decision-makers should prioritize these three issues and identify the key reasons why they are crucial for coalition relation.

(2) **Weak neutrality reducts:** The discernibility matrix $\mathbb{WM}^{\approx}$ for the weak neutral relation is initially constructed following Definition 14. The discernibility function $\triangle^{\approx}$ is then generated using Definition 15 as follows:

$$\begin{aligned} \triangle^{\approx} &= (i_1 \vee i_5) \wedge (i_1 \vee i_6) \wedge (i_1 \vee i_2 \vee i_8) \wedge (i_2 \vee i_5 \vee i_7) \wedge (i_4 \vee i_5 \vee i_9) \wedge (i_5 \vee i_6 \vee i_7) \\ &\quad \wedge (i_2 \vee i_4 \vee i_6 \vee i_9) \wedge (i_1 \vee i_3 \vee i_4 \vee i_8 \vee i_9) \\ &= (i_1 \wedge i_2 \wedge i_5) \vee (i_1 \wedge i_4 \wedge i_5) \vee (i_1 \wedge i_4 \wedge i_7) \vee (i_1 \wedge i_5 \wedge i_6) \vee (i_1 \wedge i_5 \wedge i_9) \\ &\quad \vee (i_1 \wedge i_7 \wedge i_9) \vee (i_5 \wedge i_6 \wedge i_8) \vee (i_1 \wedge i_2 \wedge i_4 \wedge i_6) \vee (i_1 \wedge i_2 \wedge i_6 \wedge i_9) \\ &\quad \vee (i_2 \wedge i_3 \wedge i_5 \wedge i_6) \vee (i_2 \wedge i_4 \wedge i_5 \wedge i_6) \vee (i_2 \wedge i_5 \wedge i_6 \wedge i_9). \end{aligned}$$

The weak neutrality reducts are determined using Property 3. The results show that there are 12 reducts for the weak neutrality relation, which are as follows:

$$\{\{i_1, i_2, i_5\}, \{i_1, i_4, i_5\}, \{i_1, i_4, i_7\}, \{i_1, i_5, i_6\}, \{i_1, i_5, i_9\}, \{i_1, i_7, i_9\}, \{i_5, i_6, i_8\},$$
$$\{i_1, i_2, i_4, i_6\}, \{i_1, i_2, i_6, i_9\}, \{i_2, i_3, i_5, i_6\}, \{i_2, i_4, i_5, i_6\}, \{i_2, i_5, i_6, i_9\}\}.$$

In the weak neutrality reducts, issues i_1, i_5, and i_6 appear most frequently, indicating that decision-makers should focus on these three issues when considering weak neutrality relations.

(3) **Weak conflict reducts:** For the weak conflict relation, the discernibility matrix $\mathbb{WM}^{\asymp}$ is first constructed according to Definition 14. The corresponding discernibility function $\triangle^{\asymp}$ is then formulated in line with Definition 15 as follows:

$$
\begin{aligned}
\triangle^{\asymp} ={}& (i_6 \vee i_9) \wedge (i_1 \vee i_2 \vee i_9) \wedge (i_1 \vee i_5 \vee i_6) \wedge (i_3 \vee i_8 \vee i_9) \wedge (i_4 \vee i_7 \vee i_9) \wedge (i_5 \vee i_8 \vee i_9) \\
& \wedge (i_1 \vee i_3 \vee i_4 \vee i_9) \wedge (i_1 \vee i_4 \vee i_5 \vee i_9) \wedge (i_2 \vee i_4 \vee i_8 \vee i_9) \wedge (i_2 \vee i_5 \vee i_6 \vee i_8) \\
& \wedge (i_2 \vee i_7 \vee i_8 \vee i_9) \wedge (i_1 \vee i_3 \vee i_4 \vee i_5 \vee i_8) \wedge (i_2 \vee i_3 \vee i_6 \vee i_7 \vee i_8) \\
& \wedge (i_1 \vee i_2 \vee i_3 \vee i_5 \vee i_7 \vee i_8) \wedge (i_3 \vee i_4 \vee i_5 \vee i_6 \vee i_7 \vee i_8) \\
={}& (i_1 \wedge i_6 \wedge i_9) \vee (i_1 \wedge i_8 \wedge i_9) \vee (i_2 \wedge i_5 \wedge i_9) \vee (i_3 \wedge i_5 \wedge i_9) \vee (i_3 \wedge i_6 \wedge i_9) \vee (i_5 \wedge i_6 \wedge i_9) \\
& \vee (i_5 \wedge i_7 \wedge i_9) \vee (i_5 \wedge i_8 \wedge i_9) \vee (i_6 \wedge i_8 \wedge i_9) \vee (i_1 \wedge i_2 \wedge i_3 \wedge i_9) \vee (i_1 \wedge i_2 \wedge i_4 \wedge i_9) \\
& \vee (i_1 \wedge i_2 \wedge i_7 \wedge i_9) \vee (i_1 \wedge i_4 \wedge i_6 \wedge i_8) \vee (i_1 \wedge i_6 \wedge i_7 \wedge i_8) \vee (i_2 \wedge i_4 \wedge i_6 \wedge i_8) \\
& \vee (i_2 \wedge i_4 \wedge i_6 \wedge i_9) \vee (i_4 \wedge i_6 \wedge i_7 \wedge i_9) \vee (i_2 \wedge i_3 \wedge i_4 \wedge i_5 \wedge i_6) \\
& \vee (i_2 \wedge i_3 \wedge i_5 \wedge i_6 \wedge i_7) \vee (i_1 \wedge i_3 \wedge i_4 \wedge i_5 \wedge i_6 \wedge i_7).
\end{aligned}
$$

Applying Property 3 produces 20 weak conflict reducts, as listed below:

$$\{\{i_1, i_6, i_9\}, \{i_1, i_8, i_9\}, \{i_2, i_5, i_9\}, \{i_3, i_5, i_9\}, \{i_3, i_6, i_9\}, \{i_5, i_6, i_9\}, \{i_5, i_7, i_9\}, \{i_5, i_8, i_9\},$$
$$\{i_6, i_8, i_9\}, \{i_1, i_2, i_3, i_9\}, \{i_1, i_2, i_4, i_9\}, \{i_1, i_2, i_7, i_9\}, \{i_1, i_4, i_6, i_8\}, \{i_1, i_6, i_7, i_8\},$$
$$\{i_2, i_4, i_6, i_8\}, \{i_2, i_4, i_6, i_9\}, \{i_4, i_6, i_7, i_9\}, \{i_2, i_3, i_4, i_5, i_6\}, \{i_2, i_3, i_5, i_6, i_7\},$$
$$\{i_1, i_3, i_4, i_5, i_6, i_7\}\}.$$

In the weak conflict reducts, either issue i_6 or i_9 appears in every reduct, suggesting that these issues play a crucial role in shaping weak conflict relations.

By analyzing the results of the alliance issue reducts, neutral issue reducts, and conflict issue reducts, several observations can be made. First, the reduct sets obtained for the three types of weak relations are all different, indicating that the importance of different issues varies across different types of relations. Second, it is observed that the number of issues contained in each reduct set ranges from three to five. This suggests that when analyzing the conflict scenario, decision-makers can simplify their consideration from the original nine issues in the incomplete three-valued situation table to just three, four, or five key issues in the reduct sets. This significantly reduces time costs and improves decision-making efficiency.

In this practical case based on an incomplete three-valued situation table, the proposed method retains the original incomplete conflict scenario without any

$$\mathbb{WM}^{=} = \begin{bmatrix} \emptyset \\ \{z_3, z_4, z_5, z_6, z_8, z_9\}\emptyset \\ \{z_2, z_5, z_9\} & \{z_5, z_6, z_9\} & \emptyset \\ \{z_5, z_6, z_7, z_8, z_9\} & \{z_5, z_6, z_8\} & \{z_2, z_5, z_6, z_7, z_9\}\emptyset \\ \{z_4, z_5, z_8, z_9\} & \{z_3, z_4, z_6, z_8\} & \{z_1, z_3, z_6, z_8\} & \{z_1, z_4, z_6, z_7, z_8\}\emptyset \\ \{z_1, z_4, z_7, z_9\} & \{z_4, z_9\} & \{z_2, z_5\} & \{z_4, z_7, z_8, z_9\} & \{z_1, z_4, z_7, z_8, z_9\}\emptyset \\ \{z_4, z_5, z_7, z_8, z_9\} & \{z_2, z_5, z_6, z_8\} & \{z_2, z_5\} & \{z_2, z_7, z_9\} & \{z_4, z_7, z_8, z_9\} & \{z_7, z_8\} & \{z_4, z_5, z_7, z_8, z_9\}\emptyset \\ \{z_2, z_6, z_9\} & \{z_3, z_4, z_5, z_6, z_8\} & \{z_2, z_5, z_6, z_9\} & \{z_2, z_5, z_8\} & \{z_4, z_6\} & \{z_2, z_4, z_5, z_6\} & \{z_2, z_4, z_5, z_6, z_8\}\emptyset \\ \{z_3, z_4, z_5, z_9\} & \{z_3, z_5, z_8\} & \{z_2, z_9\} & \{z_1, z_3, z_7\} & \{z_1, z_2, z_4, z_6\} & \{z_2, z_4, z_5, z_8, z_9\} & \{z_2, z_3, z_4, z_5\}\emptyset \\ \{z_1, z_5, z_6, z_7, z_9\} & \{z_1, z_3, z_4, z_5, z_8\} & \{z_1, z_6, z_7, z_9\} & \{z_1, z_5, z_6, z_7, z_8, z_9\}\{z_4, z_6, z_7, z_8\} & \{z_1, z_3, z_7\} & \{z_1, z_5, z_6\} & \{z_1, z_3, z_4, z_7\}\emptyset \end{bmatrix}$$

$$\mathbb{WM}^{\approx} = \begin{bmatrix} I \\ \{z_1, z_3, z_4, z_8, z_9\} \\ \{z_1, z_6, z_7\} & I \\ \{z_1, z_2, z_3, z_4, z_6, z_7, z_9\} & \{z_1, z_3, z_4, z_5, z_9\} & \{z_1, z_5, z_6, z_7\}I \\ \{z_1, z_2, z_3, z_4, z_6, z_8, z_9\} & \{z_3, z_4, z_6, z_8, z_9\} & \{z_1, z_6\} & \{z_1, z_2, z_3, z_4, z_6, z_9\} & I \\ \{z_1, z_2, z_7, z_8\} & \{z_1, z_5, z_8\} & \{z_1, z_2, z_7, z_8\} & \{z_1, z_5, z_7, z_9\}\{z_1, z_2, z_5, z_7\} & \{z_1, z_2, z_8\} & I \\ \{z_2, z_4, z_6, z_7, z_9\} & \{z_4, z_5, z_9\} & \{z_4, z_6\} & \{z_5, z_6, z_7\} & \{z_5, z_6, z_7, z_8, z_9\}\{z_2, z_4, z_6, z_9\} & \{z_2, z_5, z_7\} & I \\ \{z_1, z_2, z_3, z_4, z_5, z_6, z_8, z_9\}\{z_1, z_2, z_3, z_4, z_6, z_9\} & \{z_1, z_2, z_3, z_4, z_6, z_8, z_9\}\{z_2, z_4, z_6, z_9\} & \{z_1, z_2, z_5, z_7, z_9\}\{z_1, z_2, z_4, z_5, z_7, z_9\}\{z_1, z_2, z_4, z_8, z_9\}I \\ \{z_1, z_2, z_4, z_7, z_8, z_9\} & \{z_1, z_3, z_4, z_8, z_9\} & \{z_1, z_5, z_7\} & \{z_1, z_2, z_4, z_5, z_7, z_9\}\{z_1, z_2, z_4, z_5, z_7, z_9\} & \{z_1, z_5, z_8\} & \{z_1, z_3, z_4, z_5, z_7, z_9\}I \\ \{z_1, z_3, z_4, z_9\} & \{z_1, z_4, z_5, z_6, z_8, z_9\} & \{z_1, z_2, z_3, z_4, z_6, z_9\} & \{z_1, z_3, z_4, z_7, z_9\} & \{z_1, z_3, z_4, z_7, z_9\}\{z_1, z_3, z_4, z_7, z_8\} & \{z_1, z_5, z_8\} & \{z_1, z_3, z_4, z_5, z_9\}I \end{bmatrix}$$

$$\mathbb{WM}^{)(} = \begin{bmatrix} I \\ \{z_1, z_5, z_6\} \\ \{z_1, z_2, z_5, z_6, z_7, z_9\}\{z_1, z_6, z_9\} & I \\ \{z_1, z_2, z_3, z_4, z_5, z_8\} & \{z_1, z_5, z_6, z_8\} & \{z_1, z_5, z_6, z_7, z_8\} & I \\ \{z_2, z_3, z_4, z_7, z_8\} & \{z_2, z_4, z_6, z_7, z_9\} & \{z_2, z_3, z_7, z_8, z_9\} & I \\ \{z_1, z_2, z_4, z_7, z_9\} & \{z_2, z_4, z_6, z_8, z_9\} & \{z_2, z_6, z_7, z_9\} & \{z_2, z_4, z_7, z_9\} & \{z_2, z_4, z_5, z_8, z_9\} & I \\ \{z_2, z_5, z_6, z_8\} & \{z_4, z_6, z_8, z_9\} & \{z_2, z_5, z_6, z_9\} & \{z_2, z_7, z_8, z_9\} & \{z_2, z_4, z_6, z_7, z_8, z_9\}\{z_2, z_4, z_8, z_9\} & \{z_5, z_8, z_9\} & I \\ \{z_1, z_3, z_4, z_5, z_9\} & \{z_1, z_5, z_6, z_9\} & \{z_1, z_2, z_5, z_6, z_9\} & \{z_1, z_3, z_4, z_7\} & \{z_1, z_2, z_3, z_8, z_9\}\{z_3, z_5, z_9\} & \{z_1, z_3, z_4, z_5, z_9\}\{z_1, z_2, z_4, z_7, z_8, z_9\}\{z_1, z_3, z_4, z_5, z_8, z_9\}I \\ \{z_3, z_4, z_5, z_6, z_7, z_8\}\{z_6, z_9\} & \{z_1, z_2, z_5, z_7, z_9\} & \{z_3, z_4, z_6, z_7, z_8, z_9\}\{z_1, z_3, z_6, z_7, z_8, z_9\}\{z_5, z_6, z_7, z_8, z_9\}\{z_2, z_4, z_7, z_9\} & \{z_1, z_3, z_6, z_7, z_8, z_9\}\{z_3, z_5, z_7, z_8, z_9\}I \end{bmatrix}$$

modifications. It directly conducts issue reduction for the weak alliance, weak neutral, and weak conflict relations while accounting for the missing values. This approach enables the identification of the most critical issues for decision-makers, ensuring that the original information is preserved to the greatest extent possible.

5 Conclusion

This paper extends the research on three-way conflict analysis to the context of incomplete ternary situation tables, focusing on the issue reduction of alliance, neutrality, and conflict in such incomplete tables. First, agent pairs are classified into three categories, defining weak alliance, weak neutrality, and weak conflict relations. Subsequently, the concepts of weak alliance reduction, weak neutrality reduction, and weak conflict reduction are proposed. Based on these three types of weak relations, corresponding discernibility matrices and discernibility functions are constructed. The issue reduction results are then obtained through the calculation of discernibility functions. Finally, the proposed method is applied to NBA labor negotiations. By analyzing the results of alliance, neutrality, and conflict issue reductions, the practicality and effectiveness of the method are demonstrated.

In future research, we will further explore issue reduction in more complex conflict scenarios, such as incomplete three-valued decision situation tables, incomplete fuzzy situation tables, and dynamic three-valued situation tables, providing more effective theoretical and methodological support for complex conflict analysis.

Acknowledgements. This work is supported by the National Natural Science Foundation of China(Nos.62076040,12471431), the Scientific Research Fund of Hunan Provincial Education Department (No. 22A0233), and the Hunan Provincial Key Laboratory of Mathematical Modeling and Analysis in Engineering (No. 2018MMAEZD10).

References

1. Pawlak, Z.: On conflicts. Int. J. Man Mach. Stud. **21**(2), 127–134 (1984)
2. Pawlak, Z.: An inquiry into anatomy of conflicts. Inf. Sci. **109**, 65–78 (1998)
3. Yao, Y.Y.: Three-way decisions with probabilistic rough sets. Inf. Sci. **180**(3), 341–353 (2010)
4. Yao, Y.Y.: Three-way conflict analysis: reformulations and extensions of the Pawlak model. Knowl.-Based Syst. **180**, 26–37 (2019)
5. Lang, G.M., Miao, D.Q., Fujita, H.: Three-way group conflict analysis based on Pythagorean fuzzy set theory. IEEE Trans. Fuzzy Syst. **28**(3), 447–461 (2019)
6. Mandal, P., Samanta, S., Pal, M., Ranadive, A.S.: Regret theory based three-way conflict analysis model under q-rung orthopair fuzzy information: studies with parameter and three-way decision-making-based approaches. Artif. Intell. Rev. **56**(Suppl 3), 3417–3469 (2023)

7. Luo, J., Han, B.H., Huang, B., Geng, S.L.: Weighted three-way conflict analysis in multi-attribute decision-making perspective. Inf. Sci. **674**, 120721 (2024)
8. Hu, M.J.: Three-way conflict analysis with preference-based conflict situations. Inf. Sci. 121676 (2024)
9. Tang, J., Chen, C.F., Rao, F.: A distance-based three-way conflict analysis for fuzzy multiset. Appl. Soft Comput. **169**, 112562 (2025)
10. Gao, S., Yang, H.L., Guo, Z.L.: Three-way conflict analysis and resolution based on interval set information. Inf. Sci. 121938 (2025)
11. Li, X.N., Liang, R., Yi, H.J.: The grouping weighted averaging operator via three-way conflict analysis. Inf. Sci. 121990 (2025)
12. Pawlak, Z.: Rough sets. Int. J. Comput. Inform. Sci. **11**(5), 341–356 (1982)
13. Skowron, A., Rauszer, C.: The discernibility matrices and functions in information systems. In: Intelligent Decision Support: Handbook of Applications and Advances of the Rough Sets Theory, pp. 331–362 (1992)
14. Yao, Y.Y., Zhao, Y.: Discernibility matrix simplification for constructing attribute reducts. Inf. Sci. **179**(7), 867–882 (2009)
15. Sowkuntla, P., Prasad, P.S.: MapReduce based parallel fuzzy-rough attribute reduction using discernibility matrix. Appl. Intell. **52**(1), 154–173 (2022)
16. Wen, H.T., Xu, Y., Liang, M.S.: Parallel attribute reduction algorithm for unlabeled data based on fuzzy discernibility matrix and soft deletion behavior. Inf. Sci. **689**, 121472 (2025)
17. Kryszkiewicz, M.: Rough set approach to incomplete information systems. Inf. Sci. **112**, 39–49 (1998)
18. Dai, J.H., Wang, Z.Y., Huang, W.Y.: Interval-valued fuzzy discernibility pair approach for attribute reduction in incomplete interval-valued information systems. Inf. Sci. **642**, 119215 (2023)
19. Lang, G.M.: Three-way conflict analysis: alliance, conflict, and neutrality reducts of three-valued situation tables. Cogn. Comput. **14**(6), 2040–2053 (2022)
20. Chen, J., Zhang, X.Y.: Three-way reductions of conflict analysis based on relation matrices and integration measures. Appl. Intell. **55**(4), 315 (2025)
21. Zhang, X.Y., Chen, J.: Three-way quantitative models and three-way issue reductions of conflict analysis. Inf. Sci. **701**, 121846 (2025)
22. Xu, F., Cai, M.J., Song, H.L., Dai, J.H.: The selection of feasible strategies based on consistency measurement of cliques. Inf. Sci. **583**, 33–55 (2022)

New Models of Three-Way Conflict Analysis Based on Five Level Preferences

Haojun Liu[1,2], Qimei Xiao[1,2], Huiying Yu[1,2], and Guangming Lang[1,2(✉)]

[1] School of Mathematics and Statistics, Changsha University of Science and Technology, Changsha 410114, Hunan, China
langguangming1984@126.com
[2] Hunan Provincial Key Laboratory of Mathematical Modeling and Analysis in Engineering, Changsha University of Science and Technology, ChangSha, Hunan 410114, People's Republic of China

Abstract. Most scholars have established conflict analysis models based on the situation tables. Generally, agents tend to have imprecise assessments of issues, and their interpretations of issues are difficult to maintain consistency. The preference-based conflict analysis model aims to address these shortcomings, by employing ordered pairs over issues to replace three-value ratings in order to characterize conflict degree. However, the three-way conflict analysis models based on three-level preferences are insufficient to characterize preferences of agents. In this paper, we first define five preference relations and conflict situations based on these preference relations. Next, we assign conflict degrees to twenty-five different cases and define a conflict function for subsets of issues. Subsequently, we normalize the conflict degrees for subsets of issues and trisect the set of all pairs of agents. Finally, we validate the practicality of the model through examples. The proposed model guarantees more coherent data collection, simultaneously enhancing the explainability of preferences in three-way conflict analysis.

Keywords: Conflict analysis · Preference relation · Three-way decision · Agent relation

1 Introduction

The three-way decision model was proposed by Yao [1], which divides the universe into three regions based on Pawlak's rough set theory and offers meaningful interpretations for each region. Afterwards, three-way decision theory has already developed into an important branch of granular computing [2]. As a methodology, three-way decision theory has been widely applied in the field of artificial intelligence by integrating theories such as fuzzy sets, concept lattice, granular-ball computing and so on [3–6]. For example, Suo et al. [7] compiled a literature review on three-way decision theories, proposing various trisection structures and their applications in other theories while conducting bibliometric research on three-way decision-related articles using the three-way decision model. Xu [8] proposed a dual-universe rough set model based on ranking and reference tuples, providing interpretations for the three regions of the model and revealing the relationships with three other rough set models. Xu [9] extended the dual-universe rough set model based on ranking and reference tuples to the level of dual thresholds, demonstrating that trisections only require the calculation of a finite number of threshold pairs, and proposed a measure for determining the optimal trisection. The researches related to three-way decisions have developed rapidly in recent years.

Three-way conflict analysis embodies the idea of three-way decision, conflict has a significant impact on the development of things, and many disciplines encompass research related to conflict [10–16]. In 1983, Pawlak [17] explained the fundamental concepts of conflict and proposed a model of conflict situations. Pawlak [18] studied conflict analysis in conjunction with the classical rough set model and proposed flow distribution in the graph to examine relationships among agents. Yao [19] proposed a distance metric based on four axioms and introduced the concept of preference for the first time, defining the discrepancy between user and system rankings of documents and revealing the relationship between NPDM and other document retrieval metrics. Yao [20] proposed a trisection for agents and a trisection for pairs of agents based on a single issue. Lang [21] proposed alliance and conflict measures for three-way conflict analysis, assigning different weights to different issues. Hu [22] performed three-way conflict analysis based on preference in the conflict situation, which can be transformed into a three-value information table and a triangular fuzzy number information system.

In practical applications, the three-way conflict analysis models based on three-level preferences fail to adequately capture the preferences of agents. The contributions of this work are briefly listed as follows:

(1) We propose five types of preference relations: strong preference relation, weak preference relation, indifference relation, weak converse relation, and strong converse relation. Based on these preference relations, we define the conflict situations. Subsequently, we present the related theorems for the five types of preference relations.

(2) We consider the conflict degree between two agents for a specific pair of issues, which will later be aggregated to define the overall conflict degree for a set of issues. The conflict degree for the subset of issues can be obtained by considering all unordered pairs of issues in the subset of issues. Afterwards, we normalize the conflict degree based on the subset of issues, and obtain a specific trisection.

(3) The proposed model can be applied to multi-valued rating tables. We validate the effectiveness of the model through examples.

The rest of this paper is organized as follows. In Sect. 2, we review some basic concepts of three-way conflict analysis based on three level preferences. Section 3 proposes the three-way conflict analysis models based on five level preferences. Section 4 shows how to apply the proposed model. Section 5 concludes the work of this paper.

2   Preliminaries

In this section, we recall three-way conflict analysis based on three level preferences.

Definition 1. *(Hu [22], 2025) Suppose A is a non-empty set of agents, I is a non-empty set of issues, and $\succ_a^I$ is a preference relation for $a \in A$ and $i, j \in J \subseteq I$, preference relation, converse relation, and indifferent relation towards the subset J are defined as:*

$$\succ_a^J = \{(i, j) \in J \times J \mid i \text{ is supported more by } a \text{ than } j\}, \tag{1}$$

$$\prec_a^J = \{(j, i) \in J \times J \mid i \text{ is supported more by } a \text{ than } j\}, \tag{2}$$

$$\sim_a^J = \{(i, j) \in J \times J \mid \neg (i \succ_a^J j) \wedge \neg (i \prec_a^J j)\}. \tag{3}$$

Yao [19] pointed out in the literature that the preference relation is a weak order, satisfying properties such as asymmetry and negative transitivity, and the indifferent relation is an equivalence relation, satisfying reflexivity, symmetry, and transitivity, and $J \times J$ can be partitioned into three mutually exclusive relations: preference relation, indifference relation, and converse relation. Furthermore, the comparisons between the equivalence classes derived from the subset of issues can indicate that the former type of issues is favored over the latter.

Definition 2. *(Hu [22], 2025) Suppose A is a non-empty set of agents, I is a non-empty set of issues, $\succ_a^I$ is a preference relation for $a \in A$, let $\mathbb{R} = \{\succ_a^I \mid a \in A\}$, we can obtain a triplet $T = (A, I, \mathbb{R})$ which is characterized as a conflict situation based on three level preferences.*

The conflict situation based on three level preferences describes pairwise preferences regarding issues. Agents, issues, and the ratings of agents constitute three-way conflict analysis frameworks, which can be represented through the conflict situation based on three level preferences. In the three-way conflict analysis, the preference-based conflict situation can be transformed into a multi-valued situation table and the transformation is bidirectional.

3   Three-Way Conflict Analysis Based on Five Level Preferences

In this section, we discuss three-way conflict analysis based on five level preferences. The following definitions are given for the five preference relations.

Definition 3. *Suppose $\succ_a^I$ is a preference relation for $a \in A$ and $i, j \in J \subseteq I$, "$\succ_a^{J_s}$", "$\succ_a^{J_w}$", "$\sim_a^J$", "$\prec_a^{J_w}$" and "$\prec_a^{J_s}$" are defined as:*

$$\succ_a^{J_s} = \{(i,j) \in J \times J \mid i \text{ is more strongly supported by } a \text{ than } j\}, \quad (4)$$

$$\succ_a^{J_w} = \{(i,j) \in J \times J \mid i \text{ is more weakly supported by } a \text{ than } j\}, \quad (5)$$

$$\sim_a^J = \{(i,j) \in J \times J \mid \neg\left(i \succ_a^{J_b} j\right) \wedge \neg\left(i \prec_a^{J_b} j\right)\}, \quad b = s \text{ or } w, \quad (6)$$

$$\prec_a^{J_w} = \{(j,i) \in J \times J \mid i \text{ is more weakly supported by } a \text{ than } j\}, \quad (7)$$

$$\prec_a^{J_s} = \{(j,i) \in J \times J \mid i \text{ is more strongly supported by } a \text{ than } j\}. \quad (8)$$

Based on the five preference relations, we can obtain the conflict situation derived from five level preferences.

Definition 4. *Suppose A is a non-empty set of agents, I is a non-empty set of issues, let $\mathbb{R} = \{\succ_a^{I_b} \mid a \in A, \ b \in s, w\}$, we can obtain a triplet $T = (A, I, \mathbb{R})$, which is characterized as a conflict situation based on five level preferences.*

An example is provided to explain the conflict situation based on five level preferences.

Example 1. Consider a five level preferences-based conflict situation $T = (A, I, \mathbb{R})$. Among them, $A = \{a_1, a_2\}$, $I = \{i_1, i_2, i_3, i_4, i_5, i_6\}$. By taking $J = I$, the strong preference relations are described as follows:

$$\succ_{a_1}^{J_s} = \{(i_2, i_6), (i_2, i_5), (i_3, i_6), (i_4, i_6)\}, \qquad \succ_{a_2}^{J_s} = \{(i_3, i_4), (i_3, i_5)\}.$$

The weak preference relations are described as follows:

$$\succ_{a_1}^{J_w} = \{(i_2, i_1), (i_3, i_5), (i_4, i_5), (i_1, i_6), (i_2, i_3), (i_2, i_4), (i_3, i_1), (i_4, i_1),$$
$$(i_1, i_5), (i_5, i_6)\},$$
$$\succ_{a_2}^{J_w} = \{(i_3, i_1), (i_2, i_4), (i_6, i_4), (i_2, i_5), (i_6, i_5), (i_3, i_2), (i_3, i_6), (i_2, i_1),$$
$$(i_6, i_1), (i_1, i_4), (i_1, i_5)\}.$$

Furthermore, the weak converse relation and the strong converse relation can be derived from the previous two preference relations. We omit them here. The indifferent preference relations are described as follows:

$$\sim_{a_1}^J = \{(i_2, i_2), (i_3, i_3), (i_4, i_4), (i_1, i_1), (i_5, i_5), (i_6, i_6), (i_3, i_4), (i_4, i_3)\},$$
$$\sim_{a_2}^J = \{(i_3, i_3), (i_2, i_2), (i_6, i_6), (i_1, i_1), (i_4, i_4), (i_5, i_5), (i_2, i_6), (i_6, i_2),$$
$$(i_4, i_5), (i_5, i_4)\}.$$

The above preference relations constitute the conflict situation based on five level preferences. The following theorem holds when the preference takes five levels, which is analogous to the case where the preference takes three levels.

Theorem 1. *Suppose a subset $J \subseteq I$ and an agent $a \in A$, and "$\succ_a^{J_s}$", "$\succ_a^{J_w}$", "$\sim_a^{J}$", "$\prec_a^{J_w}$", and "$\prec_a^{J_s}$" are five preference relations for $J \subseteq I$ and $a \in A$. When the preference takes five levels, it follows that:*

$$\succ_a^{J_s} \cap \succ_a^{J_w} = \phi, \quad \succ_a^{J_s} \cap \sim_a^{J} = \phi, \quad \succ_a^{J_s} \cap \prec_a^{J_w} = \phi, \quad \succ_a^{J_s} \cap \prec_a^{J_s} = \phi,$$

$$\succ_a^{J_w} \cap \sim_a^{J} = \phi, \succ_a^{J_w} \cap \prec_a^{J_w} = \phi, \quad \succ_a^{J_w} \cap \prec_a^{J_s} = \phi, \quad \sim_a^{J} \cap \prec_a^{J_w} = \phi,$$

$$\sim_a^{J} \cap \prec_a^{J_s} = \phi, \quad \prec_a^{J_w} \cap \prec_a^{J_s} = \phi; \tag{9}$$

$$\succ_a^{J_s} \cup \succ_a^{J_w} \cup \sim_a^{J} \cup \prec_a^{J_w} \cup \prec_a^{J_s} = J \times J. \tag{10}$$

Based on the definition of the preference relations, it follows that the five preference relations are pairwise disjoint, and there are a total of "C_5^2" formulas. Additionally, $J \times J$ can be partitioned into five mutually exclusive relations: strong preference relation, weak preference relation, indifference relation, weak converse relation and strong converse relation.

The conflict degree between two agents for a pair of issues is shown in the table below in cases where the preference reaches five levels. One should note that "$j \succ_x^{J_b} i$" and "$i \prec_x^{J_b} j$" are equivalent when b takes the value s or w.

As shown in the table, "SC", "WC", "PC", "WI" and "SI" represent strong consistency, weak consistency, partial consistency, weak inconsistency and strong inconsistency, respectively. Since consistency and inconsistency are two concepts that are relative to each other, partial consistency in the table above can also be interpreted as partial inconsistency. That is, "PC" can also be replaced by "PI". By organizing the conflict degrees for different cases in the table, they can be represented by the following piecewise function.

Table 1. The conflict degree between agent x and agent y on issue i and issue j when the preference takes five levels.

Scenario	$Preference(x)$	$Preference(y)$	$Relationships$	$CD_{ij}(x,y)$
1	$i \succ_x^{J_s} j$	$i \succ_y^{J_s} j$	SC	η^{SC}
2	$i \succ_x^{J_s} j$	$i \succ_y^{J_w} j$	WC	η^{WC}
3	$i \succ_x^{J_s} j$	$i \sim_y^{J} j$	PC	η^{PC}
4	$i \succ_x^{J_s} j$	$j \succ_y^{J_w} i$	WI	η^{WI}
5	$i \succ_x^{J_s} j$	$j \succ_y^{J_s} i$	SI	η^{SI}
6	$i \succ_x^{J_w} j$	$i \succ_y^{J_s} j$	WC	η^{WC}
7	$i \succ_x^{J_w} j$	$i \succ_y^{J_w} j$	SC	η^{SC}
8	$i \succ_x^{J_w} j$	$i \sim_y^{J} j$	WC	η^{WC}
9	$i \succ_x^{J_w} j$	$j \succ_y^{J_w} i$	PC	η^{PC}
10	$i \succ_x^{J_w} j$	$j \succ_y^{J_s} i$	WI	η^{WI}
11	$i \sim_x^{J} j$	$i \succ_y^{J_s} j$	PC	η^{PC}
12	$i \sim_x^{J} j$	$i \succ_y^{J_w} j$	WC	η^{WC}

continued

Table 1. continued

Scenario	$Preference(x)$	$Preference(y)$	Relationships	$CD_{ij}(x,y)$
13	$i \sim_x^J j$	$i \sim_y^J j$	SC	η^{SC}
14	$i \sim_x^J j$	$j \succ_y^{J_w} i$	WC	η^{WC}
15	$i \sim_x^J j$	$j \succ_y^{J_s} i$	PC	η^{PC}
16	$j \succ_x^{J_w} i$	$i \succ_y^{J_s} j$	WI	η^{WI}
17	$j \succ_x^{J_w} i$	$i \succ_y^{J_w} j$	PC	η^{PC}
18	$j \succ_x^{J_w} i$	$i \sim_y^J j$	WC	η^{WC}
19	$j \succ_x^{J_w} i$	$j \succ_y^{J_w} i$	SC	η^{SC}
20	$j \succ_x^{J_w} i$	$j \succ_y^{J_s} i$	WC	η^{WC}
21	$j \succ_x^{J_s} i$	$i \succ_y^{J_s} j$	SI	η^{SI}
22	$j \succ_x^{J_s} i$	$i \succ_y^{J_w} j$	WI	η^{WI}
23	$j \succ_x^{J_s} i$	$i \sim_y^J j$	PC	η^{PC}
24	$j \succ_x^{J_s} i$	$j \succ_y^{J_w} i$	WC	η^{WC}
25	$j \succ_x^{J_s} i$	$j \succ_y^{J_s} i$	SC	η^{SC}

Definition 5. *Suppose A is a non-empty set of agents, I is a non-empty set of issues, $\succ_a^I$ is a preference relation based on five level preferences. Given $x, y \in A$, $CD(x,y) : I \times I \longrightarrow [\eta^{SC}, \eta^{WC}, \eta^{PC}, \eta^{WI}, \eta^{SI}]$ serves as a conflict measure with respect to a pair of agents x and y. For $i, j \in J \subseteq I$,*

$$
CD_{ij}(x,y) = \begin{cases}
\eta^{SC}, & \text{if } (i \succ_x^{J_b} j \wedge i \succ_y^{J_b} j) \vee (i \sim_x^J j \wedge i \sim_y^J j) \vee \\
& (j \succ_x^{J_b} i \wedge j \succ_y^{J_b} i), \quad b = s, w. \\
\eta^{WC}, & \text{if } (i \succ_x^{J_s} j \wedge i \succ_y^{J_w} j) \vee (i \succ_x^{J_w} j \wedge i \succ_y^{J_s} j) \vee \\
& (i \succ_x^{J_w} j \wedge i \sim_y^J j) \vee (i \sim_x^J j \wedge i \succ_y^{J_w} j) \vee \\
& (i \sim_x^J j \wedge j \succ_y^{J_w} i) \vee (j \succ_x^{J_w} i \wedge i \sim_y^J j) \vee \\
& (j \succ_x^{J_w} i \wedge j \succ_y^{J_s} i) \vee (j \succ_x^{J_s} i \wedge j \succ_y^{J_w} i), \\
\eta^{PC}, & \text{if } (i \succ_x^{J_s} j \wedge i \sim_y^J j) \vee (i \succ_x^{J_w} j \wedge j \succ_y^{J_w} i) \vee \\
& (i \sim_x^J j \wedge i \succ_y^{J_s} j) \vee (i \sim_x^J j \wedge j \succ_y^{J_s} i) \vee \\
& (j \succ_x^{J_w} i \wedge i \succ_y^{J_w} j) \vee (j \succ_x^{J_s} i \wedge i \sim_y^J j), \\
\eta^{WI}, & \text{if } (i \succ_x^{J_s} j \wedge j \succ_y^{J_w} i) \vee (i \succ_x^{J_w} j \wedge j \succ_y^{J_s} i) \vee \\
& (j \succ_x^{J_w} i \wedge i \succ_y^{J_s} j) \vee (j \succ_x^{J_s} i \wedge i \succ_y^{J_w} j), \\
\eta^{SI}, & \text{if } (i \succ_x^{J_s} j \wedge j \succ_y^{J_s} i) \vee (j \succ_x^{J_s} i \wedge i \succ_y^{J_s} j).
\end{cases}
\tag{11}
$$

It can be observed from the piecewise function that as the conflict degree increases, the cases where the two agents have different preferences for a pair of issues decrease. Furthermore, the number of cases where the two agents have the same preferences for a pair of issues is equal to the number of conflict degree categories.

Remark 1. The five conflict degrees reflect the differing preference scenarios of two agents regarding a pair of issues. The following provides an explanation of five conflict degrees in the context of five level preferences for clarity.

Strong Consistency: The two agents reach an agreement over issues if the following conditions are satisfied: both agents strongly prefer issue i over issue j; or both agents weakly prefer issue i over issue j; or both agents have an equal preference for issue i and issue j.

Weak Consistency: The two agents reach an agreement over issues if one agent strongly prefers issue i over issue j, and the other agent weakly prefers issue i over issue j; or the two agents reach an agreement over issues if one agent weakly prefers issue i over issue j, and the other agent has an equal preference for issue i and issue j, and so on.

Partial Consistency: The two agents reach an agreement over issues if one agent strongly prefers issue i over issue j, and the other agent has an equal preference for issue i and issue j; or the two agents have a disagreement over issues if one agent weakly prefers issue i over issue j, and the other agent weakly prefers issue j over issue i, and so on.

Weak Inconsistency: The two agents have a disagreement over issues if one agent strongly prefers issue i over issue j, and the other agent weakly prefers issue j over issue i; or the two agents have a disagreement over issues if one agent weakly prefers issue i over issue j, and the other agent strongly prefers issue j over issue i, and so on.

Strong Inconsistency: The two agents have a disagreement over issues if one agent strongly prefers issue i over issue j, and the other agent strongly prefers issue j over issue i.

According to the symmetry of the agents' positions and the interchangeability of the issues, it follows that the above results remain valid. After defining the conflict degree for five level preferences, it is also necessary to calculate the number of unordered pairs corresponding to each scenario in order to define the conflict degree for subsets of issues.

Remark 2. According to Table 1, for agents $x, y \in A$ and issues $i, j \in J \subseteq I$, when the preference takes five levels, we have:

Strong consistency: $| \succ_x^{J_s} \cap \succ_y^{J_s} | + | \succ_x^{J_w} \cap \succ_y^{J_w} | + \frac{|\sim_x^J \cap \sim_y^J| + |J|}{2}$. The number of unordered pairs corresponding to Scenario 1, and Scenario 25 are equally identical, which is denoted as $| \succ_x^{J_s} \cap \succ_y^{J_s} |$. Similarly, the number of unordered pairs corresponding to Scenario 7, and Scenario 19 are equally identical, which is denoted as $| \succ_x^{J_w} \cap \succ_y^{J_w} |$. In addition, the equivalence class of $| \sim_x^J \cap \sim_y^J |$ includes the equivalence classes of unordered pairs of the same type and unordered pairs of the different type for a pair of agents, with their quantities denoted as $|J|$ and $| \sim_x^J \cap \sim_y^J | - |J|$, respectively. Due to the symmetry of the equivalence relationship, the number of unordered pairs corresponding to Scenario 13 is represented as $\frac{|\sim_x^J \cap \sim_y^J| + |J|}{2}$.

Weak consistency: $|\succ_x^{J_s} \cap \succ_y^{J_w}|+|\succ_x^{J_w} \cap \sim_y^{J}|+|\succ_x^{J_w} \cap \succ_y^{J_s}|+|\sim_x^{J} \cap \succ_y^{J_w}|$.
The number of unordered pairs corresponding to Scenario 2, and Scenario 24 are equally identical, which is denoted as $|\succ_x^{J_s} \cap \succ_y^{J_w}|$. Similarly, the number of unordered pairs corresponding to Scenario 8 and Scenario 18 are identical, denoted as $|\succ_x^{J_w} \cap \sim_y^{J}|$. The other cases follow similarly, and so on.

Partial consistency: $|\succ_x^{J_s} \cap \sim_y^{J}|+|\succ_x^{J_w} \cap \prec_y^{J_w}|+|\sim_x^{J} \cap \prec_y^{J_s}|$. The number of unordered pairs corresponding to Scenario 3, and Scenario 23 are equally identical, which is denoted as $|\succ_x^{J_s} \cap \sim_y^{J}|$. Similarly, the number of unordered pairs corresponding to Scenario 9 and Scenario 17 are identical, denoted as $|\succ_x^{J_w} \cap \prec_y^{J_w}|$. The other cases follow similarly, and so on.

Weak inconsistency: $|\succ_x^{J_s} \cap \prec_y^{J_w}|+|\succ_x^{J_w} \cap \prec_y^{J_s}|$. The number of unordered pairs corresponding to Scenario 4, and Scenario 22 are equally identical, which is denoted as $|\succ_x^{J_s} \cap \prec_y^{J_w}|$. Similarly, the number of unordered pairs corresponding to Scenario 10 and Scenario 16 are identical, denoted as $|\succ_x^{J_w} \cap \prec_y^{J_s}|$.

Strong inconsistency: $|\succ_x^{J_s} \cap \prec_y^{J_s}|$. The number of unordered pairs corresponding to Scenario 5, and Scenario 21 are equally identical, which is denoted as $|\succ_x^{J_s} \cap \prec_y^{J_s}|$.

Given the conflict degree between two agents for a pair of issues and the number of unordered pairs corresponding to each scenario, the conflict degree for subsets of issues is then defined.

Definition 6. *Given agents $x, y \in A$ and issues $i, j \in J \subseteq I$, $\mathbb{P}^J$ denotes the set of all unordered pairs on a subset J of I, and $\langle i, j \rangle$ denotes an unordered pair belonging to $\mathbb{P}^J$. The conflict degree for a subset J of issues is defined by a function $CD_J\colon A \times A \longrightarrow \mathbb{R}$ as:*

$$CD_J(x,y) = \sum_{\langle i,j \rangle \in \mathbb{P}^J} CD_{ij}(x,y)$$

$$= \eta^{SC}(|\succ_x^{J_s} \cap \succ_y^{J_s}|+|\succ_x^{J_w} \cap \succ_y^{J_w}|+\frac{|\sim_x^{J} \cap \sim_y^{J}|+|J|}{2}) + \eta^{WC}.$$

$$(|\succ_x^{J_s} \cap \succ_y^{J_w}|+|\succ_x^{J_w} \cap \sim_y^{J}|+|\succ_x^{J_w} \cap \succ_y^{J_s}|+|\sim_x^{J} \cap \succ_y^{J_w}|)+$$

$$\eta^{PC}(|\succ_x^{J_s} \cap \sim_y^{J}|+|\succ_x^{J_w} \cap \prec_y^{J_w}|+|\sim_x^{J} \cap \prec_y^{J_s}|) + \eta^{WI}.$$

$$(|\succ_x^{J_s} \cap \prec_y^{J_w}|+|\succ_x^{J_w} \cap \prec_y^{J_s}|) + \eta^{SI}(|\succ_x^{J_s} \cap \prec_y^{J_s}|). \qquad (12)$$

The value of the conflict degree for subsets of issues is dependent on a parameter set η^{SC}, η^{WC}, η^{PC}, η^{WI} and η^{SI}. The following is a commonly used normalization method:

Definition 7. *Given a pair of agents $x, y \in A$ and a subset $J \subseteq I$, $CD_J(x, y)$ is the conflict degree for a subset J of issues, the normalized conflict degree for subset J of issues is defined as:*

$$\widehat{CD}_J(x,y) = \frac{CD_J(x,y) - min\{CD_J(x,y) \mid x,y \in A\}}{max\{CD_J(x,y) \mid x,y \in A\} - min\{CD_J(x,y) \mid x,y \in A\}}. \qquad (13)$$

The normalized conflict degree $\widehat{CD}_J(x,y)$ can attain its minimum value $\eta^{SC}|\frac{|J|(|J|+1)}{2}|$ when $|\succ_x^{J_s} \cap \succ_y^{J_s}| + |\succ_x^{J_w} \cap \succ_y^{J_w}| + \frac{|\sim_x^J \cap \sim_y^J|+|J|}{2}$ attains maximum value $\frac{|J|(|J|+1)}{2}$; the normalized conflict degree $\widehat{CD}_J(x,y)$ attains maximum value $\eta^{SI}|\frac{|J|(|J|+1)}{2}|$ when $|\succ_x^{J_s} \cap \prec_y^{J_s}|$ attains maximum value $\frac{|J|(|J|+1)}{2}$. The range of values for the normalized conflict degree $\widehat{CD}_J(x,y)$ is $[0,1]$.

Considering the actual significance of the normalized conflict degree, we can obtain trisections of pairs of agents.

Definition 8. *Given conflict situation $T = (A, I, \mathbb{R})$, the normalized conflict degree $\widehat{CD}_J(x,y)$ is the normalized conflict degree for a subset J of issues, the three types of relationships are defined as follows:*

$$R_J^{=} = \{(x,y) \in A \times A \mid \widehat{CD}_J(x,y) < \lambda\}, \tag{14}$$

$$R_J^{\approx} = \{(x,y) \in A \times A \mid \lambda \le \widehat{CD}_J(x,y) \le \mu\}, \tag{15}$$

$$R_J^{\asymp} = \{(x,y) \in A \times A \mid \widehat{CD}_J(x,y) > \mu\}. \tag{16}$$

Three agent relations $R_J^{=}$, $R_J^{\approx}$, and $R_J^{\asymp}$ are described as alliance relation, neutrality relation, and conflict relation. It is evident that $A \times A$ can be partitioned into these mutually exclusive relations. The thresholds λ and μ take values within the range $[0,1]$.

4 Application of the Model in Rating-Based Situation Tables

The preference relations based on three preference levels can be derived from the three-valued situation tables. The definition of preference relations in the five-valued situation table depends on specific transformation rules. The definition of the five-valued situation table is as follows.

Definition 9. *Suppose A is a non-empty set of agents, I is a non-empty set of issues, $r: A \times I \longrightarrow \{S^+, W^+, N, W^-, S^-\}$ is a rating function when the following conditions are satisfied:*

$$r(a,i) = \begin{cases} S^+, & \text{if } a \text{ is strongly supportive towards } i, \\ W^+, & \text{if } a \text{ is weakly supportive towards } i, \\ N, & \text{if } a \text{ is neutral towards } i, \\ W^-, & \text{if } a \text{ is weakly opposed towards } i, \\ S^-, & \text{if } a \text{ is strongly opposed towards } i. \end{cases} \tag{17}$$

The five values of rating decrease sequentially from top to bottom. Moreover, the five values can be viewed as five distinct levels. Furthermore, we can define a hierarchical function based on five-valued situation table.

Definition 10. *Suppose A is a non-empty set of agents, I is a non-empty set of issues, $r(a,i)$ represents the five possible attitudes of agent a toward an issue i. $L: \{S^+, W^+, N, W^-, S^-\} \longrightarrow \{5,4,3,2,1\}$ is a hierarchical function when the following conditions are satisfied:*

$$L(S^+) = 5, L(W^+) = 4, L(N) = 3, L(W^-) = 2, L(S^-) = 1. \tag{18}$$

For example, the difference between S^+ and W^+ is one level, and the difference between S^+ and W^- is three levels. Following the definition of the five-valued situation table, a preference relation based on the five-valued situation table can be defined as follows. For an agent $a \in A$ and issues $i, j \in J \subseteq I$,

$$\succ_a^{J_s} = \{(i,j) \in J \times J \mid |L(r(a,i)) - L(r(a,j))| = 3 \text{ or } 4\}, \tag{19}$$

$$\succ_a^{J_w} = \{(i,j) \in J \times J \mid |L(r(a,i)) - L(r(a,j))| = 1 \text{ or } 2\}, \tag{20}$$

$$\sim_a^{J} = \{(i,j) \in J \times J \mid |L(r(a,i)) - L(r(a,j))| = 0\}, \tag{21}$$

$$\prec_a^{J_w} = \{(j,i) \in J \times J \mid |L(r(a,i)) - L(r(a,j))| = 1 \text{ or } 2\}, \tag{22}$$

$$\prec_a^{J_s} = \{(j,i) \in J \times J \mid |L(r(a,i)) - L(r(a,j))| = 3 \text{ or } 4\}. \tag{23}$$

Based on the basic concepts of three-way conflict analysis and preference, the five types of preference relations when the preference takes five levels can be derived.

Theorem 2. *Given conflict situation $T = (A, I, \mathbb{R})$, the issues J can be divided into five mutually exclusive parts for $a \in A$ and $J \subseteq I$:*

$$J_a^\diamond = \{\, i \in J \mid r(a,i) = \diamond\}. \tag{24}$$

Among them, $\diamond \in \{S^+, W^+, N, W^-, S^-\}$. Strong preference relation, weak preference relation, indifference relation, converse weak preference relation and converse strong preference relation based on the five-valued situation table when the preference takes five levels are sequentially defined as:

$$\succ_a^{J_s} = J_a^{S^+} \times J_a^{S^-} \cup J_a^{S^+} \times J_a^{W^-} \cup J_a^{W^+} \times J_a^{S^-}, \tag{25}$$

$$\succ_a^{J_w} = J_a^{S^+} \times J_a^{N} \cup J_a^{W^+} \times J_a^{W^-} \cup J_a^{N} \times J_a^{S^-} \cup J_a^{S^+} \times J_a^{W^+} \cup$$
$$J_a^{W^+} \times J_a^{N} \cup J_a^{N} \times J_a^{W^-} \cup J_a^{W^-} \times J_a^{S^-}, \tag{26}$$

$$\sim_a^{J} = J_a^{S^+} \times J_a^{S^+} \cup J_a^{W^+} \times J_a^{W^+} \cup J_a^{N} \times J_a^{N} \cup J_a^{W^-} \times J_a^{W^-} \cup$$
$$J_a^{S^-} \times J_a^{S^-}, \tag{27}$$

$$\prec_a^{J_w} = J_a^{N} \times J_a^{S^+} \cup J_a^{W^-} \times J_a^{W^+} \cup J_a^{S^-} \times J_a^{N} \cup J_a^{W^+} \times J_a^{S^+} \cup$$
$$J_a^{N} \times J_a^{W^+} \cup J_a^{W^-} \times J_a^{N} \cup J_a^{S^-} \times J_a^{W^-}, \tag{28}$$

$$\prec_a^{J_s} = J_a^{S^-} \times J_a^{S^+} \cup J_a^{W^-} \times J_a^{S^+} \cup J_a^{S^-} \times J_a^{W^+}. \tag{29}$$

Furthermore, given a pair of agents $x, y \in A$ and a subset $J \subseteq I$, the number of unordered pairs corresponding to different scenarios based on the five-valued rating is defined as follows:

$$| \succ_x^{J_s} \cap \succ_y^{J_s} | = |(J_x^{S+} \times J_x^{S-} \cup J_x^{S+} \times J_x^{W-} \cup J_x^{W+} \times J_x^{S-}) \cap (J_y^{S+} \times J_y^{S-} \cup$$
$$J_y^{S+} \times J_y^{W-} \cup J_y^{W+} \times J_y^{S-})|$$
$$= |J_x^{S+} \cap J_y^{S+}| \times |J_x^{S-} \cap J_y^{S-}| + |J_x^{S+} \cap J_y^{S+}| \times |J_x^{S-} \cap J_y^{W-}| +$$
$$|J_x^{S+} \cap J_y^{W+}| \times |J_x^{S-} \cap J_y^{S-}| + |J_x^{S+} \cap J_y^{S+}| \times |J_x^{W-} \cap J_y^{S-}| +$$
$$|J_x^{S+} \cap J_y^{S+}| \times |J_x^{W-} \cap J_y^{W-}| + |J_x^{S+} \cap J_y^{W+}| \times |J_x^{W-} \cap J_y^{S-}| +$$
$$|J_x^{W+} \cap J_y^{S+}| \times |J_x^{S-} \cap J_y^{S-}| + |J_x^{W+} \cap J_y^{S+}| \times |J_x^{S-} \cap J_y^{W-}| +$$
$$|J_x^{W+} \cap J_y^{W+}| \times |J_x^{S-} \cap J_y^{S-}|. \tag{30}$$

Table 2. The five-valued situation table for three-way conflict analysis.

A	I							
	i_1	i_2	i_3	i_4	i_5	i_6	i_7	i_8
a_1	S^+	S^-	N	W^-	S^+	S^-	N	S^-
a_2	N	W^+	S^-	N	N	S^+	S^-	S^+
a_3	S^-	W^-	S^+	S^-	W^-	W^+	S^+	W^+
a_4	N	N	S^-	W^+	S^+	S^-	S^-	S^+
a_5	S^+	S^+	S^-	W^+	S^+	S^-	N	S^-

Additionally, the following calculation of the number of unordered pairs corresponding to different scenarios based on the five-valued rating is similar to the previous formula, it is not listed in detail again. The following five-valued situation table will be used for three-way conflict analysis, incorporating the theoretical framework discussed earlier.

Example 2. According to Table 2, by taking $J = I$ and using agent a_1 as an example, we can divide issues J into the following three sets:

$$J_{a_1}^{S+} = \{i_1, i_5\}, J_{a_1}^{W+} = \phi, J_{a_1}^N = \{i_3, i_7\}, J_{a_1}^{W-} = \{i_4\}, J_{a_1}^{S-} = \{i_2, i_6, i_8\}.$$

Moreover, the five preference relations for a_1 are as follows:

$$\succ_{a_1}^{J_s} = J_{a_1}^{S+} \times J_{a_1}^{S-} \cup J_{a_1}^{S+} \times J_{a_1}^{W-} \cup J_{a_1}^{W+} \times J_{a_1}^{S-}$$
$$= \{(i_1, i_2), (i_1, i_6), (i_1, i_8), (i_5, i_2), (i_5, i_6), (i_5, i_8), (i_1, i_4), (i_5, i_4)\},$$
$$\succ_{a_1}^{J_w} = J_{a_1}^{S+} \times J_{a_1}^N \cup J_{a_1}^{W+} \times J_{a_1}^{W-} \cup J_{a_1}^N \times J_{a_1}^{S-} \cup J_{a_1}^{S+} \times J_{a_1}^{W+} \cup J_{a_1}^{W+} \times J_{a_1}^N \cup$$
$$J_{a_1}^N \times J_{a_1}^{W-} \cup J_{a_1}^{W-} \times J_{a_1}^{S-}$$
$$= \{(i_1, i_3), (i_1, i_7), (i_5, i_3), (i_5, i_7), (i_3, i_2), (i_3, i_6), (i_3, i_8), (i_7, i_2), (i_7, i_6),$$
$$(i_7, i_8), (i_3, i_4), (i_7, i_4), (i_4, i_2), (i_4, i_6), (i_4, i_8)\},$$
$$\sim_{a_1}^J = J_{a_1}^{S+} \times J_{a_1}^{S+} \cup J_{a_1}^{W+} \times J_{a_1}^{W+} \cup J_{a_1}^N \times J_{a_1}^N \cup J_{a_1}^{W-} \times J_{a_1}^{W-} \cup J_{a_1}^{S-} \times J_{a_1}^{S-}$$

$$= \{(i_1, i_1), (i_1, i_5), (i_5, i_1), (i_5, i_5), (i_3, i_3), (i_3, i_7), (i_7, i_3), (i_7, i_7), (i_4, i_4),$$
$$(i_2, i_2), (i_2, i_6), (i_2, i_8), (i_6, i_2), (i_6, i_6), (i_6, i_8), (i_8, i_2), (i_8, i_6), (i_8, i_8)\},$$
$$\prec_{a_1}^{J_w} = J_{a_1}^N \times J_{a_1}^{S^+} \cup J_{a_1}^{W^-} \times J_{a_1}^{W^+} \cup J_{a_1}^{S^-} \times J_{a_1}^N \cup J_{a_1}^{W^+} \times J_{a_1}^{S^+} \cup J_{a_1}^N \times J_{a_1}^{W^+} \cup$$
$$J_{a_1}^{W^-} \times J_{a_1}^N \cup J_{a_1}^{S^-} \times J_{a_1}^{W^-}$$
$$= \{(i_3, i_1), (i_7, i_1), (i_3, i_5), (i_7, i_5), (i_2, i_3), (i_6, i_3), (i_8, i_3), (i_2, i_7), (i_6, i_7),$$
$$(i_8, i_7), (i_4, i_3), (i_4, i_7), (i_2, i_4), (i_6, i_4), (i_8, i_4)\},$$
$$\prec_{a_1}^{J_s} = J_{a_1}^{S^-} \times J_{a_1}^{S^+} \cup J_{a_1}^{W^-} \times J_{a_1}^{S^+} \cup J_{a_1}^{S^-} \times J_{a_1}^{W^+}$$
$$= \{(i_2, i_1), (i_6, i_1), (i_8, i_1), (i_2, i_5), (i_6, i_5), (i_8, i_5), (i_4, i_1), (i_4, i_5)\}.$$

Furthermore, let us use agents a_1 and a_2 as an example to calculate the conflict degree between the two agents. We can divide issues J into the following three sets with respect to a_2:

$$J_{a_2}^{S^+} = \{i_6, i_8\}, J_{a_2}^{W^+} = \{i_2\}, J_{a_2}^N = \{i_1, i_4, i_5\}, J_{a_2}^{W^-} = \phi, J_{a_2}^{S^-} = \{i_3, i_7\}.$$

Let $|J_{a_1}^\star \cap J_{a_2}^\ast|$ be represented as $u_{a_1 a_2}^{\star\ast}$. The cardinality of the intersection pairs is calculated sequentially as:

$$u_{a_1 a_2}^{S^+ S^+} = 0, \ u_{a_1 a_2}^{S^+ W^+} = 0, \ u_{a_1 a_2}^{S^+ N} = 2, \ u_{a_1 a_2}^{S^+ W^-} = 0, \ u_{a_1 a_2}^{S^+ S^-} = 0;$$
$$u_{a_1 a_2}^{W^+ S^+} = 0, \ u_{a_1 a_2}^{W^+ W^+} = 0, \ u_{a_1 a_2}^{W^+ N} = 0, \ u_{a_1 a_2}^{W^+ W^-} = 0, \ u_{a_1 a_2}^{W^+ S^-} = 0;$$
$$u_{a_1 a_2}^{N S^+} = 0, \ u_{a_1 a_2}^{N W^+} = 0, \ u_{a_1 a_2}^{N N} = 0, \ u_{a_1 a_2}^{N W^-} = 0, \ u_{a_1 a_2}^{N S^-} = 2;$$
$$u_{a_1 a_2}^{W^- S^+} = 0, \ u_{a_1 a_2}^{W^- W^+} = 0, \ u_{a_1 a_2}^{W^- N} = 1, \ u_{a_1 a_2}^{W^- W^-} = 0, \ u_{a_1 a_2}^{W^- S^-} = 0;$$
$$u_{a_1 a_2}^{S^- S^+} = 2, \ u_{a_1 a_2}^{S^- W^+} = 1, \ u_{a_1 a_2}^{S^- N} = 0, \ u_{a_1 a_2}^{S^- W^-} = 0, \ u_{a_1 a_2}^{S^- S^-} = 0.$$

The number of unordered pairs corresponding to different scenarios are calculated as follows:

$$|\succ_{a_1}^{J_s} \cap \succ_{a_2}^{J_s}| = 0, |\succ_{a_1}^{J_w} \cap \succ_{a_2}^{J_w}| = 4, |\sim_{a_1}^J \cap \sim_{a_2}^J| = 22, |\succ_{a_1}^{J_s} \cap \succ_{a_2}^{J_w}| = 0,$$
$$|\succ_{a_1}^{J_w} \cap \sim_{a_2}^J| = 11, |\succ_{a_1}^J \cap \succ_{a_2}^{J_s}| = 0, |\sim_{a_1}^J \cap \succ_{a_2}^{J_w}| = 13, |\succ_{a_1}^{J_s} \cap \sim_{a_2}^J| = 8,$$
$$|\succ_{a_1}^{J_w} \cap \prec_{a_2}^{J_w}| = 5, |\sim_{a_1}^J \cap \prec_{a_2}^{J_s}|) = 0, |\succ_{a_1}^{J_s} \cap \prec_{a_2}^{J_w}| = 6, |\succ_{a_1}^{J_w} \cap \prec_{a_2}^{J_s}| = 6,$$
$$|\succ_{a_1}^{J_s} \cap \prec_{a_2}^{J_s}| = 0.$$

The conflict degree between a_1 and a_2 is:

$$CD_J(a_1, a_2) = \eta^{SC} \times (0 + 4 + \frac{22 + 8}{2}) + \eta^{WC} \times (0 + 11 + 0 + 13) +$$
$$\eta^{PC} \times (8 + 5 + 0) + \eta^{WI} \times (6 + 6) + \eta^{SI} \times 0$$
$$= 19\eta^{SC} + 24\eta^{WC} + 13\eta^{PC} + 12\eta^{WI}.$$

The maximum and minimum values of $CD_J(a_1, a_2)$ are $36\eta^{SI}$ and $36\eta^{SC}$. The normalized for subset J of issues is calculated as:

$$\widehat{CD}_J(a_1, a_2) = \frac{CD_J(a_1, a_2) - min\{CD_J(a_1, a_2) \mid a_1, a_2 \in A\}}{max\{CD_J(a_1, a_2) \mid a_1, a_2 \in A\} - min\{CD_J(a_1, a_2) \mid a_1, a_2 \in A\}}$$

$$= \frac{-17\eta^{SC} + 24\eta^{WC} + 13\eta^{PC} + 12\eta^{WI}}{36\eta^{SI} - 36\eta^{SC}}.$$

We can compute the normalized conflict degree between multiple pairs of agents, as shown in the table. Given a pair of thresholds (λ, μ), we can determine the three-way agent relations.

Table 3. The normalized conflict degree $\widehat{CD}_J$ expressed with $\eta^{SC} = -2$, $\eta^{WC} = -1$, $\eta^{PC} = 1$, $\eta^{WI} = 2$ and $\eta^{SI} = 3$.

A	A				
	a_1	a_2	a_3	a_4	a_5
a_1	0	0.6611	0.2611	0.1611	0.0944
a_2		0	0.2444	0.05	0.1944
a_3			0	0.3278	0.4
a_4				0	0.1222
a_5					0

Table 4. Agent relations expressed with $\eta^{SC} = -2$, $\eta^{WC} = -1$, $\eta^{PC} = 1$, $\eta^{WI} = 2$, $\eta^{SI} = 3$, $\lambda = 0.15$ and $\mu = 0.3$.

A	A				
	a_1	a_2	a_3	a_4	a_5
a_1	$R_J^=$	$R_J^{\gtrless}$	$R_J^{\approx}$	$R_J^{\approx}$	$R_J^=$
a_2		$R_J^=$	$R_J^{\approx}$	$R_J^=$	$R_J^=$
a_3			$R_J^=$	$R_J^{\gtrless}$	$R_J^{\gtrless}$
a_4				$R_J^=$	$R_J^=$
a_5					$R_J^=$

We use a small example to validate the effectiveness of the trisection of agent relations. According to Table 3 and Table 4, we can draw some conclusions. For example, agent a_1 and agent a_2 have strong disagreements on some issues, agent a_4 and agent a_2 have strong agreements on some issues. In subsequent research, we will apply this model to large-scale multi-valued situation tables and subsequently propose relevant algorithms for calculating conflict degrees.

5   Conclusion

In this paper, we propose a three-way conflict analysis model based on five level preferences, which addresses the limitations of a three-way conflict analysis model based on three level preferences in characterizing agents' preferences.

Firstly, we define five types of preference relations based on five level preferences, and define a conflict function for subsets of issues. Next, we propose the conflict degree for subsets of issues is defined by determining the number of unordered pairs corresponding to each scenario. Finally, the conflict degree is standardized to establish the three-way agent relationships of alliance, conflict, and neutrality.

In the future, we will further explore how to determine the optimal threshold pairs for trisections. Furthermore, we can further subdivide five level preferences and expand them to $2k + 1$ level preferences. In addition, we can draw an analogy to the study of conflict relationships between agents and analyze the issue relationships in preference-based conflict situations.

Acknowledgements. This work is supported by the National Natural Science Foundation of China (No. 62076040), Hunan Provincial Natural Science Foundation of China (No. 2020JJ3034), the Scientific Research Fund of Hunan Provincial Education Department (No. 22A0233), the Scientific Research Fund of Chongqing Key Laboratory of Computational Intelligence (No. 2020FF04), the Graduate Research Innovation Project of Hunan Province (No. CX20220952).

References

1. Yao, Y.Y.: Three-way decision with probabilistic rough sets. Inf. Sci. **180**, 341–353 (2010)
2. Yao, Y.Y.: Three-way decision and granular computing. Int. J. Approx. Reason. **103**, 107–123 (2018)
3. Yang, J., et al.: 3WC GBNRS++: a novel three-way classifier with granular-ball neighborhood rough sets based on uncertainty. IEEE Trans. Fuzzy Syst. **32**(8), 4376–4387 (2024)
4. Li, X.N., Wang, X., Sun, B.Z., She, Y.H., Zhao, L.: Three-way decision on information tables. Inf. Sci. **545**, 25–43 (2021)
5. Yue, X.D., Liu, S.W., Qian, Q., Miao, D.Q., Gao, C.: Semi-supervised shadowed sets for three-way classification on partial labeled data. Inf. Sci. **607**, 1372–1390 (2022)
6. Luo, J.F., Hu, M.J., Lang, G.M., Yang, X., Qin, K.Y.: Three-way conflict analysis based on alliance and conflict functions. Inf. Sci. **594**, 322–359 (2022)
7. Suo, L., Yang, H., Li, Q.Y., Yang, H.L., Yao, Y.Y.: A review of three-way decision: triadic understanding, organization, and perspectives. Int. J. Approx. Reason. **173**, 109268 (2024)
8. Xu, W.Y., Jia, B., Li, X.N.: A two-universe model of three-way decision with ranking and reference tuple. Inf. Sci. **581**, 808–839 (2021)
9. Xu, W.Y., Jia, B., Li, X.N.: A generalized model of three-way decision with ranking and reference tuple. Int. J. Approx. Reason. **144**, 51–68 (2022)
10. Lang, G.: A general conflict analysis model based on three-way decision. Int. J. Mach. Learn. Cybern. **11**(5), 1083–1094 (2020). https://doi.org/10.1007/s13042-020-01100-y
11. Lang, G.M., Miao, D.Q., Cai, M.J.: Three-way decision approaches to conflict analysis using decision-theoretic rough set theory. Inf. Sci. **406–407**, 185–207 (2017)
12. Li, X.N., Yan, Y.C.: A dynamic three-way conflict analysis model with adaptive thresholds. Inf. Syst. **657**, 119999 (2024)

13. Sun, B.Z., Chen, X.T., Zhang, L.Y., Ma, W.M.: Three-way decision making approach to conflict analysis and resolution using probabilistic rough set over two universes. Inf. Sci. **807**, 809–822 (2020)
14. Yang, H.L., Wang, Y., Guo, Z.L.: Three-way conflict analysis based on hybrid situation tables. Inf. Sci. **628**, 522–541 (2023)
15. Suo, L., Yang, H.L.: Three-way conflict analysis based on incomplete situation tables: a tentative study. Int. J. Approx. Reason. **145**, 51–74 (2022)
16. Ren, R.S., Qi, J.J., Wei, L., Wei, X.S.: Tri-level conflict analysis from the angle of three-valued concept analysis. Inf. Sci. **662**, 120284 (2024)
17. Pawlak, Z.: On conflicts. Int. J. Man-Mach. Stud. **21**, 127–134 (1984)
18. Pawlak, Z.: Some remarks on conflict analysis. Eur. J. Oper. Res. **166**, 649–654 (2005)
19. Yao, Y.Y.: Measuring information retrieval effectiveness based on user preferences on documents. J. Am. Soc. Inf. Sci. **46**, 133–145 (1995)
20. Yao, Y.Y.: Three-way conflict analysis: reformulations and extensions of the Pawlak model. Knowl. Based Syst. **180**, 26–37 (2019)
21. Lang, G.M., Yao, Y.Y.: New measures of alliance and conflict for three-way conflict analysis. Int. J. Approx. Reason. **132**, 49–69 (2021)
22. Hu, M.J.: Three-way conflict analysis with preference-based conflict situations. Inf. Sci. **693**, 121676 (2025)

Three-Way Causal Structure Representation and Learning for Uncertainty

Chenglin Zhang[1], Hong Yu[1(✉)], and Guoyin Wang[1,2(✉)]

[1] Chongqing Key Laboratory of Computational Intelligence, Chongqing University of Posts and Telecommunications, Chongqing 400065, People's Republic of China
d210201035@stu.cqupt.edu.cn, yuhong@cqupt.edu.cn
[2] National Center for Applied Mathematics in Chongqing, Chongqing Normal University, Chongqing 401331, China
wanggy@cqnu.edu.cn

Abstract. Causal relationships play a crucial role in enhancing decision-making processes and optimizing system design. In real-world scenarios, different tasks introduce uncertainty in optimization objectives due to their varying requirements. Some tasks aim to discover as many causal relationships as possible, while others require highly precise causal relationships to make reliable decisions. Existing methods are confined to learning a fixed causal structure, thereby failing to adapt to these diverse requirements. To address this challenge, we propose a novel framework called Three-Way Causal Structure Representation (TW-CSR). First, we represent the framework of TW-CSR as (DCE, UCE, IN). Then, in two propositions, we discuss the relationship between precision and recall in causal structures, providing a theoretical foundation for selecting an appropriate causal graph structure to meet task requirements. Finally, based on the propositions, we propose causal structure learning methods under the TW-CSR framework for both constraint-based and score-based approaches. Experimental results demonstrate that our framework and methods effectively address the uncertainty introduced by diverse tasks.

Keywords: Uncertainty · Three-way decision theory · Causal structure

1 Introduction

Learning causal relationships is considered fundamental to artificial intelligence [5,12,13] and plays a crucial role in fields such as healthcare [3,9] and finance [7]. Researchers can learn causality through interventions and manipulations in randomized experiments. However, such approaches are often costly and may even violate ethical principles. As a result, there has been a growing focus on causal discovery from observational data, prompting the emergence of diverse causal structure learning methods.

In real-world tasks, the need for causal structures varies based on the application scenario, as shown in Fig. 1. Taking drug development as an example, researchers aim to discover as many causal relationships as possible between

© The Author(s), under exclusive license to Springer Nature Switzerland AG 2025
Q. Zhang et al. (Eds.): IJCRS 2025, LNAI 15710, pp. 127–139, 2025.
https://doi.org/10.1007/978-3-031-92741-6_10

drugs and potential side effects, even if some of these relationships may be incorrect. In contrast, in epilepsy source tracing causal analysis, precisely locating the source of abnormal discharges is critical, as it directly impacts the determination of surgical resection areas, requiring very high precision [4,15]. Obviously, the fixed causal structure learned from existing methods fails to satisfy the requirements of different tasks. Hence, a new causal structure representation framework is needed, one that can dynamically balance recall rate and precision based on the task characteristics, thus fulfilling the varied needs of different applications.

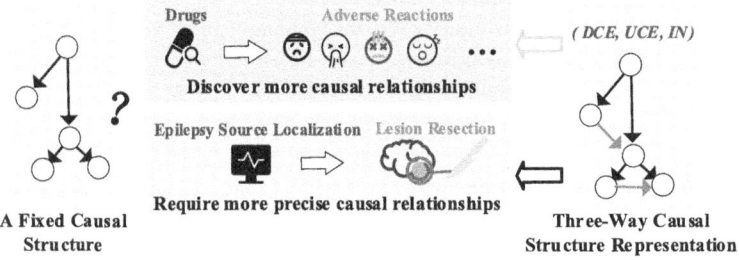

Fig. 1. Limitations of existing methods: Fixed structures fail to handle uncertainty in tasks

Inspired by three-way decision theory, we introduce a framework of TW-CSR, expressed as (DCE, UCE, IN). DCE represents definite causal edges, UCE represents uncertain causal edges, and IN represents isolated nodes. Subsequently, we propose novel causal structure learning methods under the TW-CSR framework for both constraint-based and score-based causal structure learning methods. These methods perform well in addressing the diverse needs of different tasks. Our key contributions are as follows.

- As far we know, it is the first time to introduce three-way decision theory into causal structure representation and learning, proposing a novel causal structure representation framework to address the uncertainty arising from the diverse demands of different tasks in causal structure learning.
- We discuss the relationship between the precision and recall of causal structures, and find that as new causal edges are learned, the recall of the structure either remains unchanged or improves. In addition, we investigate the conditions under which precision either improves or deteriorates.
- We propose novel methods under the TW-CSR framework for both constraint-based and score-based causal structure learning. Experimental results validate its superior performance in precision and recall.

2   Related Work

2.1   Causal Structure Representation and Learning

Causal structure learning methods can be classified into three categories: constraint-based methods, score-based methods, and hybrid methods.

(1) **Constraint-based methods:** Constraint-based methods learn causal structures through statistical tests or independence tests. The PC algorithm [17] constructs the skeleton of a causal graph by pruning conditionally independent edges and identifies edge directions using V-structures and orientation rules. However, the algorithm requires the assumption of causal sufficiency. To address the problem of unobserved hidden variables, researchers have proposed algorithms like FCI [16] and CD-NOD [6]. These methods label edges where the causal direction is uncertain, effectively capturing uncertainty. However, this labeling does not provide clear causal directions, thus failing to improve the graph's representation and not satisfying the high recall requirement. Additionally, Shimizu et al. analyzed data under the assumption of linear non-Gaussianity and introduced functional causal models such as DirectLiNGAM [14] and ParceLiNGAM [18].

(2) **Score-based methods:** Score-based methods primarily define a scoring function and employ heuristic search or deep learning to optimize for the best-scoring structure. Classical algorithms such as GES and Hill-Climbing employ heuristic search or greedy algorithms to find global or local optimal structures [2]. With the advantages of deep learning in optimization, NOTEARS [25, 26] transforms DAG constraints into continuous optimization problems, efficiently handling observed data via gradient descent.

(3) **Hybrid methods:** Hybrid causal structure learning methods integrate constraint-based and score-based methods to balance efficiency and accuracy in inferring causal relationships. Tsirlis et al. first employed conditional independence (CI) tests to reduce the search space and subsequently applied a greedy search algorithm to determine the final causal structure. [19]. Li et al. proposed a novel causal structure learning method based on a mixed structural equation model, leveraging a locally consistent score and an efficient conditional independence test for mixed-type data [8].

Although the methods mentioned above have distinct advantages, they all use fixed graphs to represent causal structures, failing to meet the varying requirements of real-world tasks. In this paper, we propose the TW-CSR framework and causal learning methods that build on it, enhancing causal structure representation by incorporating uncertain causal edges and learning adaptive structures to meet diverse task requirements. It is important to emphasize that this paper focuses exclusively on the integration of the TW-CSR framework into constraint-based and score-based methods, with an emphasis on improvement strategies.

2.2   Three-Way Decision Theory

Three-way decision theory, initially proposed by Yao [20] based on rough set theory [10, 11], provides a methodology and computational paradigm for decision making. In practical decision processes, it interprets the positive, negative and boundary regions of rough sets as acceptance, rejection and delay, respectively. Beyond the narrowly defined three-way decision, the theory also includes the broader three-way decision, such as the Perception-Cognition-Action (PCA) trilevel conceptual model [22] and the Trisecting-Acting-Outcome (Tao) model [21]. This theory and its models have been widely applied in fields such as clustering [23], time series forecasting [24], and anomaly detection [1]. These theories, methods, and applications introduce a third option, effectively reducing decision-making risks when dealing with uncertain information. Motivated by this, we introduce three-way decision theory into causal structure representation and propose a novel method for causal structure learning. In the proposed three-way causal representation, inter-variable relations are not represented as merely existing or non-existing. Rather, their relations are partitioned into three categories: definite causal edges, uncertain causal edges, and definitely non-existent causal edge. To our knowledge, this is the first time that three-way decision theory has been introduced into causal structure learning.

3   Methods

3.1   Preliminaries

Causal structure learning aims to learn a causal graph $G = (V, E)$ from a given observational dataset $D = \{X^{(i)}\}_{i=1}^{d}$, where $X^{(i)}$ represents the observed values of the i-th variable. The graph's nodes V represent the feature variables of the dataset, with $|V| = d$. The edge $\langle v_i, v_j \rangle \in E$ in the causal graph indicates a causal relationship between $v_i \in V$ and $v_j \in V$.

The methods and frameworks in this paper satisfy the following assumptions:

- The causal structure G is a directed acyclic graph.
- Causal faithfulness assumption: It states that all conditional independence relationships observed in the data are exclusively due to the causal graph's structure (e.g., d-separation), with no spurious independences arising from hidden variables or coincidental parameter values.
- Causal Markov assumption: It asserts that a variable is conditionally independent of its non-descendants in the causal graph when conditioned on its direct causes (parents), thereby linking causal structure to probabilistic independence.

3.2   Framework of Three-Way Causal Structure Representation

Existing causal structure learning methods infer a fixed graph from the data, but this uniform representation lacks the flexibility needed for varying tasks. Inspired by the three-way decision theory's Acceptance-Rejection-Delay paradigm, we introduce uncertain causal edges as a complement to the traditional binary representation of causal relationships. This addition enhances the expressiveness of causal graphs. Therefore, we partition the causal edges into three categories: (1) definite causal edges, (2) uncertain causal edges, and (3) definitely non-existent causal edges. Based on this partition, we represent the three-way causal structure representation as follows:

$$(DCE, UCE, IN). \tag{1}$$

Here, DCE represents definite causal edges, and UCE represents uncertain causal edges. Note that IN represents isolated nodes rather than definitely non-existent causal edges. In the real world, causal structures are sparse, leading to a large number of non-existent causal edges. Therefore, we do not need to explicitly represent this set. However, we can indirectly infer non-existent causal edges from the sets DCE, UCE, and IN.

Let $V(E_k) = \{v \mid \exists v_i, \langle v, v_i \rangle \in E_k \text{ or } \langle v_i, v \rangle \in E_k\}$ represents the set of all nodes in an arbitrary edge set E_k. The three-way causal structure representation (DCE, UCE, IN) has the following properties:

$$V(DCE) \cap IN = \emptyset \quad \text{and} \quad V(UCE) \cap IN = \emptyset$$
$$V(DCE) \cup V(UCE) \cup IN = V$$
$$DCE \cap UCE = \emptyset \tag{2}$$

In fact, the three-way causal structure remains a graph representation. As illustrated in Fig. 2, a tuple $(V(DCE) + IN, DCE)$ corresponds to an initial graph G_0, where causal edges are fully determined. Starting from G_0, we iteratively introduce uncertain causal edges to enrich the graph representation, generating additional graph structures. This augmentation process continues until all uncertain causal edges have been integrated, ultimately resulting in the final graph structure G_1, represented as $(V(DCE) + V(UCE) + IN, DCE + UCE)$.

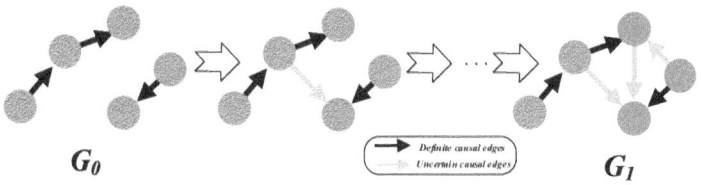

G_0 Definite causal edges / Uncertain causal edges G_1

Fig. 2. Illustration of the three-way causal structure representations.

3.3   Discussion on the Precision and Recall of Causal Structures

To better understand the suitability of G_0 and G_1 for different tasks, we first explore the relationship between precision and recall in causal structures. The calculations of precision and recall are as follows:

$$Precision = \frac{TP}{TP + FP}, \quad Recall = \frac{TP}{TP + FN}, \tag{3}$$

where TP represents the number of positive samples correctly predicted, FP represents the number of negative samples incorrectly predicted, and FN represents the number of negative samples incorrectly predicted.

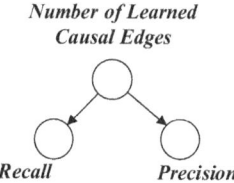

Fig. 3. The causal mechanism between *Precision* and *Recall*

It is frequently observed that precision declines while recall improves. Therefore, it is necessary to investigate the relationship between the two. From a causal perspective, the mechanism behind this change is shown in Fig. 3. The figure illustrates that variations in the number of learned causal edges act as a causal factor influencing both *Precision* and *Recall*. The inherent trade-off between these two metrics is mediated by changes in the causal edge count. We state the following propositions to describe this mechanism in detail.

Proposition 1. *Given that the causal edges n are inferred, with each edge being correct with an independent probability p, the precision of the learned causal structure follows two patterns: (1) when $p < \frac{TP}{TP+FP}$, Precision decreases; (2) when $p > \frac{TP}{TP+FP}$, Precision improves.*

Proof. We provide corresponding proofs separately in terms of precision and recall.

If one additional edge is learned, the precision of the learned new structure is represented by $(TP+1)/(TP+FP+1)$ with probability p, and by $(TP)/(TP+FP+1)$ with probability $1-p$. Therefore, the precision is updated to $p \times (TP+$

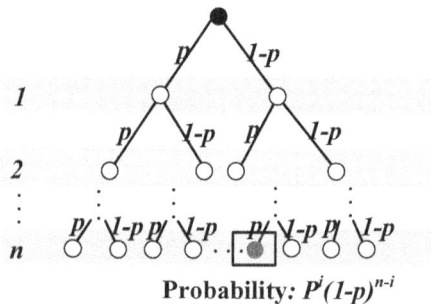

Probability: $P^i(1-p)^{n-i}$

Fig. 4. Probabilistic pathways for adding n newly learned causal edges

$1)/(TP + FP + 1) + (1 - p) \times (TP)/(TP + FP + 1)$. As shown in Fig. 4, after learning n edges, the precision is represented by $Precision'$:

$$Precision' = \frac{\sum_{i=0}^{n} \binom{n}{i} p^i (1-p)^{n-i}(TP+i)}{TP+FP+n}$$

$$= \frac{TP\sum_{i=0}^{n} \binom{i}{n} p^i (1-p)^{n-i} + \sum_{i=0}^{n} \binom{i}{n} p^i (1-p)^{n-i} i}{TP+FP+n}. \tag{4}$$

Define $S_1 = TP\sum_{i=0}^{n} \binom{i}{n} p^i (1-p)^{n-i}$ and $S_2 = \sum_{i=0}^{n} \binom{i}{n} p^i (1-p)^{n-i} i$. Using the binomial theorem, S_1 simplifies as follows:

$$S_1 = TP \times (p + 1 - p)^n = TP \times 1 = TP \tag{5}$$

For S_2, it can be viewed as the expected value $E[X]$ of a random variable $X \sim \text{Binomial}(n, p)$:

$$S_2 = np \tag{6}$$

Based on Eq. (5) and Eq. (6), we can derive Eq. (7) as follows:

$$Precision' = \frac{S_1 + S_2}{TP + FP + n} = \frac{TP + np}{TP + FP + n} \tag{7}$$

Consequently, the variation in precision can be formulated as $f(p)$:

$$f(p) = Precision' - Precision = \frac{TP + np}{TP + FP + n} - \frac{TP}{TP + FP}$$

$$= \frac{(nTP + nFP)p - nTP}{(TP + FP + n)(TP + FP)}. \tag{8}$$

Let $a = (TP + FP + n)(TP + FP)$ and $f'(p) = (nTP + nFP)p - nTP$, where $a > 0$. Since $nTP + nFP > 0$, $f'(p)$ is a monotonically increasing linear function of p in the interval $[0, 1]$, as shown in Fig. 5.

The Fig. 5 clearly shows that $f'(p)$ takes negative values ($f'(p) < 0$) in the interval $[0, \frac{TP}{TP+FP})$, which means $f(p) < 0$. Meanwhile, $f'(p)$ assumes non-negative values ($f'(p) \geq 0$) in the interval $[\frac{TP}{TP+FP}, 1]$, which means $f(p) \geq 0$. Hence, the proposition is proven.

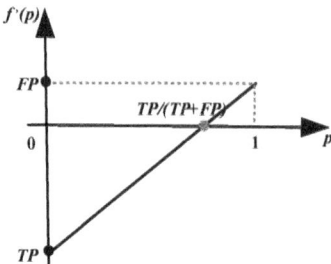

Fig. 5. Graph of the function $f'(p)$ with $f'(p) = 0$ at $p = \frac{TP}{TP+FP}$.

Proposition 2. *Given that the causal edges n are inferred, with each edge being correct with an independent probability p, the recall of the new learned causal structure will remain unchanged or improve.*

Proof. Similar to the previous proof, after the addition of n new edges through learning, the updated recall rate $Recall'$ is expressed as:

$$Recall' = \frac{\sum_{i=0}^{n} \binom{n}{i} p^i (1-p)^{n-i}(TP+i)}{TP+FN}$$
$$= \frac{TP+np}{TP+FN}.$$

$$(9)$$

According to the $Recall = TP/(TP+FN)$, we express the changing recall rate as $g(p)$:

$$g(p) = Recall' - Recall = \frac{TP+np}{TP+FN} - \frac{TP}{TP+FN}$$
$$= \frac{np}{TP+FN}.$$

$$(10)$$

Given that $g(p) \geq 0$ holds for all cases, we can infer that the recall rate will remain unchanged or increase when the newly learned n edges are added. Therefore, we proved our proposition.

From the above proposed propositions, it can be inferred that by adding new causal edges, we can achieve a switch from high precision to high recall requirements. The graph G_1 is constructed by incrementally adding causal edges to G_0. Consequently, the recall of G_1 will be no less than that of G_0, while its precision may decrease, especially when the precision of G_0 is relatively high. Therefore, the tuple $(V(DCE) + IN, DCE)$ representing G_0 is suited for tasks requiring high precision, while the tuple $(V(DCE) + V(UCE) + IN, DCE + UCE)$ representing G_1 is suited for tasks demanding high recall.

3.4   Three-Way Causal Structure Learning Method

Causal structure learning methods include both constraint-based and score-based approaches. We separately discuss these two categories of methods and propose a causal structure learning approach based on the three-way causal structure representation.

In constraint-based methods, conditional independence testing plays a key role. In conditional independence testing, we can use different significance levels to indicate whether variables are conditionally independent, thus controlling the number of causal edges learned. Under strict conditions, the significance level is set to 0.01, while under relaxed conditions, the significance level is set to 0.1. Accordingly, the causal learning method within our framework is given in Algorithm 1.

Algorithm 1. The constraint-based casual structure learning method within the framework of TW-CSR

Input: The observed data D;
Output: (DCE, UCE, IN);
1: Initial $DCE \leftarrow \emptyset$, $UCE \leftarrow \emptyset$ and $IN \leftarrow \emptyset$;
2: When the significance level is fixed at 0.01, the constraint-based method derives the causal graph (V, E_l) from the data D;
3: When the significance level is fixed at 0.1, the constraint-based method derives the causal graph (V, E_u) from the data D;
4: $DCE \leftarrow E_l$, $UCE \leftarrow E_u - E_l$ and $IN \leftarrow V - V(E_l) \cup V(E_u)$;
5: **return** (DCE, UCE, IN);

For score-based causal structure learning methods, we control the number of learned causal edges by adding an L_2 norm constraint to the score function. Without constraints, the maximum number of causal edges learned corresponds to $(V(DCE) + V(UCE) + IN, DCE + UCE)$. With the sparse constraint, the number of causal edges learned is reduced, corresponding to $(V(DCE) + IN, DCE)$.

4   Experiment

In this section, we conducted experiments on synthetic datasets with node sizes of 50, 100, 150, 200. The synthetic data generation process comprehensively includes both linear and nonlinear mechanisms. To validate the effectiveness of the causal structure learning method under the TW-CSR framework, we selected comparative methods, including constraint-based methods such as the PC [17] and FCI [16] algorithm and score-based methods such as NOTEARS [25,26]. The comparison metrics of the mtheds contain two items: Precision P and Recall R. The precision is the percentage of correct causal relationships among learned causal relationships between two feature variables. The recall calculates the percentage of correct causal relationships among true causal relationships between

Algorithm 2. The score-based casual structure learning method within the framework of TW-CSR

Input: The observed data D;
Output: (DCE, UCE, IN);
 1: Initial $DCE \leftarrow \emptyset$, $UCE \leftarrow \emptyset$ and $IN \leftarrow \emptyset$;
 2: The constraint-based method derives the causal graph (V, E) from the data D;
 3: With the sparse constraint, the causal edges of the causal graph (V, E) are discretized, yielding the causal graph (V, E_l);
 4: Without constraints, the causal edges of the causal graph (V, E) are discretized, yielding the causal graph (V, E_u);
 5: $DCE \leftarrow E_l$, $UCE \leftarrow E_u - E_l$ and $IN \leftarrow V - V(E_l) \cup V(E_u)$;
 6: **return** (DCE, UCE, IN)

Table 1. Comparison of methods on linear synthetic data for different node sizes

Method	Nodes = 50		Nodes = 100		Nodes = 150		Nodes = 200	
	P	R	P	R	P	R	P	R
PC [17]	**16.25%**	54.62%	**16.31%**	62.59%	**11.47%**	71.95%	**12.89%**	60.60%
PC (TW-CSR)	13.82%	**67.77%**	12.58%	**73.26%**	11.36%	**82.57%**	11.09%	**71.63%**
FCI [16]	**18.08%**	63.81%	**18.63%**	72.99%	**18.47%**	86.11%	**14.91%**	73.81%
FCI (TW-CSR)	17.76%	**76.15%**	17.89%	**83.34%**	16.93%	**92.78%**	13.47%	**82.88%**
NOTEARS [25]	99.26%	88.66%	95.85%	85.00%	96.69%	84.66%	99.23%	86.16%
NOTEARS (TW-CSR)	**100.00%**	**92.66%**	**100.00%**	**91.33%**	**99.28%**	**89.77%**	**100.00%**	**95.66%**

Table 2. Comparison of methods on nonlinear synthetic data for different node sizes

Method	Nodes = 50		Nodes = 100		Nodes = 150		Nodes = 200	
	P	R	P	R	P	R	P	R
PC [17]	27.51%	71.82%	24.88%	73.27%	22.14%	78.76%	19.24%	79.89%
PC (TW-CSR)	**35.76%**	**79.90%**	**33.04%**	**81.21%**	**29.71%**	**80.73%**	**28.30%**	**84.75%**
FCI [16]	28.85%	83.90%	23.68%	83.55%	27.16%	87.73%	18.76%	88.95%
FCI (TW-CSR)	**34.43%**	**87.19%**	**34.93%**	**89.26%**	**32.22%**	**88.98%**	**29.93%**	**92.49%**
NOTEARS [25]	86.79%	90.00%	87.95%	91.67%	96.69%	84.66%	99.23%	86.16%
NOTEARS (TW-CSR)	**94.57%**	**94.00%**	**94.92%**	**93.00%**	**91.28%**	**87.73%**	**91.33%**	**95.66%**

two feature variables. All experiments were performed on an i7-9700 CPU @ 3.00 GHz with 16 GB of memory.

Tables 1 and 2 show the comparison results on linear and nonlinear synthetic data. Under our framework, constraint-based and score-based causal structure learning methods achieve good performance in both precision and recall, except for some results of the PC and FCI algorithms on linear data. We analyze the reasons for the unsatisfactory results of the PC and FCI algorithms.

Using the causal network with 50 nodes as an example, we plot the changes in precision and recall with the number of causal edges for the PC algorithm on linear and nonlinear data. In Fig. 6(a), we observe that the precision of the

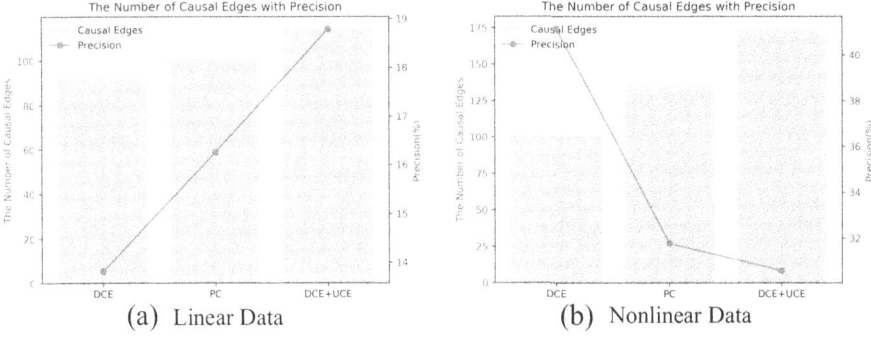

Fig. 6. Relationship between the number of causal edges and precision on linear and nonlinear synthetic data

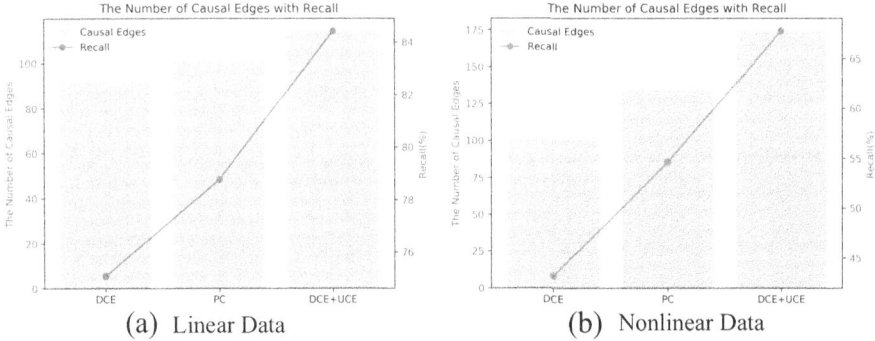

Fig. 7. Relationship between the number of causal edges and recall on linear and nonlinear synthetic data

$(V(DCE) + IN, DCE)$ learned on linear data is only around 10%. According to Proposition 1, If the correctness of newly added causal edges surpasses 10%, the precision of the causal graph will increase considerably. Clearly, this is an easily achievable condition. Ultimately, the structure determined by our method using the tuple $(V(DCE) + IN, DCE)$ has lower precision than the structure obtained by the original method.

However, this situation can be improved by the algorithm's performance. In Fig. 6(b), when the precision of $(V(DCE) + IN, DCE)$ reaches 20% or higher, increasing precision through the addition of uncertain causal edges is not easy. Thus, under our framework, our method achieves the best precision results. For recall, Proposition 2 states that the worst-case scenario is no change, so the graph structure corresponding to $(V(DCE) + V(UCE) + IN, DCE + UCE)$ under our framework achieves ideal recall. The continuous increase in recall in Fig. 7 also supports this conclusion.

5   Conclusion

In this paper, we address the challenge of uncertainty in causal structure learning by proposing a novel framework called three-way causal structure representation (TW-CSR). To the best of our knowledge, this is the first work to introduce the theory of three-way decision into the representation and learning of causal structures. In the framework of TW-CSR, we propose new causal structure methods for both constraint-based and score-based approaches.

However, our proposed three-way causal structure representation framework still requires certain manual configurations. For example, in constraint-based approaches, we empirically set lenient and strict thresholds for conditional independence tests, which generate the graph structures $DCE(V(DCE)+IN, DCE)$ and $(V(DCE)+V(UCE)+IN, DCE+UCE)$, respectively. In score-based methods, L_2 regularization is selected without exploring alternative regularization norms. In the future, we will consider reducing subjective manual configurations by adopting a data-driven approach based on causal identifiability in the data. Additionally, we plan to incorporate the three-way causal structure representation framework into hybrid causal learning methods.

Acknowledgments. This study was supported in part by the National Natural Science Foundation of China (62221005, 62136002, 62233018).

Disclosure of Interests. The authors have no competing interests to declare that are relevant to the content of this article.

References

1. Chen, G.S., et al.: Three-way unsupervised anomaly detection of sequential patterns. Int. J. Mach. Learn. Cybern. (2025). https://doi.org/10.1007/s13042-025-02533-z
2. Chickering, D.M.: Optimal structure identification with greedy search. J. Mach. Learn. Res. **3**, 507–554 (2002)
3. Cross-disorder group of the psychiatric genomics consortium and others: identification of risk loci with shared effects on five major psychiatric disorders: a genome-wide analysis. Lancet **381**(9875), 1371–1379 (2013)
4. Eom, T.H.: Electroencephalography source localization. Clin. Exp. Pediatr. **66**(5), 201 (2022)
5. Glymour, C., Zhang, K., Spirtes, P.: Review of causal discovery methods based on graphical models. Front. Genet. **10**, 1–15 (2019)
6. Huang, B., et al.: Causal discovery from heterogeneous/nonstationary data. J. Mach. Learn. Res. **21**(1), 3482–3534 (2020)
7. Imbens, G.W.: Nonparametric estimation of average treatment effects under exogeneity: a review. Rev. Econ. Stat. **86**(1), 4–29 (2004)
8. Li, Y., Xia, R., Liu, C., Sun, L.: A hybrid causal structure learning algorithm for mixed-type data. In: Proceedings of the AAAI Conference on Artificial Intelligence, pp. 7435–7443 (2022)
9. Mani, S., Cooper, G.F.: Causal discovery from medical textual data. In: Proceedings of the AMIA Symposium, pp. 542–546 (2000)

10. Pawlak, Z.: Rough sets. Int. J. Comput. Inf. Sci. **11**, 341–356 (1982)
11. Pawlak, Z.: Rough Sets: Theoretical Aspects of Reasoning About Data, vol. 9. Springer, Dordrecht (2012)
12. Pearl, J.: Causality. Cambridge University Press, Cambridge (2009)
13. Pearl, J., Mackenzie, D.: The Book of Why: The New Science of Cause and Effect. Basic books (2018)
14. Shimizu, S., et al.: Directlingam: a direct method for learning a linear non-gaussian structural equation model. J. Mach. Learn. Res. **12**(4), 1225–1248 (2011)
15. Spinelli, L., et al.: Semiautomatic interictal electric source localization based on long-term electroencephalographic monitoring: a prospective study. Epilepsia **64**(4), 951–961 (2023)
16. Spirtes, P., Meek, C., Richardson, T.: Causal discovery from medical textual data. In: Proceedings of the Eleventh Conference on Uncertainty in Artificial Intelligence, pp. 499–506 (1995)
17. Spirtes, P., Glymour, C., Scheines, R.: Causation, Prediction, and Search. MIT Press, Cambridge (2001)
18. Tashiro, T., Shimizu, S., Hyvärinen, A., Washio, T.: Parcelingam: a causal ordering method robust against latent confounders. Neural Comput. **26**(1), 57–83 (2014)
19. Tsirlis, K., Lagani, V., Triantafillou, S., Tsamardinos, I.: On scoring maximal ancestral graphs with the max-min hill climbing algorithm. Int. J. Approximate Reasoning **102**, 74–85 (2018)
20. Yao, Y.: Three-way decisions with probabilistic rough sets. Inf. Sci. **180**(3), 341–353 (2010)
21. Yao, Y.: An outline of a theory of three-way decisions. In: Proceedings of International Conference on Rough Sets and Current Trends in Computing, pp. 1–17. Springer (2012)
22. Yao, Y.: Tri-level thinking: models of three-way decision. Int. J. Mach. Learn. Cybern. **11**(5), 947–959 (2020)
23. Yu, H., Wang, X., Wang, G., Zeng, X.: An active three-way clustering method via low-rank matrices for multi-view data. Inf. Sci. **507**, 823–839 (2020)
24. Yu, H., Wang, Z., Xie, Y., Wang, G.: A multi-granularity hierarchical network for long-and short-term forecasting on multivariate time series data. Appl. Soft Comput. **157**, 111537 (2024)
25. Zheng, X., Aragam, B., Ravikumar, P., Xing, E.P.: DAGs with NO TEARS: continuous optimization for structure learning. In: Proceedings of Advances in Neural Information Processing Systems (2018)
26. Zheng, X., Dan, C., Aragam, B., Ravikumar, P., Xing, E.P.: Learning sparse nonparametric DAGs. In: Proceedings of International Conference on Artificial Intelligence and Statistics (2020)

A Sequential Three-Way Decision Model Based on Uncertainty Measurement

Hongguang Xia[1,2] and Jun Hu[1,2(✉)]

[1] Chongqing Key Laboratory of Computational Intelligence, Chongqing University
of Posts and Telecommunications, Chongqing, People's Republic of China
[2] School of Computer Science and Technology, Chongqing University of Posts and
Telecommunications, Chongging, People's Republic of China
hujun@cqupt.edu.cn

Abstract. As an approach of granular computing, the sequential three-way decision(S3WD) model has been widely studied in practical applications. Fruitful results have been achieved in existing research on improving the accuracy of S3WD model. However, the positive and negative regions obtained by a pair of probability thresholds (α, β) introduced by the decision-theoretic rough set model will inevitably lead to the misclassification of some objects. To improve the accuracy of the S3WD model, this paper proposes a S3WD model from the perspective of uncertainty. Firstly, the uncertainty of the equivalence class is defined, and a pair of thresholds are constructed according to the shadowed set theory to measure the uncertainty of the equivalence class, and the equivalence class with large uncertainty in the positive and negative region near both sides of the boundary region is screened out. Finally, the S3WD model based on uncertainty measurement is proposed. The experimental results show that the proposed model has better performance in classification ability compared with the classical S3WD model.

Keywords: sequential three-way decision · equivalence class ·
uncertainty · threshold · shadowed set

1 Introduction

With the rapid development of the Internet, accompanied by the emergence of massive data, data mining, as a data analysis technology, stands out in the era of big data. Among them, the S3WD model [18,21], as a common data mining technique, has received extensive attention and research. It can effectively deal with uncertain information, and has been deeply studied in many disciplines, such as attribute reduction [5,12,14], pattern recognition [10], neural networks [11,13], image analysis [17,22], and cluster analysis [2,15], etc. Compared with the three-way decision-making model [16,20], the S3WD model proposed has better versatility and scalability. The main idea is to divide the entire universe into three disjoint regions through known information, namely positive region, negative region, and boundary region, and then refine the boundary region of each layer by continuously adding attributes, and finally make decisions on all the objects.

Q. Zhang et al. (Eds.): IJCRS 2025, LNAI 15710, pp. 140–153, 2025.
https://doi.org/10.1007/978-3-031-92741-6_11

Through a gradual approach, the S3WD model realizes the decision-making from coarse to fine granularity. In fuzzy decision systems, Qian et al. [9] introduced a cost-sensitive S3WD model, emphasizing the integration of cost considerations. Chen et al. [1] further improved the model by restructuring the granularity to reduce decision costs in multi-class scenarios. Muraleedharan et al. [6] used the Chi-Square statistic to assess multiple attributes, enhancing decision efficiency. Luo et al. [4] extended S3WD to group decision-making, addressing vagueness and hesitation in collaborative contexts. Additionally, Qian et al. [8] incorporated regret theory into a hierarchical multi-attribute decision-making method, demonstrating its stability and effectiveness. Zhang et al. [23] proposed a S3WD model with autonomous error correction (S3WD-AEC) from the perspective of the subdivision of granules. These studies collectively highlight the adaptability of the S3WD model in various fields, from fuzzy systems and multi-attribute problems to group and financial decision-making.

In existing S3WD models, the positive region, negative region, and boundary region are determined based on a pair of probability thresholds (α, β). That is, once the positive region and negative region have been decided, no further processing is done on them, and only the boundary region is used as the universe for further subdivision. This inevitably leads to some misclassification of objects. The reason for this result is that the equivalence classes in the positive and negative regions have certain uncertainties.

To address the above issue, this paper proposes a novel S3WD model based on uncertainty measurement, with the following main contributions.

1. Positive region and negative region based on uncertainty measurement is defined. The uncertainty of equivalence classes is characterized by conditional probability, and a pair of decision thresholds is constructed based on shadowed set theory to measure the uncertainty of equivalence classes.
2. A novel S3WD model based on uncertainty measurement is proposed. Using shadowed set-based thresholds, equivalence classes with high uncertainty in the positive region and negative region are filtered and reclassified into the boundary region, where they are further subdivided at finer granularities, thus improving the accuracy of the S3WD model.

2   Preliminaries

To facilitate the discussion in this paper, this section provides a brief review of the basic concepts of three-way decision and sequential three-way decision.

Definition 1. *Given a decision information system $S = (U, AT = C \cup D, \nu, f)$, let $\Omega = \{[X, \neg X]\}$ be a state set, where X represents the target concept; $\neg X$ represents the non-target concept; State set σ indicates whether an object belongs to X or not. The universe U is divided into three pairwise disjoint regions: Positive region $POS(X)$, Boundary region $BND(X)$ and Negative region $NEG(X)$. $A = \{a_P, a_B, a_N\}$ be an action set, where a_P denotes the acceptance action, indicating that object x is assigned to the positive region $POS(X)$; a_B denotes the non-commitment action, indicating that object x is assigned to the boundary* region $BND(X)$; and a_N denotes the rejection action, indicating that object x is assigned to the negative region $NEG(X)$. Considering that different actions lead to different cost, let $\lambda_{PP}, \lambda_{BP}, \lambda_{NP}$ be the cost when object x belongs to the target concept X and actions a_P, a_B, a_N are taken, respectively. Similarly, using $\lambda_{PN}, \lambda_{BN}, \lambda_{NN}$, denote the cost when object x does not belong to X, and actions a_P, a_B, a_N are taken, respectively.

Therefore, for an object x , the expected cost $R(a \mid [x])$ resulting from taking the corresponding action can be expressed as:

$$R(a_P \mid [x]) = \lambda_{PP} \Pr(X \mid [x]) + \lambda_{PN} \Pr(\neg X \mid [x])$$
$$R(a_B \mid [x]) = \lambda_{BP} \Pr(X \mid [x]) + \lambda_{BN} \Pr(\neg X \mid [x]) \qquad (1)$$
$$R(a_N \mid [x]) = \lambda_{NP} \Pr(X \mid [x]) + \lambda_{NN} \Pr(\neg X \mid [x])$$

where $[x]$ denotes the equivalence class in universe U; $\Pr(X \mid [x])$ represents the conditional probability of X given equivalence class $[x]$; $\Pr(X \mid [x]) = \frac{|X \cap [x]|}{||[x]||}$

In the Bayesian decision procedure, the action that results in the least cost is considered the optimal decision action, and the following decision rule as follows:

(P) If $\Pr(X \mid [x]) \geq \alpha$ and $\Pr(X \mid [x]) \geq \gamma$, decide $x \in POS(X)$;

(B) If $\Pr(X \mid [x]) \leq \alpha$ and $\Pr(X \mid [x]) \geq \beta$, decide $x \in BND(X)$;

(N) If $\Pr(X \mid [x]) \leq \beta$ and $\Pr(X \mid [x]) \leq \gamma$, decide $x \in NEG(X)$.

where

$$\alpha = \frac{(\lambda_{PN} - \lambda_{BN})}{(\lambda_{PN} - \lambda_{BN}) + (\lambda_{BP} - \lambda_{PP})}$$

$$\beta = \frac{(\lambda_{BN} - \lambda_{NN})}{(\lambda_{BN} - \lambda_{NN}) + (\lambda_{NP} - \lambda_{BP})} \qquad (2)$$

$$\gamma = \frac{(\lambda_{PN} - \lambda_{NN})}{(\lambda_{PN} - \lambda_{NN}) + (\lambda_{NP} - \lambda_{PP})}.$$

By considering a reasonable kind of cost functions with $\lambda_{PP} \leq \lambda_{BP} < \lambda_{NP}$ and $\lambda_{NN} \leq \lambda_{BN} < \lambda_{PN}$, the following condition is further satisfied: $\frac{(\lambda_{BP} - \lambda_{PP})}{(\lambda_{PN} - \lambda_{BN})} < \frac{(\lambda_{NP} - \lambda_{BP})}{(\lambda_{BN} - \lambda_{NN})}$, then $0 \leq \beta < \gamma < \alpha \leq 1$. In this case, the following simplified decision rules can be obtained:

(P) If $\Pr(X \mid [x]) \geq \alpha$, decide $x \in POS(X)$;

(B) If $\beta < \Pr(X \mid [x]) < \alpha$, decide $x \in BND(X)$;

(N) If $\Pr(X \mid [x]) \leq \beta$, decide $x \in NEG(X)$.

Definition 2. *Given a decision information system* $S = (U, AT = C \cup D, V, f)$, *and a nested sequence of attribute sets* $C_1 \subset C_2 \subset \cdots \subset C_n \subseteq C$, *which can induce a nested sequence of equivalence relations* $IND(C_n) \subset IND(C_{n-1}) \subset \cdots \subset IND(C_1)$. *The multi-layer granularity structure based on the set of conditional attributes is denoted as* GL, *the granularity structure at the* $i(i = 1, 2, \ldots, n)$ *-th layer is denoted as* GL_i. *Then,* GL *and* GL_i *are defined as follows:*

$$GL = (GL_1, GL_2, \dots, GL_n)$$
$$GL_i = (U_i, AT_i = C_i \cup D_i, V_i, f_i), i = 1, 2, \dots, n \tag{3}$$

where C_i represents the subset of conditional attributes in the i-th layer of the granularity structure; U_i represents the universe of the i-th layer of the granularity structure; AT_i is the union of the current conditional attribute subset C_i and the decision attribute set D.

In the sequential process, the positive region, boundary region, and negative region of each layer can be defined as:

$$POS(X_i) = \{x \in U_i \mid \Pr(X_i \mid [x]_{C_i}) \geq \alpha\};$$
$$BND(X_i) = \{x \in U_i \mid \beta \leq \Pr(X_i \mid [x]_{C_i}) \leq \alpha\}; \tag{4}$$
$$NEG(X_i) = \{x \in U_i \mid \Pr(X_i \mid [x]_{C_i}) \leq \beta\},$$

The S3WD model, in situations with limited information, progressively acquires new information to carry out an incremental decision-making process [19]. However, the positive and negative regions obtained through the pair of probability thresholds (α, β), introduced by the decision-theoretic rough set model inevitably lead to classification errors for some objects.

Example 1. Given a decision information system $S = (U, AT = C \cup D, V, f)$, as shown in Table 1, the non-empty finite universe is $U = \{x_1, x_2, x_3, x_4, x_5, x_6, x_7, x_8, x_9, x_{10}, x_{11}, x_{12}\}$, where the condition attribute subsets are $C_1 = \{a_1\}, C_2 = \{a_1, a_2\}, \dots, \quad C_4 = \{a_1, a_2, a_3, a_4\}$. The decision attribute set is denoted as D, and the set $X = \{x \mid D(x) = 1\}$ represents the set where the decision attribute equals 1. Suppose the threshold pair (α, β) derived from the Bayesian decision criterion is $(\alpha, \beta) = (0.7, 0.35)$. The S3WD is shown in Table 2.

Table 1. Decision information system

U	a_1	a_2	a_3	a_4	D
x_1	1	0	0	1	1
x_2	1	0	0	0	2
x_3	0	1	1	0	1
x_4	0	1	0	0	2
x_5	1	0	0	0	1
x_6	0	1	1	1	1
x_7	1	0	0	0	2
x_8	0	0	0	1	1
x_9	1	1	1	0	2
x_{10}	1	1	0	0	1
x_{11}	1	0	0	0	2
x_{12}	0	0	0	0	2

Table 2. Sequential three-way decisions results.

Layer	POS	BND	NEG
1	$\emptyset$	$\{x_1, x_2, x_3, x_4, x_5, x_6, x_7, x_8, x_9, x_{10}, x_{11}, x_{12}\}$	$\emptyset$
2	$\{x_3, x_4, x_6\}$	$\{x_1, x_2, x_5, x_7, x_8, x_9, x_{10}, x_{11}, x_{12}\}$	$\emptyset$
3	$\{x_3, x_4, x_6, x_{10}\}$	$\{x_1, x_2, x_8, x_{12}\}$	$\{x_5, x_7, x_9, x_{11}\}$
4	$\{x_1, x_3, x_4, x_6, x_8, x_{10}\}$	$\emptyset$	$\{x_2, x_5, x_7, x_9, x_{11}, x_{12}\}$

The equivalence class divisions at each layer are as follows:

$Layer$ 1: $U_1/\{C_1\} = \{\{x_1, x_2, x_5, x_7, x_9, x_{10}, x_{11}\}, \{x_3, x_4, x_6, x_8, x_{12}\}\}$.

$Layer$ 2: $U_2/\{C_2\} = \{\{x_1, x_2, x_5, x_7, x_{11}\}, \{x_9, x_{10}\}, \{x_3, x_4, x_6\}, \{x_8, x_{12}\}\}$.

$Layer$ 3: $U_3/\{C_3\} = \{\{x_1, x_2\}, \{x_5, x_7, x_{11}\}, \{x_9\}, \{x_{10}\}, \{x_8, x_{12}\}\}$.

$Layer$ 4: $U_4/\{C_4\} = \{\{x_1\}, \{x_2\}, \{x_8\}, \{x_{12}\}\}$.

In the $Layer$ 2, the equivalence class x_3, x_4, x_6 has a conditional probability of $2/3 > 0.65$, so it is classified into the positive negative. However, it can be seen that the uncertainty is still relatively large, which causes x_4 to be misclassified. Similarly, in the $Layer$ 2, misclassification occurs for objects with high uncertainty. From this multi-granularity structure of the decision information table, it can also be observed that as new attributes are added, objects that were misclassified in the previous layer due to insufficient information can be corrected in the current layer. Therefore, we can use a new pair of thresholds to filter equivalence classes with high uncertainty in the positive region and negative region, and reassign them to the boundary region for further refinement at a finer granularity.

3   A Sequential Three-Way Decision Model Based on Uncertainty Measurement

This section first analyzes the uncertainty of equivalence classes in S3WD; Subsequently, a new pair of threshold values is designed using shadowed set theory, and the positive region and negative region based on uncertainty measurement is defined. Building upon this foundation, a S3WD model based on uncertainty measurement is proposed.

3.1   Uncertainty of Equivalence Classes in Sequential Three-Way Decision

In the S3WD model, the uncertainty of equivalence classes is often characterized by conditional probability. That is, for the target concept X, if the conditional probability of an equivalence class is closer to 1 or 0, it means that the uncertainty of the equivalence class based on X is smaller.

Definition 3. *Given a decision information system $S = (U, AT = C \cup D, V, f)$. Suppose $[x]_A$ is the equivalence class formed under the equivalence relation* *induced by the attribute subset A. Then, the uncertainty of $[x]_A$ can be expressed as:*

$$Unc([x]_A) = \frac{|X \cap [x]|}{||[x]||} \tag{5}$$

There are many methods for measuring uncertainty, among which the shadowed set can efficiently handle uncertainty information. Compared to the threshold pair proposed in three-way decision theory [19], the classic shadowed set model can optimize based on the uncertainty of objects, without considering cost parameters based on Bayesian decision theory. This eliminates the need to manually assign loss functions to construct threshold pairs, thus ensuring that the uncertainty loss during the object partitioning process is minimized. Therefore, a new pair of threshold values can be derived through shadowed set theory to perform uncertainty-based filtering of equivalence classes.

Definition 4. *Suppose $GL = (GL_1, GL_2, \ldots, GL_n)$ is an n-layer granularity structure, where $GL_i = (U_i, AT_i = C_i \cup D_i, V_i, f_i)$ represents the i-th layer granularity structure partitioned by the subset of conditional attributes C_i. Suppose that threshold pair (α_i', β_i') is determined at the i-th granularity layer using shadowed set theory. Then, the degree of membership of object x for the target concept X_i can be defined by the conditional probability as follows:*

$$\mu_{X_i}(x) = \Pr(X_i \mid [x]_{C_i}) \tag{6}$$

In shadowed set theory, an optimal threshold pair can be obtained based on the degree of membership of objects in the universe [7]. Specifically, this can be achieved through the following optimization function:

$$V_{(\alpha_i', \beta_i')}(\mu_{X_i}) = | \sum_{\mu_{X_i}(x) \geq \alpha_i'} (1 - \mu_{X_i}(x)) + \sum_{\mu_{X_i}(x) \leq \beta_i'} \mu_{X_i}(x) - $$
$$card(\{x \in U | \beta_i' < \mu_{X_i}(x) < \alpha_i'\} | \tag{7}$$

where $\alpha_i' + \beta_i' = 1$; $card(\cdot)$ represents the cardinality of the set.

By minimizing Eq. 7, the optimal threshold pair (α_i', β_i') can be obtained.

Theorem 1. *For Eq. 7, let $\beta_i' = \tau_i, \alpha_i' = 1 - \tau_i$. Suppose that in the sequential decision process, the threshold is updated by minimizing $V_{(\tau_i)}(\mu_{X_i})$ via the mapping $f : [0, 0.5] \to [0, 0.5], \tau_{i+1} = f(\tau_i)$, and assume that: (1) The mapping f is continuous on $[0, 0.5]$; (2) f is a contraction; i.e., there exists a constant $0 < L < 1$ such that for any $\tau_i, \tau_{i+1} \in [0, 0.5], |f(\tau_i) - f(\tau_{i+1})| \leq L|\tau_i) - f(\tau_{i+1})|$. Then, the sequence $\{\tau_i\}$ converges to a unique fixed point $\tau^* \in [0, 0.5]$ such that $\lim_{i \to \infty} \tau_i = \tau^*$ and $f(\tau^*) = \tau^*$. Consequently, the decision threshold pair $(\alpha_i', \beta_i') = (\tau^*, 1 - \tau^*)$ is stable and convergent.*

Proof. Consider the closed interval $S = [0, 0.5]$. With the standard absolute value metric, S is a complete metric space. By assumption, for any $\tau_i, \tau_{i+1} \in X, |f(\tau_1) - f(\tau_2)| \leq L|\tau_1 - \tau_2|$, where $0 < L < 1$. By the Banach Fixed Point Theorem, every contraction mapping on a complete metric space S has a unique fixed point $\tau^* \in S$ satisfying $f(\tau^*) = \tau^*$. Moreover, for any initial value $\tau_0 \in X$, the iterative sequence $\tau_{i+1} = f(\tau_i)$. Therefore, in the sequential decision-making process, if at each step the threshold is updated by minimizing the loss function $\Omega(\tau)$ and the resulting update mapping f is a contraction, then the sequence $\{\tau_n\}$ converges to the unique fixed point τ^*. Consequently, the decision threshold pair $(\alpha_i', \beta_i') = (\tau^*, 1 - \tau^*)$.

Definition 5. *Suppose $GL = (GL_1, GL_2, \ldots, GL_n)$ is an n-layer granularity structure, where $GL_i = (U_i, AT_i = C_i \cup D_i, V_i, f_i)$ represents the i-th layer granularity structure partitioned by the subset of conditional attributes C_i. Let (α_i', β_i') be the threshold pair based on shadowed set on the i-th layer. Then, the positive region and negative region based on uncertainty measurement at the i-th layer can be defined as:*

$$
\begin{aligned}
Ecu(POS(X_i)) &= \{x \in POS(X_i) \mid \beta_i' < \Pr(X_i \mid [x]_{C_i}) < \alpha_i'\} \\
Ecu(NEG(X_i)) &= \{x \in NEG(X_i) \mid \beta_i' < \Pr(X_i \mid [x]_{C_i}) < \alpha_i'\}
\end{aligned}
\tag{8}
$$

3.2 The Algorithm of Sequential Three-Way Decision Model Based on Uncertainty Measurement

Based on the above discussion, after obtaining the positive, negative, and boundary regions through the pair of probability thresholds from the decision-theoretic rough set model, it is necessary to measure the uncertainty of the equivalence classes using the threshold pair based on shadowed set. The equivalence classes with larger uncertainty are selected and reclassified into the boundary region, where further decisions are made at a finer granularity, which can improve classification accuracy.

Definition 6. *Given a decision information system $S = (U, AT = C \cup D, V, f)$, C is the set of conditional attributes. Let C_i be a subset of conditional attributes, where $i = 1, 2, \ldots, n$. Suppose $GL = (GL_1, GL_2, \ldots, GL_n)$ is an n-layer granularity structure, where $GL_i = (U_i, AT_i = C_i \cup D_i, V_i, f_i)$ represents the i-th layer granularity structure partitioned by the subset of conditional attributes C_i. Then, the universe of each layer can be expressed as:*

$$
U_i = \begin{cases}
U, & i = 1 \\
Ecu(POS(X_{i-1})) \cup Ecu(NEG(X_{i-1})) \cup BND(X_{i-1}), & i > 1
\end{cases}
\tag{9}
$$

when $i = 1$, the initial universe is U; when $i > 1$, due to the presence of equivalence classes with larger uncertainty in the positive and negative regions of the previous layer, the universe at this layer consists of the boundary region of the previous layer and the equivalence classes with larger uncertainty selected through the threshold pair based on shadowed set.

As shown in Fig. 1, it is the framework of the proposed model. According to the framework, the model proposed in this paper can also be implemented by Algorithm 1.

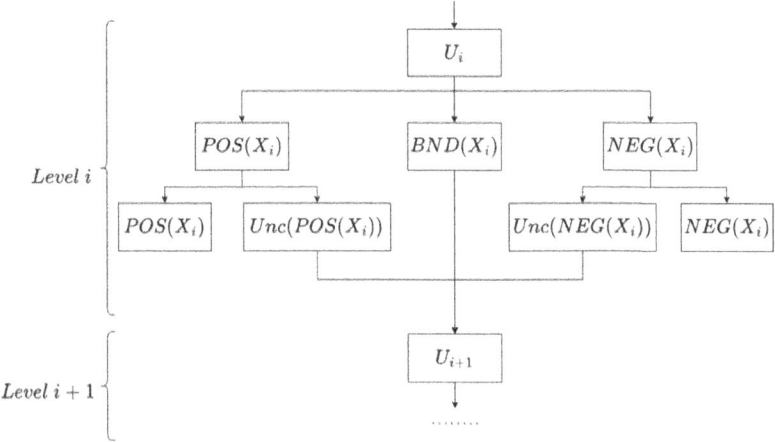

Fig. 1. The framework of the proposed model

The time complexity of Algorithm 1 is $T(n) = O\left(\frac{(1+m)\cdot m}{2} \cdot n + 50 \cdot m \cdot n\right)$. Since the number of objects in the universe n decreases continuously during the sequential process, the actual time complexity will be much lower than $T(n)$. Because the proposed model performs uncertainty partitioning on equivalence classes and calculates the threshold pair based on shadowed set, its time complexity is slightly higher than that of the classical S3WD model.

4   Experimental Comparison and Analysis

To prove the effectiveness of the proposed model, this section selects 8 datasets from UCI and compares the proposed model with classical sequential three-way decision model(S3WD) [19]. The experimental environment is a computer running Windows 10, equipped with 16 GB RAM, a 3.0 GHz CPU, and the programming language used is MATLAB.

4.1   Experimental Setup

In the selected 8 datasets, some have more than 2 decision classes, while three-way decision are mainly introduced for the two-class decision problem. Therefore, in the experiments, only the decision class with the first and second highest number of objects was selected. The basic information of the datasets is shown in Table 3.

Algorithm 1: The algorithm of sequential three-way decision model based on uncertainty measurement

---

Input: Decision information system $S = (U, AT = C \cup D, V, f)$, cost function matrix $\Lambda = \{\lambda_{PP}, \lambda_{BP}, \lambda_{NP}, \lambda_{PN}, \lambda_{BN}, \lambda_{NN}\}$.

Output: Positive region $POS(X)$ and negative region $NEG(X)$ based on the target concept X.

1 $POS = \emptyset, NEG = \emptyset$;
2 **for** $i = 1$ **to** $n - 1$ **do**
3 $\quad POS_i = \emptyset, BND_i = \emptyset, NEG_i = \emptyset$;
4 $\quad$ **if** $i = 1$ **then**
5 $\quad\quad U_i = U$;
6 $\quad$ Calculate $\Pr(X_i| [x]_{C_i})$;
7 $\quad$ Calculate (α_i, β_i);
8 $\quad$ **if** $\Pr(X_i| [x]_{C_i}) \geq \alpha_i$ **then**
9 $\quad\quad POS_i = POS_i \cup [x]_{C_i}$;
10 $\quad$ **if** $\beta_i < \Pr(X_i| [x]_{C_i}) < \alpha_i$ **then**
11 $\quad\quad BND_i = BND_i \cup [x]_{C_i}$;
12 $\quad$ **if** $\Pr(X_i| [x]_{C_i}) \leq \beta_i$ **then**
13 $\quad\quad NEG_i = NEG_i \cup [x]_{C_i}$;
14 $\quad$ Calculate (α_i', β_i');
15 $\quad$ Calculate $Ecu(POS(X_i))$ and $Ecu(NEG(X_i))$;
16 $\quad$ Update the positive region $POS_i = POS_i - Ecu(POS(X_i))$;
17 $\quad$ Update the negative region $NEG_i = NEG_i - Ecu(NEG(X_i))$;
18 $\quad POS = POS \cup POS_i$;
19 $\quad NEG = NEG \cup NEG_i$;
20 $\quad$ Update the universe $U_i = Ecu(POS(X_{i-1})) \cup Ecu(NEG(X_{i-1}))\cup$
21 $\quad BND(X_{i-1})$;
22 **if** $i = n$ **then**
23 $\quad$ Calculate $\Pr(X_i| [x]_{C_i})$;
24 $\quad$ **if** $\Pr(X_i| [x]_{C_i}) \geq \gamma$ **then**
25 $\quad\quad POS_i = POS_i \cup [x]_{C_i}$;
26 $\quad$ **if** $\Pr(X_i| [x]_{C_i}) < \gamma$ **then**
27 $\quad\quad NEG_i = NEG_i \cup [x]_{C_i}$;
28 **return** *POS,NEG*;

Since the threshold conditions for three-way decision are determined by the cost functions based on Bayesian decision theory, in order to better demonstrate the versatility of the model, five sets of cost functions were randomly generated, as shown in Table 4. For classification, comparisons are often made using the two metrics: *Accuracy* and F_1 [3].

Table 3. Description of the datasets

ID	Datasets	Objects	Attributes
1	Breast Cancer Wisconsin	699	9
2	Tic-Tac-Toe	985	9
3	Contraceptive Method Choice	1473	8
4	Mammographic mass	961	5
5	Car evaluation	1728	6
6	Nursery	12960	8
7	Website Phishing	1353	10
8	Congressional Voting Records	435	16

Table 4. Cost functions

ID	λ_{PP}	λ_{BP}	λ_{NP}	λ_{PN}	λ_{BN}	λ_{NN}
Case 1	0	2	10	10	2	0
Case 2	0	2	8	13	1	0
Case 3	0	2	7	10	1	0
Case 4	0	2	7	4	1	0
Case 5	0	1	4	5	2	0

Accuracy is defined as:

$$Accuracy = \frac{|D \cap POS(D)| + |\neg D \cap NEG(D)|}{|U|} \tag{10}$$

where $|D \cap POS(D)|$ represent the number of objects assigned to the positive region that originally belong to the decision attribute D; $|\neg D \cap NEG(D)|$ represents the number of objects assigned to the negative region that originally do not belong to the decision attribute D; $|U|$ represents the total number of objects in the universe.

F_1 is the harmonic mean of *Precision* and *Recall*, taking both metrics into account. *Precision*, *Recall*, and F_1 can be defined as:

$$Precision = \frac{|D \cap POS(D)|}{|\neg D \cap POS(D)| + |D \cap POS(D)|} \tag{11}$$

$$Recall = \frac{|D \cap POS(D)|}{|D \cap NEG(D)| + |D \cap POS(D)|} \tag{12}$$

$$F_1 = \frac{2 \cdot Precision \cdot Recall}{Precision + Recall} \tag{13}$$

4.2   Experimental Results

The experimental results on *Accuracy* are shown in Fig. 2, and the experimental results on F_1 are shown in Fig. 3.

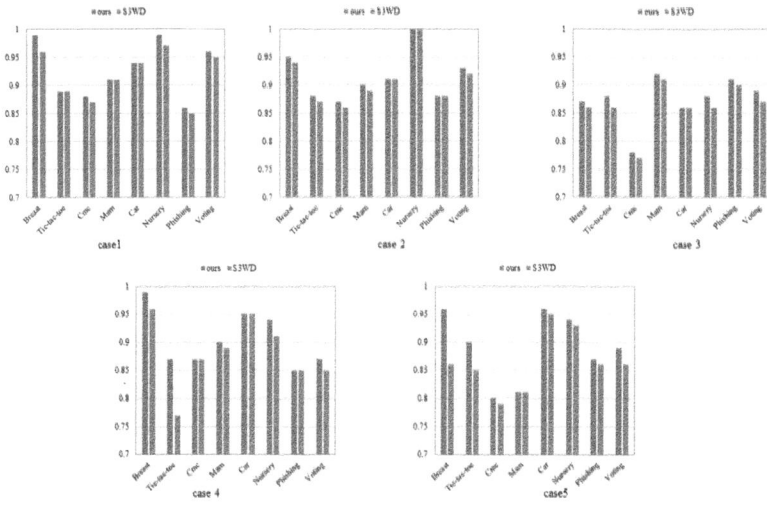

Fig. 2. The results on *Accuracy*

As can be clearly seen in Fig. 2, for the eight datasets, the proposed model achieves *accuracy* that is equal to or higher than classical S3WD in terms of *Accuracy*. From a general perspective of classification models, the higher the *Accuracy*, the better the classification performance of the model. For some datasets, the *Accuracy* in some case were equal under different loss functions. For example, for the Website Phishing dataset, in case 4, the *Accuracy* are equal. This is because the information contained in the original dataset's attributes was insufficient. Even though the threshold pair based on shadowed set was used to filter out the equivalence classes with high uncertainty and further processed at a finer granularity, the uncertainty of the equivalence classes remained high, resulting in no improvement in classification performance compared to classical S3WD. Similarly, when examining the F_1, it is clearly evident from Fig. 3 that the proposed model outperforms classical S3WD. The reason for these is that the proposed model further filters some of the equivalence classes near the boundary

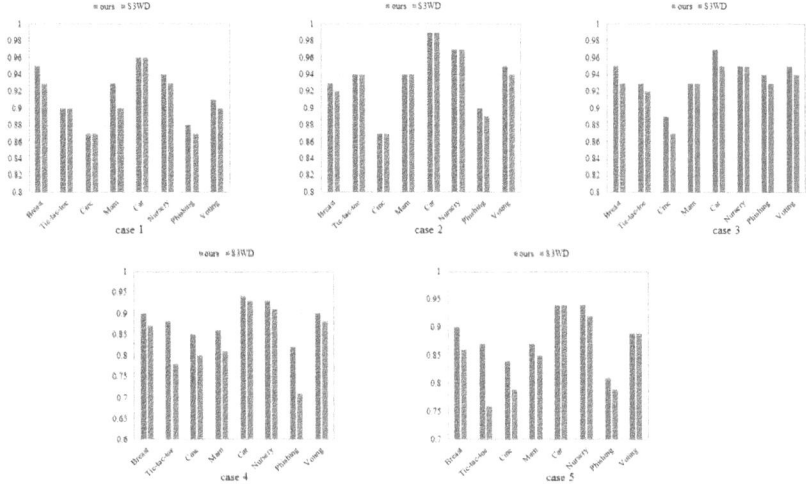

Fig. 3. The results on F_1.

region, placing those with higher certainty into a finer layer of consideration, which effectively improves the accuracy of decisions.

In summary, the experimental results demonstrate that under the conditions of the 8 datasets and five loss functions, the proposed model shows improved accuracy on both the *Accuracy* and F_1. It can be concluded that the proposed model, by considering the selection of equivalence classes when handling datasets with high uncertainty, provides better classification performance in practical applications.

5   Conclusion

This paper proposes a S3WD model based on uncertainty measurement. The model, from the perspective of uncertainty, measures the equivalence classes in S3WD, selecting equivalence classes with highest uncertainty near the positive and negative regions on both sides of the boundary region. These classes are then refined at a finer granularity, reducing the uncertainty of the equivalence classes while improving the accuracy of the S3WD. In future study, we will attempt to explore other methods to characterize the uncertainty of equivalence classes, as well as to find the optimal threshold pairs.

References

1. Chen, W., Zhang, Q., Dai, Y.: Sequential multi-class three-way decisions based on cost-sensitive learning. Int. J. Approximate Reasoning **146**, 47–61 (2022)
2. Du, M., Zhao, J., Sun, J., Dong, Y.: M3w: multistep three-way clustering. IEEE Trans. Neural Netw. Learn. Syst. **35**(4), 5627–5640 (2022)
3. Kaiwen, G., Hongliang, P., Alin, H.: Classification algorithm based on feature selection and clustering. J. Jilin Univ. (Sci. Ed.) **28**(2), 395–398 (2018)
4. Luo, N., Zhang, Q., Xie, Q., Wang, Y., Yin, L., Wang, G.: Sequential three-way group decision-making for double hierarchy hesitant fuzzy linguistic term set. Inf. Sci. **687**, 121403 (2025)
5. Ma, X.A., Yao, Y.: Three-way decision perspectives on class-specific attribute reducts. Inf. Sci. **450**, 227–245 (2018)
6. Muraleedharan, R.K., Nair, L.R.: A novel approach for sequential three-way decision using chi-square statistic as the assessment metric. Revue d'Intelligence Artificielle **38**(1) (2024)
7. Pedrycz, W.: Shadowed sets: representing and processing fuzzy sets. IEEE Trans. Syst. Man Cybern. Part B (Cybern.) **28**(1), 103–109 (1998)
8. Qian, J., Lu, Y., Yu, Y., Zhou, J., Miao, D.: Hierarchical sequential three-way multi-attribute decision-making method based on regret theory in multi-scale fuzzy decision systems. IEEE Trans. Fuzzy Syst. (2024)
9. Qian, W., Zhou, Y., Qian, J., Wang, Y.: Cost-sensitive sequential three-way decision for information system with fuzzy decision. Int. J. Approximate Reasoning **149**, 85–103 (2022)
10. Savchenko, A.V.: Sequential three-way decisions in multi-category image recognition with deep features based on distance factor. Inf. Sci. **489**, 18–36 (2019)
11. Savchenko, A.V.: Fast inference in convolutional neural networks based on sequential three-way decisions. Inf. Sci. **560**, 370–385 (2021)
12. Wang, Z., Shi, C., Wei, L., Yao, Y.: Tri-granularity attribute reduction of three-way concept lattices. Knowl.-Based Syst. **276**, 110762 (2023)
13. Wu, Y., Cheng, S., Li, Y., Lv, R., Min, F.: STWD-SFNN: sequential three-way decisions with a single hidden layer feedforward neural network. Inf. Sci. **632**, 299–323 (2023)
14. Xia, D., et al.: Interactive fuzzy knowledge distance-guided attribute reduction with three-way accelerator. Knowl.-Based Syst. **279**, 110943 (2023)
15. Xu, Y., Niu, G.: Research on multi-view clustering algorithm based on sequential three-way decision. Appl. Soft Comput. **158**, 111590 (2024)
16. Yang, X., Li, T., Fujita, H., Liu, D., Yao, Y.: A unified model of sequential three-way decisions and multilevel incremental processing. Knowl.-Based Syst. **134**, 172–188 (2017)
17. Yang, X., Zhou, X., Huang, B., Li, H., Wang, T.: A three-way decision method on multi-scale single-valued neutrosophic decision systems. Artif. Intell. Rev. **57**(4), 109 (2024)
18. Yao, Y.: Three-way decisions with probabilistic rough sets. Inf. Sci. **180**(3), 341–353 (2010)
19. Yao, Y.: The superiority of three-way decisions in probabilistic rough set models. Inf. Sci. **181**(6), 1080–1096 (2011)
20. Yao, Y.: An outline of a theory of three-way decisions. In: International Conference on Rough Sets and Current Trends in Computing, pp. 1–17 (2012)

21. Yao, Y.: Granular computing and sequential three-way decisions. In: International Conference on Rough Sets and Knowledge Technology, pp. 16–27. Springer (2013)
22. Ye, X., Liu, D.: An interpretable sequential three-way recommendation based on collaborative topic regression. Expert Syst. Appl. **168**, 114454 (2021)
23. Zhang, Q., Huang, Z., Wang, G.: A novel sequential three-way decision model with autonomous error correction. Knowl.-Based Syst. **212**, 106526 (2021)

Granular-Ball Three-Way Decision for Robust Text Classification

Jie Zhang, Yanhua Li, Xiaocao Ouyang, Lingfei Ren, and Xin Yang[✉]

School of Computing and Artificial Intelligence, Southwestern University of Finance and Economics, Chengdu 611130, China
{oyxiaocao,renlf,yangxin}@swufe.edu.cn

Abstract. Intent classification plays a crucial role in the development of dialogue systems. Most existing research assumes that training data is correctly labeled; however, real-world annotated data often contains noise due to the time-consuming and labor-intensive labeling process. Noisy training data degrades model representations, making noise-robust intent classification essential. Existing approaches for handling noisy data primarily rely on coarse-grained semantic representations, failing to capture the true semantic distribution. Moreover, they do not account for the uncertainty in noise recognition and label correction, instead directly classifying a sample as either clean or noisy. To address these challenges, we propose a Robust Granular Ball Three-Way Decision (RGB3WD) method. Specifically, we leverage granular ball clustering to represent samples with multiple granular balls, capturing fine-grained semantic structures within clusters. For noise recognition, we introduce a label-consistency-based three-way decision rule, categorizing samples into positive, boundary, and negative regions. Finally, we design region-specific strategies and develop a loss function to learn noise-robust representations. Extensive experiments on multiple datasets demonstrate the effectiveness of the proposed method.

Keywords: Granular-Ball · Three-way Decision · Robust Intent Classification

1 Introduction

Accurate user intent classification plays a crucial role in online service systems, especially in dialog systems, as it significantly enhances user satisfaction. Most existing text classification methods [27] assume that the training data is accurately labeled. However, real-world data often contains mislabeled samples due to the time-consuming and labor-intensive nature of data annotation, and such noisy data can negatively impact the accuracy of text classification models.

Previous research has demonstrated significant progress in developing robust text classification to mitigate the impact of noisy data [1,2]. These approaches typically focus on enhancing feature representations by recognizing noise, followed by training the model with the processed clean data [3,14]. Although

© The Author(s), under exclusive license to Springer Nature Switzerland AG 2025
Q. Zhang et al. (Eds.): IJCRS 2025, LNAI 15710, pp. 154–168, 2025.
https://doi.org/10.1007/978-3-031-92741-6_12

these methods have proven effective, they still exhibit the following limitations. **Issue 1**: These methods often treat all samples within a class as a homogeneous entity during representation learning, failing to capture the fine-grained semantic information inherent within each class. **Issue 2**: Directly categorizing samples as either noisy or clean neglects the uncertainty associated with noise types, increasing the risk of miss-recognition and consequently compromising the effectiveness of downstream training. **Issue 3**: Threshold-based noise identification methods lack robustness, as pre-defined thresholds are not adaptive, potentially leading to poor model generalization and excessive dependence on specific datasets.

To address the aforementioned issues, we propose a Robust Granular-Ball Three-Way Decision (RGB3WD) method for text classification. Specifically, for Issue 1, a fine-grained representation learning method is designed to capture the underlying data structure. Inspired by the concept of granular computing, granular balls with varying size are utilized to represent all datasets. For Issue 2, we propose a three-way decision rule based on label-consistency, which categorizes all samples within a granular ball into three domains: the positive domain (POS), the boundary domain (BND), and the negative domain (NEG). The positive domain consists of samples that are label-consistent with the granular ball, allowing for confident label correction. The boundary domain represents samples whose true labels remain uncertain concerning the granular ball's label. The negative domain includes samples that are label-inconsistent with the granular ball. For Issue 3, we propose a neighbor label-consistency rule to further divide samples in the boundary region to either the positive or negative region. Additionally, we design a novel loss function that pulls label-consistent samples closer while pushing label-inconsistent samples away, thereby enhancing representation learning and improving model robustness.

The contributions of this paper are shown as follows:

- This paper proposes a noise-robust text classification method based on granular-ball three-way decision, which adapts to data distribution while effectively addressing the uncertainty of true-label.
- Adaptive granular-ball clustering is utilized to obtain the fine-grained representation of intent to show class data distribution.
- We design a label-consistency-based three-way decision rule to determine the consistency between samples and the granular ball label, categorizing them into consistent, inconsistent, and uncertain groups. This approach to handling uncertainty reduces decision-making risks.
- We design distinct strategies for the three regions, incorporating a neighbor label-consistency rule for the boundary region and introducing a new loss function to enhance representation learning.

2   Preliminaries

In this section, we review some foundational works relevant to our study, including granular-ball computing and three-way decisions.

2.1 Granular-Ball Computing

Granular-ball computing is an efficient, robust, and interpretable method for adaptive multi-granularity representation and computation [18]. Research on granular-ball computing include granular-ball clustering [6,19], granular-ball classification [17,20], granular-ball representation in feature selection [4] and text adversarial defense [15] and others.

In text classification, granular-ball representation enables the effective capture of semantic patterns by clustering similar data points into compact groups, thereby improving classification performance. Furthermore, the adaptive nature of granular-ball generation is crucial for handling noisy data, minimizing the impact of outliers.

Granular-ball representation inherently offers advantages for handling noisy label learning, yet only a limited number of related studies have been conducted [16], none of which address the uncertainty of noise types. In fact, the natural centroid and radius of granular balls can assist in analyzing the uncertainty of sample noise types.

2.2 Three-Way Decision

The three-way decision [22,24] is proposed as an extension of the commonly used binary decision model by introducing a third option. In general, the three-way decision divides the universe into three regions: positive, negative, and boundary regions, which represent the areas of acceptance, rejection, and noncommitment, respectively. Three regions are defined as follows:

$$POS_{\alpha,\beta} = \{x \in U \mid \mu_{\tilde{A}}(x) \geq \alpha\},$$
$$BND_{\alpha,\beta} = \{x \in U \mid \alpha < \mu_{\tilde{A}}(x) < \beta\}, \tag{1}$$
$$NEG_{\alpha,\beta} = \{x \in U \mid \mu_{\tilde{A}}(x) \leq \beta\},$$

where $\mu_{\tilde{A}}(x)$ is the membership function of the fuzzy set $\tilde{A}$, mapping each element x in the universe U to a value in [0, 1]. These regions represent:$POS_{\alpha,\beta}$: The set of elements x with $\mu_{\tilde{A}}(x) \geq \alpha$ (positive).$NEG_{\alpha,\beta}$: The set of elements x with $\mu_{\tilde{A}}(x) \leq \beta$ (negative).$BND_{\alpha,\beta}$: The set of elements x where $\alpha < \mu_{\tilde{A}}(x) < \beta$ (boundary).

Due to its significant advantages in addressing uncertainty, ambiguity, and incomplete information, three-way decision has gained increasing popularity in recent years across various decision-making models [11,23]. Compared to the traditional binary decision-making model, the three-way decision provides a more flexible representation of the three states-acceptance, rejection, and uncertainty-by introducing boundary regions. This allows for more refined and diversified solutions to complex decision-making problems. This feature makes the three-way decision highly effective in fields such as risk management [9], fuzzy decision [25,26], and supply chain optimization [12]. Furthermore, it demonstrates its unique advantages in handling ambiguous and dynamically evolving situations.

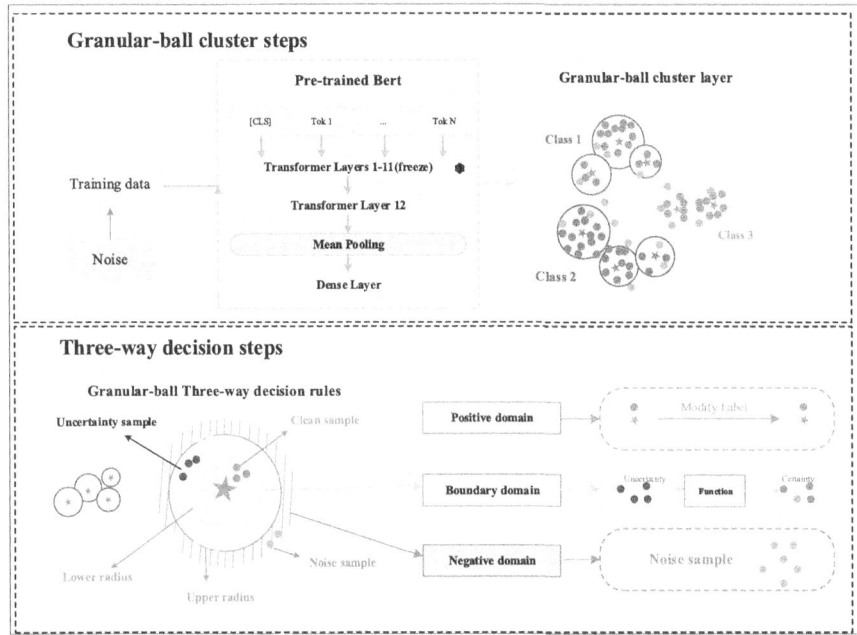

Fig. 1. The framework of RGB3WD.

Three-way decision theory has demonstrated outstanding performance across various fields, particularly in tackling complex problems and decision-making processes. It is particularly effective in managing uncertainty, ambiguity, and incomplete information. Therefore, we propose the use of three-way decision-making to address the uncertainty inherent in noise type.

3   Methodology

3.1   Problem Statement and Method Overview

Noise learning plays a critical role in text classification tasks, particularly in situations where the task relies on labeled data that may be prone to errors. In many real-world applications, the process of labeling text data is often susceptible to human mistakes, inconsistencies, or ambiguities, which can significantly impact the performance of classification models (Fig. 1).

The proposed noise-robust granular-ball three-way decision method consists of three main modules: (a) Granular-ball clustering, (b) Granular-Ball three-way noise recognition based on label-consistency, and (c) Three-Way strategies for representation learning. Specifically, (a) we utilize a pre-trained BERT model [7] to extract feature representations and apply the granular-ball clustering method to group all samples into multiple granular balls. (b) Based on the features of the granular balls and the spatial distribution of the samples, all samples are divided into three regions: positive, boundary, and negative. (c) Different strategies are applied to samples in the three regions to correct their labels, and a loss function is designed for representation learning.

3.2 Granular-Ball Clustering

We first use a pre-trained BERT model to extract text features, while freezing the parameters of all layers except the final one. To capture the fine-grained semantic structure of the text and facilitate noise identification, we aim to uncover sub-class information within each category. Traditional clustering methods can address this issue, but they often require the number of sub-clusters to be predefined. In contrast, granular-ball clustering does not require a predefined number of clusters. It is an efficient and adaptive clustering method that more accurately reflects the distribution characteristics of the data.

We use the **BERT** model to extract deep intent features. Given the ith input sentence s_i, we get all its token embeddings $[CLS, T_1, \cdots, T_N] \in \mathbb{R}^{(N+1) \times H}$ from the last hidden layer of BERT, where CLS is the vector for text classification, N is the sequence length and H is the hidden layer size. We perform mean-pooling [10] on these token embeddings to synthesize high-level semantic features in a sentence, and then obtain an average representation $x_i \in \mathbb{R}^H$. To further strengthen feature extraction capability, we feed x_i to a dense layer h to get the intent representation $z_i \in \mathbb{R}^D$:

$$z_i = h(x_i) = \sigma(W_h x_i + b_h), \tag{2}$$

where D is the dimension of the intent representation, σ is a *ReLU* activation function, $W_h \in \mathbb{R}^{H \times D}$ and $\mathbf{b}_h \in \mathbb{R}^D$ respectively denote the weights and the bias term of layer h.

Next, we proceed with the granular ball clustering phase. In this phase, we begin by initializing the features of each sample to form the initial granular ball. Let $X_i = \{(z_i, y_i), \ldots, (z_N, y_N)\}$ be the intent examples with their corresponding labels. we cluster the samples into multiple granular balls. The objective of granular ball clustering is to group similar samples together, thereby forming a set of granular balls based on the similarity of samples in the feature space.

A granular ball gb_j, which contains a set of similar samples, is defined by three key attributes: the number of samples n_j, representing the total number of samples in the granular ball; the center of mass O_j, which is the average representation of all the samples within gb_j, computed as follows:

$$O_j = \frac{1}{n_j} \sum_{i=1}^{n_j} z_i, \tag{3}$$

where z_i is the representation of the i-th sample in the granular ball; and the radius r_j, which is the average distance from the center of mass to all the samples, calculated as

$$r_j = \frac{1}{n_j} \sum_{i=1}^{n_j} \|z_i - O_j\|, \tag{4}$$

where $\|z_i - O_j\|$ denotes the Euclidean distance between the sample z_i and the center of mass O_j. The label l_j of the granular sphere gb_j is the category with the largest number of samples within it. The process of adaptive granular ball clustering is iterative. Initially, all samples are placed into a single granular sphere. The purity p_j of a granular sphere indicates the proportion of samples with the most common label in the granular sphere, and it is computed as

$$p_j = \frac{\text{count}(l_j)}{n_j}, \tag{5}$$

where $\text{count}(l_j)$ is the number of samples with the label l_j in the granular sphere.

Based on a purity threshold p_l and a sample count threshold n_l, the granular ball undergoes a split under the following conditions: when the granular ball is split, it is partitioned into multiple new granular balls, each corresponding to a distinct label in the original label set L_j. Specifically, a sample from each unique label is randomly selected as the pseudo-center of mass, and the remaining samples are assigned to the closest granule sphere based on the Euclidean distance between the sample and the pseudo-center of mass. After several iterations, the granule ball clustering will result in a set of granules $G = \{gb_1, gb_2, \ldots, gb_m\}$, where each granule gb_j is assigned a label $l_j \in \{1, 2, \ldots, K\}$, with K representing the total number of categories. To mitigate the impact of noisy samples and prevent overfitting, only those granules whose purity p_j exceeds a predefined threshold p_t and whose sample size n_j is greater than a threshold n_t are retained as valid granules. The final set of clustered granules, reflecting the intrinsic structure of the data, will serve as inputs for model training and evaluation.

3.3 Granular-Ball Three-Way Noise Recognition Based on Label-Consistency

Existing noise recognition methods typically classify samples as noisy or clean based on thresholds or other criteria. This deterministic decision-making process increases the risk of incorrect noise type classification, which can negatively impact subsequent model training. In this section, we propose a method based on the granular-ball three-way decision approach. This method classifies the samples belonging to a granular ball into three regions: Positive, Boundary, and Negative by the relationship between the sample's distance from the granular ball center and the ball's radius. For boundary samples, we employ a delayed decision strategy, which reduces the risk of noise classification errors.

We define two radii thresholds, the lower radius r_j^- and the upper radius r_j^+, to represent the distribution boundaries of a granular ball, where $r_j^- < r_j^+$ and both are determined based on the granular ball's radius r_j. The samples belonging to the granular ball gb_j are then divided into three regions: positive, boundary, and negative.

Specifically, (1) samples that are closer to the centroid of the granular ball typically share the same label as the granular ball. Thus, we assign samples within the lower radius r_j^- to the positive region, meaning they are label-consistent with the label of the granular ball gb_j. (2) Samples that lie outside the upper radius r_j^+ are classified into the negative region, as their true labels are likely inconsistent with the label of the granular ball due to their distance from the centroid. (3) For samples that fall between the lower radius r_j^- and the upper radius r_j^+, we cannot confidently determine whether their true labels align with the label of the granular ball. Therefore, these samples are considered label-uncertain and are assigned to the boundary region.

In total, all samples belonging to the granular ball gb_j are divided into three regions: positive, boundary, and negative, corresponding to label-consistent, label-uncertain, and label-inconsistent samples with respect to gb_j, respectively. The decision rules are as follows:

$$
\begin{aligned}
POS(gb_j) &= \{z_i \in gb_j | \|z_i - O_j\| < r_j^- \}, \\
BND(gb_j) &= \{z_i \in gb_j | r_j^- < \|z_i - O_j\| < r_j^+ \}, \\
NEG(gb_j) &= \{z_i \in gb_j | r_j^+ < \|z_i - O_j\| \},
\end{aligned}
\tag{6}
$$

where $\|z_i - O_j\|$ denote the Euclidean distance between sample z_i and centroid O_j. Based on the aforementioned rules, we divide all samples in training data into three regions and apply different strategies for samples in each region.

3.4 Three-Way Strategies for Representation Learning

The advantage of the three-way decision lies in its ability to handle uncertainty through a three-region partitioning strategy. In this section, we apply different strategies to samples in each of the three regions to mitigate the impact of noise and design a loss function to guide model training.

(1) For samples in the $POS(gb_j)$, since they are considered to be label-consistent with the granular ball, we directly modify their labels to match the label of the granular ball, regardless of their original labels.
(2) For samples in the $NEG(gb_j)$, as they are deemed label-inconsistent with the granular ball and their true labels remain unknown, we exclude them from training to prevent any adverse effects.
(3) For samples in the $BND(gb_j)$, we propose a neighbor label-consistency strategy to further analyze these samples. Specifically, for each sample in the boundary region, we obtain the labels of its n nearest neighbors within the same granular ball $l = \{l_i^1, \ldots, l_i^n\}$ (we set $n=5$). If a certain proportion (above p_r, we set p_r as o.6) of these neighboring samples share the same label as the granular ball, the sample is label-consistent with gb_j and be reassigned to the positive region. Otherwise, it is assigned to the negative region. The decision rules are formulated as follows:

$$
\begin{aligned}
POS_U(gb_j) &= \{z_i \in BND(gb_j) | \frac{\sum_{i=1}^{n} \mathbb{I}[l_i^n = l_j]}{n} > p_r \}, \\
NEG_U(gb_j) &= \{z_i \in BND(gb_j) | \frac{\sum_{i=1}^{n} \mathbb{I}[l_i^n = l_j]}{n} \leq p_r \},
\end{aligned}
\tag{7}
$$

where $\mathbb{I}[\cdot]$ is an indicator function returning 1 if true, 0 otherwise. Similar to the label correction approach applied to samples initially assigned to the positive and negative regions, we update the labels of samples in $POS_U(gb_j)$ to match the label of gb_j, while discarding the samples in $NEG_U(gb_j)$.

Next, we design a loss function to train the model representation. For samples from regions $POS(gb_j)$ and $POS_U(gb_j)$, we first compute their cross-entropy loss, formulated as follows:

$$\mathcal{L}_{CE} = -\sum_{i=1}^{C} y_i \log \hat{y}_i, \tag{8}$$

where C is the number of total intent categories, y_i represents the correct label, and $\hat{y}_i$ denotes the predicted probability distribution.

Furthermore, for samples in $POS_U(gb_j)$ and $NEG_U(gb_j)$, we design a distance-based loss function to enhance representation learning. Specifically, for samples in $POS_U(gb_j)$, which are considered label-consistent with the granular ball, we encourage them to move closer to the centroid of gb_j. Conversely, for samples in $NEG_U(gb_j)$, which are regarded as label-inconsistent, we aim to push them farther from the centroid. Thus, we propose the following loss function for samples in $POS_U(gb_j)$ and $NEG_U(gb_j)$:

$$\mathcal{L}_{POS_U(gb_j)} = \sum_{z_i \in POS_U(gb_j)} \|z_i - O_j\|^2, \tag{9}$$

$$\mathcal{L}_{NEG_U(gb_j)} = \sum_{z_i \in NEG_U(gb_j)} \frac{1}{\|z_i - O_j\|^2}. \tag{10}$$

The overall training objective is written as follows:

$$\mathcal{L} = \mathcal{L}_{CE} + \mathcal{L}_{POS_U(gb_j)} + \mathcal{L}_{NEG_U(gb_j)}. \tag{11}$$

4   Experiments

4.1   Dateset

To evaluate the effectiveness of our approach, we conducted experiments on two widely used benchmark datasets. **StackOveflow** [21] is a text-based dataset containing 3,370,528 programming questions across 20 categories. **BANKING** [5] is an online banking inquiry dataset consisting of 13,080 instances and 77 intent classes. **CLINC** [8] contains 15,000 user queries covering 150 intent categories. This dataset can be evaluated using our method after introducing noise. For analysis, we randomly selected an equal number of examples from each category as a subset. Table 1 presents detailed statistics of these datasets.

Algorithm 1. Granular-Ball Three-Way Decision for Noise Handling
***
Require: Training dataset D, granular balls $\{gb_j\}$
Ensure: Processed dataset with label corrections
1: **for** each granular ball gb_j **do**
2:     Compute centroid O_j and radius r_j
3:     Determine lower and upper radius thresholds r_j^- and r_j^+
4:     **for** each sample $z_i \in gb_j$ **do**
5:         Compute Euclidean distance $d_i = \|z_i - O_j\|$
6:         **if** $d_i < r_j^-$ **then**
7:             Assign z_i to $POS(gb_j)$
8:         **else if** $d_i > r_j^+$ **then**
9:             Assign z_i to $NEG(gb_j)$
10:         **else**
11:             Assign z_i to $BND(gb_j)$
12:         **end if**
13:     **end for**
14:     **for** each sample $z_i \in BND(gb_j)$ **do**
15:         Retrieve labels of n nearest neighbors in gb_j
16:         Compute proportion p of neighbors sharing gb_j's label
17:         **if** $p > p_r$ **then**
18:             Assign z_i to $POS_U(gb_j)$
19:         **else**
20:             Assign z_i to $NEG_U(gb_j)$
21:         **end if**
22:     **end for**
23: **end for**
24: **for** each sample $z_i \in POS(gb_j) \cup POS_U(gb_j)$ **do**
25:     Modify label of z_i to match gb_j
26: **end for**
27: **for** each sample $z_i \in NEG(gb_j) \cup NEG_U(gb_j)$ **do**
28:     Discard z_i from training set
29: **end for**
30: Compute cross-entropy loss:

$$\mathcal{L}_{CE} = -\sum_{i=1}^{C} y_i \log \hat{y}_i$$

31: Compute distance-based loss:

$$\mathcal{L}_{POS_U(gb_j)} = \sum_{z_i \in POS_U(gb_j)} \|z_i - O_j\|^2$$

$$\mathcal{L}_{NEG_U(gb_j)} = \sum_{z_i \in NEG_U(gb_j)} \frac{1}{\|z_i - O_j\|^2}$$

32: Final loss function:

$$\mathcal{L} = \mathcal{L}_{CE} + \mathcal{L}_{POS_U(gb_j)} + \mathcal{L}_{NEG_U(gb_j)}$$

Table 1. Statistics of datasets.

Dataset	#Class	#Train	#Valid	#Test	Length
StackOverflow	20	12,000	2,000	6,000	9.18
BANKING	77	9,003	1,000	3,080	11.91
CLINC	150	15,000	3,000	5,700	3.37

4.2 Setting

In the representation learning phase, we employ a pre-trained BERT model [7], with the parameters of all layers kept constant except for the final layer. To facilitate fine-tuning of the last layer, we set the training batch size to 128 and the learning rate to 2×10^{-5}. The adaptive granular ball clustering is controlled through two properties: a purity limit, denoted as p_l, and a sample size limit, denoted as n_l. All hyperparameters are set according to the default configuration, minimizing the impact of hyperparameter tuning on the experimental results. The experiments are conducted on a system equipped with an Intel(R) Core(TM) i5-12400F processor and an NVIDIA GeForce RTX 4070 Ti GPU.

4.3 Main Results

To validate the superiority of our proposed RGB3WD method, we conduct comparative experiments with several state-of-the-art intent classification methods. These baselines are shown as follows:

- **SelfMix** [13]: A self-supervised learning-based intent classification method that enhances model generalization through a data augmentation strategy by mixing different samples. SelfMix leverages a combination of contrastive learning and data augmentation to facilitate semantic representation learning, thereby improving classification performance.
- **GBRAIN** [16]: A neural network-based intent classification method that incorporates an attention mechanism to construct global semantic relationships and utilizes graph structures to enhance intent recognition. GBRAIN captures both local and global dependencies in the input text, improving classification accuracy, particularly in low-resource scenarios.

We report the Accuracy (Acc) and F1-score (F1), which reflect the effectiveness of text classification under noisy labels, with the best-performing results highlighted in **bold**. Experiments are conducted on three datasets with noise-containing ratios of 10%, 20%, and 30%. The results are shown in Table 2.

Based on Acc and F1 as evaluation metrics, our RGB3WD method shows significant advantages on all three datasets under different noise levels. As the noise ratio increases, the RGB3WD method consistently outperforms the other two baseline methods. Specifically, when confronted with highly noisy data, RGB3WD not only maintains a high classification accuracy but also outperforms the other baseline methods in terms of F1. This indicates its robustness

Table 2. Results of intent classification across StackOverflow, BANKING, and CLINC with different noise ratios (10%, 20%, and 30%).

	Methods	StackOverflow		BANKING		CLINC	
		Acc	F1	Acc	F1	Acc	F1
10%	SelfMix	57.20	60.47	45.26	55.18	64.47	60.78
	GBRAIN	56.25	64.44	51.25	50.58	68.46	63.66
	RGB3WD	**85.40**	**76.30**	**83.95**	**84.08**	**77.28**	**81.05**
20%	SelfMix	66.80	66.59	41.88	53.61	45.44	57.19
	GBRAIN	50.00	58.84	46.56	40.56	59.62	56.76
	RGB3WD	**79.60**	**73.09**	**70.26**	**71.89**	**69.12**	**74.20**
30%	SelfMix	72.93	70.13	51.97	56.25	43.55	54.27
	GBRAIN	37.50	47.78	43.44	36.70	46.15	35.93
	RGB3WD	**77.53**	**72.40**	**65.13**	**70.21**	**68.07**	**72.64**

and stability in dealing with complex noisy environments. These results highlight that the RGB3WD method effectively enhances model performance when dealing with noise interference, providing more reliable outcomes.

4.4   Ablation Study

To further investigate the effectiveness of our proposed approach, we conducted a series of ablation experiments on three benchmark datasets: StackOverflow, BANKING, and CLINC. The main innovation of this paper lies in the granular ball representation and the three-way decision to mitigate the impact of noise on text classification. To validate the effectiveness of the proposed method, we conducted a series of experiments by removing these critical modules. Initially, we removed the three-way decision module (named "w/o 3WD") and directly utilized cross-entropy learning for representation. Subsequently, we further eliminated both the granular ball representation and three-way decision modules (named "w/o GB&3WD") to examine the changes in classification performance.

We introduced varying levels of noise into the datasets, specifically at 10%, 20%, and 30% noise ratios, to assess the robustness of our method. Table 3 presents the results of our ablation study. From the results, we observe a significant performance degradation when noise is introduced without any additional module. The "w/o GB&3WD " method consistently yields the lowest accuracy and F1 scores across all noise levels and datasets, demonstrating the negative impact of noise on classification performance. By incorporating granular ball clustering ("w/o 3WD"), performance improves considerably, indicating its effectiveness in mitigating noise to some extent. However, the combination of granular ball clustering with the three-way decision ("RGB3WD") achieves the highest accuracy and F1 scores across all datasets and noise levels, underscoring the complementary benefits of both techniques.

Table 3. Ablation study across StackOverflow, BANKING, and CLINC with different noise ratios (10%, 20%, and 30%).

	Methods	StackOverflow		BANKING		CLINC	
		Acc	F1	Acc	F1	Acc	F1
10%	**RGB3WD**	**85.40**	**76.30**	**83.95**	**84.08**	**77.28**	**81.05**
	w/o 3WD	83.07	75.36	73.03	75.40	67.03	70.44
	w/o GB&3WD	57.87	44.92	55.39	42.30	59.91	63.58
20%	**RGB3WD**	**79.60**	**73.09**	**70.26**	**71.89**	**69.12**	**74.20**
	w/o 3WD	61.73	57.70	61.69	65.54	51.93	57.07
	w/o GB&3WD	39.40	24.47	39.61	24.33	35.26	40.10
30%	**RGB3WD**	**77.53**	**72.40**	**65.13**	**70.21**	**68.07**	**72.64**
	w/o 3WD	57.33	55.31	50.79	55.58	48.95	54.40
	w/o GB&3WD	20.00	6.67	10.53	2.06	25.61	31.74

Furthermore, as the noise ratio increases, the gap between "RGB3WD" and other methods becomes more pronounced, highlighting the robustness of our approach in handling high-noise scenarios. These findings validate the necessity of both granular ball clustering and three-way decisions in achieving superior classification performance under noisy conditions.

4.5 Parametric Analysis

In this section, we will analyze all hyperparameters, the following hyperparameter analysis experiments are performed on the Stackoverflow dataset with a noise ratio of 20%.

The upper radius r_j^+ and lower radius r_j^- in our method are set to be $1\times$ the granular ball's radius r_j and $0.5\times$ the granular ball's radius r_j, respectively. To evaluate the impact of these parameters, we conduct experiments with varying upper and lower radii, and the results are presented in Fig. 2. The results indicate that a significantly reduced upper radius leads to a decline in performance. Similarly, the performance decreases when the lower radius is either excessively

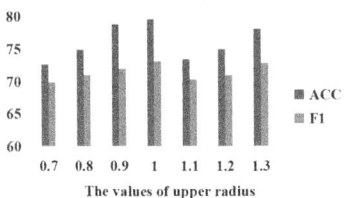 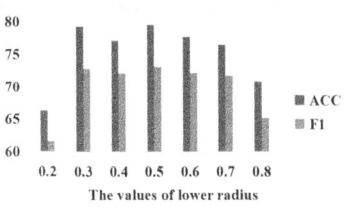

Fig. 2. The upper and lower radii are taken as granular ball's radius with different multiplicities.

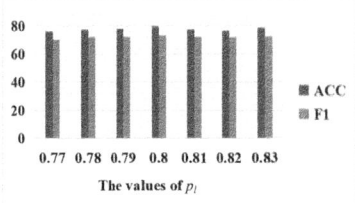 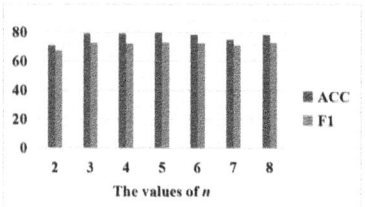

Fig. 3. Experimental results for different p_l and n.

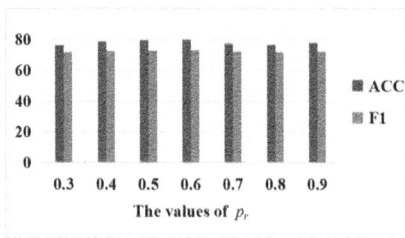

Fig. 4. Experimental results for different values of p_r.

large or too small. Based on these observations, we select a higher-level parameter setting to achieve optimal performance.

Next, we analyze the purity p_l and n. To evaluate the impact of these parameters, we conduct experiments using different values, and the corresponding results are presented in Fig. 3. The results indicate that the performance remains relatively stable across varying parameter values, demonstrating the robustness of the proposed method.

Finally, we analyze the parameter p_r, and the corresponding experimental results are presented in Fig. 4. The results indicate that the performance remains stable across different values of p_r, further demonstrating the robustness of the proposed method.

5   Conclusion

In this paper, we propose an intent classification method: Robust Granular Ball Three-Way Decision. Specifically, different sizes of granular balls are generated by adaptive granular ball clustering to reflect the real semantic structure during the representation learning process. In addition, considering the uncertainty of textual noise, we introduce three-way decision on the basis of granular ball clustering as a way to analyze the noise in a more refined way. Finally, the experiment proves the superiority of the RGB3WD method.

In future work, we will further enhance the RGB3WD method by exploring more adaptive and efficient clustering techniques to improve the accuracy of intent classification. Additionally, we aim to incorporate deep learning techniques, such as transformer-based models, to optimize feature representation learning.

References

1. Ali, M.N., Falavigna, D., Brutti, A.: Time-domain joint training strategies of speech enhancement and intent classification neural models. Sensors **22**(1), 374 (2022)
2. Ali, M.N., Schmalz, V.J., Brutti, A., Falavigna, D.: A speech enhancement front-end for intent classification in noisy environments. In: 2021 29th European Signal Processing Conference (EUSIPCO), pp. 471–475. IEEE (2021)
3. Amin, S.U., Alsulaiman, M., Muhammad, G., Mekhtiche, M.A., Hossain, M.S.: Deep learning for EEG motor imagery classification based on multi-layer CNNS feature fusion. Future Gener. Comput. Syst. **101**, 542–554 (2019)
4. Cao, X., Yang, X., Xia, S., Wang, G., Li, T.: Open continual feature selection via granular-ball knowledge transfer. IEEE Trans. Knowl. Data Eng. **36**(12), 8967–8980 (2024)
5. Casanueva, I., Temčinas, T., Gerz, D., Henderson, M., Vulić, I.: Efficient intent detection with dual sentence encoders. In: Proceedings of the Workshop on Natural Language Processing for Conversational AI, pp. 38–45 (2020)
6. Cheng, D., Liu, S., Xia, S., Wang, G.: Granular-ball computing-based manifold clustering algorithms for ultra-scalable data. Expert Syst. Appl. **247**, 123313 (2024)
7. Devlin, J., Chang, M.W., Lee, K., Toutanova, K.: Bert: Pre-training of deep bidirectional transformers for language understanding. In: Proceedings of the 2019 Conference of the North American Chapter of the Association for Computational Linguistics: Human Language Technologies, volume 1 (long and short papers), pp. 4171–4186 (2019)
8. Larson, S., et al. An evaluation dataset for intent classification and out-of-scope prediction. arXiv preprint arXiv:1909.02027 (2019)
9. Li, H., Zhou, X.: Risk decision making based on decision-theoretic rough set: a three-way view decision model. Int. J. Comput. Intell. Syst. **4**(1), 1–11 (2011)
10. Lin, T. E., Xu, H., Zhang, H.: Discovering new intents via constrained deep adaptive clustering with cluster refinement. In: Proceedings of the AAAI Conference on Artificial Intelligence, vol. 34, pp. 8360–8367 (2020)
11. Liu, D., Liang, D., Wang, C.: A novel three-way decision model based on incomplete information system. Knowl. Based Syst. **91**, 32–45 (2016)
12. Mondal, A., Roy, S.K: Behavioural three-way decision making with fermatean fuzzy mahalanobis distance: application to the supply chain management problems. Appl. Soft Comput. **151**, 111182 (2024)
13. Qiao, D., et al.: Selfmix: robust learning against textual label noise with self-mixup training. arXiv preprint arXiv:2210.04525 (2022)
14. Wang, F., et al. Residual attention network for image classification. In: Proceedings of the IEEE Conference on Computer Vision and Pattern Recognition, pp. 3156–3164 (2017)
15. Wang, Z., Li, J., Xia, S., Lin, L., Wang, G.: Text adversarial defense via granular-ball sample enhancement. In: Proceedings of the International Conference on Multimedia Retrieval, pp. 348–356 (2024)
16. Wang, Z., Zhang, T., Xia, S., Lin, L., Wang, G.: GBRAIN: Combating textual label noise by granular-ball based robust training. In: Proceedings of the International Conference on Multimedia Retrieval, pp. 357–365 (2024)

17. Xia, S., Dai, X., Wang, G., Gao, X., Giem, E.: An efficient and adaptive granular-ball generation method in classification problem. IEEE Trans. Neural Netw. Learn. Syst. **35**(4), 5319–5331 (2022)
18. Xia, S., Liu, Y., Ding, X., Wang, G., Yu, H., Luo, Y.: Granular ball computing classifiers for efficient, scalable and robust learning. Inf. Sci. **483**, 136–152 (2019)
19. Xia, S., Shi, B., Wang, Y., Xie, J., Wang, G., Gao, X. GBCT: efficient and adaptive clustering via granular-ball computing for complex data. IEEE Trans. Neural Netw. Learn. Syst. (2025)
20. Xie, Q., et al.: GBG++: a fast and stable granular ball generation method for classification. IEEE Trans. Emerg. Top. Comput. Intell. **8**(2), 2022–2036 (2024)
21. Xu, JM., et al.: Short text clustering via convolutional neural networks. In: Proceedings of the 1st Workshop on Vector Space Modeling for Natural Language Processing, pp. 62–69 (2015)
22. Yang, X., Li, Y.H., Li, T.R.: A review of sequential three-way decision and multigranularity learning. Int. J. Approximate Reasoning **152**, 414–433 (2023)
23. Yao, Y.Y.: Three-way decisions with probabilistic rough sets. Inf. Sci. **180**(3), 341–353 (2010)
24. Yao, Y.Y.: The superiority of three-way decisions in probabilistic rough set models. Inf. Sci. **181**(6), 1080–1096 (2011)
25. Ye, J., Zhan, J,M., Xu, Z.S.: A novel decision-making approach based on three-way decisions in fuzzy information systems. Inf. Sci. **541**, 362–390 (2020)
26. Zhan, J.M., Ye, J., Ding, W.P., Liu, P.E.: A novel three-way decision model based on utility theory in incomplete fuzzy decision systems. IEEE Trans. Fuzzy Syst. **30**(7), 2210–2226 (2021)
27. Zhang, H., Xu, H., Lin, T.E.: Deep open intent classification with adaptive decision boundary. In: Proceedings of the AAAI Conference on Artificial Intelligence, pp. 14374–14382 (2021)

Medicine and Health Data Mining

Interpretable Disease Progression Path for Next Admission Diagnosis Event Prediction in Healthcare Data via Hybrid Rule-Transformer Network

Fanxin Xu[1,2], Hong Yu[1,2], Zuqiang Su[1,2], Ping Zhang[4],
and Guoyin Wang[1,2,3(✉)]

[1] Chongqing Key Laboratory of Computational Intelligence, Chongqing University of
Posts and Telecommunications, Chongqing 400065, People's Republic of China
d230201041@stu.cqupt.edu.cn, {yuhong,suzq}@cqupt.edu.cn,
wanggy@cqnu.edu.cn
[2] Key Laboratory of Cyberspace Big Data Intelligent Security, Ministry of
Education, Chongqing, China
[3] National Center for Applied Mathematics in Chongqing, Chongqing Normal
University, Chongqing 401331, China
[4] Sichuan Huhui Software Co., LTD, Mianyang, China
zhangping@schhsw.com

Abstract. Next admission diagnosis event prediction is one of the core
tasks based on electronic health records. Existing researches on this task
mainly focus on learning accurate disease representations and then fusing
these disease representations into deep learning models for learning and
prediction. This approach implicitly models the disease progressing path
between diseases in the model, while ignoring the explicit disease progression path between diseases that can be established by the data itself.
Therefore, it is impossible to accurately and explicitly represent the disease progression path to achieve the purpose of interpreting the disease
progression path. For this problem, this paper propose the Hybrid Rule-
Transformer Network, it first represent the interpretable disease progression paths as positive region-based decision rules by Pawlak Rough
Set, then fuse the original disease embedding and the rule embedding
by a dynamic gating fusion strategy, and finally realize the extraction of
patient time series modeling information through the Transformer layer.
Extensive experiments on two real EHR datasets show that the model
this paper established has achieved state-of-the-art in terms of F1-score
and Recall, and provides process interpretability of the disease progression paths between visits.

Keywords: Disease Progression Path · Rough Set · Next Admission
Diagnosis Event Prediction · Transformer

1 Introduction

The advancement of Electronic Health Records (EHR)-driven predictive
analytics enables cross-institutional frameworks for forecasting subsequent

Q. Zhang et al. (Eds.): IJCRS 2025, LNAI 15710, pp. 171–183, 2025.
https://doi.org/10.1007/978-3-031-92741-6_13

hospitalization diagnoses, offering enhanced capabilities to tailor patient-centric interventions through data-informed clinical intelligence. The realization of a universal next admission diagnosis event prediction model through EHR data analysis is conductive to personalized care for patients [1,2]. In EHR data, admission diagnosis events are the most fine-grained medical concepts and event units that reflect the patient's health status. They can reflect the health information of the patient's current visit status. Usually, a single visit of a patient contains a single or multiple diagnosis events. Based on the medical concept representation learned from the diagnosis event data of multiple visits of the patient, many researches have been conducted in the next admission diagnosis event prediction.

From the perpective of the evolutionary procession path between disease events in different visits, the existing next admission diagnosis event prediction researches can be generally divided into four categories: the first category is based on the ontology graph hierarchy between disease events [2–5,7], which use the prior hierarchical graph structure information of disease events to enhance the visit embedding; the second category is based on the co-occurrence graph structure of the co-occurrence relationship between disease events [5–8], which use the co-occurrence characteristics between disease events to construct a co-occurrence graph and then enhance the visit embedding; the third category is based on the disease progression function in the same single disease event [9], which is to introduce the time factor to learn the degree of development of the same disease event over time, and then integrate it into the evolution of embedding between visits; the fourth category is based on the indepent node of events [10–12], which directly learns the evolution of the embedding between visits based on the original multi-hot embedding. In these researches, the disease progression paths in different visits are not explicitly modeled.

Aiming at the problem of how to explicitly model the disease progression paths, this paper represents the interpretable disease progression paths as positive region-based decision rules by Pawlak Rough Set [22], and represents them as the form of antecedents and consequents. Based on the disease progression path rules, this paper proposes a Hybrid Rule-Transformer Network, and successfully applies it to the next admission diagnosis event prediction task. The main contributions of this paper are as follows:

1. Designing a **Hybrid Rule-Transformer Network** (HRTN) that combines the rule base and Transformer for the next admission diagnosis event prediction.
2. Representing the interpretable disease progression paths as positive region-based decision rules by Pawlak Rough Set, and representing them as a rule base composed of rules in the form of antecedents and consequents.
3. A dynamic gated fusion strategy is designed to dynamically fuse the original diagnosis events embedding and the rule matched events embedding, and then the Transformer encoder is used to process the fused embedding of variable length visits.
4. Experiments on two real EHR datasets show that compared with existing models, the model proposed in this paper is either the best or the second best in terms of Recall and F1-score, reaching state-of-the-art (SOTA).

2 The Next Admission Diagnosis Event Prediction Problem Formulation

EHR data contain temporal patient records of visits to medical institutions. This section introduces the background knowledge of EHR data and the problem formulation of predicting the next admission diagnosis event task.

2.1 Diagnosis Event Codes

A key type in EHR is diagnosis event. Each admission diagnosis event can be represented as a set of unique diagnosis event codes, denoted as $C = \{c_1, c_2, \ldots, c_{|C|}\}$, where $|C|$ is the size of unique diagnosis event codes. Each code represents a diagnosis. The diagnosis event codes are typically predefined by modern disease classification systems, such as ICD-9-CM or ICD-10.

2.2 EHR Dataset

The EHR of each patient can be represented as time-sequenced array of diagnosis event codes. For a patient $u \in U$, where $|U|$ is the number of patients in EHR dataset, the diagnoses in the t-th visit are defined as a multi-hot vector $x_t \in \{0,1\}^{|C|}$. Then the historical visit information of patient u can be represented as $X^u = [x_1, x_2, \cdots, x_{T^u}], t = 1, \ldots, T^u$ indicates the index of visit, and T^u denotes the total visit times of patient u. The EHR dataset is defined as $D = \{X^u | u \in U\}$. To simplify, this paper explain the algorithms for one patient, omitting the superscript u for clear.

2.3 Disease Progression Path

Suppose the set of diagnosis events for the patient's t-th visit is $C_t = \{c_i | x_{t,i} = 1\} \subseteq C$, where $x_t \in \{0,1\}^{|C|}$ is a multi-hot vector representation, and $x_{t,i} = 1$ indicates that the t-th visit contains the diagnosis code c_i. The disease progression path mapping between two consecutive visits is defined as a binary relationship: $M_t : \mathcal{P}(C_t) \to \mathcal{P}(C_{t+1})$, where $\mathcal{P}(C_t)$ is the power set of C_t, representing any combination (single or multiple) of diagnosis events in the previous visit, $\mathcal{P}(C_{t+1})$ is the possible combination of diagnosis events in the next visit.

The specific mapping can be expressed as: $\forall S_t \subseteq C_t, M_t(S_t) = \cup_{c \in S_t} \phi(c) \cup \psi(S_t)$, where $\phi(c) : C \to \mathcal{P}(C)$ represent disease progression path of single diagnosis event (such as complications, outcome relationships), $\psi(S_t) : \mathcal{P}(C) \to \mathcal{P}(C)$ represent disease progression path of multi diagnosis events (e.g., comorbidities triggering new diseases).

2.4 Next Admission Diagnosis Event Prediction

Given an EHR dataset D, a patient u and u's previous diagnosis events X, next admission diagnosis event prediction is to predict the diagnosis events y^{T+1} of the future visit $T+1$. For example, the ground-truth of next admission diagnosis event prediction in the visit $T + 1$ is the diagnosis events $y^{T+1} \in \{0,1\}^{|C|}$.

3   Methodology

This section mainly introduce the method for positive region-based decision rules extraction, followed by details of the proposed Hybrid Rule-Transformer Network (HRTN). As illustrated in Fig. 1, the overall framework of HRTN comprises four core components: multi-hot vector representation, positive region-based rules matching, dynamic gate fusion for multi-embedding, and transformer-enhanced temporal modeling. The input EHR data for HRTN model is first constructed into multi-hot vector representation in each visit. Then the input of visit is matched with the positive region-based rules to generate matched events. Subsequently fusing the original diagnosis events embedding and the rule matched events embedding as the input for the transformer layer by dynamic gate strategy. Finally, the output of the last visit processed by the Hybrid Rule-Transformer Network is input into the predictor.

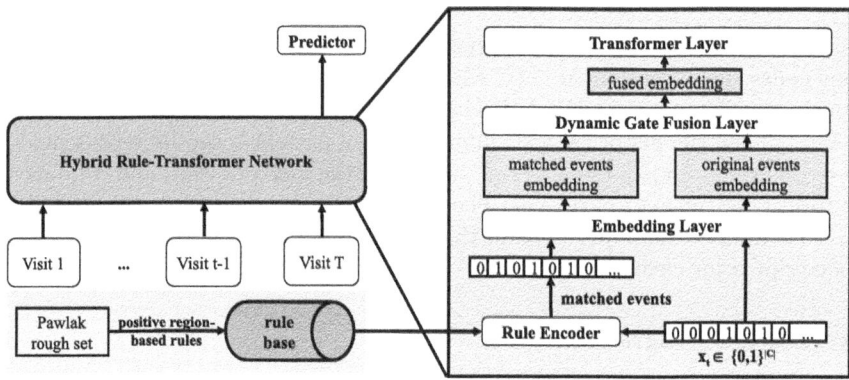

Fig. 1. The overall framework of the proposed Hybrid Rule-Transformer Network.

3.1   Positive Region-Based Decision Rules Extraction

In the definition of Sect. 2.3, this paper defines the disease progression path as the mapping of any combination (single or multiple) of diagnosis events in the previous visit to the possible combination of diagnosis events in the next visit. Pawlak Rough Set is naturally suitable for reliable rule extraction in this problem. Representing the interpretable disease progression paths as positive region-based decision rules by Pawlak Rough Set.

Algorithm 1. Positive Region-based Decision Rules Extraction Algorithm

---

Require: Raw EHR dataset $P = \{[x_1, \ldots, x_n, y]\}$
Ensure: Rule base $\mathcal{R}_{\det}$
1: **Step 1: Construct Decision Table**
2: **for all** patient $p \in P$ **do**
3:     **for** $t = 1$ to $n - 1$ **do**
4:         $\mathcal{CA}_t^p \leftarrow x_t$ {Current visit events}
5:         $\mathcal{DA}_t^p \leftarrow x_{t+1}$ {Next visit events}
6:         $UT \leftarrow UT \cup \{(\mathcal{CA}_t^p, \mathcal{DA}_t^p)\}$
7:     **end for**
8: **end for**
9: **Step 2: Iterate Through Decision Events**
10: **for all** decision event $d \in \mathcal{DA}$ **do**
11:     **Compute Positive Region:**
12:     $\mathrm{POS}_{\mathcal{CA}}(d) \leftarrow \bigcup\{E \in UT/\mathcal{CA} \mid E \subseteq [d]_{\mathcal{DA}}\}$
13:     **Step 3: Attribute Reduction**
14:     $R \leftarrow \emptyset$
15:     **while** $\exists a \in \mathcal{CA} \setminus R$ such that $\gamma_{R \cup \{a\}}(d) > \gamma_R(d)$ **do**
16:         Select attribute a^* that maximizes $\gamma_{R \cup \{a\}}(d)$
17:         $R \leftarrow R \cup \{a^*\}$
18:     **end while**
19:     **Step 4: Generate Deterministic Rules**
20:     **if** $\gamma_R(d) = \gamma_{\mathcal{CA}}(d)$ **and** $\mathrm{conf}(R \to d) \geq \tau_{\mathrm{conf}}$ **then**
21:         $\mathcal{R}_{\det} \leftarrow \mathcal{R}_{\det} \cup \{R \to d\}$
22:     **end if**
23: **end for**

Suppose the set of diagnosis events for the patient's t-th visit is $C_t = \{c_i | x_{t,i} = 1\} \subseteq C$, where $x_t \in \{0,1\}^{|C|}$ is a multi-hot vector representation, and $x_{t,i} = 1$ indicates that the t-th visit contains the diagnosis code c_i. The set of diagnosis events for the patient's $t + 1$-th visit is $C_{t+1} = \{c_i | x_{t+1,i} = 1\} \subseteq C$. Naturally, the two visits can form a decision table consisting of conditional attributes and decision attributes in the Pawlak Rough Set, defined as $UT = \{(X_t, Y_t) \mid X_t = C_t, \ Y_t = C_{t+1}\}_{t=1}^{n-1}$. Each condition-decision pair represents the disease processing path from a single visit event to the next event of the patient. This paper first converts the decision table into the representation of $UT = (\mathcal{CA}, \mathcal{DA})$, where $\mathcal{CA}$ represents the condition attribute and $\mathcal{DA}$ represents the decision attribute, which is convenient for subsequent understanding.

For each decision event $d \in \mathcal{DA}$, defining its decision class as $[d]_{\mathcal{DA}} = \{x \in UT \mid x_d = 1\}$. The detailed postive region-based decision rules extraction algorithm can be seen in Algorithm 1. It mainly traverses the decision events and calculates the positive region of each decision event, defined as $\mathrm{POS}_{\mathcal{CA}}(d) = \bigcup\{E \in UT/\mathcal{CA} \mid E \subseteq [d]_{\mathcal{DA}}\}$, indicating that the conditional attribute can uniquely determine the sample set of decision event d. The positive region is determined by the positive region coverage $\gamma_{\mathcal{CA}}(d) = \frac{|\mathrm{POS}_{\mathcal{CA}}(d)|}{|UT|}$. Based on the determined positive region set, the minimum subset $R \subseteq \mathcal{CA}$ is found from the conditional attribute set $\mathcal{CA}$ so that the positive region coverage remains unchanged, the optimization goal can be formulated as $\min |R| s.t. \gamma_{\mathcal{CA}}(d) = \gamma_R(d)$. Based on the simplified attribute set R, the deterministic rules is generated, which satisfied $\frac{|\mathrm{POS}_R(d) \cap [d]_{\mathcal{DA}}|}{|\mathrm{POS}_R(d)|} \geq \tau_{\mathrm{conf}}$.

3.2   Hybrid Rule-Transformer Network

Rule Encoder. The core task of the Rule Encoder designed in this paper is to encode the rule base into a computable tensor form to achieve fast matching of rules with input data. The detailed algorithm can be seen in Algorithm 2. This component mainly consists of three subcomponents: 1. Initialization: parsing the rules in the rule base into rule antecedent matrix and rule consequent matrix. 2. Rule Matching: achieving accurate matching of input data with the rules in the rule base. 3. Event Prediction: Outputing the contribution of the successfully matched rules and the probability distribution of the diagnosis event level. The contribution of the rules is used to reflect the contribution of each rule to the current input, and the probability distribution of the diagnosis event level represents the possibility of the target diagnosis event derived by the rules.

Algorithm 2. Rule Encoder

Require: $\mathbf{X} \in \{0,1\}^{|C|}$ {input event vector}, $\mathcal{R}_{\mathrm{det}}$ {Rule base}
Ensure: $\mathbf{S}_{\mathrm{weighted}} \in [0,1]^{|\mathcal{R}_{\mathrm{det}}|}$ {Weighted rule scores}, $\mathbf{E} \in \{0,1\}^{|C|}$ {matched event vector}
1: **Initialization:**
2: $\mathbf{M} \leftarrow \mathrm{BinaryMask}(\mathcal{R}_{\mathrm{det}})$ {Rule antecedent matrix $\in \{0,1\}^{|\mathcal{R}_{\mathrm{det}}| \times |C|}$}
3: $\mathbf{T} \leftarrow \mathrm{TargetMatrix}(\mathcal{R}_{\mathrm{det}})$ {Rule consequent matrix $\in \{0,1\}^{|\mathcal{R}_{\mathrm{det}}| \times |C|}$}
4: $\mathbf{W} \leftarrow [\mathrm{conf}(r) \forall r \in \mathcal{R}_{\mathrm{det}}]$ {Learnable weight vector $\in \mathbb{R}^{|\mathcal{R}_{\mathrm{det}}|}$}
5: **Rule Matching:**
6: $\mathbf{S} \leftarrow \mathbf{X} \cdot \mathbf{M}^{\top}$ {Raw matching scores $\in \mathbb{N}^{|\mathcal{R}_{\mathrm{det}}|}$}
7: $\mathbf{M}_{\mathrm{match}} \leftarrow (\mathbf{S} \geq \mathrm{sum}(\mathbf{M}, \dim = 1) - \epsilon)$ {Exact match matrix, $\epsilon = 10^{-6}$}
8: $\mathbf{S}_{\mathrm{weighted}} \leftarrow \mathbf{M}_{\mathrm{match}} \odot \mathbf{W}^{\top}$ {Confidence weighting}
9: **Event Prediction:**
10: $\mathbf{E} \leftarrow \sigma(\mathbf{S}_{\mathrm{weighted}} \cdot \mathbf{T} - \tau)$ {Threshold activation}

11: where $\sigma(x) = \begin{cases} 1 & x \geq 0 \\ 0 & \text{otherwise} \end{cases}$

Dynamic Gated Fusion Strategy. The core task of the Dynamic Gated Fusion Strategy designed in this paper is to fuse the original diagnosis events embedding and the rule matched events embedding through dynamic weight allocation. The detailed algorithm can be seen in Algorithm 3. This component consists of three subcomponents: 1. Embedding Concatenation: it converts the original event and the matching event into corresponding embeddings and then concatenates them to obtain a joint representation. 2. Gating Weight Calculation: it generates gating weights through fully connected layers and nonlinear activation functions. The gating weights reflect the importance of data-driven embedding in the current visit. 3. Weighted Embedding Fusion: it uses gated weights to perform a weighted sum of the two embeddings to obtain a fused embedding, so that the embedding obtained by data-driven learning and the embedding obtained by rule matching can be adaptively fused in the same embedding space to achieve complementary advantages between the two.

Algorithm 3. Dynamic Gated Fusion Strategy

Require: $\mathbf{M} \in \mathbb{R}^{|C|}$ {original events vector}, $\mathbf{R} \in \mathbb{R}^{|C|}$ {Rule-based matched events vector}, d {embedding size}
Ensure: Fused embedding $\mathbf{F} \in \mathbb{R}^d$
  1: **Step 1: Eembedding Concatenation**:
  2: $\mathbf{M} = \text{Embedding}(\mathbf{M})$ $(\mathbf{M} \in \mathbb{R}^d)$
  3: $\mathbf{R} = \text{Embedding}(\mathbf{R})$ $(\mathbf{R} \in \mathbb{R}^d)$
  4: $\mathbf{H} = \text{Concat}(\mathbf{M}, \mathbf{R})$ $(\mathbf{H} \in \mathbb{R}^{2d})$
  5: **Step 2: Gating Weight Calculation**:
  6: $\mathbf{G} = \sigma\left(\mathbf{W}_g \mathbf{H} + \mathbf{b}_g\right)$ $(\mathbf{W}_g \in \mathbb{R}^{2d \times d}, \sigma = \text{Sigmoid})$
  7: **Step 3: Weighted Embedding Fusion**:
  8: $\mathbf{F} = \mathbf{G} \odot \mathbf{M} + (\mathbf{1} - \mathbf{G}) \odot \mathbf{R}$ $(\odot = \text{element-wise multiplication})$
  9: **return F**

Transformer for Temporal Information Extraction. Based on the fusion embedding formed by Algorithm 3, HRTN uses Transformer for temporal information extraction. In this model, Transformer uses a 1-layer encoder to extract the global dependencies of the patient's visit sequence. The multi-head self-attention mechanism is used to dynamically calculate the association weights between historical diagnosis events, and finally through the effective time step feature extraction (taking the last visit hidden state), the long-term temporal context is aggregated for the final Predictor.

4   Experiments

4.1   Experimental Setting

Datasets and Preprocessing. This paper use MIMIC-III [20] and MIMIC-IV [21] to validate the predictive results of HRTN. MIMIC-III contains 7,493 patients with multiple visits ($T \geq 2$) from 2001 to 2012. MIMIC-IV contains 85,155 patients with multiple visits from 2008 to 2019. To avoid temporal overlap with MIMIC-III, this paper randomly sample 10000 patients from MIMIC-IV from 2013 to 2019. Spliting the two datasets based on patients into training/validation/test sets randomly, which contain 6,000/493/1,000 patients for MIMIC-III and 8,000/1,000/1,000 for MIMIC-IV, respectively. See Table 1 for dataset statistics. The number of the positive region-based decision rules in MIMIC-III is 1529, in MIMIC-IV is 1772.

Table 1. Dataset Statistics Comparison between MIMIC-III and MIMIC-IV

Metric	MIMIC-III	MIMIC-IV
# Patients	7,493	10,000
– Max. # visits	42	77
– Avg. # visits	2.66	3.70
# Codes	4,880	5995
– Max. # codes per visit	39	39
– Avg. # codes per visit	13.06	13.49

Baselines. To compare HRTN with state-of-the-art models, selecting the following methods as baselines:

– multi-hot embedding-based model: RETAIN [10], Deepr [11], Dipole [12].
– ontology graph-based model: GRAM [3], KAME [13], KNOWRISK [15], G-BERT [16], CGL [4].
– co-occurrence relationship-based model: DG-RNN [14], GCNN [18], Chet [6], BioDynGraph [19].
– disease progression function-based model: Timeline [9], HiTANet [17].

Parameter Settings. The embedding size d is 48. The confidence for τ_{conf} is set to 0.99 to obtain the deterministic rules of Pawlak Rough Set, and is not set to 1 to account for possible errors in floating-point calculations. The batch size is 32. The epochs is setting as 50. The optimizer is AdamW. The learning rate is set as 0.005, reduced to 0.001 at the 15th epoch, and reduced to 0.0005 at the 30th epoch. The proposed model is implemented using Python 3.11.5 and PyTorch 2.6.0 on a machine with i7-1255U CPU, 16 GB memory, and T550 Laptop GPU.

Evaluation Metrics. Evaluation metrics for next admission diagnosis event prediction are weighted F1 score (F1-score) and top-k recall (R@k). F1-score is the weighted sum of the F1 scores of all diagnosis events. It measures the overall effectiveness of predictions across all diagnosis events. R@k is the average ratio of required diagnosis events in the top-k predictions to the total number of required diagnosis events per visit. It measures the accuracy of predictions for a subset of diagnosis events.

4.2   Experimental Results

This paper compared the prediction performance of the next admission diagnosis event of different models on the MIMIC-III and MIMIC-IV datasets. See Table 2 fore detailed compared results, See Figure 2 and 3 for detailed comprehensive comparison of evaluation metrics in MIMIC-IV and MIMIC-III datasets.

Table 2. Next Admission Diagnosis Event Prediction results on the MIMIC-III and MIMIC-IV datasets of best parameter settings are presented in terms of R@k (%) and F1 score (%). The bold ones are the best values, and the underlined ones are the second-best values.

Model	MIMIC-III			MIMIC-IV		
	R@10	R@20	F1-score	R@10	R@20	F1-score
RETAIN (2016) [10]	26.13	35.08	20.69	28.02	34.46	24.71
Deepr (2017) [11]	24.74	33.47	18.87	26.29	33.93	24.08
Dipole (2017) [12]	24.98	34.02	19.35	27.38	35.48	23.69
Timeline (2018) [9]	25.75	34.83	20.46	29.00	37.13	25.26
HiTANet (2020) [17]	26.02	35.97	21.15	27.45	36.37	24.92
GRAM (2017) [3]	26.51	35.80	21.52	27.29	36.36	23.50
KAME (2018) [13]	24.97	33.99	20.10	25.10	34.85	21.88
KNOWRISK (2019) [15]	25.11	34.95	20.88	27.78	36.10	25.11
G-BERT (2019) [16]	25.86	35.31	19.88	27.16	35.86	24.49
CGL (2021) [4]	26.64	36.72	21.92	28.52	37.15	25.41
DG-RNN (2019) [14]	26.75	36.81	22.01	27.92	37.10	25.75
GCNN (2021) [18]	27.18	37.52	23.11	28.91	37.11	25.97
Chet (2022) [6]	**28.64**	37.87	22.63	<u>30.28</u>	38.69	26.35
BioDynGraph (2024) [19]	28.15	<u>38.10</u>	**25.21**	30.13	<u>38.95</u>	<u>27.09</u>
HRTN(**Ours**)	<u>28.35</u>	**38.56**	<u>24.83</u>	**31.32**	**40.19**	**28.03**

From the perspective of best results, in the MIMIC-III dataset, Chet [6] ranks the best at R@10 in this dataset with an R@10 of 28.64%, while the HRTN model better than it with an R@20 of 38.56% and an F1-score of 24.83%, respectively; BioDynGraph [19] ranks the best at F1-score in this dataset with an F1-score of 25.21%, while the HRTN model better than it with an R@10 of 28.35% and an R@20 of 38.56% respectively. In the MIMIC-IV dataset, the HRTN model has made breakthroughs across the board: R@10 reaches 31.32%, which ranks first; R@20 reaches 40.19%, which ranks first; and the F1-score reaches 28.03%, which ranks first. The two datasets together show that HRTN is the SOTA model in comprehensive comparison of evaluation metrics.

From the perspective of disease progression path, in MIMIC-IV dataset, the F1-scores of the multi-hot embedding-based model such as RETAIN [10], Deepr [11], Dipole [12] were all below 25%, indicating that simple multi-hot embedding struggles to capture complex disease relationships. the ontology graph-based models such as GRAM [3], KAME [13], KNOWRISK [15], enhanced the embedding by medical ontology graph, these models achieved improved R@20. However, their limited ability to model temporal relationships to further performance gains. The disease progression function-based models such as Timeline [9], HiTANet [17], achieved strong R@10 performance(29% on MIMIC-IV), underscoring the critical role of the temporal path. The co-occurrence relationship-based models such as Chet [6] and BioDynGraph [19] were greater than 25%, their focus on statistical correlations may overlook temporal dependencies in disease progression. In MIMIC-III dataset, the proposed disease progression path-based model HRTN is the greatest model than the ontology graph-based models, co-occurrence relationship-based models and the disease progression function-based models in comprehensive evaluation metrics. The newly proposed HRTN not only leads in the core metrics of the two datasets, but also shows stronger adaptability in the larger-scale MIMIC-IV.

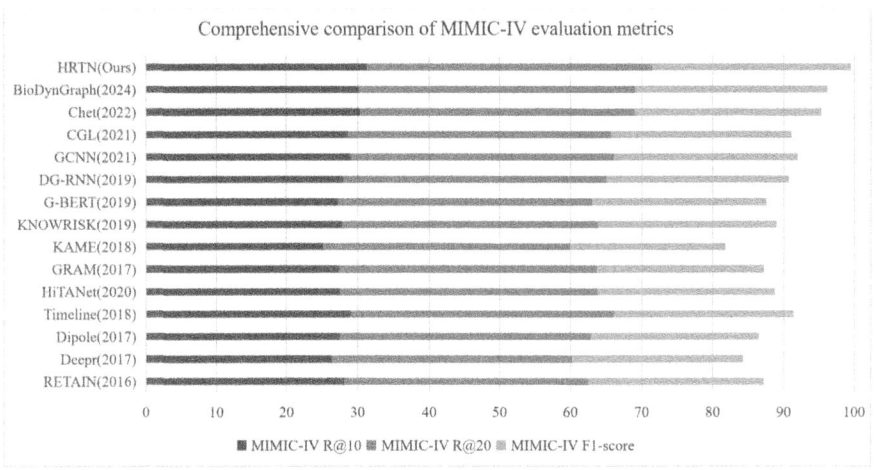

Fig. 2. The comprehensive comparison of MIMIC-IV evaluation metrics.

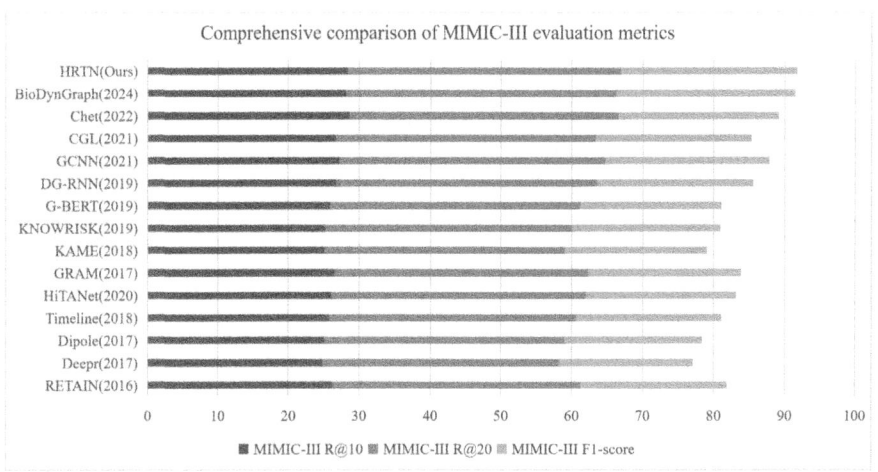

Fig. 3. The comprehensive comparison of MIMIC-III evaluation metrics.

4.3   Ablation Study

To further analyze the effectiveness of the Rule Encoder module in HRTN, we compare the HRTN variant without Rule Encoder with HRTN on the MIMIC-III and MIMIC-IV datasets:

– HRTN(R-): Removing the Rule Encoder module with the Dynamic Gate Fusion module in HRTN.

Table 3. Next Admission Diagnosis Event Prediction results for the HRTN variant without Rule Encoder on the MIMIC-III and MIMIC-IV datasets.

Models	MIMIC-III			MIMIC-IV		
	R@10	R@20	F1-score	R@10	R@20	F1-score
HRTN(R-)	26.71	36.91	23.24	29.29	39.11	25.38
HRTN	**28.35**	**38.56**	**24.83**	**31.32**	**40.19**	**28.03**

Table 3 shows the results of next admission diagnosis event prediction for HRTN and HRTN(R-) on the MIMIC-III and MIMIC-IV datasets. Experimental results show that the removal of the rule encoder significantly reduces the model performance. In the MIMIC-III dataset, R@10 decreased by 5.78% (from 28.35 to 26.71), R@20 decreased by 4.28% (from 38.56 to 36.91), and F1-score decreased by 6.40% (from 24.83 to 23.24); In the MIMIC-IV dataset, the performance degradation was more obvious, R@10 decreased by 6.48% (from 31.32 to 29.29), R@20 decreased by 2.69% (from 40.19 to 39.11), and F1-score decreased by 9.45% (from 28.03 to 25.38). These quantitative results strongly demonstrate the key role of the Rule Encoder module with the Dynamic Gate Fusion module in the model's predictive ability. The results confirm that decision rules and temporal dynamics are interdependent: rules compensate for data-driven models' inability to capture rare diagnostic logic, while the fusion mechanism enables context-aware synergy. The severe performance degradation of HRTN(R-) validates that these components are non-redundant and essential for diagnostic reasoning capabilities.

5   Conclusion

In this paper, to address the problem that the disease progression paths is not explicitly modeled in the next admission diagnosis event prediction, the proposed HRTN by representing the interpretable disease progression paths as positive region-based decision rules and designing a rule encoder to match the rules for explicitly showing that which disease progression path is activated. Transformer layer is adopting to extract the fused information between original events and matched events. Experimental results on two real EHR datasets demonstrate the effectiveness of the proposed method on the next admission diagnosis event prediction task. In the future, hoping that more researchers can enhance the interpretability of explicit modeling of disease progression paths from different perspectives.

Acknowledgments. This work has been supported by the National Key Research and Development Program of China under grant 2021YFF0704100.

References

1. Xu, Y., et al.: VecoCare: visit sequences-clinical notes joint learning for diagnosis prediction in healthcare data. In: Proceedings of the 32nd International Joint Conference on Artificial Intelligence (IJCAI 2023), pp. 4921–4929 (2023)
2. Lv, H., Chen, Z., Yang, Y., Ma, G., Tao, Y., Yang, C.: BoxCare: a box embedding model for disease representation and diagnosis prediction in healthcare data. In: Companion Proceedings of the ACM Web Conference 2024, pp. 1130–1133 (2024)
3. Choi, E., Bahadori, M.T., Song, L., Stewart, W.F., Sun, J.: GRAM: graph-based attention model for healthcare representation learning. In: Proceedings of the 23rd ACM SIGKDD International Conference on Knowledge Discovery and Data Mining (KDD 2017), pp. 787–795 (2017)
4. Lu, C., Reddy, C.K., Chakraborty, P., Kleinberg, S., Ning, Y.: Collaborative graph learning with auxiliary text for temporal event prediction in healthcare. In: Proceedings of the 30th International Joint Conference on Artificial Intelligence (IJCAI 2021), pp. 3529–3535 (2021)
5. Lu, C., Reddy, C.K., Ning, Y.: Self-supervised graph learning with hyperbolic embedding for temporal health event prediction. IEEE Trans. Cybern. **53**(4), 2124–2136 (2021)
6. Lu, C., Han, T., Ning, Y.: Context-aware health event prediction via transition functions on dynamic disease graphs. In: Proceedings of the 36th AAAI Conference on Artificial Intelligence (AAAI 2022), vol. 36, no. 4, pp. 4567–4574 (2022)
7. Sun, Z., Yang, X., Feng, Z., Xu, T., Fan, X., Tian, J.: EHR2HG: modeling of EHRs data based on hypergraphs for disease prediction. In: 2022 IEEE International Conference on Bioinformatics and Biomedicine (BIBM), pp. 1730–1733. IEEE (2022)
8. Tan, Y., et al.: Enhancing personalized healthcare via capturing disease severity, interaction, and progression. In: 2023 IEEE International Conference on Data Mining (ICDM), pp. 1349–1354. IEEE (2023)
9. Bai, T., Zhang, S., Egleston, B.L., Vucetic, S.: Interpretable representation learning for healthcare via capturing disease progression through time. In: Proceedings of the 24th ACM SIGKDD International Conference on Knowledge Discovery & Data Mining (KDD 2018), pp. 43–51 (2018)
10. Choi, E., Bahadori, M.T., Sun, J., Kulas, J., Schuetz, A., Stewart, W.: RETAIN: an interpretative predictive model for healthcare using reverse time attention mechanism. Adv. Neural. Inf. Process. Syst. **29**, 1–9 (2016)
11. Nguyen, P., Tran, T., Wickramasinghe, N., Venkatesh, S.: Deepr: a convolutional net for medical records. IEEE J. Biomed. Health Inform. **21**(1), 22–30 (2017)
12. Ma, F., Chitta, R., Zhou, J., You, Q., Sun, J., Gao, J.: DIPOLE: diagnosis prediction in healthcare via attention-based bidirectional recurrent neural networks. In: Proceedings of the 23rd ACM SIGKDD International Conference on Knowledge Discovery and Data Mining (KDD 2017), pp. 1903–1911 (2017)

13. Ma, F., You, Q., Xiao, H., Chitta, R., Zhou, J., Gao, J.: KAME: knowledge based attention model for diagnosis prediction in healthcare. In: Proceedings of the 27th ACM International Conference on Information and Knowledge Management (CIKM 2018), pp. 743–752 (2018)
14. Yin, C., Zhao, R., Qian, B., Lv, X., Zhang, P.: Domain knowledge guided deep learning with electronic health records. In: 2019 IEEE International Conference on Data Mining (ICDM 2019), pp. 738–747. IEEE (2019)
15. Zhang, X., Qian, B., Li, Y., Yin, C., Wang, X., Zheng, Q.: KnowRisk: an interpretable knowledge-guided model for disease risk prediction. In: 2019 IEEE International Conference on Data Mining (ICDM 2019), pp. 1492–1497. IEEE (2019)
16. Shang, J., Ma, T., Xiao, C., Sun, J.: Pre-training of graph augmented transformers for medication recommendation. In: Proceedings of the 28th International Joint Conference on Artificial Intelligence (IJCAI 2019), pp. 5953–5959. IJCAI (2019)
17. Luo, J., Ye, M., Xiao, C., Ma, F.: HiTANet: hierarchical time-aware attention networks for risk prediction on electronic health records. In: Proceedings of the 26th ACM SIGKDD International Conference on Knowledge Discovery & Data Mining (KDD 2020), pp. 647–656 (2020)
18. Ramirez, R., et al.: Prediction and interpretation of cancer survival using graph convolution neural networks. Methods **192**, 120–130 (2021)
19. Li, Q., You, T., Chen, J., Zhang, Y., Du, C.: BioDynGrap: biomedical event prediction via interpretable learning framework for heterogeneous dynamic graphs. Expert Syst. Appl. **244**, 122964 (2024)
20. Johnson, A.E., et al.: MIMIC-III, a freely accessible critical care database. Sci. Data **3**(1), 1–9 (2016)
21. Johnson, A.E., et al.: MIMIC-IV, a freely accessible electronic health record dataset. Sci. Data **10**(1), 1 (2023)
22. Pawlak, Z.: Rough sets. Int. J. Comput. Inf. Sci. **11**(5), 341–356 (1982)

ConformalRefiner: Retinal Vessel Topology Reconstruction via Conformal Risk Control

Xiaolong Pang[1] , Zhipeng Wei[2] , Jie Shi[2] , and Xiaodong Yue[1,2,3(✉)]

[1] Artificial Intelligence Institute of Shanghai University, Shanghai, China
{pxlames,yswantfly}@shu.edu.cn
[2] School of Computer Engineering and Science, Shanghai University, Shanghai, China
{weizhipeng,jieshi}@shu.edu.cn
[3] VLN Lab, NAVI MedTech Co., Ltd., Shanghai, China

Abstract. Retinal vessel segmentation has been widely applied in ophthalmic disease diagnosis. However, current deep-learning-based vessel segmentation methods still suffer from disconnected vessel structures. They struggle with noise interference and low signal-to-noise ratios in difficult-to-separate regions. We typically set fixed threshold values for class probabilities output by the network to separate the results, but this can lead to ignoring vessels in ambiguous regions. Furthermore, current threshold selection methods do not take into account the overall distribution of the sample population. To address these issues, we propose a plug-and-play vessel segmentation reconstruction network, ConformalRefiner, which employs Conformal Risk Control. Firstly, we use a threshold calibration method based on conformal risk control theory to alleviate the uncertainty in selecting an initial threshold, thereby obtaining a significant threshold value that includes more challenging vessels in the calibrated result. Additionally, we design a dual-input reconstruction network that utilizes topological calibration outcomes to guide the reconstruction of the initial segmentation mask. Finally, to tackle the issue of noise introduced by the calibration, we employ vascular topology priors to further enhance performance. Experimental results show our approach outperforms state-of-the-art methods on DRIVE and FIVE datasets, especially in topological connectivity metrics.

Keywords: Retinal vessel segmentation · Conformal risk control ·
Topology prior · Deep learning

1 Introduction

The structure of the retinal vascular system contains important information and changes in its morphological features are associated with various diseases [1]. Fundus retinal vascular segmentation is vital in diagnosing retinopathy and helps ophthalmologists detect and diagnose various retinopathies [2].

Q. Zhang et al. (Eds.): IJCRS 2025, LNAI 15710, pp. 184–198, 2025.
https://doi.org/10.1007/978-3-031-92741-6_14

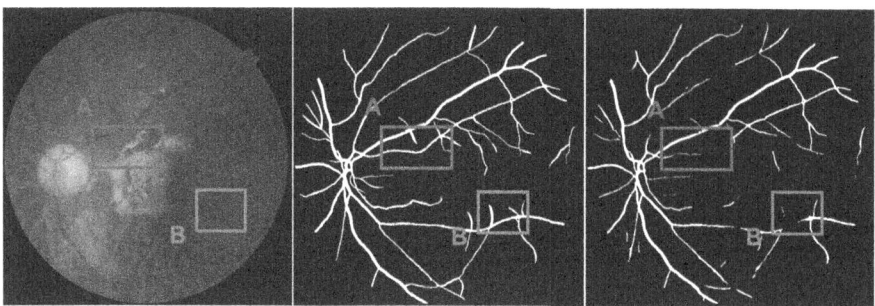

Fig. 1. Example images and corresponding segmentations obtained using the UNet model, illustrating two key challenges: (A) vessel segmentation impaired by lesion-induced noise that interferes with accurate blood vessel recognition, and (B) topological discontinuities in low signal-to-noise regions where fine vascular structures fail to be properly identified.

While deep learning has made significant breakthroughs in the field of medical image segmentation, vessel segmentation still faces serious challenges [3]. As shown in Fig. 1, these challenges primarily manifest in two aspects: on the one hand, lesion-induced noise interferes with blood vessel recognition, making it difficult to accurately identify vascular structures in affected areas; on the other hand, in low signal-to-noise ratio regions, blood vessel identification becomes challenging, often leading to topological discontinuities in the segmentation results. Traditional fixed-threshold methods exhibit significant limitations, failing to adequately segment vessels in these difficult regions.

Traditional segmentation tasks focus too much on improving overlap accuracy, making it difficult for models to pay attention to subtle topological structures. In recent years, various methods have been introduced to address the challenges of segmentation topological errors. The existing methods can be broadly categorized into three types: (1) Network-based methods [4–6] adopt nested network structures that can capture features of both large and fine blood vessels simultaneously or improve sensitivity to vessels by introducing deformable convolution structures. However, due to the tiny proportion of vessels in the entire image, these methods inevitably face the issue of severe category imbalance, causing the model to tend to learn background category features during the process, significantly reducing the accuracy of vessel recognition. (2) Loss-based methods [7–9] model the connectivity, boundaries, and topological features of vessels and introduce these topological constraints during the optimization process. However, such methods are still susceptible to interference from noise in low signal-to-noise ratio regions. (3) Post-processing segmentation methods [10–13] play a specific role in repairing vessel interruptions. However, this usually relies heavily on the segmentation results of the first stage, which is incapable of reconstructing the missing vessels from the first phase.

The above methods demonstrate limitations in addressing topological connectivity, and their fixed threshold selection strategy is not well-adapted to this task. To address these issues, we propose ConformalRefiner, a post-processing framework independent of the initial segmentation model. This method draws on Conformal Risk Control (CRC) theory by allocating a partial calibration set with FNR as the risk control function, obtains a segmentation threshold with a guaranteed significance level across the overall sample. Furthermore, we apply vessel fractal dimension topological priors to filter the calibrated results, reducing noise proportion and effectively increasing the vascular foreground samples overlooked in the initial segmentation. Finally, we break through the limitations of prior post-processing models by designing a network with a dual-input fusion mechanism that takes both the original segmentation prediction and calibrated results as complementary feature inputs to the reconstruction network, while introducing a topology-aware loss function (TopoLoss) to guide the reconstruction network in learning to maintain vessel structure connectivity. Our proposed model can be seamlessly integrated as a plug-and-play component into any existing segmentation model.

Our contributions can be summarized as follows:

- For the first time, we innovatively apply conformal risk control theory to vessel segmentation and the calibration results were filtered through their own topological prior.
- We propose a novel post-processing framework that incorporates CRC calibrated vessel structures as enhanced feature priors into the reconstruction network, which significantly improves the topological segmentation performance of the model.
- Experimental validation on multiple publicly available retinal vascular datasets demonstrates that our method has significant advantages over existing post-processing techniques in repairing broken vessel connectivity.

2   Related Work

2.1   Retinal Vessel Segmentation and Topology-Aware Methods

Retinal vessel segmentation, a critical task in medical image analysis, has witnessed remarkable progress with the advancement of deep learning technologies in recent years. Early research primarily focused on classic network architectures such as UNet [14], which effectively preserved multi-scale features through encoder-decoder structures and skip connections, establishing a foundation for precise segmentation. Subsequently, researchers proposed various improved architectures [15–17] by introducing attention mechanisms or integrating different feature extraction strategies, further enhancing segmentation accuracy. However, these methods mainly emphasize pixel-level accuracy while often neglecting the topological characteristics of vascular structures.

To address the challenge of topology preservation, researchers have developed various topology-aware loss functions. Hu et al. [7] proposed a loss based on persistent homology theory, Chen et al. [18] utilized Euler numbers from topology as constraints, and Hu et al. [19] constructed loss functions based on topological homotopy transformations. The Discrete Morse Theory (DMT) Loss introduced by [9] identifies critical topological structures and incorporates them into the training process. Centerline dice coefficient (Cldice) [8] measures topological similarity by computing the intersection between segmentation masks and their skeletons. Recently, Zhang et al. [20] presented a strategy based on hard pixel mining, enhancing topological accuracy while maintaining computational efficiency.

Another category of methods focuses on improving network architectures. Qi et al. [6] introduced dynamic snake convolution, adaptively focusing on elongated and tortuous structures. Kong et al. [21] designed tree-structured convolutional gated recurrent units that explicitly model the topological structure of coronary arteries. DoubleUNet [4] implicitly enhances topology preservation capabilities through a dual U-Net structure.

Post-processing methods have also been extensively studied. DVAE-refiner [22] utilizes denoising variational autoencoders to map initial segmentations to a compressed latent space before reconstruction, significantly enhancing connectivity. ErrorNet [23] generates synthetic label biases through a noise-injection VAE, helping networks capture error patterns that are then fused with original results. Additionally, researchers have developed universal post-processing frameworks such as morphological operations, conditional random fields [10], CascadePSP [12], SegRefiner [11] and DeepClosing [13]. These methods are not dependent on specific frontend models, offering flexible options for optimizing tubular structure segmentation.

2.2   Uncertainty-Based Segmentation Methods

Despite significant advances in deep learning-based image segmentation methods, these models often lack reliable uncertainty quantification, which is crucial for clinical applications where erroneous predictions may lead to harmful consequences. Conformal Prediction (CP) has emerged as a powerful framework to address this limitation with statistical guarantees.

Angelopoulos et al. [24] established theoretical foundations for distribution-free uncertainty quantification. Building on this, Angelopoulos et al. [25] introduced the "learn then test" method, separating model training from uncertainty quantification. The Conformal Risk Control (CRC) framework [26] provides prediction regions with controlled risk, offering a principled approach that extends from classification to segmentation problems.

CP has demonstrated significant value in various medical image segmentation tasks. The conformal confidence sets developed by [27] provide prediction regions with preset confidence levels for biomedical image segmentation, effectively addressing challenges posed by complex anatomical structures and variations in image quality. Nguyen et al. [28] applied CP to prostate MRI segmentation, generating pixel-level uncertainty maps. Koch et al. [29] proposed a method enabling automatic quality assessment of segmentation results. For retinal applications, Ghoshal et al. [30] demonstrated CP could enhance diagnostic model confidence. Inspired by these works, we adopted CRC to assist threshold selection in the first segmentation stage, increasing through calibration the proportion of vessel foreground samples that were initially overlooked during segmentation.

3   Method

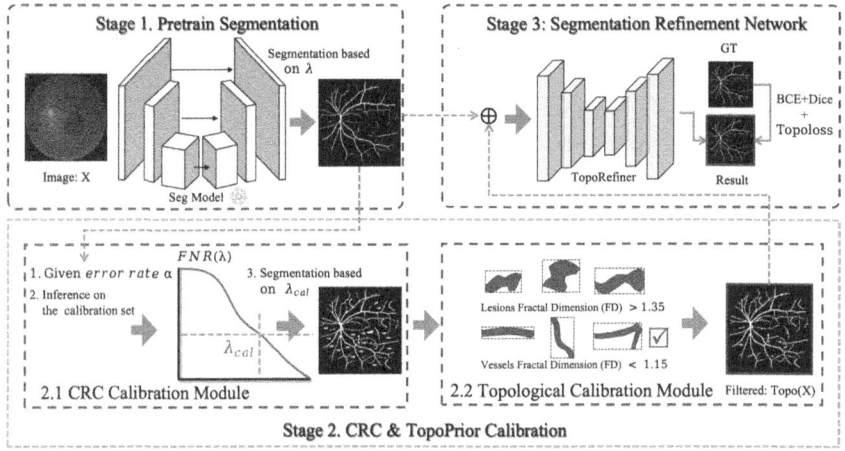

Fig. 2. An illustration of the proposed ConformalRefiner framework.

The ConformalRefiner method proposed in this paper consists of a segmentation threshold calibration module based on CRC, a topology calibration module based on a vascular self-similarity prior, and a reconstruction network. As shown in Fig. 2, The method first calibrates the class probability threshold for the pretrained segmentation model through conformal risk control theory, obtaining adaptive threshold values that preserve more elusive vascular information. Next, it utilizes the self-similarity of vessels for topology calibration, effectively distinguishing vasculature in newly added regions from noise. Finally, the original prediction results and the calibrated segmentation results are jointly input into the reconstruction network to enhance the topological connectivity of the vascular structure. This framework boasts good model generality and can be easily migrated to any segmentation model.

3.1   Segmentation Calibration Based on Conformal Risk Control

In this subsection, we propose using CRC theory to calibrate segmentation results. Traditional segmentation models typically set fixed threshold values for class probabilities to obtain masked results, but this faces a key challenge: there is significant uncertainty in the selection of segmentation thresholds, making it difficult to effectively balance false negatives (missed vessels) and false positives (over-segmentation) during threshold adjustment. Leading to an inability to determine a segmentation threshold that achieves optimal performance across the entire sample distribution. Inspired by the CRC [26] framework, we have developed a method to address this challenge with statistical guarantees. CRC, as a risk management approach based on statistical theory, constructs rigorous statistical guarantees to control prediction risk through a given significance level. It is defined as follows:

Let D = (X, Y) = $\{(x_1, y_1), \ldots, (x_n, y_n)\}$ denote the training dataset, and $\ell(\hat{y}, y)$ represent the loss function, where $\hat{y}$ is the predicted output and y is the ground truth label. CRC aims to construct a prediction set $\mathcal{C}(X_{\text{test}})$ for new test data $(X_{\text{test}}, Y_{\text{test}})$, ensuring that for a specified error rate $\alpha \in [0, 1]$ the expected risk of the prediction set remains bounded by α [24].

$$\mathbb{E}[\ell(\mathcal{C}(X_{\text{test}}), Y_{\text{test}})] \leq \alpha, \tag{1}$$

where, the expected risk $\mathbb{E}[\ell(\cdot)]$ measures the average deviation between the model prediction $\mathcal{C}(X_{test})$ and the true label Y_{test}. In the retinal vessel segmentation setting, we work with $M \times N \times 3$ images and predict $M \times N$ binary masks, where "1" indicates a vessel pixel. Assuming there exists a pre-trained segmentation model $\hat{f}$ that outputs an $M \times N$ grid representing the estimated probability of each pixel being a vessel pixel. We define $\mathcal{C}_\lambda(X) = \{(m, n) : \hat{f}(X)_{(m,n)} \geq \lambda\}$, where (m, n) represents pixel coordinates and λ serves as the segmentation threshold.

In this method, to obtain segmentation results on the entire test set with a false negative rate (FNR) lower than α, resulting in more significant segmented vascular regions while maintaining a confidence level of $1 - \alpha$, we employ FNR as the risk control function, as shown in Eq. 2.

$$\ell_i^{\text{FNR}}(\lambda) = 1 - \frac{|Y_i \cap \mathcal{C}_\lambda(X_i)|}{|Y_i|} \tag{2}$$

During the calibration process, we partition an independent and identically distributed calibration set $(X_{cal}, Y_{cal}) = \{(X_j, Y_j)\}_{j=1}^n$ that is separate from the training set. By optimizing Eq. 3, we obtain the threshold parameter λ_{cal}.

$$\lambda_{cal} = \inf\left\{\lambda : \hat{R}(\lambda) \leq \alpha - \frac{B - \alpha}{n}\right\} \tag{3}$$

Here, B represents a finite value that serves as the upper bound of the loss function, which is set to 1 in our application. The term n denotes the size of the calibration set, and $\hat{R}(\lambda) = \frac{\ell(C_\lambda(X_1), Y_1) + \cdots + \ell(C_\lambda(X_n), Y_n)}{n}$ represents the empirical risk on the calibration data. After calibration, the prediction set $C_{\lambda_{cal}}(X_{\text{test}})$ satisfies Eq. 1.

Theorem 1 *(Conformal coverage guarantee [26]). Suppose $(X_i, Y_i)_{i=1,...,n}$ and (X_{test}, Y_{test}) are i.i.d. sample from some distribution. Further, suppose ℓ is a monotone function of λ. Then the following holds:*

$$P\left(Y_{test} \in C(X_{test})\right) \geq 1 - \alpha \qquad (4)$$

According to the theorem 1, by setting the threshold $\alpha = 0.05$, we can ensure with 95% confidence that our prediction results capture all accurate vascular structures. This provides a more comprehensive result set and establishes a reliable threshold with statistical guarantees.

3.2   Topological Calibration Based on Vascular Self-Similarity Prior

Retinal images contain blood vessels and pathological lesions like hemorrhages and hard exudates [31]. For computational models, blood vessels and these noise elements are often confusable, especially in low signal-to-noise regions. Existing post-processing methods use pixel-level closing operations or connected component analysis to remove small noise elements, but lack precision in element selection. Pathological lesions typically appear as irregular patches, while vessels show linear patterns [32]. To address this, [33] introduced fractal dimension theory, where objects with different morphological structures have distinct fractal dimensions for differentiation. Furthermore, [34] proposed using fractal dimensions to distinguish between actual blood vessels and pathological noise. Inspired by this approach, we integrate this prior knowledge to topological calibration. The fractal dimension is defined as follows [33,35].

Definition 1. *Let A be a bounded set in n-dimensional Euclidean space. If A can be expressed as the union of N_r non-overlapping and non-covering subsets that are similar to A (with scaling ratio r), then A is said to be self-similar. The fractal dimension D of A is defined as:*

$$D = \frac{\log N(r)}{\log(1/r)} \qquad (5)$$

where r represents a uniform scaling factor across all coordinates, this mathematical definition, though theoretically rigorous, has practical limitations. Researchers have developed various methods to estimate fractal dimensions, with the box-counting method becoming dominant in digital image analysis for measuring fractal characteristics of geometric structures. The following expression can formally describe the box-counting dimension:

$$D_{box} = \lim_{r \to 0} \frac{\log N(r)}{\log(1/r)} \qquad (6)$$

where N_r represents the number of geometric units required to completely cover the target structure under the condition that the radius does not exceed r.

In our task, Algorithm 1 is the entire algorithm workflow, consisting of three key steps: (1) We perform segmentation calibration using CRC to obtain

$C_{\lambda_{cal}}(X)$, while generating initial predictions $C_{\lambda_{0.5}}(X)$ using the standard 0.5 threshold. (2) We identify uncertainty regions by computing $C_{\lambda_{cal}}(X) - C_{\lambda_{0.5}}(X)$ and filter small connected components lacking sufficient sampling points for reliable analysis, as fractal dimension estimation requires multi-scale characteristics. The filtered uncertainty set is denoted as $U(X)$. (3) We select actual blood vessels using fractal dimension threshold method [34], which established values of 1.15 for vessels and 1.35 for lesions. We classify uncertain regions in $U(X)$ as follows: components with fractal dimensions below 1.15 are identified as vessels, those exceeding 1.35 are categorized as lesions, and regions with values ranging from 1.15 to 1.35 are classified according to their topological adjacency to previously identified vessels, following the principle of vascular anatomical continuity. Through this method, we obtain the topology-informed calibrated result $\text{Topo}(X)$.

Algorithm 1: Segmentation Calibration and Refinement Module

Require: Input image X, Confidence threshold λ_{cal}, Standard threshold
      0.5

Ensure : Topologically calibrated segmentation result $\text{Topo}(X)$

$C_{\lambda_{cal}}(X) \leftarrow$ Conformal Risk Control(X, λ_{cal});

$C_{\lambda_{0.5}}(X) \leftarrow$ Standard Threshold Inference$(X, 0.5)$;

$U(X) \leftarrow C_{\lambda_{cal}}(X) - C_{\lambda_{0.5}}(X)$;

$U_{\text{filtered}}(X) \leftarrow$ PrefilterSmallComponents$(U(X))$;

$\text{Vessels}(X) \leftarrow \emptyset, \text{Topo}(X) \leftarrow C_{\lambda_{0.5}}(X)$;

for _component $c \in U_{filtered}(X)$_ **do**

    fractal_dim $\leftarrow$ ComputeFractalDimension(c);

    if _fractal_dim < 1.15 **or** ($1.15 \leq$ fractal_dim ≤ 1.35 **and**_
    IsTopologicallyAdjacentTo(c, Vessels(X))) **then**

        $\text{Vessels}(X) \leftarrow \text{Vessels}(X) \cup \{c\}$;

        $\text{Topo}(X) \leftarrow \text{Topo}(X) \cup \{c\}$;

return _Topo(X)_

3.3   Segmentation Refinement Network

Finally, we adopt a traditional U-shaped network as the second-stage segmentation model, denoted as TopoRefiner, but deliberately remove the skip connections. This well-justified design choice effectively prevents the network from directly learning an identity mapping, as skip connections often cause the network to simply copy the input segmentation mask rather than learn deeper feature representations.

We use $\mathcal{C}_{\lambda_{0.5}}(X)$ and $\text{Topo}(X)$ as new inputs, as shown in Eq. 5.

$$Y_{final} = \text{TopoRefiner}([\mathcal{C}_{\lambda_{0.5}}(X) \oplus \text{Topo}(X)]) \tag{7}$$

where $\mathcal{C}_{\lambda_{0.5}}(X)$ represents the binarized mask with a threshold of 0.5, and $\oplus$ denotes the feature concatenation operation. In addition to using Binary Cross-Entropy (BCE) and Dice loss functions for supervised learning, we introduce

Topoloss as a topological constraint to guide the model in learning mutually reinforcing feature representations. The complete loss function is shown as follows:

$$\mathcal{L}_{\text{total}} = \mathcal{L}_{\text{BCE}} + \mathcal{L}_{\text{Dice}} + \lambda \mathcal{L}_{\text{Topoloss}} \tag{8}$$

The weighting terms λ are set to 0.0001, following the optimal parameter configuration determined in [7].

4   Experiments

4.1   Datasets

To evaluate the effectiveness of the proposed segmentation method, we selected three recognized benchmark datasets for retinal vessel segmentation for experimental validation: the DRIVE [36] and FIVE [37]. The FIVE contains 800 image samples, which is a significantly larger sample size than the other datasets; hence, we chose this dataset for detailed testing and ablation experiments.

4.2   Evaluation Metrics

Based on and in conjunction with previous researchers [4,7,8,14,16], using evaluation metrics, we employed two types of metrics to evaluate the performance of our approach. The first set of metrics consisted of overlap-based metrics, some of which included Dice, Precision, and Recall. Additionally, we introduced the LSRecall [38] metric based on local vessel salience, which specifically evaluates the algorithm's segmentation performance in low-salience vessel regions and better reflects the ability to preserve vascular topological structures.

The second set of indicators is topological indicators, including the Cldice and Betti error (β_0, β_1). Cldice evaluates topological structure preservation by calculating the similarity between skeletonized predictions and ground truth. Based on the work in [7], we introduce Betti numbers for local topological assessment, where β_0 counts connected components to reflect branch integrity, and β_1 counts loops to evaluate connection correctness.

4.3   Implementation and Details

We adopt a U-shaped encoder-decoder architecture as the refined segmentation network for Method 3, which shares similarities with UNet in its overall structure but eliminates skip connections. The network training uses the Adam optimizer with an initial learning rate set to 0.001. The training process is conducted on a single RTX 3090 GPU with a batch size of 8 and an input image cropping size of 256 × 256. For other key parameters in network training, the error rate α_1 for CRC is set to 0.05, and the window size in the topological loss function is set to 73 as referenced in [7]. The configuration for the first segmentation model follows the best parameter settings from the original work.

Table 1. Quantitative Performance Evaluation of the Proposed Methods on DRIVE and FIVE Datasets

Method	Dice↑	Precision↑	Recall↑	lsRecall↑	ClDice↑	β_0 ↓	β_1 ↓
DRIVE							
UNet [14]	**0.809**	**0.812**	0.796	0.759	0.811	1.207	0.531
+ ours	0.806	0.807	**0.814**	**0.763**	**0.813**	**1.049**	**0.524**
ResUnet [16]	**0.812**	**0.817**	0.803	0.762	0.813	1.203	0.502
+ ours	0.808	0.740	**0.815**	**0.768**	**0.817**	**0.983**	**0.496**
TopoLoss [7]	**0.810**	**0.826**	0.794	0.743	0.812	1.073	0.476
+ ours	[8]	0.817	**0.813**	**0.751**	**0.814**	**0.921**	**0.458**
Cldice [8]	**0.809**	**0.804**	0.810	0.736	0.820	0.963	0.474
+ ours	0.806	0.802	**0.821**	**0.741**	**0.824**	**0.826**	**0.450**
DoubleUNet [4]	**0.816**	**0.806**	0.815	0.786	0.821	0.952	0.502
+ ours	0.815	0.803	**0.822**	**0.797**	**0.824**	**0.816**	**0.489**
FIVE							
UNet	**0.896**	**0.902**	0.864	0.790	0.914	0.429	0.233
+ ours	0.883	0.874	**0.889**	**0.802**	**0.918**	**0.391**	**0.216**
ResUnet	**0.901**	**0.904**	0.896	0.846	0.916	0.417	0.233
+ ours	0.894	0.882	**0.902**	**0.848**	**0.924**	**0.413**	**0.232**
TopoLoss	**0.851**	**0.877**	0.830	0.782	0.871	0.372	0.207
+ ours	0.850	0.875	**0.886**	**0.806**	**0.893**	**0.334**	**0.201**
Cldice	**0.901**	**0.912**	0.904	0.848	0.917	0.368	0.202
+ ours	0.882	0.906	**0.905**	**0.849**	**0.920**	**0.361**	0.202
DoubleUNet	**0.908**	**0.915**	0.920	0.882	0.553	**0.364**	**0.187**
+ ours	0.905	0.913	**0.927**	**0.893**	**0.434**	0.367	0.192

Table 2. Comparison of Different Post-processing Methods Based on the FIVE Dataset

Method	Dice↑	Recall↑	ClDice↑	β_0 ↓
UNet [14]	**0.896**	0.864	0.914	0.429
Closing (k = 3)	0.893	0.862	0.914	0.416
ConvCRF (k = 1) [10]	0.852	0.833	0.891	0.426
CascadePSP [12]	0.702	0.698	0.687	0.431
SegRefiner [11]	0.873	0.844	0.892	0.415
DeepClosing [13]	**0.896**	0.871	0.917	0.393
Ours	0.883	**0.889**	**0.918**	**0.391**

4.4   Main Results

Our proposed post-processing method is generic and can be seamlessly integrated into various segmentation models. In our experiments, a diverse set of representative segmentation methods were selected as benchmarks, including [4, 14, 16], and [7, 8] with topology-aware capabilities. Table 1 shows quantitative results for three datasets. Our method significantly outperforms existing approaches in terms of the Cldice topology indicator, owing to its exceptional capacity for repairing topological structures. The significant improvement in the recall rate validates the effectiveness of the method based on CRC and topological prior calibration to maximize the replacement of missing vessel segments. Although there is a slight decrease in precision and Dice coefficient, this over-segmentation does not introduce noise and is fully acceptable in clinical applications.

For a comprehensive evaluation, we performed comparative experiments using UNet as the initial segmentation model Table 2. The comparison methods include traditional post-processing methods, the classic Closing operation, conditional random field-based ConvCRF [10], and several general-purpose deep-

Table 3. Ablation study of the key components.

Method	CRC	TP	Refiner	TopoLoss	Dice↑	ClDice↑	β_0↓
UNet					**0.896**	0.914	0.429
Ours	✔				0.701	0.824	0.581
	✔	✔			0.832	0.851	0.506
	✔	✔	✔		0.873	0.878	0.432
	✔	✔	✔	✔	0.883	**0.918**	**0.391**

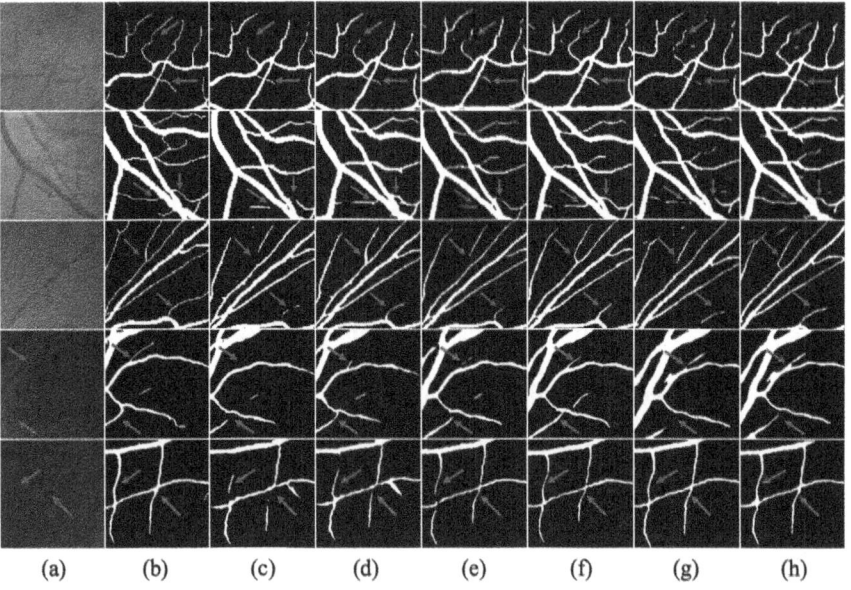

| (a) | (b) | (c) | (d) | (e) | (f) | (g) | (h) |

Fig. 3. Example of blood vessel segmentation results on two datasets. (a) Original images, (b) Ground truth, (c) Unet, (d) Unet+Ours, (e) Cldice, (f) Cldice+Ours, (g) DoubleUnet, (h) DoubleUnet+Ours. The first three rows are the results of the DRIVE, and the last two rows are the results of the FIVE dataset.

learning-based image segmentation methods [11–13]. The results showed that our method achieved a significant improvement in topological metrics.

4.5   Analysis

Ablation Study of the Key Components. To evaluate each module, we conducted a series of ablation studies on the proposed method by removing certain components. We present the ablation studies of components, including CRC segmentation calibration, topological prior calibration, the reconstruction network Refiner, and topological constraints during training (TopoLoss). The results are shown in Table 3. It can be observed that all proposed components contribute to performance improvement.

Visualization. As shown in Fig. 3, the visualization of segmentation results on DRIVE and FIVE. Compared to any baseline model, our method significantly enhances topological connectivity, ensuring the integrity and continuity of critical vessel structures.

5   Conclusion

In this paper, we introduce the ConformalRefiner, a retinal vascular segmentation reconstruction network based on CRC topology calibration, which innovatively combines the theory of CRC with vascular self-similarity priors, significantly enhancing the topological integrity of retinal vessel segmentation. Our main contributions include introducing CRC theory into retinal vessel segmentation for adaptive threshold calibration, constructing an effective topology calibration based on vessel self-similarity features, and designing a dual-input fusion reconstruction mechanism to successfully repair vascular breakpoints. Experimental validation shows that our method significantly outperforms existing advanced methods in terms of CLDice and β_0, among other topology evaluation indicators, while maintaining excellent model generality on public datasets such as DRIVE and FIVE. Although the current calibration process operates independently of model training, in the future we will work to integrate the CRC calibration process with optimization, explore more efficient reconstruction network architectures, and extend this framework to other medical image segmentation tasks.

Acknowledgments. This work was supported by the National Natural Science Foundation of China (Serial Nos. 61991410, 61976134), OpenProject Foundation of Intelligent Information Processing Key Laboratory of Shanxi Province, China (No. CICIP2021001), Natural Science Foundation of Shanghai (No. 21ZR1423900), and Shanghai Science and Technology Innovation Action Plan (22511101903).

References

1. Ricci, E., Perfetti, R.: Retinal blood vessel segmentation using line operators and support vector classification. IEEE Trans. Med. Imaging **26**(10), 1357–1365 (2007)
2. Qureshi, I., Ma, J., Abbas, Q.: Recent development on detection methods for the diagnosis of diabetic retinopathy. Symmetry **11**(6), 749 (2019)
3. Olubunmi Omobola Sule: A survey of deep learning for retinal blood vessel segmentation methods: taxonomy, trends, challenges and future directions. IEEE Access **10**, 38202–38236 (2022)
4. Jha, D., Riegler, M.A., Johansen, D., Halvorsen, P., Johansen, H.D.: Doubleu-net: a deep convolutional neural network for medical image segmentation. In: 2020 IEEE 33rd International Symposium on Computer-based Medical Systems (CBMS), pp. 558–564. IEEE (2020)
5. Yang, X., Li, Z., Guo, Y., Zhou, D.: DCU-net: a deformable convolutional neural network based on cascade u-net for retinal vessel segmentation. Multimedia Tools Appl. **81**(11), 15593–15607 (2022)
6. Qi, Y., He, Y., Qi, X., Zhang, Y., Yang, G.: Dynamic snake convolution based on topological geometric constraints for tubular structure segmentation. In: Proceedings of the IEEE/CVF International Conference on Computer Vision, pp. 6070–6079 (2023)
7. Hu, X., Li, F., Samaras, D., Chen, C.: Topology-preserving deep image segmentation. In: Advances in Neural Information Processing Systems, vol. 32 (2019)
8. Shit, S., et al.: cldice-a novel topology-preserving loss function for tubular structure segmentation. In: Proceedings of the IEEE/CVF Conference on Computer Vision And Pattern Recognition, pp. 16560–16569 (2021)
9. Hu, X., Wang, Y., Fuxin, L., Samaras, D., Chen, C.: Topology-aware segmentation using discrete morse theory. arXiv preprint (2021). arXiv:2103.09992
10. Sutton, C., McCallum, A.: An introduction to conditional random fields for relational learning. Introduction Stat. Relational Learn. **2**, 93–128 (2006)
11. Wang, M., Ding, H., Liew, J.H., Liu, J., Zhao, Y.,Wei, Y.: SegRefiner: towards model-agnostic segmentation refinement with discrete diffusion process. arXiv preprint (2023). arXiv:2312.12425
12. Cheng, H.K., Chung, J., Tai, Y.W., Tang, C.K.: Cascadepsp: toward class-agnostic and very high-resolution segmentation via global and local refinement. In: Proceedings of the IEEE/CVF Conference on Computer Vision and Pattern Recognition, pp. 8890–8899 (2020)
13. Wu, Q., Chen, Y., Liu, W., Yue, X., Zhuang, X.: Deep closing: enhancing topological connectivity in medical tubular segmentation. IEEE Trans. Med. Imaging (2024)
14. Ronneberger, O., Fischer, P., Brox, T.: U-Net: convolutional networks for biomedical image segmentation. In: Navab, N., Hornegger, J., Wells, W.M., Frangi, A.F. (eds.) MICCAI 2015. LNCS, vol. 9351, pp. 234–241. Springer, Cham (2015). https://doi.org/10.1007/978-3-319-24574-4_28
15. Oktay, O., et al.: Attention u-net: learning where to look for the pancreas. arxiv preprint (2018). arXiv:1804.03999
16. Zhang, Z., Liu, Q., Wang, Y.: Road extraction by deep residual u-net. IEEE Geosci. Remote Sens. Lett. **15**(5), 749–753 (2018)
17. Chen, J., et al.: Transunet: transformers make strong encoders for medical image segmentation. arXiv preprint (2021). arXiv:2102.04306

18. Chen, X., Luo, X., Zhao, Y., Zhang, S., Wang, G., Zheng, Y.: Learning euler's elastica model for medical image segmentation. arXiv preprint (2020). arXiv:2011.00526

19. Xiaoling, H.: Structure-aware image segmentation with homotopy warping. Adv. Neural. Inf. Process. Syst. **35**, 24046–24059 (2022)

20. Zhang, G., Dong, C., Li, Y.: Topology-preserving hard pixel mining for tubular structure segmentation. In: BMVC, pp. 846–853 (2023)

21. Kong, B., et al.: Learning tree-structured representation for 3D coronary artery segmentation. Comput. Med. Imaging Graph. **80**, 101688 (2020)

22. Araújo, R.J., Cardoso, J.S., Oliveira, H.P.: A deep learning design for improving topology coherence in blood vessel segmentation. In: Shen, D., Liu, T., Peters, T.M., Staib, L.H., Essert, C., Zhou, S., Yap, P.-T., Khan, A. (eds.) MICCAI 2019. LNCS, vol. 11764, pp. 93–101. Springer, Cham (2019). https://doi.org/10.1007/978-3-030-32239-7_11

23. Tajbakhsh, N., Lai, B., Ananth, S.P., Ding, X.: Errornet: learning error representations from limited data to improve vascular segmentation. In: 2020 IEEE 17th International Symposium on Biomedical Imaging (ISBI), pp. 1364–1368. IEEE (2020)

24. Angelopoulos, A.N., Bates, S.: A gentle introduction to conformal prediction and distribution-free uncertainty quantification. arXiv preprint (2021). arXiv:2107.07511

25. Angelopoulos, A.N., Bates, S., Candès, E.J., Jordan, M.I., Lei, L.: Learn then test: calibrating predictive algorithms to achieve risk control. arXiv preprint (2021). arXiv:2110.01052

26. Angelopoulos, A.N., Bates, S., Fisch, A., Lei, L., Schuster, T.: Conformal risk control. arXiv preprint (2022). arXiv:2208.02814

27. Davenport, S.: Conformal confidence sets for biomedical image segmentation. arXiv preprint (2024). arXiv:2410.03406

28. Nguyen, K.M., Fernandez-Quilez, A.: Segmentation uncertainty with statistical guarantees in prostate MRI. In: 2024 32nd European Signal Processing Conference (EUSIPCO), pp. 1636–1640. IEEE (2024)

29. Koch, L.M., Baumgartner, C.F.: Conformal performance range prediction for segmentation output quality control. In: Uncertainty for Safe Utilization of Machine Learning in Medical Imaging: 6th International Workshop, UNSURE 2024, Held in Conjunction with MICCAI 2024. Marrakesh, 10 Oct 2024, Proceedings, vol. 15167, p. 81. Springer, Cham (2025)

30. Ghoshal, B., et al. Making deep learning models clinically useful-improving diagnostic confidence in inherited retinal disease with conformal prediction. In: International Workshop on Uncertainty for Safe Utilization of Machine Learning in Medical Imaging, pp. 47–58. Springer, Cham (2024)

31. Khojasteh, P., Aliahmad, B., Kumar, D.K.: Fundus images analysis using deep features for detection of exudates, hemorrhages and microaneurysms. BMC Ophthalmol. **18**, 1–13 (2018)

32. Masters, B.R.: Fractal analysis of the vascular tree in the human retina. Annu. Rev. Biomed. Eng. **6**(1), 427–452 (2004)

33. Cannon, J.W.: The fractal geometry of nature. by Benoit B. Mandelbrot. Am. Math. Monthly **91**(9), 594–598 (1984)

34. Wang, W., Wang, W., Zhangping, H.: Retinal vessel segmentation approach based on corrected morphological transformation and fractal dimension. IET Image Proc. **13**(13), 2538–2547 (2019)

35. Sarkar, N., Chaudhuri, B.B.: An efficient differential box-counting approach to compute fractal dimension of image. IEEE Trans. Syst. Man Cybern. **24**(1), 115–120 (1994)
36. Staal, J., Abràmoff, M.D., Niemeijer, M., Viergever, M.A., Ginneken, B.V.: Ridge-based vessel segmentation in color images of the retina. IEEE Trans. Med. Imaging **23**(4), 501–509 (2004)
37. Jin, K., et al.: Fives: a fundus image dataset for artificial intelligence based vessel segmentation. Sci. data **9**(1), 475 (2022)
38. Parella, J.P., da Silva, M.V., Comin, C.H.: A new approach for evaluating and improving the performance of segmentation algorithms on hard-to-detect blood vessels. arXiv preprint (2024). arXiv:2406.13128

Multi-view Fusion Enhanced Social Text Representation for Depression Detection

Jie Chen[1,2,3], Xuetong You[1,2,3], Shu Zhao[1,2,3(✉)], Jingxuan Xiao[4],
and Yanping Zhang[1,2,3]

[1] Key Laboratory of Intelligent Computing and Signal Processing, Ministry
of Education, Hefei 230601, Anhui Province, People's Republic of China
[2] School of Computer Science and Technology, Anhui University, Hefei 230601,
Anhui Province, People's Republic of China
[3] Information Materials and Intelligent Sensing Laboratory of Anhui Province,
Hefei 230601, Anhui Province, People's Republic of China
zhaoshuzs2002@hotmail.com
[4] Faculty of Innovation Engineering, Department of Computer Science,
Macau University of Science and Technology, Macau, People's Republic of China

Abstract. Depression has become a major global public health challenge. Social media texts reflect users' emotional and psychological states, providing a valuable data source for depression detection through natural language analysis. However, current research still faces limitations in modeling the collaborative representation of user texts and dynamic behaviors. To address this, we propose MVDep, a model that integrates multi-view features to enhance social text representation for depression detection. Specifically, we enhance text representation by investigating depression from multi-view, including social media texts, user profiles, and social behaviors, enabling more effective depression detection. First, we conduct initial filtering of social media texts using the Diagnostic and Statistical Manual of Mental Disorders (DSM-5), and then screen out the uncertain domain of data samples based on the three-way decision (3WD). Next, we extract quantified user profiles and social behavioral features to enhance the representation of depression tendencies for uncertain domain samples. Finally, we fuse the multi-view features derived from social media texts, user profiles and social behaviors, improving the model's ability to detect depression. We evaluate our model on the SWDD dataset, demonstrating its superiority over state-of-the-art baseline models in depression detection tasks.

Keywords: Depression Detection · Social Media Texts · Multi-View
Feature Fusion · 3WD

1 Introduction

Depression is a severe global health issue, affecting over 340 million people worldwide, with a prevalence rate of 4.4% [1]. If untreated, it can lead to severe consequences, including suicide, which claims more than 700,000 lives annually [2]. Despite available treatments, 76%–85% of individuals in low- and middle-income

Q. Zhang et al. (Eds.): IJCRS 2025, LNAI 15710, pp. 199–212, 2025.
https://doi.org/10.1007/978-3-031-92741-6_15

countries lack access to effective care due to stigma and resource limitations [3], highlighting the urgent need for early detection and intervention. The widespread use of platforms like Twitter and Sina Weibo has enabled users to share personal views and emotions [4], providing a rich data source for analyzing mental states through user-generated content [5]. Social media texts often reflect depression-related emotional characteristics, such as worthlessness and self-hatred, offering reliable and actionable insights for detection [6,7]. These textual features, combined with advanced natural language processing techniques, make social media a robust platform for depression detection.

Recent studies on depression detection in social media have primarily focused on text-based features, utilizing both classical machine learning approaches, such as SVM and LR [8], and deep learning techniques, including LSTM [9], and GNNs [10]. These methods have been effective in identifying depression-related language patterns. Additionally, research has increasingly emphasized the importance of incorporating social context and user interactions [8,10,11], highlighting the potential of integrating multiple data sources for a more comprehensive analysis. However, current depression detection research still has significant limitations. Existing research lacks a comprehensive representation of the interplay between text features, user profiles, and social behavioral features. Specifically, it fails to model the interactions and dynamic evolution of these features, which restricts the understanding of the semantic depth of depressive texts.

To solve the above-mentioned problems, this paper proposes MVDep, a model that enhances the representation of social media texts by incorporating user social network features. MVDep employs the DSM-5 [12] and 3WD theory to filter out uncertain data samples from user texts. Then, we extract social network features from user profiles and social behaviors. Finally, we fuse multi-view features from social media texts, user profiles, and social behaviors to comprehensively capture negative emotion patterns. Our key contributions are as follows:

(1) We propose a MVDep model that achieves a collaborative representation of user texts and social network features, thereby enhancing text representation for depression detection.
(2) Based on 3WD, we screen out texts into certain and uncertain domains. Texts in the uncertain domain are enhanced by integrating comprehensive social network features derived from quantified user profiles and social behavioral features.
(3) We conduct extensive experiments on the SWDD dataset, achieving 0.9686 classification accuracy and outperforming state-of-the-art baseline models.

2   Related Work

In this section, we summarize some closely related works. Current methods for online depression detection mainly include two directions: Text-based Depression Detection and User Social Network Feature-enhanced Depression Detection.

2.1 Text-Based Depression Detection

Multiple studies have shown a correlation between linguistic expression on social media and depression. Eichstaedt et al. [13] further pointed out that individuals with depression are more likely to use first-person singular pronouns, such as "I" and "my", and frequently employ words associated with distress, depressive symptoms, and rumination. Chiong et al. [14] proposed 90 features based on emotional lexicons and text content, which were used as inputs for a depression detection classifier.

Detecting depression-related content from users' historical texts presents a significant challenge due to the limited proportion of social media data directly relevant to mental health. To address this issue, Saha et al. [15] proposed a joint modeling framework that extracts topic-based and psycholinguistic features from texts, enabling the classification of mental health-related online communities. Building on this, Zogan et al. [16] utilized the BART model to generate condensed summaries of user text sequences, effectively reducing noise and redundancy by compressing extensive historical texts into a single, informative summary.

2.2 User Social Network Feature-Enhanced Depression Detection

User social network features include user profile and social behavior aspects. User profile features, such as gender and age, offer valuable insights into mental health, while social behavioral features, such as midnight texting frequency and mention frequency, play a key role in assessing psychological states.

User Profile Features. Studies by Negriff et al. [17] found a negative correlation between depressive symptoms and the number of Facebook friends. This means that the more pronounced the depressive symptoms, the fewer friends individuals tend to have on Facebook. De Choudhury et al. [8] found that depressed adults have fewer followers and follow fewer people compared to non-depressed adults. They explored potential user behaviors to perform user-oriented depression detection. By measuring behavioral attributes such as social engagement, emotion, language, and linguistic style on Twitter, they identified useful signals for depressive characteristics.

User Social Behavioral Features. Razak et al. [18] proposed a machine-learning method based on content and activity features to detect depression from Twitter texts. They trained and evaluated classifiers by extracting users' network behaviors and text features to distinguish depressed users from non-depressed ones. Zhang et al. [19] constructed user behavior graphs and used GNNs to model the contextual dependencies of user behaviors. They also employed LSTM to capture the dynamic evolution of behavioral sequences and integrated temporal behavior and thematic semantic representations through cross-perspective collaborative learning, thereby enhancing depression detection performance.

Based on the aforementioned studies, this paper proposes a comprehensive model that integrates text-based features, user profiles, and social behavioral features to identify individuals at risk of depression. This comprehensive approach enhances the semantic understanding of textual content, significantly improving the performance of the detection model.

3   Methodology

3.1   Problem Formulation

In this paper, we aim to assess the risk of depression in individuals based on their text history. For a user u, we can collect his/her historical texts, denoted as $P = \{p_1, p_2, \ldots, p_n\}$. Each of the texts has a corresponding timestamp, denoted as $t_1, t_2, \ldots, t_n$. Our task is formalized as determining the depression label y for user u, given their history of texts P, where $y \in \{0, 1\}$.

3.2   Model Architecture

As shown in Fig. 1, our MVDep model consists of three main components: the Reliable Filtering Module, the Social Network Feature Extraction Module, and the Multi-view Feature Fusion Module. To provide a clearer and more intuitive understanding of our proposed model, Algorithm 1 presents the pseudocode for the complete MVDep model.

Reliable Filtering Module (RFM). streamlines the initial assessment of social media texts by applying the DSM-5 and 3WD theory, categorizing content into positive, negative, and uncertain domains, thus focusing on texts indicative of depression and minimizing irrelevant data interference.

Social Network Feature Extraction Module (SNFEM). enhances depression detection by extracting social network features, including user profiles and social behavioral features, from social media. User profiles, such as the ratio of original to private texts, reflect emotional expression and social engagement. Social behavioral features, including late-night posting frequency and the frequency of @mentions, provide deeper insights into users' emotional trajectories and interaction patterns.

Multi-view Feature Fusion Module (MVFFM). integrates the extracted social network and textual features, enabling a more precise evaluation of users' emotional states. Employing the IT [20] architecture, a sophisticated deep CNN ensemble, enhances the model's ability to represent features robustly.

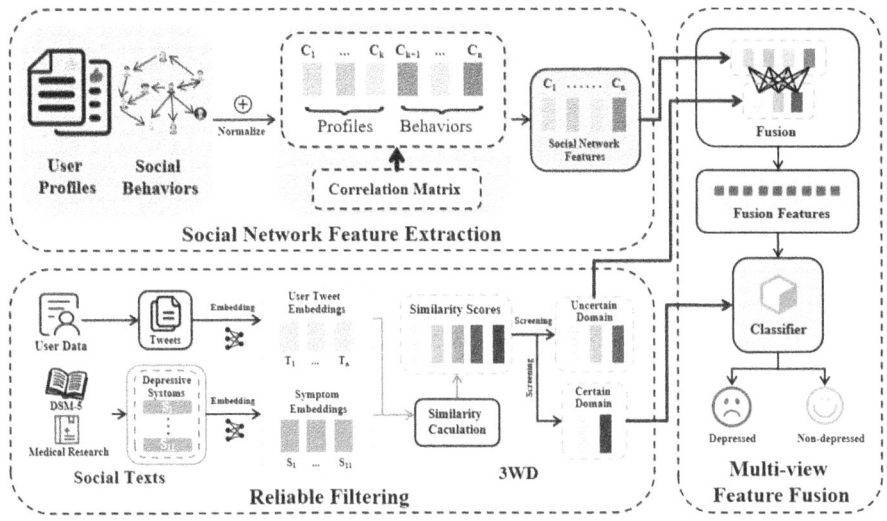

Fig. 1. The Architecture of our Proposed MVDep.

Overall, by leveraging the RFM for initial data filtering and refinement, extracting comprehensive user profiles and social behavioral features through the SNFEM, and fusing social network and textual features with the MVFFM, the model provides a multi-view feature fusion approach, significantly improving the accuracy of depression detection in social media users.

3.3   Reliable Filtering Module

To improve efficiency and model performance, we propose a initial filtering step in the Reliable Filtering Module. Using the DSM-5, we identify 11 key depressive symptoms and extract the top K most relevant texts for each user. These texts are categorized into positive, negative, and uncertain domains based on 3WD theory.

Preliminary Screening (PS). For the preliminary screening, we use the DSM-5 as a template to evaluate the relevance of each text. Based on the DSM-5 and relevant medical research, we propose a set of depression symptom descriptions tailored to online social media vocabulary, denoted as S. The text matrix T_S of depression symptoms consists of the following descriptions:

$$T_S = \{S_1, S_2, \ldots, S_m\} \tag{1}$$

where S_i represents the textual description of a specific depression symptom, and m is the total number of symptoms.

Algorithm 1. MVDep Model for Depression Detection

1: **Input:** User texts U, Depression template T, Upper threshold θ_1, Lower threshold θ_2

2: **Output:** *prediction*

3: **procedure** RELIABLEFILTERING(U, T, θ_1, θ_2)

4:     **for** each text u_i in U **do**

5:         $sim_i \leftarrow$ CALCULATECOSINESIMILARITY(u_i, T)

6:         **if** $sim_i > \theta_1$ **then**

7:             $P_{\text{positive}} \leftarrow P_{\text{positive}} \cup \{u_i\}$

8:         **else if** $sim_i < \theta_2$ **then**

9:             $P_{\text{negative}} \leftarrow P_{\text{negative}} \cup \{u_i\}$

10:        **else**

11:            $P_{\text{boundary}} \leftarrow P_{\text{boundary}} \cup \{u_i\}$

12:        **end if**

13:    **end for**

14:    **return** P_{boundary}

15: **end procedure**

16: **procedure** SOCIALNETWORKFEATUREEXTRACTION(U)

17:    $F_{\text{user}} \leftarrow \left(\frac{N_{\text{original}}}{N_{\text{total}}}, \frac{N_{\text{private}}}{N_{\text{total}}} \right)$

18:    $F_{\text{behavior}} \leftarrow \left(\frac{N_{\text{midnight}}}{N_{\text{total}}}, \frac{N_{\text{mention}}}{N_{\text{total}}} \right)$

19:    $F_{\text{social}} \leftarrow \alpha F_{\text{user}} + \beta F_{\text{behavior}}$

20:    **return** F_{social}

21: **end procedure**

22: **procedure** MULTI-VIEWFEATUREFUSION(U)

23:    $P_{\text{boundary}} \leftarrow$ RELIABLEFILTERING(U, T, θ_1, θ_2)

24:    $F_{\text{social}} \leftarrow$ SOCIALNETWORKFEATUREEXTRACTION(U)

25:    $C \leftarrow$ CALCULATECOSINESIMILARITY(P_{boundary}, T)

26:    $S \leftarrow w_1 \cdot C + w_2 \cdot F_{\text{social}}(u_i), \quad w_1 + w_2 = 1$

27:    $F_{\text{combined}}(u_i) \leftarrow F_{\text{certain}}(u_i) \cup F_{\text{social}}(u_i)$

28:    $prediction \leftarrow$ CLASSIFIER($F_{\text{combined}}(u_i)$)

29:    **return** *prediction*

30: **end procedure**

To compute depression symptom scores from user texts, we first extract the embeddings for each depressive symptom description and transform the text matrix T_S into an embedding matrix E_S:

$$E_S = \{f(s_1), f(s_2), \ldots, f(s_m)\} = \{s_e^1, s_e^2, \ldots, s_e^m\} \tag{2}$$

where s_e^i represents the embedding of depressive symptom description S_i, and f refers to the pre-trained multilingual embedding model. For each user text t, we extract its embedding t_e and use E_s to compute depression symptom scores S_d for the text:

$$\begin{aligned} S_d &= f_x(t_e, E_s) \\ &= \{f_s(t_e, s_e^1), f_s(t_e, s_e^2), \ldots, f_s(t_e, s_e^m)\} \\ &= \{\theta_1, \theta_2, \ldots, \theta_m\}, s_c^m \in E_S \end{aligned} \tag{3}$$

where θ_i is the depression symptom score of text t, and f_s is the similarity function. In this paper, we use cosine similarity, denoted as cosine-sim:

$$C = \text{cosine-sim}(u, v) = \frac{\sum_{i=1}^{n} u_i v_i}{\sqrt{\sum_{i=1}^{n} u_i^2}\sqrt{\sum_{i=1}^{n} v_i^2}}, \tag{4}$$

where u and v are vectors, u_i and v_i are their respective components, and n is the dimensionality.

A higher similarity score with the symptom descriptions indicates a stronger match between the text and the symptom, signifying a more evident tendency towards depression. We then extract the top K texts with the highest depression tendencies for each user, represented as $[P_{i,1}, P_{i,2}, \ldots, P_{i,K}]$.

Text Classification with 3WD. Based on the similarity scores computed from the preceding process, we apply 3WD theory to classify texts into certain and uncertain domains, thereby identifying texts that warrant further analysis.

$$\begin{aligned} POS &= \{x \subset P \mid H(x) > \theta_1\} \\ NEG &= \{x \subset P \mid H(x) < \theta_2\} \\ BND &= \{x \subset P \mid \theta_1 \geq H(x) \geq \theta_2\} \end{aligned} \tag{5}$$

where P represents all texts obtained after template filtering, and H is the scoring function based on cosine similarity. θ_1 and θ_2 are the upper and lower thresholds, respectively. Texts in the POS and NEG domains show clear tendencies toward depression or non-depression, and are directly fed into the classifier. Texts in the BND domain, which exhibit ambiguous emotional tendencies, are enhanced through the integration of quantified social network features within the multi-view fusion module, thereby strengthening their textual representations for depression detection. The boundary domain text set is represented as:

$$B = [P_{i,1}, P_{i,2}, ..., P_{i,b}] \tag{6}$$

where b is the number of texts filtered by 3WD Theory.

3.4   Social Network Feature Extraction Module

This module enhances the identification of depressive states by analyzing user social network features on social media. It incorporates both user profile features and social behavioral features to better understand emotional patterns and interaction behaviors in the online environment.

User Profile Features. User profile features include the proportion of original texts, private texts, as well as the number of followers and followings. These features reflect users' levels of social activity, self-expression tendencies, and emotional states. For instance, a high proportion of original texts may indicate a tendency to express psychological states, whereas a high proportion of private texts might reflect emotional avoidance or self-isolation.

To quantitatively capture these tendencies, we define several key metrics. As an example, the proportion of private texts is calculated to measure the user's tendency toward emotional avoidance or self-isolation. $N_{\text{private}}(u_i)$ represent the number of private texts and $N_{\text{total}}(u_i)$ denotes the total number of texts. The proportion of private texts, denoted as $F_{\text{social}}(u_i)$, is given by:

$$F_{\text{user}} = \frac{N_{\text{private}}}{N_{\text{total}}} \tag{7}$$

User Social Behavioral Features. User social behavioral features include text comment frequency, midnight texting frequency and mention frequency, along with other engagement-related indicators. These features reveal users' interaction patterns and emotional expressions on social media. For example, a high frequency of midnight texting may be associated with sleep disorders, and a low repost frequency indicate weaker emotional expression within the social network.

Social Network Feature Construction. To capture a comprehensive understanding of the user's emotional state, we define the social network feature vector, denoted as $F_{\text{social}}(u_i)$, which integrates both user profile features $F_{\text{user}}(u_i)$ and social behavioral features $F_{\text{behavior}}(u_i)$:

$$F_{\text{social}}(u_i) = \alpha F_{\text{user}}(u_i) + \beta F_{\text{behavior}}(u_i), \quad \alpha + \beta = 1 \tag{8}$$

In the case of texts in the uncertain domain, these social network features are used to enhance textual representations, providing a more comprehensive view of the user's emotional tendencies. Here, α and β are balancing coefficients that control the contributions of user profiles and social behavioral features.

3.5   Multi-view Feature Fusion Module

In this subsection, we propose to enhance text representation through the effective integration of multi-view features, encompassing social network features and textual semantics. By fusing these perspectives, our strategy facilitates a more nuanced and accurate assessment of users' emotional states, enabling the detection of subtle yet clinically relevant depression-related signals.

Fusion of Social Network and Textual Features. For texts classified in the uncertain domain, we calculate the emotional state score S by combining the cosine similarity score C, which quantifies the alignment between the text and depressive symptoms, with user social network features. The emotional state score is computed as the sum of these two components as follows:

$$S = w_1 C + w_2 F_{\text{social}}(u_i), \quad w_1 + w_2 = 1 \tag{9}$$

where C is the cosine similarity score between the text and depressive symptom lexicon, and $F_{\text{social}}(u_i)$ is the combined feature vector from the social network information. The weights w_1 and w_2 determine the relative contribution of each feature type.

Fusion of Certain and Uncertain Domains. For texts in the certain domain, the feature vector $F_{\text{certain}}(u_i)$ consists of the cosine similarity score C. For texts in the uncertain domain, $F_{\text{social}}(u_i)$ is the enhanced feature vector that incorporates both textual and social network features, representing the emotional state score S. The unified feature vector for user u_i, denoted as $F_{\text{combined}}(u_i)$, is calculated as follows:

$$F_{\text{combined}}(u_i) = F_{\text{certain}}(u_i) \cup F_{\text{social}}(u_i) \tag{10}$$

To evaluate the performance of the feature fusion process, we employ IT [20], a deep CNN specifically designed for time-series data. The ensemble architecture of IT enables it to capture complex, hierarchical patterns in both textual and social network data, thereby enhancing depression detection through efficient processing of multi-view input features.

4   Experiments

4.1   Experiment Setup

Dataset. We evaluate our model on the SWDD [21] dataset. Statistical details of the SWDD dataset are summarized in Table 1. Since SWDD is an imbalanced dataset, we randomly selected 3,500 depressed users and 3,500 non-depressed users for the experimental dataset. The dataset is divided into three subsets: 60% for training, 20% for validation, and 20% for testing. The number of depressed and non-depressed users in each test set is balanced.

Table 1. The statistics of the dataset SWDD.

Category	Users	Texts
Depressed	3,711	785,689
Non-depressed	19,526	4,068,732
Total	**23,237**	**4,854,421**

Baseline. To verify the effectiveness of our feature extraction method in online depression detection, we compare the performance of our proposed method with five models.

X-A-BiLSTM [22] is a deep learning approach that combines XGBoost for data filtering and an Attention-BiLSTM classifier for detecting depression in imbalanced datasets.

SS3 [23] is an incremental and early classification text classifier that generates confidence vectors by progressively simplifying document chunks.

MFFN [24] uses a multi-task learning framework to integrate heterogeneous modal information for predicting depressive users on social networks.

DSTS-IT [21] is a depression detection method based on the multivariate time series features of users' depressive symptoms.

EmoNeZha [25] enhances depression detection by constructing user behavior graphs and applying cross-view collaborative learning to capture the dynamic evolution of user behaviors.

Evaluation Metrics. To ensure result reliability, we employed a 10-fold cross-validation approach to mitigate potential bias and randomness. The results are averaged across accuracy, recall, precision, and F1 score metrics.

4.2   Experimental Results and Parameter Analysis

This subsection presents the experimental results, comparing our model's performance with existing baseline models and analyzing the impact of user text volume and the percentage of depressed users on detection accuracy.

Table 2. Performance Comparison of Different Models.

Model	Accuracy	Precision	Recall	F1-Score
X-A-BiLSTM [22]	0.7614	0.7638	0.7614	0.7609
SS3 [23]	0.7832	0.8379	0.7382	0.7741
MFFN [24]	0.8921	0.8925	0.8921	0.8921
DSTS_IT [21]	0.9202	0.9203	0.9202	0.9202
EmoNeZha [25]	**0.9714**	0.9673	0.9411	0.9502
Our Model	0.9687	**0.9694**	**0.9687**	**0.9686**

Experimental Results. As presented in Table 2, our model demonstrates competitive performance, achieving an accuracy of 0.9687, along with superior precision (0.9694), recall (0.9687), and F1-score (0.9686). Although EmoNeZha achieves a marginally higher accuracy (0.9714), the balanced performance of our model across all evaluation metrics emphasizes its robustness. This performance is a result of the effective integration of textual features and multi-view social network features, which facilitates precise and nuanced depression detection. During our experiments, we set the hyperparameters as follows: $\alpha = 0.1$, $\beta = 0.9$, $w_1 = 0.7$, and $w_2 = 0.3$.

As presented in Table 2, our model demonstrates competitive performance, achieving an accuracy of 0.9687, along with superior precision (0.9694), recall (0.9687), and F1-score (0.9686). This performance advantage stems from the

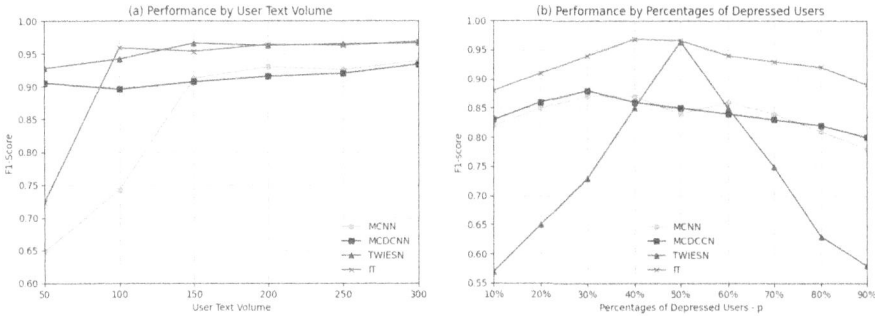

Fig. 2. Parameter Analysis of Model Performance.

effective integration of textual features and multi-view social network information, enabling our model to capture subtle depressive signals that may be missed
by models relying solely on text. Although EmoNeZha achieves slightly higher
accuracy (0.9714), its performance across other metrics is less consistent, potentially due to overfitting to specific dataset patterns. In contrast, our model leverages dynamic feature fusion, enhancing its adaptability to diverse user behaviors.
Therefore, despite not having the highest accuracy, our model delivers more balanced and robust performance, particularly in scenarios where both precision
and recall are critical. During our experiments, we set the hyperparameters as
follows: $\alpha = 0.1$, $\beta = 0.9$, $w_1 = 0.7$, and $w_2 = 0.3$.

User Text Volume Analysis. We investigate the impact of the number of user
texts on depression detection results. As shown in Fig. 2(a), the performance of
all models improves significantly as the user text volume increases. However, once
the number of texts exceeds 100, the performance of each classifier experiences
slight fluctuations with further increases in text volume. This suggests that while
a larger volume of data generally improves model performance, the marginal
benefit diminishes beyond a certain threshold.

Depressed User Percentage Analysis. In real-world scenarios, the proportion of depressed users is typically low. To evaluate the performance of the models
under different conditions, we tested them with varying proportions of depressed
users (denoted as ρ), ranging from 10% to 90% in 10% increments. The results
are shown in Fig. 2(b). The figure demonstrates that, except for the TWIESN
model, which performs best when trained on a balanced dataset, other models
perform optimally on slightly unbalanced datasets. Specifically, the MCNN and
MCDCNN models achieve the best performance when trained on datasets with
30% of depressed users, while the IT model performs best with 40% of depressed
users.

4.3 Ablation Study

In this subsection, we systematically evaluate the effectiveness of each module in the proposed model and their contributions to depression detection performance through ablation experiments. Table 3 presents the results of the ablation study. In the table, "PS" stands for preliminary screening using the DSM-5, "3WD" indicates filtering texts with 3WD theory, "SNFEM" represents Social Network Feature Extraction Module, which includes user profiles and social behavioral features. Our results demonstrate that each module plays a crucial role in enhancing the overall detection accuracy.

Table 3. Experimental results (F1-Score) on Ablation Study.

	PS	3WD	SNFEM	Accuracy	Precision	Recall	F1-Score
Base				0.8770	0.8807	0.8770	0.8767
1	✓			0.8873	0.8892	0.8873	0.8872
2	✓		✓	0.9380	0.9419	0.9380	0.9380
3	✓	✓	✓	**0.9687**	**0.9694**	**0.9687**	**0.9686**

The ablation study demonstrates that the Reliable Filtering Module and Social Network Feature Extraction Module play pivotal roles in the proposed model. The Reliable Filtering Module improves data quality and significantly enhances the model's noise robustness, while the Social Network Feature Extraction Module provides additional contextual information, enriching the representation of the user's psychological state.

5 Conclusion

This paper proposes a model named MVDep, which integrates multi-view features, including social media texts, user profiles, and social behaviors, to enhance text representation for depression detection. Experimental results demonstrate that the MVDep algorithm exhibits outstanding performance, verifying its effectiveness and robustness in the task of depression detection. This study not only provides new technical approaches for mental health detection in online social networks but also explores a new direction for applying 3WD theory in the field of mental health. Future work will focus on further optimizing the model's efficiency, exploring its generalization ability in multilingual and multicultural contexts, and investigating how to apply this model to the detection of other mental health issues.

Acknowledgments. Our work is supported by the National Natural Science Foundation of China (62476003), Anhui Province Excellent Scientific Research and Innovation Team (2024AH010004), Anhui Provincial Natural Science Foundation - Water Science

Joint Fund (2408055US006), the University Synergy Innovation Program of Anhui Province (GXXT-2023-050), and SMP-Zhipu. Al Large Model Cross-Disciplinary Fund (SMP-Zhipu20240210). We also acknowledge the support from Zhipu AI-Anhui University Joint Research Center, and the High-Performance Computing Platform of Anhui University.

References

1. Zhang, T., Yang, K., Ananiadou, S.: Sentiment-guided transformer with severity-aware contrastive learning for depression detection on social media. In: The 22nd Workshop on Biomedical Natural Language Processing and BioNLP Shared Tasks, pp. 114–126 (2023)
2. World Health Organization: Depressive disorder (depression). https://www.who.int/news-room/fact-sheets/detail/depression. Accessed June 2023
3. Iyortsuun, N.K., Kim, S.-H., Jhon, M., Yang, H.-J., Pant, S.: A review of machine learning and deep learning approaches on mental health diagnosis. In: Healthcare, vol. 11, no. 3, p. 285. MDPI (2023)
4. Yao, M., Chelmis, C., Zois, D.-S.: Cyberbullying ends here: towards robust detection of cyberbullying in social media. In: The World Wide Web Conference, pp. 3427–3433 (2019)
5. Chatzakou, D., et al.: Detecting cyberbullying and cyberaggression in social media. ACM Trans. Web (TWEB) 13(3), 1–51 (2019)
6. Tejaswini, V., Sathya Babu, K., Sahoo, B.: Depression detection from social media text analysis using natural language processing techniques and hybrid deep learning model. ACM Trans. Asian Low-Resource Lang. Inf. Process. 23(1), 1–20 (2024)
7. Kerasiotis, M., Ilias, L., Askounis, D.: Depression detection in social media posts using transformer-based models and auxiliary features. Soc. Netw. Anal. Min. 14(1), 196 (2024)
8. De Choudhury, M., Gamon, M., Counts, S., Horvitz, E.: Predicting depression via social media. In: Proceedings of the International AAAI Conference on Web and Social Media, vol. 7, no. 1, pp. 128–137 (2013)
9. Ghosh, S., Anwar, T.: Depression intensity estimation via social media: a deep learning approach. IEEE Trans. Comput. Soc. Syst. 8(6), 1465–1474 (2021)
10. Mihov, I., Chen, H., Qin, X., Ku, W.-S., Yan, D., Liu, Y.: Mentalnet: heterogeneous graph representation for early depression detection. In: 2022 IEEE International Conference on Data Mining (ICDM), pp. 1113–1118. IEEE (2022)
11. Pirayesh, J., Chen, H., Qin, X., Ku, W.-S., Yan, D.: Mentalspot: effective early screening for depression based on social contagion. In: Proceedings of the 30th ACM International Conference on Information & Knowledge Management, pp. 1437–1446 (2021)
12. Regier, D.A., Kuhl, E.A., Kupfer, D.J.: The DSM-5: classification and criteria changes. World Psychiat. 12(2), 92–98 (2013)
13. Eichstaedt, J.C., et al.: Facebook language predicts depression in medical records. Proc. Natl. Acad. Sci. 115(44), 11203–11208 (2018)
14. Chiong, R., Budhi, G.S., Dhakal, S.: Combining sentiment lexicons and content-based features for depression detection. IEEE Intell. Syst. 36(6), 99–105 (2021)
15. Saha, B., Nguyen, T., Phung, D., Venkatesh, S.: A framework for classifying online mental health-related communities with an interest in depression. IEEE J. Biomed. Health Inf. 20(4), 1008–1015 (2016)

16. Zogan, H., Razzak, I., Jameel, S., Xu, G.: Depressionnet: a novel summarization boosted deep framework for depression detection on social media. arXiv preprint arXiv:2105.10878 (2021)
17. Negriff, S.: Depressive symptoms predict characteristics of online social networks. J. Adolesc. Health **65**(1), 101–106 (2019)
18. Razak, C., Zulkarnain, M.A., Hamid, S., Anuar, N.B., Jali, M.Z., Meon, H.: Tweep: a system development to detect depression in twitter posts. In: Alfred, R., Lim, Y., Haviluddin, H., On, C.K. (eds.) Computational Science and Technology. LNEE, vol. 603, pp. 543–552. Springer, Singapore (2020). https://doi.org/10.1007/978-981-15-0058-9_52
19. Zhu, J., Zhang, Z., Guo, Z., Li, Z.: Sentiment classification of anxiety-related texts in social media via fusing linguistic and semantic features. IEEE Trans. Comput. Soc. Syst. (2024)
20. Ismail Fawaz, H., et al.: Inceptiontime: finding alexnet for time series classification. Data Min. Knowl. Disc. **34**(6), 1936–1962 (2020)
21. Cai, Y., Wang, H., Ye, H., Jin, Y., Gao, W.: Depression detection on online social network with multivariate time series feature of user depressive symptoms. Expert Syst. Appl. **217**, 119538 (2023)
22. Cong, Q., Feng, Z., Li, F., Xiang, Y., Rao, G., Tao, C.: XA-BiLSTM: a deep learning approach for depression detection in imbalanced data. In: 2018 IEEE International Conference on Bioinformatics and Biomedicine (BIBM), pp. 1624–1627. IEEE (2018)
23. Burdisso, S.G., Errecalde, M., Montes-y-Gómez, M.: A text classification framework for simple and effective early depression detection over social media streams. Expert Syst. Appl. **133**, 182–197 (2019)
24. Wang, Y., Wang, Z., Li, C., Zhang, Y., Wang, H.: Online social network individual depression detection using a multitask heterogenous modality fusion approach. Inf. Sci. **609**, 727–749 (2022)
25. Zhang, Z., Li, Z., Zhu, J., Guo, Z., Shi, B., Hu, B.: Enhancing user sequence representation with cross-view collaborative learning for depression detection on Sina Weibo. Knowl.-Based Syst. **293**, 111650 (2024)

An Interpretable Framework for Pulmonary Nodule Malignancy Prediction

Ruqi Wang[1,2], Guoyin Wang[1,2,3]([✉]), Yunfei Zhou[4], and Qun Liu[1,2]

[1] Key Laboratory of Cyberspace Big Data Intelligent Security, Ministry of Education, Chongqing University of Posts and Telecommunications, Chongqing 400065, China
d200201018@stu.cqupt.edu.cn, liuqun@cqupt.edu.cn
[2] Chongqing Key Laboratory of Computational Intelligence, Chongqing University of Posts and Telecommunications, Chongqing 400065, China
[3] National Center for Applied Mathematics in Chongqing, Chongqing Normal University, Chongqing 401331, China
wanggy@cqnu.edu.cn
[4] Department of Neurosurgery, The First Affiliated Hospital of Xi'an Jiaotong University, Xi'an 71004, China

Abstract. Achieving the reliable and automatic pulmonary nodule malignancy prediction is a critical step in reducing the workload of physicians and advancing towards intelligent medical treatment. The 'black-box' of deep neural networks is a major obstacle in achieving reliable prediction. An interpretable neural network framework for pulmonary nodule malignancy prediction is proposed to ensure reliable predictions. It mainly includes image generation module, mapping network module, reversible network module and reliable classifier module. The latent space constructed based on the image generation module can completely and accurately represent the features of pulmonary nodules. By mapping the input images to the latent space, the accurate representations of pulmonary nodules can be obtained, and then they are mapped to the feature space through the reversible network module. Finally, based on data-driven knowledge in feature space and knowledge-driven medical knowledge, the reliable malignancy prediction is achieved. The experiment shows that our model can accurately extract the features of pulmonary nodules and reliably predict malignancy based on those specific features and medical knowledge. In addition, this model can be used to analyze the difference between data-driven knowledge and knowledge-driven knowledge, which helps enrich and refine medical knowledge while interpreting the basis for prediction.

Keywords: interpretable · pulmonary nodule malignancy prediction · reliable classifier · latent space · medical knowledge

1 Introduction

Due to the high incidence and mortality, lung cancer has become one of the common diseases affecting human health and life expectancy [3]. The early diagnosis

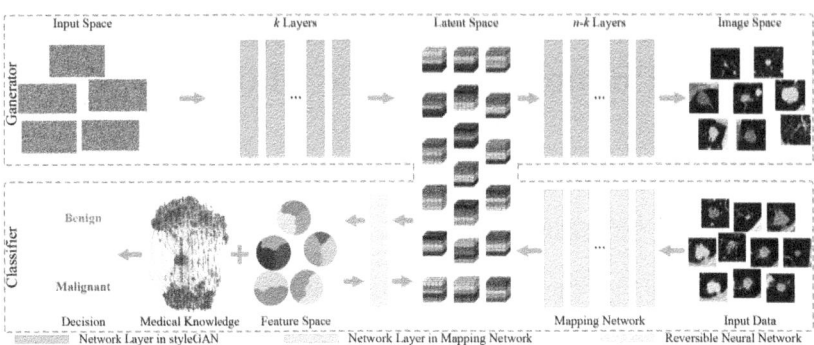

Fig. 1. The interpretable neural network framework for nodule malignancy prediction.

of lung cancer achieved through the screening with chest computed tomography(CT) can effectively reduce the mortality rate of lung cancer [3]. The malignant degree of pulmonary nodules, the "spots" less than or equal to 3cm in diameter detected by chest CT, is an important basis for the early diagnosis of lung cancer [9]. Accurately predicting the benign or malignant pulmonary nodules according to CT images requires profound medical knowledge and rich clinical experience, which brings a significant workload to doctors. How to help physicians to efficiently and reliably diagnose pulmonary nodules represents a key challenge in enhancing the accuracy of early lung cancer detection and advancing towards intelligent medical treatment.

In recent years, with the widespread application of deep neural networks in machine learning, many researchers have introduced them into the field of pulmonary nodule malignancy prediction. The accuracy [12], false positive rate [15] and robustness [9] of pulmonary nodule prediction have been continuously improved. Semi-supervised learning [5] and few-shot learning [11] in pulmonary nodule prediction have also effectively overcome the difficulties in data acquisition. These efforts have achieved remarkable results. However, the predictions of pulmonary nodule malignancy require extremely high credibility. Due to the "black-box" and data-driven nature of traditional deep neural networks, it is unknown whether the feature space used for classification is complete and accurate. It is necessary to construct an interpretable feature space to improve the reliability of the pulmonary nodule malignancy prediction.

Since the physicians mainly diagnose the malignant degree of pulmonary nodules by analyzing their distinguishing features in CT images, interpreting and representing the different features of pulmonary nodules in feature space is the key to making a reliable diagnosis. In this paper, an interpretable neural network framework for pulmonary nodule malignancy prediction is proposed to improve the reliability of diagnostic results by combining data-driven and knowledge-driven approaches as shown in Fig. 1. Firstly, the latent space of styleGAN [7], which can accurately represent and realistically generate the pulmonary nodule images, is used to construct a complete feature space of pulmonary nodule.

Then, data-driven knowledge is obtained by analyzing the correlation between malignancy and features of pulmonary nodules. Finally, the reliable malignancy prediction of input pulmonary nodule is diagnosed jointly with data-driven knowledge and knowledge-driven medical knowledge. The main contributions of this paper are as follows:

- We propose a high-performance interpretable framework for pulmonary nodule malignancy prediction based on the data-driven and knowledge-driven.
- Interpretable and complete feature spaces based on styleGAN can accurately represent different features of pulmonary nodules and their relationships.
- The introduction of medical knowledge helps the model to improve the accuracy and reliability of malignancy prediction.
- Experiments show that our model can give reliable malignancy prediction along with predictive evidence.

2   Related Work

In recent years, deep neural networks have greatly improved the performance of pulmonary nodule malignancy prediction. To improve the accuracy of diagnosis, Causey et al. proposed NoduleX [2] to predict the malignancy of pulmonary nodules from CT images based on convolutional neural networks and prior knowledge. Literature [18] used the Denoising Auto Encoder (DAEK) and ensemble learning method to comprehensively extract and analyze the morphological texture features (TF) and deep semantic features. A novel architecture Res-trans network [10] was proposed to cope with various forms of pulmonary nodules and accurately predict the malignancy degree of pulmonary nodules from 3 to 30 mm. Due to the fuzzy features of pulmonary nodules in CT images, Liao et al. proposed a multi-view 'divide-and-rule' (MV-DAR) model [9] to learn features from both reliable and ambiguous annotations, which achieves the superior performance in CT images with noise. The classifier of pulmonary nodules (DCNN) in literature [15] automatically obtained the feature based on an end-to-end DCNN architecture, which effectively reduces the false positives. A Fuse-Long Short-Term Memory-Convolutional Neural Network (F-LSTM-CNN) [12] was proposed to distinguish the benign and malignant nodules by fusing visual features and deep features from affine transformation. Yiet al. [20] realized multilabel classification of pulmonary nodules on unbalanced data using multi-label softmax loss and reduced the ranking errors between the labels and within the labels during training. Ahmed et al. [1] proposed an automated system for classifying and detecting pulmonary nodules reducing the False Positive Rate (FPR).

Because deep neural networks can not be understood by researchers, their predictions have low credibility. In order to improve the reliability of pulmonary nodule malignancy prediction model based on deep neural network. Shen et al. [14] proposed a novel interpretable deep hierarchical semantic convolutional neural network (HSCNN) to interpret the basis of malignant degree of pulmonary nodules by inputting the low-level semantic features. Xu et al. introduced the medical knowledge into convolutional neural networks though the Hyper-Attention Mechanism(HAM) [19] to obtain crucial contextual information and make reliable decisions. Roy et al. [13] introduced a self-attention attribute module and a self-attention spatial module to build an attribute-driven Generative Adversarial Network (ADGAN), which can reasonably predict the malignant degree according to the attributes of nodules. A convolutional neural network (CNN)-based MTL model [4] was proposed to accurately represent the visual attributes of pulmonary nodule, it can be used to reveal the inter-attribute dependencies for clinical interpretation in pulmonary nodule malignancy prediction.

In addition, due to the difficulty in obtaining pulmonary nodule data and the high cost of labeling, researchers also explored pulmonary nodule malignancy prediction in semi-supervised learning and few-shot learning. Mobiny et al. proposed a memory-augmented capsule network [11] to adequately mining the knowledge from the target domains. This model can achieve competitive results in the case of heavy noise, artifacts, and adversarial attacks using only a few annotated samples. In order to reduce the dependence on abundant training data, Fu et al. proposed a reverse adversarial classification network (RACN) [5] based on semi-supervised learning to mining uncertain information and distinguishing features.

3   Framework

This section details the construction of our framework as shown in Fig. 1. We first train the styleGAN [7] to generate the realistic images of pulmonary nodules, and use semantic-level interpretation method to obtain the knowledge in its latent space. Then, the reversible neural network module are used to map these knowledge into the complete feature space. After mapping the input image into the feature space, the reliable classifier combines the data-driven knowledge in the feature space and the knowledge-driven medical knowledge to accurately predict malignancy.

3.1   Latent Space in Pulmonary Nodule Generator

The latent space of the generated model can generate images realistically, therefore, its feature representation of the input data is stronger than that of the classification model constructed only by the classification loss function. The styleGAN, an advanced generative model capable of mapping noise data to any image space, is chosen as the pulmonary nodule generator. Researches on the interpretability of styleGAN show that its W space $W \subseteq \mathbb{R}^d$, one of the latent space, in generator can accurately represent the feature information. Based on our previous work [17], the feature of pulmonary nodule in latent space can be described as a high-dimensional Gaussian distribution, as shown in Definition 1.

For any latent variable $\boldsymbol{w} \in W$, its membership degree $n(\boldsymbol{w})$ for the feature distribution F is defined as:

$$n(\boldsymbol{w}, F) = \frac{1}{d} \sum_{i}^{d} \exp(-\frac{(\boldsymbol{w}[i] - \boldsymbol{\mu}[i])^2}{2\boldsymbol{\sigma}[i]^2}). \tag{1}$$

Definition 1. $F = (\boldsymbol{\mu}, \boldsymbol{\sigma}, h(\cdot))$ *is called a 3-tuple representing the distribution of feature in d-dimensional latent space where:* $\boldsymbol{\mu} = (\mu_1, \mu_2, \cdots, \mu_d)$ *denotes a vector of feature center for each dimension.* $\boldsymbol{\sigma} = (\sigma_1, \sigma_2, \cdots, \sigma_d)$ *denotes a vector of the feature variance for each dimension.* $h(\boldsymbol{w}) = \frac{1}{\sqrt{2\pi}\sigma} \exp(-\frac{(w-\mu)^2}{2\sigma^2})$ *is feature probability density distribution in d-dimensional latent space, where* $h_i(\boldsymbol{w}[i])$ *is the sub-component of* $h(\boldsymbol{w})$.

Because the generator can only calculate forward, it cannot obtain the latent variable corresponding to real images. In order to obtain the distribution details of different features in the latent space, we designed a mapping network $m(\cdot)$ to map the pulmonary nodule images into the latent space, as shown in Fig. 1. The input of the mapping network is the pulmonary nodule images (64×64) and the output is a d-dimensional vectors. The mapping network consists of 5 convolutional layers, 5 Batch Normalization layers, and 1 fully connected layer. The generator in well-trained styleGAN is used to generates a large amount of pulmonary nodule images $\boldsymbol{I}_{g_1}, \boldsymbol{I}_{g_2}, \cdots, \boldsymbol{I}_{g_n}$ and their corresponding latent variable $\boldsymbol{w}_1, \boldsymbol{w}_2, \cdots, \boldsymbol{w}_n$ in the latent space as the temporary dataset for training mapping network. Then, the mean-square error (MSE) loss is used to train mapping network to make it map arbitrary images to the latent space, as shown in

$$L_{mse} = ||m(\boldsymbol{im}) \odot \boldsymbol{w}||^2. \tag{2}$$

The images $\boldsymbol{I}_{t_1}, \boldsymbol{I}_{t_2}, \cdots, \boldsymbol{I}_{t_n}$ in the training dataset of pulmonary nodules can be mapped to the latent space by the well-trained mapping network, and their corresponding latent variables $\boldsymbol{w}_1, \boldsymbol{w}_2, \cdots, \boldsymbol{w}_n$ can also be obtained. Then, according to the label information in the training dataset, the obtained latent variables are divided into different sets. Different sets contain latent variables with corresponding feature. For example, the latent variables containing the feature of Linear-Sphericity and Round-Sphericity are classified into sets $\boldsymbol{S}_{s_l}$ and $\boldsymbol{S}_{s_r}$, respectively. The Gaussian Mixture Model (GMM) is used to analyze the distribution mechanism of latent variables in different sets, so as to obtain the distribution F of each feature in the latent space.

According to the properties of Gaussian distribution [17], the average Hellinger distance can be used to measure the distance between different feature distribution. For arbitrary distributions F_i and F_j, their average Hellinger distance in the d-dimensional latent space is defined in Eq. (3).

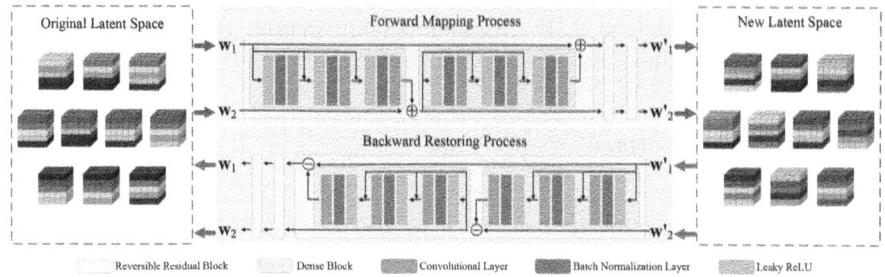

Fig. 2. The structure of reversible network.

$$D(F_i, F_j) = \frac{1}{d} \sum_{k=1}^{d} \frac{1}{\sqrt{2}} ||h_{ik}(\boldsymbol{w}[k]) - h_{jk}(\boldsymbol{w}[k])||_2$$

$$= \frac{1}{d} \sum_{k=1}^{d} (1 - \sqrt{\frac{2\sigma_{ik}\sigma_{jk}}{\sigma_{ik}^2 + \sigma_{jk}^2}} e^{-\frac{(\mu_{ik}-\mu_{jk})^2}{4(\sigma_{ik}^2+\sigma_{jk}^2)}}), \tag{3}$$

where $h_{ik}(\cdot)$, μ_{ik} and σ_{ik} are the parameters of the feature distribution F_i on the dimension k, $h_{jk}(\cdot)$, μ_{jk} and σ_{jk} are the parameters of the feature distribution F_j on the dimension k. The greater the distance, the more distant the relationship between features, and the closer the distance, the smaller the relationship between features. In this way, the interpretable latent space with feature representation and realistic generation is obtained.

3.2   Reversible Neural Network

Since the latent space of styleGAN is mainly used to generate pulmonary nodule images, features in the latent space are usually highly entangled [14], which hinders the distinction of different features in the latent space. Therefore, by introducing the reversible neural network module [6] to construct a new feature space, we can make the features in the feature space have obvious differentiation. Because the reversible neural network can establish the bijection relationship between the original latent space and the new feature space, which gives the model the ability of both classification and generation as shown in Fig. 1. The structure of the reversible neural network is shown in Fig. 2, which consists of 3 reversible residual blocks, each of which contains 3 convolutional layers and 3 Batch Normalization layers. Forward and reverse calculations of residual blocks can be realized by different connections of different residual blocks. Two different residual blocks in the k-th reversible residual block are defined as $fr_k(\cdot)$ and $gr_k(\cdot)$, respectively, and the input data are $\boldsymbol{w}_1$ and $\boldsymbol{w}_2$. The forward output $\boldsymbol{w}_1'$ and $\boldsymbol{w}_2'$ of the reversible residual block are shown in Eq. (4).

$$\begin{cases} \boldsymbol{w}_1' = \boldsymbol{w}_1 + fr_k(\boldsymbol{w}_2); \\ \boldsymbol{w}_2' = \boldsymbol{w}_2 + gr_k(\boldsymbol{w}_1). \end{cases} \tag{4}$$

When the connection of two residual blocks is changed, the input data w_1 and w_2 can be reconstructed according to the outputs w'_1 and w'_2, as shown in the Eq. (5).

$$\begin{cases} w_2 = w'_2 - gr_k(w'_1); \\ w_1 = w'_1 - fr_k(w_2). \end{cases} \tag{5}$$

Three reversible residual blocks constitute a complete reversible neural network module, denoted by $r(\cdot)$, which has an output w.

In order to make the features in the new feature space more distinguishable, we calculate the distribution of features in initial new feature space based on Definition 1 before training. The parameters of these feature distributions are used as prior knowledge to constrain the training of reversible neural network module. Specifically, based on the attribute of features, the feature distributions $F_1, F_2, \cdots, F_n$ can be divided into different feature subspaces S_{sub}, and the feature distributions within the same feature subspaces share the same category. For example, the Sphericity feature subspace contains features Linear and Round. In feature subspace S_{sub_k}, for the latent variable $w' = r(w)$, obtained by reversible neural network module from the latent space, its membership degree of feature $F_{k_i}, F_{k_i} \in S_{sub_k}$ is represented as $y_{pre_{k_i}} = n(w, F_{k_i})$ based on Eq. (1). The membership degree of all features constitutes the membership vector $y_{pre} = \{y_{pre_{k_1}}, y_{pre_{k_2}}, \cdots, y_{pre_{k_n}}\}$. Since the latent variables can contain only one feature in the same feature subspace. The reversible neural network module can be trained with the cross entropy loss, as shown in Eq. (6).

$$L_{rev} = \frac{1}{M} \sum_{j=1}^{M} \sum_{i=1}^{n} y_{pre_{k_i}} \log y_i, \tag{6}$$

where M represents the number of samples, n represents the number of features in the feature subspace, y_i represents the real label. Each feature subspace needs to a set of parameters trained by Eq. (6), and all feature subspaces form the complete feature space S. The distinction between the different feature distributions based on Definition 1 in the trained feature space is more obvious, which is more conducive to the classification task.

3.3   Reliable Classifier with Medical Knowledge

For any input pulmonary nodule images im, mapping network and reversible neural network modules can be used to map them into the feature space, and then their malignancy prediction is obtained based on the knowledge of data-driven and knowledge-driven. The knowledge of data-driven is composed of the pulmonary nodules attribute "malignancy" alone, which mainly relies on the training dataset iterative training to predict the malignant degree. The data-driven prediction $p_{data}(im)$ is defined as Eq. (7).

$$p_{data}(im) = \begin{cases} 0, \text{ if } p_{ben}^{data} \geq p_{mal}^{data}; \\ 1, \text{ otherwise}, \end{cases} \tag{7}$$

Algorithm 1. The algorithm of training the interpretable neural network framework for pulmonary nodule malignancy prediction

---

Input: training dataset, label y

Output:

1: Training styleGAN with training dataset.
2: Sampling 10K generated pulmonary nodule images and its latent variables from styleGAN.
3: **while** $epoch < e_m$ **do**
4:     Training the mapping network with sample data by loss function l_{mse}.
5: **end while**
6: **while** $epoch < e_r$ **do**
7:     Training the reversible neural network with training dataset and label y by loss function l_{rev}.
8: **end while**
9: Constructing classifier based on Eq. (7), Eq. (9) and Eq. (10)
10: **return** reliable classifier

---

where $p_{ben}^{data} = n(r(m(\boldsymbol{im})), F_{benign})$ and $p_{mal}^{data} = n(r(m(\boldsymbol{im})), F_{malignancy})$ represents the membership degree of benign and malignant based on their distributions respectively, $m(\cdot)$ represents the mapping network, $r(\cdot)$ represents the reversible neural network, $n(\cdot, F_i)$ represents the membership of the input to feature F_i, 0 and 1 represent benign and malignant pulmonary nodules, respectively.

In the malignancy prediction of pulmonary nodules, the data-driven module is more prone to misjudgment when the membership of benign and malignant nodules is similar. This requires the model to make reliable predictions based on medical knowledge. Based on the medical literatures of pulmonary nodules [16] and the guidance of professional physicians, we assign correlation weights between feature "Benign" and other features F_i as $c_{b,i}$. A higher correlation weight indicates a higher correlation between the feature and benign. Similarly, we assign correlation weights between feature "Malignant" and other features F_i as $c_{m,i}$. Since the correlation between attributes and malignancy is difficult to judge, we assume that they are equally correlated. Then the benign and malignant average correlation weights can be respectively expressed as

$$
\begin{cases}
p_{benign}^{kno} = \frac{1}{A} \sum_{k=1}^{A} \sum_{i=1}^{n} c_{b,ki} n(r(m(\boldsymbol{im})), F_{ki}); \\
p_{maly}^{kno} = \frac{1}{A} \sum_{k=1}^{A} \sum_{i=1}^{n} c_{m,ki} n(r(m(\boldsymbol{im})), F_{ki}),
\end{cases}
\tag{8}
$$

where p_{benign}^{kno} and p_{maly}^{kno} respectively represent the benign and malignant average correlation. A is the number of feature subspaces, n represents the number of features in the feature subspace, F_{ki} is the i-th feature in k-th feature subspace S_{sub_k}. Therefore, the knowledge-driven prediction $p_{kno}(\boldsymbol{im})$ can be constructed as Eq. (9).

$$
p_{kno}(\boldsymbol{im}) = \begin{cases}
0, \text{if} p_{ben}^{kno} \geq p_{maly}^{kno}; \\
1, \text{ortherwise}.
\end{cases}
\tag{9}
$$

0 and 1 represent benign and malignant pulmonary nodules in knowledge-driven module, respectively.

The complete prediction $p_{all}(\boldsymbol{im})$ of malignancy pulmonary nodule is shown in the Eq. (10).

$$p_{all}(\boldsymbol{im}) = \begin{cases} p_{data}, \text{ if } |p_{ben}^{data} - p_{mal}^{data}| > \phi; \\ p_{kno}, \text{ otherwise,} \end{cases} \qquad (10)$$

where ϕ is a hyperparameter used to judge the similarity between the membership of benign and malignant pulmonary nodules. The reliable pulmonary nodule malignancy prediction is obtained by considering both data-driven and knowledge-driven predictions.

3.4 Training

The training of the interpretable neural network framework for pulmonary nodule malignancy prediction proposed in this paper includes styleGAN generator training, mapping network training and reversible neural network training in three stages. The styleGAN generator with original setting is first trained to generate realistic images of pulmonary nodule. The mapping network is then trained e_m times based on the loss function L_{mse} so that it can map real pulmonary nodule images to the latent space of the styleGAN generator. Finally, the reversible neural network is trained e_r times to construct new feature space that can be used for feature classification based on the original latent space. The classification stage of pulmonary nodule malignancy prediction based on knowledge of data-driven and knowledge-driven is constructed by Eq. (7), Eq. (9) and Eq. (10) without additional training. The complete training process is shown in the Algorithm 1.

4 Experiment

4.1 Implementation Details

In this paper, LIDC-IDRI dataset is used to verify the proposed model. Due to the uncertainty of the unclear features, we only focus on the influence of clear features on pulmonary nodule malignancy prediction [8], different pulmonary nodules and their explicit features are shown in Table 1. The feature space $\boldsymbol{S}$ we constructed contains 8 feature subspaces, namely 'Margin' feature subspace with 2 features, 'Sphericity' feature subspace with 3 features, 'Spiculation' feature subspace with 2 features, 'Texture' feature subspace with 3 features, 'Calcification' feature subspace with 5 features, 'Internal structure' feature subspace with 2 features, 'Lobulation' feature subspace with 2 features and 'Malignancy' feature subspace with 3 features. Note that, the feature 'Benign' in 'Malignancy' feature subspace is comprised of 'Highly unlikely' and 'Moderately unlikely' in the original data, the feature 'Malignant' in 'Malignancy' feature subspace is made up of 'Highly suspicious' and 'Moderately suspicious' in the original data.

Table 1. Features and correlation weights of pulmonary nodules in medical knowledge.

Attribute	Feature	Correlation c_b	Correlation c_m
Margin	Poorly defined	0.83	0.17
	Sharp	0.17	0.83
Sphericity	Linear	0.56	0.11
	Ovoid	0.33	0.33
	Round	0.11	0.56
Spiculation	Marked	0.17	0.83
	None	0.83	0.17
Texture	Non-solid	0.56	0.11
	Part Solid	0.33	0.33
	Solid	0.11	0.56
Calcification	Popcorn	0.33	0.07
	Laminated	0.27	0.13
	Solid	0.20	0.20
	Non-central	0.07	0.33
	Central	0.13	0.27
Internal structure	Soft tissue	0.43	0.57
	Fluid	0.57	0.43
Lobulation	Marked	0.17	0.83
	None	0.83	0.17
Malignancy	Benign	–	–
	Indeterminate	–	–
	Malignancy	–	–

The 'Malignancy' feature subspace is used for the data-driven malignancy prediction module. The other seven feature subspaces are used for the knowledge-driven malignancy prediction module, and the correlations assigned according to medical knowledge are shown in Table 1. The original CT images are segmented into (64×64) images by using the annotations in the dataset, and the pulmonary nodules are ensured to be located in the center of the images.

The generated image size of styleGAN is changed to (64×64), other parameters are same as the public code of styleGAN. The mapping network is iteratively trained for $e_m = 500$ iterations with a learning rate 0.003. For each feature subspace, the reversible network with the same structure is iteratively trained for $e_r = 200$ iterations with learning rate 0.001.

Table 2. The objective evaluation of styleGAN and mapping network.

	IS	FID	MSE(latent)	MSE(image)
styleGAN	2.82	22.02	–	–
Mapping Network	–	–	2.56E-02	4.91E-02

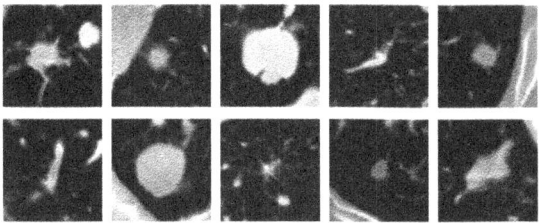

Fig. 3. The subjective evaluation of generator.

Table 3. Accuracy of feature classification in different feature subspaces. Internal is Internal structure.

Model	Margin	Sphericity	Spiculation	Texture	Calcification	Internal	Lobulation
HSCNN	0.725	0.552	–	0.834	**0.908**	–	–
Ours	**0.885**	**0.898**	**0.772**	**0.864**	0.871	**0.954**	**0.951**

4.2   The Performance of Generator and Mapping Network

In this section, we evaluate the generation ability of styleGAN and the mapping capability of the mapping network. After iterative training, the generated effect of styleGAN on pulmonary nodules is shown in Fig. 3. It can generate a variety of pulmonary nodules with different features. In addition, we objectively evaluated styleGAN using Inception Score (IS) [17] and Fréchet Inception Distance (FID) [17], two methods specifically designed to evaluate the generation of generative adversarial networks, as shown in Table 2. For the mapping network, we only require it to accurately map the real image to the latent space, so the mean square error (MSE) is used to test its mapping ability, as shown in Table 2. MSE(latent) represents the error between the original latent variables obtained by sampling and the mapping latent variables obtained by mapping from sampled images. MSE(image) represents the error between the real pulmonary nodule images and their reconstructed images from the mapping latent variables. Figure 3 and Table 2 show that the generative and reconstruction ability of styleGAN and mapping networks can meet our needs.

4.3   Pulmonary Nodule Malignancy Prediction

Since the proposed model is mainly based on the feature space for classification, the feature classification performance in each feature subspace is first compared. Since it can output both pulmonary nodule malignancy prediction and pulmonary nodule feature prediction, the HSCNN [14] is used to compare the classification accuracy of each feature subspace with the proposed model, as shown in Table 3. The classification accuracy of our model is higher than the HSNN model in most feature subspaces, and it has the ability to classify in more feature subspaces than HSCNN model.

Table 4. Comparison of classification performance in malignancy pulmonary nodules.

Model	NoduleX	HSCNN	DAEK	DCNN	ADGAN	Ours(kno)	Ours(data)	Ours(all)
ACC	0.932	0.842	0.931	**0.978**	0.928	0.807	0.826	0.939
SPE	0.985	0.889	0.839	**0.972**	0.951	0.768	0.800	0.890
SEN	0.879	0.705	0.818	0.971	0.892	**0.979**	0.850	0.978

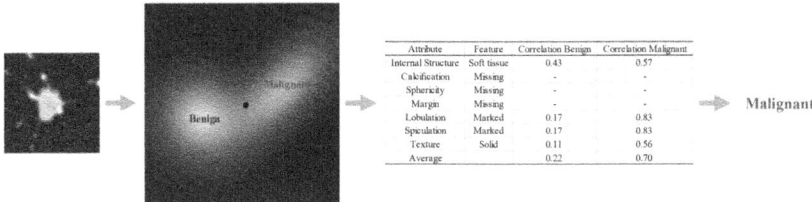

Fig. 4. The interpretable malignant prediction consisting of conclusion and basis.

For the task of pulmonary nodule malignancy prediction, the accuracy (ACC), specificity (SPE) and sensitivity (SEN) are used to measure the performance of the models, their calculations are shown in Eq. (11).

$$\begin{cases} ACC = \frac{TP+TN}{TP+TN+FP+FN} \\ SPE = \frac{TN}{FP+TN} \\ SEN = \frac{TP}{TP+FN} \end{cases} \tag{11}$$

where TP represents true positive, TN represents true negative, FP represents false positive, and FN represents false negative.

The comparison results are shown in Table 4. Ours(kno) represents the performance of the knowledge-driven classification module on the testing dataset. Ours(data) represents the performance of the data-driven classification module on the testing dataset, Ours(all) represents the performance of the data-driven classification module on part of the testing dataset obtained by hyperparameter $\phi \geq 0.001$. Experiments show that the result of Ours(kno) has a high SEN value, indicating that it can accurately predict all malignant nodules, which is also in line with the strategy of physicians to avoid missing diagnosis of malignant nodules. But this can also lead to a large number of benign nodules being misjudged as malignant. The results of Ours(data) cannot achieve superior accuracy. These indicates that the use of knowledge-driven or data-driven alone cannot achieve optimal diagnosis. Ours(all) combined with knowledge-driven or data-driven achieves excellent results, indicating that the knowledge-driven module effectively overcomes the misjudgment when the membership of benign and malignant nodules is similar in data-driven module. For any input image of pulmonary nodule, our model can make the interpretable malignancy predictions and their basis. As shown in the Fig. 4, the input sample is located between 'Benign' and 'Malignant' distributions in the feature space, which is difficult for

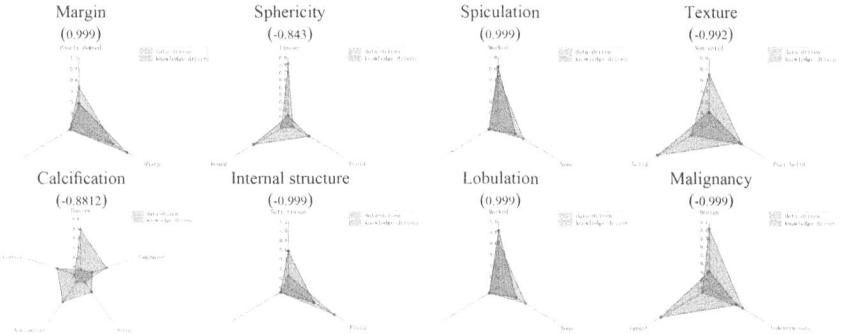

Fig. 5. The correlation between data-driven and knowledge-driven knowledge acquisition.

the model to distinguish. Therefore, medical knowledge is used to make a credible diagnosis based on the detailed features of the samples. This further proves the reliability of our classifier. The reasons for the lower prediction accuracy than DCNN may be the errors in feature classification and biases in embedded medical knowledge. Because our prediction has higher reliability, its comprehensive evaluation is better than other comparison models.

4.4   Latent Space Analysis

Since the styleGAN is mainly trained on the training dataset, the knowledge about pulmonary nodule in its latent space is obtained by data-driven. The correlation weights $c_{m,i}$ between the feature F_i and the malignant pulmonary nodules in Table 1 are obtained by knowledge-driven from medical journals and physicians. We measure the real distance $D(F_{mal}, F_i)$ between the 'Malignant' feature and other features based on the mean Hellinger distance defined in Eq. (3). The relative distance $D_r(F_{mal}, F_i)$ is obtained by normalizing the real distance. The $D_r(F_{mal}, F_i)$ and c_m are used to construct the radar chart to analyze the differences between data-driven and knowledge-driven knowledge, as shown in Fig. 5. Note that, for feature subspaces containing only two features, an extra feature with a value of 0 is added to construct the radar chart, which does not affect the original relationship. Then, by analyzing the Pearson correlation coefficient between $D_r(F_{mal}, F_i)$ and c_m, the consistency of the data-driven knowledge and the knowledge-driven knowledge is obtained. Since the distance and the correlation weight represent the opposite relationship, the negative correlation between $D_r(F_{mal}, F_i)$ and c_m means the knowledge of data-driven knowledge and knowledge-driven are consistent, otherwise, it is inconsistent. As shown in Fig. 5, data-driven and knowledge-driven knowledge are consistent except for feature subspace 'Margin', feature subspace 'Spiculation' and feature subspace 'Lobulation'. In medical knowledge, these three attributes have a strong reference value for malignancy, and in order to reduce the risk of misdiagnosing malignancy as benign, physicians will conservatively assume that all abnormal features are malignant. This also shows that relying solely on data-driven or knowledg-driven is not credible. Our reliable classifier can combine the two kinds of knowledge and make full use of their respective advantages to improve the performance and reliability of diagnosis.

5   Conclusion

To make the predictions based on deep neural network reliable, An interpretable neural network framework for pulmonary nodule malignancy prediction is proposed. The interpretable and complete feature space is constructed to accurately represent the pulmonary nodules based on the trained styleGAN. The mapping relationship between pulmonary nodule images and latent space is established through mapping network, and the representation of pulmonary nodule is obtained in latent space. The representation is mapped into the feature space by a reversible network, and the features of pulmonary nodules are determined according to the relationship between the mapping results and feature distribution. Finally, the reliable malignancy predictions of pulmonary nodules is obtained combining with data-driven and knowledge-driven approaches. Experiments on the LIDC-IDRI dataset show that this framework can accurately predict malignant pulmonary nodules and provide the reliable basis for prediction. The construction of classification model based on a generative model also provides a new solution for achieving reliable classification model. In the future, we will focus on optimizing the structure of the classifier module to improve the prediction performance and achieve high performance and reliable pulmonary nodule malignancy prediction, and try to apply it to clinical data.

Acknowledgment. This work is supported by The National Natural Science Foundations of China (61936001 and 62221005), Natural Science Foundations of Chongqing (cstc2021ycjh-bgzxm0013) and The Key Cooperation Project of Chongqing Municipal Education Commission(HZ2021008). We would like to thank editor and reviewers for insightful comments and advice.

References

1. Ahmed, I., Chehri, A., Jeon, G., Piccialli, F.: Automated pulmonary nodule classification and detection using deep learning architectures. IEEE/ACM Trans. Comput. Biol. Bioinf. **20**(4), 2445–2456 (2023)
2. Causey, J.L., et al.: Highly accurate model for prediction of lung nodule malignancy with ct scans. Sci. Rep. **8**(1), 9286 (2018)
3. Del Ciello, A., Franchi, P., Contegiacomo, A., Cicchetti, G., Bonomo, L., Larici, A.R.: Missed lung cancer: when, where, and why? Diagn. Interv. Radiol. **23**(2), 118 (2017)
4. Fu, X., Bi, L., Kumar, A., Fulham, M., Kim, J.: An attention-enhanced cross-task network to analyse lung nodule attributes in ct images. Pattern Recogn. **126**, 108576 (2022)

5. Fu, Y., Xue, P., Xiao, T., Zhang, Z., Zhang, Y., Dong, E.: Semi-supervised adversarial learning for improving the diagnosis of pulmonary nodules. IEEE J. Biomed. Health Inf. **27**(1), 109–120 (2023)
6. Gomez, A.N., Ren, M., Urtasun, R., Grosse, R.B.: The reversible residual network: backpropagation without storing activations. Adv. Neural Inf. Process. Syst. **30** (2017)
7. Karras, T., Laine, S., Aila, T.: A style-based generator architecture for generative adversarial networks. In: Proceedings of the IEEE/CVF Conference on Computer Vision and Pattern Recognition, pp. 4401–4410 (2019)
8. Li, Y., Wu, X., Yang, P., Jiang, G., Luo, Y.: Machine learning for lung cancer diagnosis, treatment, and prognosis. Genom. Proteomics Bioinf. **20**(5), 850–866 (2022)
9. Liao, Z., Xie, Y., Hu, S., Xia, Y.: Learning from ambiguous labels for lung nodule malignancy prediction. IEEE Trans. Med. Imaging **41**(7), 1874–1884 (2022)
10. Liu, D., Liu, F., Tie, Y., Qi, L., Wang, F.: Res-trans networks for lung nodule classification. Int. J. Comput. Assist. Radiol. Surg. , 1–10 (2022). https://doi.org/10.1007/s11548-022-02576-5
11. Mobiny, A., et al.: Memory-augmented capsule network for adaptable lung nodule classification. IEEE Trans. Med. Imaging **40**(10), 2869–2879 (2021)
12. Qiao, J., Fan, Y., Zhang, M., Fang, K., Li, D., Wang, Z.: Ensemble framework based on attributes and deep features for benign-malignant classification of lung nodule. Biomed. Signal Process. Control **79**, 104217 (2023)
13. Roy, R., Mazumdar, S., Chowdhury, A.S.: Adgan: attribute-driven generative adversarial network for synthesis and multiclass classification of pulmonary nodules. IEEE Trans. Neural Netw. Learn. Syst. **35**(2), 2484–2495 (2024)
14. Shen, S., Han, S.X., Aberle, D.R., Bui, A.A., Hsu, W.: An interpretable deep hierarchical semantic convolutional neural network for lung nodule malignancy classification. Expert Syst. Appl. **128**, 84–95 (2019)
15. Suresh, S., Mohan, S.: Nroi based feature learning for automated tumor stage classification of pulmonary lung nodules using deep convolutional neural networks. J. King Saud Univ.-Comput. Inf. Sci. **34**(5), 1706–1717 (2022)
16. Thai, A.A., Solomon, B.J., Sequist, L.V., Gainor, J.F., Heist, R.S.: Lung cancer. Lancet **398**(10299), 535–554 (2021)
17. Wang, R., Wang, G., Gu, L., Liu, Q., Liu, Y., Guo, Y.: Intuitively interpreting gans latent space using semantic distribution. Knowl.-Based Syst., 112894 (2025)
18. Xiao, N., Qiang, Y., Bilal Zia, M., Wang, S., Lian, J.: Ensemble classification for predicting the malignancy level of pulmonary nodules on chest computed tomography images. Oncol. Lett. **20**(1), 401–408 (2020)
19. Xu, W., Wang, K., Lin, J., Lu, Y., Huang, S., Zhang, X.: Knowledge-guided and hyper-attention aware joint network for benign-malignant lung nodule classification. In: 2020 IEEE International Conference on Image Processing (ICIP), pp. 310–314. IEEE (2020)
20. Yi, L., Zhang, L., Xu, X., Guo, J.: Multi-label softmax networks for pulmonary nodule classification using unbalanced and dependent categories. IEEE Trans. Med. Imaging **42**(1), 317–328 (2023)

CMIF: A Cross-Modal Information Fusion Approach for Molecular Property Prediction

Maoyuan Zang[1,2], Xu Gong[1,2], Rui Han[1,2], and Qun Liu[1,2(✉)]

[1] Key Laboratory of Cyberspace Big Data Intelligent Security, Ministry of Education, Chongqing, China
{s230232045,d210201006,d230201010}@stu.cqupt.edu.cn
[2] Chongqing Key Laboratory of Computational Intelligence, Chongqing University of Posts and Telecommunications, Chongqing 400065, China
liuqun@cqupt.edu.cn

Abstract. Molecular property prediction is vital in bioinformatics, as accurate and efficient predictions can greatly speed up drug discovery and lower R&D costs. In recent years, the development of deep learning has driven the innovation of data-driven molecular representation learning, which has derived sequence-based, graph-based, and geometry-based molecular representations. Many methods have been proposed to fuse molecular representations of different modalities to improve the performance of molecular property prediction. However, most recent studies have focused on one or two modes and combined them in a simple way, they also neglected the alignment of heterogeneous features. To address these challenges, we propose CMIF, a cross-modal information fusion framework for molecular property prediction, which integrates the three modal representations of sequence, graph, and geometry. The proposed CMIF encodes each modal feature separately through a specific neural network and innovatively employs a cross-attention mechanism for deep feature fusion to avoid the limitations of traditional linear operations. To further solve the problem of heterogeneous feature alignment, CMIF introduces a dual self-supervised strategy of contrast learning and cross-modal matching to strengthen the inter-modal semantic consistency. We evaluated the performance of CMIF on seven molecular datasets, and the results show that our method outperforms baseline models in most cases, validating the effectiveness of CMIF in comprehensively capturing molecular features.

Keywords: Molecular Property Prediction · Molecular Representation Learning · Cross-Modal · Self-Supervised Learning

1 Introduction

Molecular property prediction is crucial for drug discovery, as accurate molecular feature representations facilitate drug discovery [3,8], drug-target interaction prediction [4], and drug-drug interaction prediction. Recently, the advancement of deep learning has driven the learning of chemical information based on large-scale data. Molecular representations in deep learning can be categorized into

Q. Zhang et al. (Eds.): IJCRS 2025, LNAI 15710, pp. 228–241, 2025.
https://doi.org/10.1007/978-3-031-92741-6_17

three main types: sequence-based, graph-based, and geometry-based representations.

Sequence representations, such as SMILES [18], encode molecules as strings but lack spatial and topological information, limiting their performance. Molecular graphs, on the other hand, can represent topological structures and connectivity and are widely used for molecular property prediction, but they do not provide spatial information or certain specific properties, such as atomic chirality. Geometrical representations include features like atomic positions, bond angles, and torsional angles, addressing the limitations of molecular graphs. As a result, methods like GEM [2] have been proposed to extract geometric features. However, a single representation cannot fully capture molecular features, leading to the rise of multi-modal strategies. Among these, bimodal strategies include combining sequence and topological information [5,19], as well as combining molecular graphs and 3D coordinates [9,15,23]. Although bimodal methods improve representation capabilities, relying on two types of representations still has limitations and cannot fully capture molecular features. Some studies have attempted to integrate three modalities [13,22], but they often use weighted summation or concatenation, which may lead to the loss of critical information and face challenges in aligning heterogeneous information.

Considering the limitations of existing methods, we propose CMIF, a Cross-Modal Information Fusion framework for molecular property prediction. CMIF is capable of learning representations from three modalities: SMILES sequences, molecular topological graphs, and molecular geometric features, by encoding each modality through specific networks. Unlike simple feature concatenation, CMIF employs a cross-attention mechanism for feature fusion, effectively leveraging the characteristics of each modality to obtain a comprehensive molecular representation. Additionally, to promote the alignment of heterogeneous features, CMIF employs two self-supervised learning approaches, contrast learning, and cross-modal matching, to ensure inter-modal consistency. Our contributions can be summarized as follows:

- We propose a molecular representation learning framework for information fusion across three modalities, integrating sequence, graph, and geometric modalities into molecular property prediction.
- We introduced the cross-attention mechanism as a feature fusion strategy and employed two self-supervised learning methods to facilitate the alignment of different modal features.
- We conducted extensive experiments on seven molecular datasets, and CMIF outperformed all baseline models, achieving a relative improvement of 7.6%.

2   Related Work

2.1   Uni-Modal Molecular Representation Learning

Due to SMILES encoding technology is compatible with NLP techniques, many studies have proposed RNN-based SMILES molecular representation learning methods. For instance, SMILES-Transformer [6] and SMILES-BERT [17] were pre-trained on large-scale molecular data and subsequently fine-tuned for downstream tasks. However, SMILES cannot capture the spatial structural information of molecules or the complex relationships between atoms and chemical bonds, whereas graph neural networks (GNNs) are better suited for extracting topological structure from molecular graphs [7]. For example, the Communicative Message Passing Neural Network (CMPNN) [14] sends aggregated bond information to a communicative kernel to enhance atom representations, and CD-MVGNN [10] employs MPNN on atom- and bond-oriented graph structures for multi-view molecular representation learning. Although GNNs excel in extracting topological information, they still face limitations in capturing molecular spatial structures, such as atomic positions, bond lengths, and bond angles. To address this issue, GEM [2] introduced a neural network architecture capable of capturing molecular geometric information such as bond angles and bond lengths, while LineEvo [12] utilizes atomic feature evolution, bond evolution, and position evolution to capture spatial information like bond angles and torsion angles.

2.2   Multi-modal Molecular Representation Learning

Recent multi-modal molecular representation learning methods have focused on designing complex multi-modal learning frameworks [13] and interaction mechanisms between modalities. For example, MMSG [19] enhances molecular semantic features by cross-utilizing chemical features from molecular graphs and SMILES sequences. GraSeq [5] combines sequence models and graph neural network models for molecular reconstruction. Other studies have explored the integration of graph and geometric information. For instance, GraphMVP [9] employs self-supervised learning to reveal the relationship between 2D topology and 3D geometry, while Zhu et al. [23] proposed a unified pretraining framework for both 2D and 3D molecular representations using a masking strategy to leverage both 2D molecular graphs and 3D conformations. Additionally, Premu-Net [22] and MolLM [16] further investigate the fusion of sequence, graph, and geometric features. However, the feature fusion strategies in these approaches primarily rely on simple concatenation of representations, which may limit the effective utilization of the inherent characteristics of each modality. Moreover, they lack any feature alignment strategies and fail to fully explore the potential of self-supervised learning to improve performance. To address these limitations, this paper proposes CMIF, which integrates sequence, topology, and geometry modalities to learn a comprehensive molecular representation. It employs an advanced cross-attention mechanism as the fusion strategy and incorporates two self-supervised learning techniques to align heterogeneous features effectively.

3   Methods

In this section, the problem definition of the research question is first introduced, and then the implementation methods and functions of each module of the cross-modal learning framework CMIF are described in detail. As shown in Fig. 1, CMIF is mainly composed of a Multi-Modal Encoder Module, an Alignment Module, and a Prediction Module.

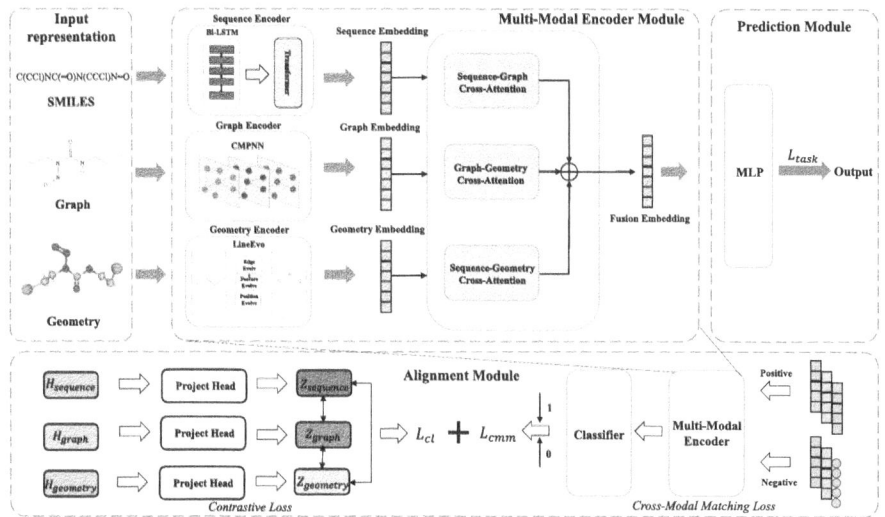

Fig. 1. Overview of CMIF.

3.1   Problem Definition

The goal of molecular property prediction is to predict a particular physicochemical property of a molecule based on its structure. The molecular representation is learned using a trained high-quality neural network model to obtain an accurate molecular representation that contains information about the molecular structure. Formally, the molecular property prediction problem can be defined as:

$$H = f(X), \qquad (1)$$

$$y = P(H), \qquad (2)$$

where X is the input after molecular encoding, f is the neural network model used to obtain the molecular representation, H is the molecular representation, P is the molecular property predictor, and y is the target property.

3.2   Multi-modal Encoder Module

The multi-modal encoder of CMIF consists of three separate encoders and a feature fusion layer based on cross-attention, the three independent encoders are Sequence Encoder, Graph Encoder, and Geometry Encoder. The details of the multi-modal encoder are given below.

Sequence Encoder. First, a molecular SMILES sequence can be denoted as $S = \{s_i | i = 1, ..., T\}$, where s_i is the symbol and T is the sequence length. Then it is mapped to a feature vector $X = \{x_t | t = 1, ..., T\}$ using one-hot encoding and the contextual dependencies are captured by Bi-LSTM. Finally, the forward and backward hidden states are concatenated to obtain the final hidden state h_t, which can be denoted as:

$$h_t = Concat(\overrightarrow{LSTM}(x_t, \overrightarrow{h_{t-1}}), \overleftarrow{LSTM}(x_t, \overleftarrow{h_{t-1}}). \tag{3}$$

The entire sequence is represented as an $H_{seq} = \{h_1, ..., h_T\}$, and finally the H_{seq} is fed into the Transformer to learn the long-distance dependencies hidden in the sequence.

Graph Encoder. A molecule can be described as $G = (V, E)$, where V and E refer to the set of atoms and edges. CMIF uses CMPNN [14] as a graph encoder to obtain a molecular graph representation. CMPNN contains two core components, $AGGREGATE$ and $COMMUNICATE$. Among them, $AGGREGATE$ is used to aggregate the representations from neighboring edges as messages, and $COMMUNICATE$ updates the nodes by passing in the messages obtained from $AGGREGATE$ aggregation and the last moment node representation to update the node. The initial node hiding state $h^0(v) = x(v)$ and the initial edge hiding state $h^0(e_{u,v}) = x(e_{u,v})$. The t-th propagation can be expressed as:

$$m^t(v) = AGGREGATE(h^{t-1}(e_{u,v})), \tag{4}$$

$$h^t(v) = COMMUNICATE(m^t(v), h^{t-1}(v)), \tag{5}$$

$$m^t(e_{u,v}) = h^t(v) - h^{t-1}(e_{u,v}), \tag{6}$$

$$h^t(e_{u,v}) = \sigma(h^0(e_{u,v}) + W \cdot m^t(e_{u,v})), \tag{7}$$

where $m^t(v)$ is the message of node v, $h^t(v)$ and $h^t(e_{u,v})$ are their hidden states at time step t, W is the weight matrix, and σ is the activation function. After T iterations, the final node representations and messages are obtained by the following steps:

$$m(v) = AGGREGATE(h^T(e_{u,v})), \tag{8}$$

$$h(v) = COMMUNICATE(m(v), h^T(v), x(v)). \tag{9}$$

Geometry Encoder. CMIF uses the LineEvo [12] layer as the geometry encoder to obtain geometric features, which is a pluggable module in GNNs. The core idea of the LineEvo layer is based on a line graph transformation strategy, which converts the fine-grained molecular graph into a coarse-grained graph, thereby generating new edge connections, atomic features, and atomic positions. Specifically, LineEvo transforms the edges in the original graph into nodes of the line graph, with edges sharing common nodes in the original graph being connected in the new graph. It then evolves atomic features using the attention mechanism of GATv2 [1] and generates new atomic positions by computing the midpoints of atomic pairs:

$$p'_{ij} = (p_i + p_j)/2, \tag{10}$$

where p_i and p_j denote the atomic position coordinates, with the evolved position p_{ij} representing the midpoint between atoms i and j. The evolved edges correspond to the neutral line of a three-atom system, with a length equal to half of the third edge, facilitating the capture of long-range graph information and additional spatial features such as bond angles and dihedral angles. In CMIF, multiple LineEvo layers are stacked after the GATv2 layer, where GATv2 first extracts molecular representations and the subsequent LineEvo layers progressively transform these representations into multi-granular information, enabling hierarchical spatial information learning from the atomic level to coarser granularities.

Cross-Attention Feature Fusion. Since the attention mechanism has shown remarkable performance in extracting various information, this work proposes a multi-cross-attention strategy for feature fusion that captures multi-modal feature information and enhances feature representation. Three modal information embeddings can be obtained after processing three modal encoders: H_{seq}, H_{graph}, H_{geo}. Then two-by-two cross-attention computations are performed on the three modal embeddings in the following order: graph-geometry, graph-sequence, and geometry-sequence, and then three representations after cross-attention can be obtained: $Attn_{graph-geo}$, $Attn_{graph-seq}$, and $Attn_{geo-seq}$. The process can be formulated as:

$$Attn_{graph-geo} = CrossAttention(f_Q(H_{geo}), f_K(H_{graph}), f_V(H_{graph})), \tag{11}$$

$$Attn_{graph-seq} = CrossAttention(f_Q(H_{graph}), f_K(H_{seq}), f_V(H_{seq})), \tag{12}$$

$$Attn_{geo-seq} = CrossAttention(f_Q(H_{seq}), f_K(H_{geo}), f_V(H_{geo})), \tag{13}$$

$$CrossAttention(Q, K, V) = Softmax(\frac{QK^T}{\sqrt{C/d}}) \cdot V, \tag{14}$$

where Q, K, and V denote the query vector, key vector, and value vector, respectively, C and d denote the dimension of the embedding matrix and the number of attention heads, f is the projection function, and $Attn_{graph-geo}$, $Attn_{graph-seq}$, and $Attn_{geo-seq}$ are the representations of the graph-geometry, graph-sequence, and geometry-sequence, respectively, after the cross-attention computation.

The final fusion representation H_{fusion} is obtained by stitching the above three representations:

$$H_{fusion} = Concat(Attn_{graph-geo}, Attn_{graph-seq}, Attn_{geo-seq}). \tag{15}$$

3.3   Alignment Module

There are subspace distances between the representations of different modalities of a molecule, and the current challenge is how to align these modal data. To this end, this work introduces two self-supervised learning strategies, contrast learning and cross-modal matching, to align the information of the three modalities into a unified potential space.

Contrastive Loss. To ensure consistency between modalities, the model was trained to maximize the similarity of the representations of the same molecules while minimizing the similarity of different molecules by introducing a similarity-based loss of contrast:

$$\mathcal{L}_{CL} = \mathcal{L}_{cl}(z_{seq}, z_{graph}) + \mathcal{L}_{cl}(z_{seq}, z_{geo}) + \mathcal{L}_{cl}(z_{graph}, z_{geo}), \tag{16}$$

$$\mathcal{L}_{cl}(a, b) = -log \frac{exp(sim(a_i, b_i)/T)}{\sum_{j=1, i \neq j}^{N} exp(sim(a_i, b_j)/T)}, \tag{17}$$

where z_{seq}, z_{graph} and z_{geo} are the outputs obtained by mapping H_{seq}, H_{graph} and H_{geo} into the space of the contrastive loss, $sim(,)$ is the similarity calculation function and T is the contrast loss rate.

Cross-Modal Matching Loss. Cross-Modal Matching (CMM) aims to facilitate alignment between features by predicting whether representations of sequences, graphs, and geometries correspond to the same molecule, and generates negative samples by disrupting multi-modal inputs in small batches. The CMM loss can be expressed as:

$$\mathcal{L}_{CMM} = \sum_{(S, G, H) \in B} cross\_entropy(y(S, G, H), p(\mathcal{M}(h_S, h_G, h_H))), \tag{18}$$

where B is a small batch of disrupted samples, y is whether the data in the sample corresponds to the same molecule, p is a predictor consisting of a fully connected layer, and $softmax$ activation function, and $\mathcal{M}$ is the multi-modal encoder.

3.4   Prediction Module

The fusion representation H_{fusion} is fed into the predictor P to predict molecular properties and trained using different loss functions for the tasks. The binary cross-entropy (BCE) loss is used for the classification task, and the mean square error (MSE) loss is used for the regression task. Formally, it can be formulated as:

$$y_{pred} = P(H_{fusion}), \tag{19}$$

$$\mathcal{L}_{comb} = \mathcal{L}_{CL} + \mathcal{L}_{CMM}, \tag{20}$$

$$\mathcal{L}_{cls} = \mathcal{L}_{BCE}(y_{pred}, y) + \lambda \cdot \mathcal{L}_{comb}, \tag{21}$$

$$\mathcal{L}_{reg} = \mathcal{L}_{MSE}(y_{pred}, y) + \lambda \cdot \mathcal{L}_{comb}, \tag{22}$$

where P consists of a fully connected layer and $\mathcal{L}_{comb}$ is the sum of $\mathcal{L}_{CL}$ and $\mathcal{L}_{CMM}$ as the combination loss. Given the difference between the combination loss and the prediction task loss, a scaling factor λ is introduced in this model to balance these two types of losses, thus enhancing the stability of the training process.

4   Experiments

4.1   Benchmark Datasets

To evaluate the performance of the CMIF framework, we conducted classification and regression task experiments on seven public benchmark datasets provided by MoleculeNet [20]. Table 1 shows all the information about the datasets used, including the dataset size, the number and type of tasks, and the description of the properties involved.

Table 1. Detailed information on the baseline dataset used.

Dataset	Tasks	Compounds	Task Type	Metric	Property
BACE	1	1513	Classification	ROC-AUC	inhibition of human β-secretase 1
BBBP	1	2039	Classification	ROC-AUC	ability to penetrate the blood-brain barrier
ClinTox	2	1478	Classification	ROC-AUC	toxicity
HIV	1	41127	Classification	ROC-AUC	HIV inhibition
ESOL	1	1128	Regression	RMSE	water solubility
FreeSolv	1	642	Regression	RMSE	hydration free energy in water
Lipophilicity	1	4200	Regression	RMSE	octanol/water distribution coefficient

4.2   Baselines

We compared CMIF with nine competitive baseline methods to demonstrate the effectiveness of our model, including sequence-based, graph-based, geometry-based, and multi-modal methods:

- Seq2SeqFP [21]: An unsupervised molecular representation method based on gated recurrent unit (GRU).
- SMILES Transformer [6]: A Transformer-based sequential language model for processing molecular fingerprints via unsupervised learning.
- CMPNN [14]: Its full name is communication message passing neural network (CMPNN), which enhances the message interaction between nodes and edges through the communicative kernel.
- GATv2 [1]: An improved graph attention network that enhances the graph data representation by introducing a learnable weighting mechanism to enhance the modeling of relationships between nodes.

- GEM [2]: A method for molecular representation learning via geometric enhancement.
- GraSeq [5]: A method for molecular representation learning by combining graphs and sequences.
- GraphMVP [9]: A pre-training method for learning molecular representations by utilizing 2D topology and 3D geometric views.
- 3D Infomax [15]: A method for molecular property prediction by maximizing the mutual information between the learned 3D vectors and the graph representation.
- FTMMR [13]: A molecular property prediction method that integrates three dimensions of molecular representation and uses Transformer for feature fusion.

4.3 Experimental Settings

To evaluate the learning capability of CMIF, we randomly split the datasets and set the ratio of train, validation, and test sets to 8:1:1. For evaluation metrics, we used ROC-AUC for classification tasks and RMSE for regression tasks. To minimize the impact of random data splits, we averaged the results over 5 random seeds for each task. During model training, we used the Adam optimizer with an initial learning rate of 1e–3 or 1e–4, a maximum learning rate of 2e–3 or 2e–4, and a batch size of 32. The hidden layer dimensions for the three independent encoders were set to 256, with 4 attention heads in the Transformer sequence encoder. The weight factor λ for loss balancing was set to 0.08, and the contrast learning rate T was set to 0.1. All experiments were conducted on a Ubuntu Server with a GPU (NVIDIA GeForce RTX 3090).

4.4 Experimental Results and Analysis

To validate the effectiveness of our proposed Cross-Modal Information Fusion (CMIF) framework, we conducted a comparison with nine baseline models across four classification tasks and three regression tasks, with the results shown in Table 2. The optimal results are highlighted in bold, the second-best results are underlined, and "–" indicates that the model did not undergo the corresponding experimental test. Table 2 provides the following observations: 1) CMIF achieves state-of-the-art performance across all baseline datasets by simultaneously integrating sequence representations, graph representations, and geometry representations. Specifically, CMIF outperforms the second-best models by an average of 4.4% in classification tasks and 10.9% in regression tasks, validating its strong capability in capturing comprehensive molecular representations. 2) Compared to GraSeq, CMIF introduces a geometric feature learning module and a cross-attention-based feature fusion module, resulting in an average performance improvement of 33.9%. Compared to 3D Infomax and GraphMVP, CMIF incorporates sequence feature learning, enabling the capture of molecular chirality and chemical scaffolds, leading to a 36.0% performance improvement. Compared to

FTMMR, CMIF introduces a Transformer module when learning SMILES and uses two self-supervised learning methods to facilitate feature alignment, achieving a relative improvement of 19.2%. 3) CMIF significantly improves prediction performance on the ClinTox, ESOL, and FreeSolv datasets, indicating that our method excels in small-scale datasets and can still capture critical information from molecules even when sample sizes are limited.

Table 2. The molecular property prediction results of different baseline models and CMIF on seven datasets from MoleculeNet. For different metrics, "↑" means the higher the better, and "↓" means the lower the better.

Method	Classification (ROC-AUC ↑)				Regression (RMSE ↓)		
	BBBP	BACE	ClinTox	HIV	ESOL	FreeSolv	Lipophilicity
Seq2SeqFP	0.889	0.753	0.904	–	1.248	2.865	–
SMILES Transformer	0.906	0.739	0.917	0.683	1.081	2.376	1.169
CMPNN	0.935	0.884	0.894	0.782	0.845	1.349	0.614
GATv2	0.894	0.849	0.893	–	0.633	0.948	0.605
GEM	0.725	0.849	0.911	0.806	0.789	1.815	0.664
GraSeq	0.942	0.763	0.628	0.772	1.255	2.739	0.648
GraphMVP	0.735	0.857	0.803	0.770	1.047	1.867	0.769
3D Infomax	0.695	0.785	0.611	0.754	0.877	2.256	1.226
FTMMR	0.757	0.869	0.923	0.784	0.780	1.480	0.610
CMIF(ours)	**0.978**	**0.925**	**0.985**	**0.827**	**0.565**	**0.773**	**0.582**

4.5 Ablation Studies

To validate the effectiveness of the modules, we conducted ablation experiments to investigate the impact of different modalities on prediction performance. The experimental results in Table 3 show that removing any modality leads to a performance decline, further confirming that CMIF, by integrating sequence, graph, and geometric information, captures a more comprehensive molecular representation. Specifically, removing the graph modality caused a significant performance drop in the BACE, FreeSolv, and Lipophilicity datasets, indicating that graph representations are better at capturing features related to molecular properties in these datasets. On the other hand, removing the sequence modality resulted in a notable performance decrease on the ClinTox dataset, suggesting that the model is more capable of capturing chemical information related to molecular toxicity from SMILES representations. Notably, the geometric modality contributed less to the performance, likely because molecular coordinates are generated by chemical tools. Coordinates generated by chemical tools, such as RDKit, may not fully reflect the true conformation of the molecule, especially for complex or flexible molecules. Additionally, they may introduce noise or errors, thereby weakening the role of geometric features like bond angles and torsional angles in prediction tasks. Overall, CMIF benefits from the integration of all three modalities, leading to improved performance.

Table 3. Ablation study on the performance of different modalities on seven benchmark datasets. w/o Sequence, w/o Graph, and w/o Geometry indicate the removal of the sequence, graph, and geometry feature modules in CMIF.

Method	Classification (ROC-AUC ↑)				Regression (RMSE ↓)		
	BBBP	BACE	ClinTox	HIV	ESOL	FreeSolv	Lipophilicity
-w/o Sequence	0.969	0.887	0.905	0.813	0.573	0.915	0.598
-w/o Graph	0.944	0.712	0.952	0.762	0.658	1.787	1.165
-w/o Geometry	0.969	0.905	0.973	0.809	0.584	0.791	0.593
CMIF(ours)	**0.978**	**0.925**	**0.985**	**0.827**	**0.565**	**0.773**	**0.582**

Additionally, we investigated the impact of removing the cross-attention module (w/o cross-attention) and the feature alignment module (w/o alignment) on the model's performance. We replaced the cross-attention-based feature fusion method with a simple concatenation approach, and the experimental results are shown in Fig. 2. CMIF showed performance improvements across all four classification datasets, highlighting the importance of the cross-attention fusion module and the alignment module. However, on the ESOL and Lipophilicity datasets, feature concatenation outperformed cross-attention, possibly due to the loss of some original features during cross-attention. Overall, the cross-attention mechanism is effective in capturing key features from different modalities, and contrast learning and cross-modal matching also facilitate feature alignment, all of which contribute positively to the improvement of model performance.

4.6   Molecular Representation Visualization

To visually demonstrate the effectiveness of CMIF in integrating molecular information from three modalities, we applied t-SNE [11] for dimensionality reduction and visualization of the learned molecular representations on the BBBP dataset. We compared the representations learned by the three modality encoders and our CMIF framework. As shown in Fig. 3, both the Graph Encoder and CMIF achieve significant separation of positive and negative samples; however, CMIF exhibits a clearer classification boundary and demonstrates more concentrated clustering compared to the graph encoder, further validating the effectiveness of CMIF in capturing more comprehensive molecular representations. The Sequence Encoder, which can only learn global information from SMILES, shows less distinct classification results. Notably, the Geometry Encoder almost fails to effectively distinguish positive and negative samples, likely because relying solely on geometric information is insufficient to capture the features related to molecular properties. However, the Geometry Encoder is capable of small-scale clustering of molecules based on different physical and chemical properties, proving its ability to effectively learn and differentiate molecular structures. In summary, CMIF's integration of Sequence, Graph, and Geometry modalities demonstrates strong foresight and achieves state-of-the-art performance.

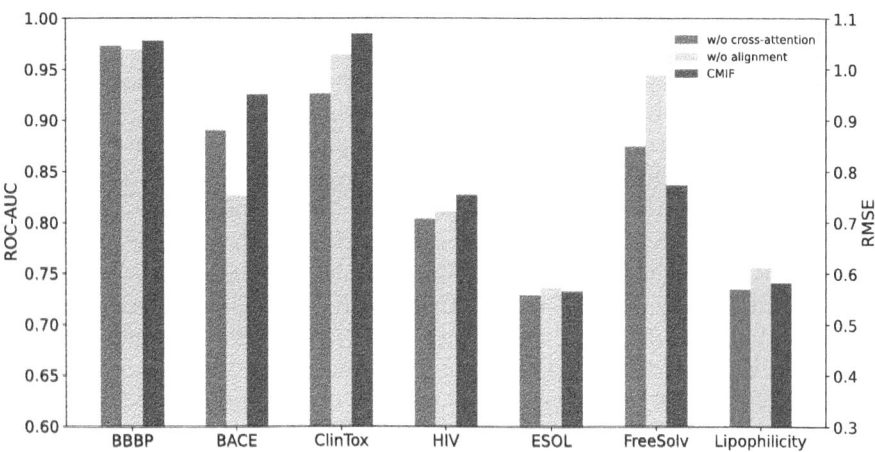

Fig. 2. Performance comparison of the CMIF model with ablation of cross-attention and alignment modules. The x-axis represents different datasets, the left y-axis represents ROC-AUC values, and the right y-axis represents RMSE values.

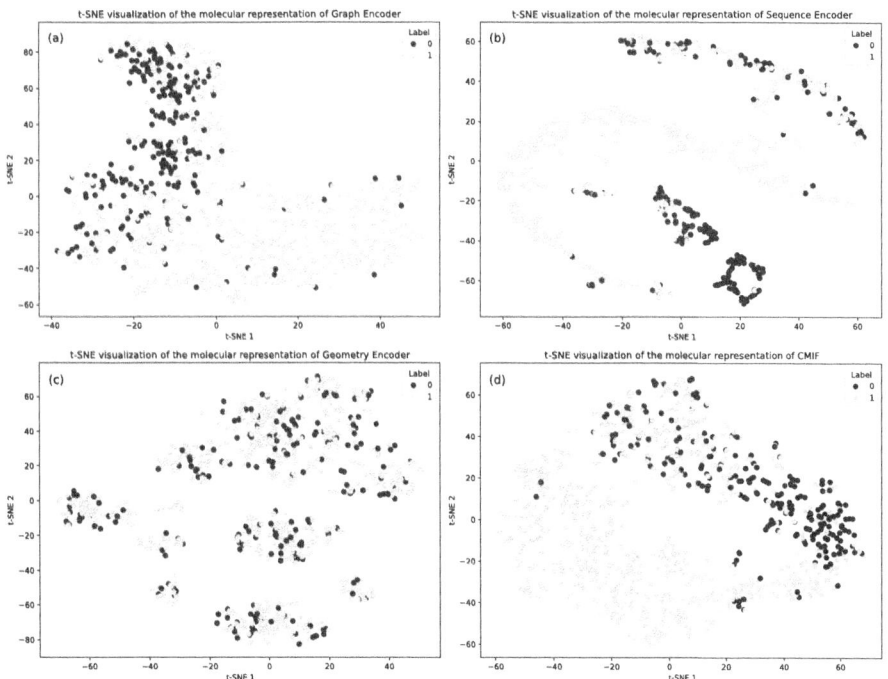

Fig. 3. t-SNE visualization of molecular representation on the BBBP dataset. (**a**), (**b**), (**c**) represent the visualization of the molecular representation of the Graph Encoder, Sequence Encoder, and Geometric Encoder, and (**d**) represent the visualization results of the molecular representation of our proposed CMIF.

5 Conclusions

In this paper, we propose a novel cross-modal information fusion molecular representation learning framework, called CMIF, for molecular property prediction. CMIF integrates features from the molecular sequence, graph, and geometry modalities using a Multi-modal Encoder Module, an Alignment Module, and a Prediction Module. It employs a dual self-supervised learning strategy to align heterogeneous features and introduces a cross-attention mechanism to focus on key features, thereby enhancing the quality of the learned representations. Extensive experimental results demonstrate that CMIF outperforms existing baseline models on multiple datasets, validating its effectiveness in molecular property prediction. However, there is still room for improvement in our approach. For instance, the cross-attention mechanism does not always outperform feature concatenation in certain cases. Our future work will explore more effective fusion mechanisms, optimize modality encoders, enhance model prediction speed, and design experiments to investigate which missing raw features contribute to performance degradation, while further validating the model's generalization ability on additional datasets.

Acknowledgements. This work is supported by The National Natural Science Foundations of China (61936001), Natural Science Foundations of Chongqing (cstc2021ycjh-bgzxm0013), The Key Cooperation Project of Chongqing Municipal Education Commission (HZ2021008). We would like to thank editor and reviewers for insightful comments and advice.

References

1. Brody, S., Alon, U., Yahav, E.: How attentive are graph attention networks? In: International Conference on Learning Representations (2022)
2. Fang, X., et al.: Geometry-enhanced molecular representation learning for property prediction. Nat. Mach. Intell. **4**(2), 127–134 (2022)
3. Gong, X., Liu, Q., Han, R., Guo, Y., Wang, G.: MIFS: an adaptive multipath information fused self-supervised framework for drug discovery. Neural Netw., 107088 (2025)
4. Gong, X., Liu, Q., He, J., Guo, Y., Wang, G.: Multigrandti: an explainable multi-granularity representation framework for drug-target interaction prediction. Appl. Intell. **55**(2), 1–19 (2025)
5. Guo, Z., Yu, W., Zhang, C., Jiang, M., Chawla, N.V.: Graseq: graph and sequence fusion learning for molecular property prediction. In: Proceedings of the 29th ACM International Conference on Information & Knowledge Management, pp. 435–443 (2020)
6. Honda, S., Shi, S., Ueda, H.R.: Smiles transformer: pre-trained molecular fingerprint for low data drug discovery. arXiv preprint arXiv:1911.04738 (2019)
7. Li, Z., Jiang, M., Wang, S., Zhang, S.: Deep learning methods for molecular representation and property prediction. Drug Disc. Today **27**(12), 103373 (2022)

8. Liu, M., Yang, Y., Liu, Q., Liu, L., Wang, G.: A knowledge-driven self-supervised approach for molecular generation. IEEE/ACM Trans. Comput. Biol. Bioinf. **21**(5), 1579–1590 (2024)
9. Liu, S., Wang, H., Liu, W., Lasenby, J., Guo, H., Tang, J.: Pre-training molecular graph representation with 3D geometry. In: International Conference on Learning Representations (2022)
10. Ma, H., et al.: Cross-dependent graph neural networks for molecular property prediction. Bioinformatics **38**(7), 2003–2009 (2022)
11. Van der Maaten, L., Hinton, G.: Visualizing data using t-SNE. J. Mach. Learn. Res. **9**(11), 2579–2605 (2008)
12. Ren, G.P., Wu, K.J., He, Y.: Enhancing molecular representations via graph transformation layers. J. Chem. Inf. Model. **63**(9), 2679–2688 (2023)
13. Son, Y.H., Shin, D.H., Kam, T.E.: FTMMR: fusion transformer for integrating multiple molecular representations. IEEE J. Biomed. Health Inf. **28**(7), 4361–4372 (2024)
14. Song, Y., Zheng, S., Niu, Z., Fu, Z.H., Lu, Y., Yang, Y.: Communicative representation learning on attributed molecular graphs. In: 29th International Joint Conference on Artificial Intelligence and the 17th Pacific Rim International Conference on Artificial Intelligence (IJCAI-PRICAI2020), pp. 2831–2838 (2020)
15. Stärk, H., et al.: 3D infomax improves GNNS for molecular property prediction. In: International Conference on Machine Learning, pp. 20479–20502 (2022)
16. Tang, X., Tran, A., Tan, J., Gerstein, M.B.: Mollm: a unified language model for integrating biomedical text with 2d and 3d molecular representations. Bioinformatics **40**(Supplement_1), i357–i368 (2024)
17. Wang, S., Guo, Y., Wang, Y., Sun, H., Huang, J.: Smiles-bert: large scale unsupervised pre-training for molecular property prediction. In: Proceedings of the 10th ACM International Conference on Bioinformatics, Computational Biology and Health Informatics, pp. 429–436 (2019)
18. Weininger, D.: Smiles, a chemical language and information system. 1. introduction to methodology and encoding rules. J. Chem. Inf. Comput. Sci. **28**(1), 31–36 (1988)
19. Wu, T., Tang, Y., Sun, Q., Xiong, L.: Molecular joint representation learning via multi-modal information of smiles and graphs. IEEE/ACM Trans. Comput. Biol. Bioinf. **20**(5), 3044–3055 (2023)
20. Wu, Z., et al.: Moleculenet: a benchmark for molecular machine learning. Chem. Sci. **9**(2), 513–530 (2018)
21. Xu, Z., Wang, S., Zhu, F., Huang, J.: Seq2seq fingerprint: an unsupervised deep molecular embedding for drug discovery. In: Proceedings of the 8th ACM International Conference on Bioinformatics, Computational Biology, and Health Informatics, pp. 285–294 (2017)
22. Zhang, H., Wu, J., Liu, S., Han, S.: A pre-trained multi-representation fusion network for molecular property prediction. Inf. Fusion **103**, 102092 (2024)
23. Zhu, J., et al.: Unified 2D and 3D pre-training of molecular representations. In: Proceedings of the 28th ACM SIGKDD Conference on Knowledge Discovery and Data Mining, pp. 2626–2636 (2022)

Adversarial Transfer Learning for Predicting Drug Sensitivity in Single-Cell Data

Huawei Zhang[1,2], Shaoshuai Zhu[1,2], Fei Lin[3], Qinghua Zhang[4], Fan Yang[2(✉)], and Qingke Zhang[1(✉)]

[1] School of Information Science and Engineering, Shandong Normal University, Jinan, China
tsingke@sdnu.edu.cn
[2] School of Public Health, Cheeloo College of Medicine, Shandong University, Jinan, China
fanyang@sdu.edu.cn
[3] School of Continuing Education, Shandong University, Jinan, China
[4] Chongqing University of Posts and Telecommunications, Chongqing, China

Abstract. Drug resistance remains a primary cause of cancer treatment failure, while single-cell-level heterogeneity within the tumor microenvironment poses significant challenges for predicting drug sensitivity. This study proposes a framework for single-cell drug sensitivity prediction based on adversarial transfer learning. Our approach addresses the scarcity of annotated single-cell data by transferring drug response knowledge from bulk cell-line data (source domain) to single-cell data (target domain) through adversarial domain adaptation. We first establish a drug response classification model via supervised learning in the source domain, followed by adversarial alignment of feature distributions between domains. The experimental results demonstrate that our framework surpasses existing models in cross-domain transfer learning tasks, achieving 94% classification accuracy.

Keywords: adversarial domain adaptation · transfer learning · single cell · drug sensitivity

1 Introduction

Drug resistance and the resulting ineffectiveness of drug treatments lead to up to 90% of cancer-related deaths [1]. In addition to intrinsic resistance in tumor subpopulations, cancer cells may develop resistance through multiple mechanisms including pathway activation [2], alternative splicing [3], and drug efflux [4]. Within the tumor microenvironment, distinct cancer subpopulations exhibit marked heterogeneity in gene expression profiles and functional states, directly influencing drug response diversity. Consequently, there is an urgent need to predict drug sensitivity at single-cell resolution and identify causal genetic determinants of therapeutic outcomes [5]. Single-cell RNA sequencing (scRNA-seq) enables transcriptomic profiling at cellular resolution, providing new insights into

Q. Zhang et al. (Eds.): IJCRS 2025, LNAI 15710, pp. 242–256, 2025.
https://doi.org/10.1007/978-3-031-92741-6_18

tumor heterogeneity and drug resistance mechanisms. The field encounters challenges in acquiring drug sensitivity annotations at single-cell resolution, primarily due to the substantial experimental material requirements and prohibitively high equipment costs associated with single-cell sequencing technologies.

In transfer learning applications, negative transfer [6] represents an important challenge, characterized by performance degradation when transferring knowledge between domains. This phenomenon may arise from source domain overfitting (leading to poor target domain generalization) or inappropriate feature selection (transferring irrelevant or counterproductive features).

To address these challenges, this study proposes an adversarial transfer learning framework for single-cell drug sensitivity prediction. The proposed architecture employs a domain discriminator for adversarial training to enable domain-invariant feature learning, while simultaneously aligning feature distributions between domains using Maximum Mean Discrepancy (MMD) loss in the target domain encoder. This dual mechanism effectively mitigates negative transfer effects. Leveraging available drug sensitivity labels from bulk cell line data, our framework demonstrates successful knowledge transfer through adversarial domain adaptation, enabling accurate drug response predictions at single-cell resolution. Experimental validation shows that our framework achieves 94% classification accuracy in single-cell drug sensitivity prediction, significantly outperforming conventional transfer learning approaches.

The primary contributions of this work are as follows.

- This study introduces novel application of adversarial transfer learning to single-cell drug sensitivity prediction, effectively addressing the challenge of missing drug response annotations at cellular resolution. The adversarial learning mechanism in the target domain enhances transfer generalization through enriched feature encoding.
- In our target domain encoder, the Maximum Mean Discrepancy (MMD) distance is used to reduce the distributional disparities between the source and target domains, thereby facilitating the learning of domain invariance.
- Asymmetric feature mapping enables more accurate simulation of low-level feature discrepancies between domains, facilitating improved cross-domain feature space alignment (Fig. 1).

2 Related Work

The cellular heterogeneity unveiled through single-cell data analysis is crucial for improving the accuracy of drug sensitivity prediction and significantly promotes the design of combination drug strategies [7–9]. Although single-cell technology is relatively new, researchers have increasingly adopted these approaches to investigate disease heterogeneity. Single-cell RNA sequencing (scRNA-seq) has proven particularly valuable in uncovering transcriptional variations associated with drug resistance mechanisms [10]. It has been demonstrated in a study that scRNA-seq technology is capable of effectively characterizing and predicting the

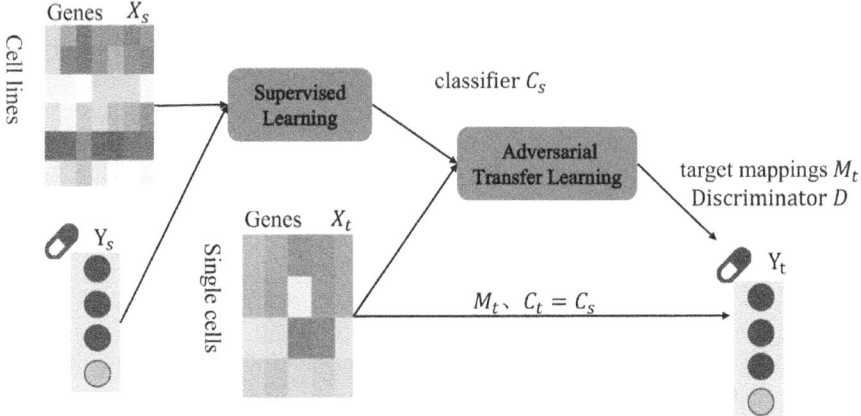

Fig. 1. Overall framework. The model obtains a classifier C_s through supervised learning on X_s (source domain), undergoes adversarial training on X_t (target domain) to obtain a target domain mapping M_t, and finally migrates the classifier C_s to single-cell data (the target domain) to predict drug sensitivity.

most aggressive tumor subpopulations in lung adenocarcinoma [11]. Comparative analyses between single-cell and conventional bulk sequencing approaches reveal that transcriptional signatures of drug-resistant subpopulations are frequently obscured by dominant cell populations, resulting in critical biological insights being masked. These findings underscore the necessity of transitioning from bulk tissue analysis to single-cell resolution when studying cellular heterogeneity [12]. The scope of single-cell research has expanded from simply identifying cell types to elucidating the biological mechanisms that lead to drug resistance in resistant subpopulations. Drug resistance can arise from both genetic [13] and non-genetic [14] factors. Genetic drug resistance usually stems from heritable mutations in cellular DNA, which make cells and their descendants resistant to existing drug treatments. The task of this paper is to predict drug sensitivity based on single-cell transcriptional data.

In the academic field of domain transfer learning, there have been a large number of previous research achievements [15–21]. For unlabeled target domains, the prevailing paradigm involves minimizing feature distribution discrepancies between domains to guide representation learning [22]. To achieve this goal, some methods have adopted the Maximum Mean Discrepancy (MMD) loss [23]. MMD measures the distribution difference by calculating the norm of the difference between the means of the two domains. The Deep Domain Confusion (DDC) method [17] introduces MMD on the basis of the conventional source domain classification loss to learn representations that are both discriminative and domain-invariant. The Deep Adaptation Networks (DAN) [24]apply MMD to layers embedded in the reproducing kernel Hilbert space, effectively matching the higher-order statistics of the two distributions. A novel architecture known as the Variational Auto-Encoder-Long-Short-Term-Memory Network-Local Weighted

Deep Sub-Domain Adaptation Network (VLSTM-LWSAN) [25] has been proposed for RUL prediction. Through the local weighted deep sub-domain adaptation network, this architecture compresses the input data into an interpretable latent space and aligns the fine-grained features across subdomains. A novel metric that integrates MMD and CORAL is introduced to amplify domain confusion [26]. Other methods choose adversarial losses to minimize domain shift and learn representations that can distinguish source labels but not domains. Literature [22] proposed adding a domain classifier that predicts the binary domain labels of the input (a single fully connected layer) and designed a domain confusion loss to encourage its predictions to be as close as possible to the uniform distribution of binary labels. The ReverseGrad algorithm [27] similarly views domain invariance as a binary classification task. However, it achieves this by inverting the gradients of the domain classifier, thereby directly increasing the classifier's loss. Researchers proposed CoCoGAN [28], a conditional coupled generative adversarial network, to capture the joint distribution of dual-domain samples for domain adaptation. It uses source-domain samples from a relevant task (RT) to provide high-level concepts approximating the target domain, and dual-domain samples from an irrelevant task (IRT) to capture shared correlations between the two domains. Recent studies [29] propose a novel framework that uses a reconstruction loss to align source and target distributions while preserving target domain labels. Due to the sparsity and heterogeneity of single-cell gene expression data, reconstruction loss may not be able to accurately measure the performance of the model. It is difficult to have a relatively stable and predictable structure as in image data.

3   Preliminaries

3.1   Transfer Learning

Before defining transfer learning, we first revisit the concepts of a domain and a task [30].

Domain: A domain $\mathcal{D}$ comprises two components: a feature space $\mathcal{X}$ and a marginal distribution $P(X)$. Formally, $\mathcal{D} = \{\mathcal{X}, P(X)\}$. Here, $\mathcal{X}$ represents an instance set, defined as $\mathcal{X} = \{\mathbf{x}_i \in \mathcal{X}, i = 1, \ldots, n\}$. There are two domains in this paper: the source domain $\mathcal{D}_s$ (cell line data) and the target domain $\mathcal{D}_t$ (single-cell data).

Task: A task $\mathcal{T}$ consists of a label space $\mathcal{Y}$ and a decision function f, expressed as $\mathcal{T} = \{\mathcal{Y}, f\}$. In my drug sensitivity classification prediction model, the output is the predicted conditional probability distribution of drug sensitivity classes for each instance. The decision function f is implicit and is learned from sample data. In this case, $f(\mathbf{x}_j) = \{P(y_k|\mathbf{x}_j)|y_k \in \mathcal{Y}, k = 1, \ldots, |\mathcal{Y}|\}$, where y_k represents the labels of the drug sensitivity class and $\mathbf{x}_j$ is the feature vector for the j-th instance.

In practice, a domain is typically observed through a set of instances, which may or not include label information. In the context of this paper, drug sensitivity labels for cell line data are easily obtainable $\mathcal{D}_S = \{(\mathbf{x}_i, y_i) | \mathbf{x}_i \in \mathcal{X}^S, y_i \in \mathcal{Y}^S, i = 1, \ldots, n^S\}$, while drug sensitivity labels for single-cell data are missing due to cost and other issues.

Transfer Learning: Given the source domain $\mathcal{D}_S = \{(\mathbf{x}_i^S, y_i^S)\}_{i=1}^{n_S}$, where $\mathbf{x}_i^S \in \mathcal{X}^S$ is the feature vector of the source domain, and $y_i^S \in \mathcal{Y}^S$ is the corresponding label; the target domain $\mathcal{D}_T = \{(\mathbf{x}_j^T, y_j^T)\}_{j=1}^{n_T}$, where $\mathbf{x}_j^T \in \mathcal{X}^T$ is the feature vector of the target domain, and $y_j^T \in \mathcal{Y}^T$ is the corresponding label (which may be partially unknown). The goal of transfer learning is to use the knowledge from the source domain $\mathcal{D}_S$ to learn a model for the target domain $\mathcal{D}_T$, thereby maximizing the predictive performance on the target domain.

3.2   Adversarial Losses

The adversarial losses train the adversarial discriminator using a standard classification loss $\mathcal{L}_{\mathrm{adv}_D}$. However, they differ in the loss used to train the mapping $\mathcal{L}_{\mathrm{adv}_M}$.

$$\mathcal{L}_{\mathrm{adv}_M} = -\mathcal{L}_{\mathrm{adv}_D}. \tag{1}$$

This optimization aligns with the true minimax objective for generative adversarial networks. Nevertheless, this objective can pose challenges, as the discriminator tends to converge rapidly in the early stages of training, leading to vanishing gradients. Creating two distinct optimization objectives, one for the generator component and one for the discriminator component. In this case, $\mathcal{L}_{\mathrm{adv}_D}$ stays the same, while $\mathcal{L}_{\mathrm{adv}_M}$ is modified to:

$$\mathcal{L}_{\mathrm{adv}_M}(\mathbf{X}_s, \mathbf{X}_t, D) = -\mathbb{E}_{\mathbf{x}_t \sim \mathbf{x}_t}[\log D(M_t(\mathbf{x}_t))]. \tag{2}$$

This objective shares the same fixed-point characteristics with the minimax loss function, yet it offers more robust gradients for the target mapping. It is important to note that in this scenario, separate mappings are employed for the source and target domains, and M_t is trained adversarially. This method is similar to the GAN [31] framework, in which the source distribution stays unchanged while the generator's distribution is trained to match it.

The GAN loss function is typically used when the generator aims to replicate a static distribution. However, when both distributions are dynamic, this goal may cause instability—once the mapping reaches its optimal state, the discriminator may simply invert its prediction. To address this, a study [22] introduced the domain confusion objective. With this objective, the mapping is trained using a cross-entropy loss function based on a uniform distribution.

$$\mathcal{L}_{\mathrm{adv}_M}(\mathbf{X}_s, \mathbf{X}_t, D) =$$
$$- \sum_{d \in \{s,t\}} \mathbb{E}_{\mathbf{x}_d \sim \mathbf{x}_d} \left[\frac{1}{2} \log D(M_d(\mathbf{x}_d)) + \frac{1}{2} \log(1 - D(M_d(\mathbf{x}_d))) \right]. \tag{3}$$

This loss function plays a crucial role in ensuring that the adversarial discriminator perceives the two domains as identical.

4   Adversarial Transfer Learning

The framework of this paper consists of two main components: (1) Supervised learning, where a model is established to predict response label classification at the source domain batch level (cell-lines). (2) Adversarial learning, the discriminator determines whether a sample is from the source domain or the target domain, and the target mapping ensures the mapping space of the target domain as close as possible to the mapping space of the source domain. The classifier are transferred to the single-cell level for label prediction (Fig. 2).

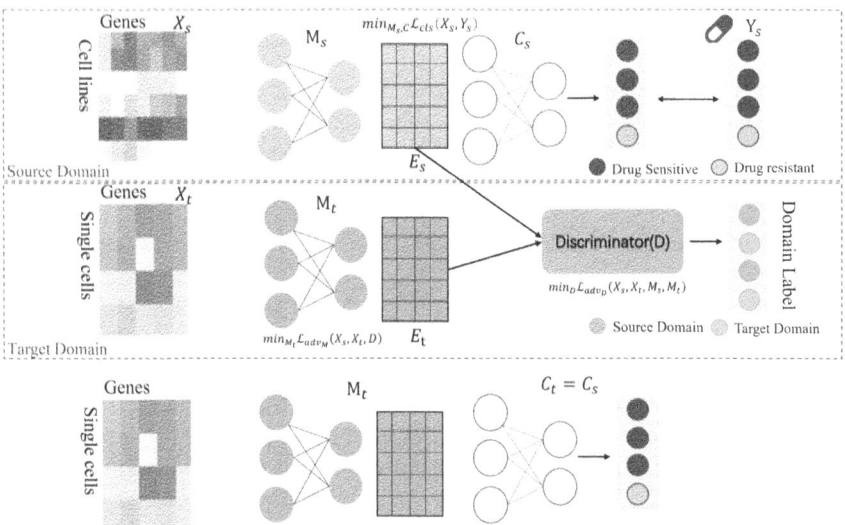

Fig. 2. Adversarial transfer learning framework. During the adversarial learning phase, the target domain encoder learns to generate mappings that are as similar as possible to the source domain feature space, while the discriminator serves to distinguish the origin of the features, identifying them as coming from either the source domain or the target domain. Both are trained adversarially until the discriminator can no longer distinguish.

In unsupervised and adaptive learning, it is assumed that cell line gene expression data X_s and drug sensitivity labels Y_s are extracted from the source domain distribution $p_s(x, y)$, and single-cell gene expression data X_t is extracted from the target distribution $p_t(x, y)$, with no label observations. The objective is to develop a target representation M_t and classifier C_t to classify target data into 2 classes, only sensitive and insensitive cases. Given that direct supervised learning on the target domain is infeasible, our method learns a source representation mapping M_s and a source classifier C_s. Subsequently, it adapts this model to fit the target domain.

In the adversarial domain adaptation approach, the primary objective is to regulate the learning process of the source mappings M_s and target mappings M_t. This study opts for a design where the source and target mappings are independent, achieved through the use of non-shared weights. This strategy offers greater flexibility, as it enables the model to learn more domain-specific feature extraction techniques. Given that the target domain lacks labels, and without weight sharing, improper initialization or training could lead the target model to quickly converge on suboptimal solutions. To address this, the study employs a pre-trained source model as the initial basis for the target representation space. During adversarial training, the source model is kept fixed, while M_t makes efforts to minimize the distance between the empirical distributions of the source mappings $M_s(X_s)$ and target mapping $M_t(X_t)$.

Supervised Learning in the Source Domain. The source classification model C_s is applicable to the target representation directly, eliminating the necessity to develop a distinct target classifier, so it is set that $C = C_s = C_t$.

$$min_{M_s,C} \mathcal{L}_{cls}(X_s, Y_s) = \mathbb{E}_{(x_s,y_s) \sim (X_s,Y_s)} - \sum_{k=1}^{K} 1_{[k=y_s]} log C(M_s(x_s)). \quad (4)$$

Adversarial Learning in the Target Domain. The domain discriminator D discriminates whether the data point is from the source domain or the target domain, while the target mapping tries to make the target domain data as close as possible to the source domain data mapping, and they are updated through adversarial training. Therefore, it is optimized according to the standard supervised loss $\mathcal{L}_{adv_D}(X_s, X_t, M_s, M_t)$, and M_t is optimized according to the loss $\mathcal{L}_{adv_M}(X_s, X_t, D)$, defined as follows:

$$\min_{D} \mathcal{L}_{adv_D}(X_s, X_t, M_s, M_t) =$$
$$- \mathbb{E}_{x_s \sim X_s}[\log D(M_s(X_s))] - \mathbb{E}_{x_t \sim X_t}[\log(1 - D(M_t(X_t)))]. \quad (5)$$

$$\mathcal{L}_{adv_M}(\mathbf{X}_s, \mathbf{X}_t, D) =$$
$$- \sum_{d \in \{s,t\}} \mathbb{E}_{\mathbf{x}_d \sim \mathbf{X}_d} \left[\frac{1}{2} \log D(M_d(\mathbf{x}_d)) + \frac{1}{2} \log(1 - D(M_d(\mathbf{x}_d))) \right] + MMD(X_s, X_t).$$
$$(6)$$

The distance between two distributions is defined as:

$$MMD(X,Y) = \left\| \frac{1}{n} \sum_{i=1}^{n} \Phi(x_i) - \frac{1}{m} \sum_{j=1}^{m} \Phi(y_j) \right\|_H^2 . \tag{7}$$

The target representation M_t and classifier C_t, obtained through the afore-mentioned steps, are applied to the single-cell RNA sequencing data X_t to derive Y_t, which represents the predicted results of single-cell drug sensitivity.

5   Data and Preprocessing

Data: The dataset from the GDSC database[1] includes drug response anno-tations such as area under the dose-response curve (AUC) and half-maximal inhibitory concentration (IC50). Additionally, gene expression data for cell lines (RMA-normalized base expression profiles) is also available on the GDSC web-site. This paper additionally gathered the CCLE cell line expression profiles and the PRISM cell line viability assays[2]. The GDSC database and the CCLE database both provide gene expression data and drug sensitivity information. To ensure the consistency and comparability of the data when integrating the two databases, it is crucial to retain shared genes as well as cell lines and drugs that have no missing values. Although some information may be lost during the integration process, the proportion of this lost information is relatively small in the overall dataset, and its impact on the experimental results is limited. Over-all, this paper collected drug response data with 1280 cancer cell-lines, 15,962 genes and 83 drugs. All scRNA-seq data can be obtained from Gene Expression Omnibus (GEO)[3] (Table 1).

Table 1. Description of Datasets.

Data	Samples	Features	Introduction
GSE110894	1504	25921	Leukemia cells treated with the I.ET.762 inhibitor
GSE149383	2740	18380	PC9 lung cancer cells treated with ERLOTINIB
GSE140440	324	60555	Prostate cancer cells treated with DOCETAXEL

Preprocessing [32]: The classification of drug sensitivity for cell lines in batch datasets is based on AUC values obtained from the GDSC and CCLE databases. For each drug, the AUC scores are converted into binary labels to indicate whether the cell lines are sensitive or resistant to the drug. Cell lines that respond positively to a particular drug are assigned a label of 1, whereas those that are

---

[1] https://www.cancerrxgene.org/.
[2] https://depmap.org/portal/.
[3] https://www.ncbi.nlm.nih.gov/geo/query/acc.cgi.

resistant to the drug are assigned a label of 0. To organize this data, a waterfall approach is employed to arrange the AUC values of the cell lines from highest to lowest, creating an AUC-cell line curve. In this graphical representation, the x-axis corresponds to the different cell lines, while the y-axis indicates the AUC values. The AUC cutoff value is established using two distinct methods: First, for linear curves—characterized by a Pearson correlation coefficient of the regression line exceeding 0.95—the AUC cutoff between sensitivity and resistance is set at the median AUC value across all cell lines. Second, for non-linear curves, the cutoff is determined by identifying the AUC value of a boundary data point that lies farthest from the line segment connecting the data points with the highest and lowest AUC values.

Using SCANPY53 [33], we conducted quality control and preprocessing of the scRNA-seq data. Specifically, the "filter cells" and "filter genes" functions are used to filter out cells with less than 200 detected genes (indicating no cells in the droplet) and genes that can only be detected in less than 3 cells. Cells with a mitochondrial gene expression percentage higher than 10% are removed. Then, the expression values are scaled using the "preprocessingMinMaxScaler" from the sklearn [34] package. After preprocessing and filtering, the GSE110894 dataset retained 1,135 cells and 3,935 genes; the GSE149383 dataset retained 1,947 cells and 3,097 genes; the GSE140440 dataset retained 324 cells and 10,906 genes.

6   Results

To ensure the robustness of our evaluation, we employed 5-fold cross-validation [35]. In this method, the dataset is randomly divided into five equal parts. For each iteration, one part is held out as the validation set while the remaining four parts are used for training the model. This process is repeated five times, with each of the five parts used exactly once as the validation set. The performance metrics reported are the average values obtained from these five iterations, providing a more reliable and generalized assessment of our model's performance. This paper uses ACC (Accuracy), AUC (Area Under Curve), Pre (Precision), Recall, and F1 Score as metrics to evaluate model performance. The calculation formulas for each evaluation metrics are as follows.

$$ACC = \frac{TP + TN}{TP + TN + FP + FN}, \tag{8}$$

where TP denotes the number of samples that are truly sensitive and correctly predicted as sensitive. TN indicates the number of samples that are truly drug-resistant and correctly predicted as drug-resistant. FP refers to the number of samples that are actually drug-resistant but incorrectly predicted as sensitive. FN signifies the number of samples that are truly sensitive but incorrectly predicted as drug-resistant. Accuracy measures the ratio of correctly classified samples to the total number of samples.

The Area Under the Curve (AUC) serves as a metric for evaluating the performance of classification models across different threshold settings. It quantifies a model's capacity to differentiate between classes and is visually represented by the area beneath the Receiver Operating Characteristic (ROC) curve. The ROC curves of our method on the three datasets are shown in Fig. 3. The ROC curve is constructed by plotting the true positive rate (TPR) against the false positive rate (FPR) at various thresholds. An AUC of 1.0 signifies a flawless model, whereas an AUC of 0.5 indicates a model that performs no better than random guessing. Unlike the other two datasets, GSE140440 has only 324 samples and 10,906 features post - processing. This high dimensionality and small sample size somewhat restrict the model's performance. Nevertheless, comparative experiments demonstrate that our model still surpasses others.

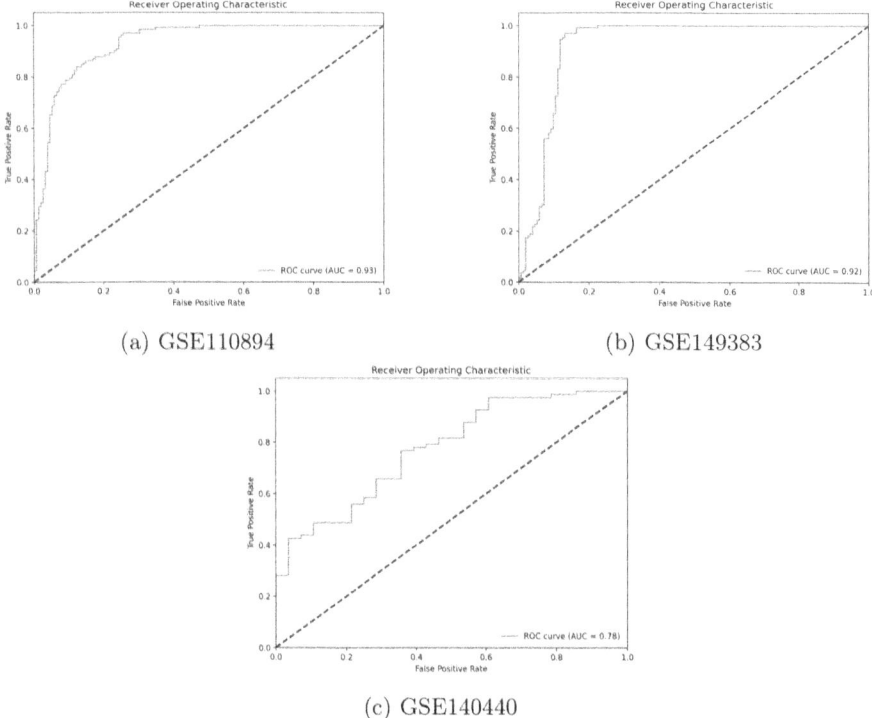

(a) GSE110894

(b) GSE149383

(c) GSE140440

Fig. 3. The ROC curve of the model for GSE110894, GSE149383 and GSE140440.

$$Pre = \frac{TP}{TP + FP}. \tag{9}$$

Precision describes the proportion of positive identifications that were actually correct. It measures the accuracy of the positive predictions made by the model.

$$Recall = \frac{TP}{TP + FN}. \tag{10}$$

Recall describes the proportion of actual positives that were correctly identified. It measures the model's ability to find all relevant cases within the data.

$$\text{F1 Score} = 2 \times \frac{\text{Pre} \times \text{Recall}}{\text{Pre} + \text{Recall}}. \tag{11}$$

The F1 Score is the harmonic mean of Precision and Recall, used to measure a test's accuracy. It provides a single measure that balances both precision and recall, and is particularly useful when dealing with imbalanced datasets (Tables 2, 3 and 4).

Table 2. Results of different model using GSE110894.

Model	ACC	AUC	Pre	Recall	F1score
Source only	0.68	0.66	0.80	0.40	0.54
DANN	0.73	0.81	0.74	0.69	0.71
Domain Confusion	0.90	0.92	0.91	0.89	0.89
our proposed	0.94	0.94	0.93	0.95	0.94

Table 3. Results of different model using GSE149383.

Model	ACC	AUC	Pre	Recall	F1score
Source only	0.58	0.63	0.60	0.56	0.50
DANN	0.80	0.89	0.81	0.76	0.77
Domain Confusion	0.81	0.88	0.80	0.78	0.79
our proposed	0.93	0.93	0.90	0.95	0.92

Table 4. Results of different model using GSE140440.

Model	ACC	AUC	Pre	Recall	F1score
Source only	0.45	0.44	0.41	0.35	0.38
DANN	0.64	0.63	0.60	0.56	0.58
Domain Confusion	0.73	0.76	0.74	0.69	0.69
our proposed	0.80	0.78	0.80	0.78	0.79

To verify the effectiveness of the method proposed in this paper, we compared it with three other models: source only model, DANN, and Domain Confusion, across three different target domain datasets. The results indicate that our method significantly outperforms the other models in all performance metrics.

Taking the GSE110894 dataset as an example, the accuracy (ACC) of our method reached 0.94, which is significantly higher than that of source only model at 0.68 and the DANN model at 0.73, with improvements of 0.26 and 0.21 respectively ($p < 0.01$). Additionally, compared to the Domain Confusion model, our method also scored 0.04 higher ($p < 0.05$). The p-values were calculated using Welch's t-test.

(1) Source only: The training of this model relies solely on data sourced from the source domain. It does not attempt to make any adjustments or optimizations for the data in the target domain, so it can serve as a benchmark for evaluating the performance of other domain adaptation methods. Therefore, the performance on the GSE140440 dataset is relatively poor, while it is comparatively better on the GSE110894 dataset, due to the latter's data dimension being closer to that of the source domain. (2) DANN (Domain Adversarial Training of Neural Networks) [15]: The main idea of DANN is the ReverseGrad algorithm, which treats domain invariance as a binary classification problem, but by reversing the gradient of the domain discriminator, it directly maximizes the loss of the domain classifier. Adversarial training itself is relatively complex; a simple gradient reversal essentially provides convenience in coding, but it is extremely unstable during the training process. (3) Domain confusion: The Deep Domain Confusion (DDC) method [17] introduces the MMD on the basis of the conventional source domain classification loss to learn representations that are both discriminative and domain-invariant, mapping the features of the source and target domains together into a new shared space for symmetric mapping. The DANN and Domain Confusion models can alleviate the issues caused by dimensionality differences to some extent, thereby enhancing the model's performance on the target domain. However, on the GSE140440 dataset, due to the significant dimensionality differences, the improvement in performance is limited. (4) Ours: This paper's method first learns discriminative representations using labels in the source domain, and then maps the target data to the source feature space using an asymmetric mapping learned through domain adversarial loss, which can better simulate the differences in low-level features than symmetric mapping. The method proposed in this paper demonstrates the best performance across all datasets, indicating that the method has good adaptability to differences in the number of samples and genes. It can effectively handle high-dimensional, small-sample data, thereby achieving good results on various datasets.

Parameters. This paper maps single-cell gene expression data and cell line gene expression data to a 512-dimensional latent space, which is passed to the discriminator for adversarial learning. Gene expression data is high-dimensional, which makes computations complex and prone to overfitting. A dimensionality of 512 reduces the complexity of the data while retaining sufficient feature information, which helps the model learn the essential characteristics of the data. This dimension has demonstrated good model performance and stability in preliminary experiments. During the training process, this dimension allows the model to effectively learn the feature representations of both the source and target domain data, and achieve feature distribution alignment through adversarial learning, thereby improving the model's predictive accuracy. The dimensions of the layers of the source domain neural network are [4096,2048,1024,512], and a pre-trained source model is used as the initialization of the target representation space. And the source model is fixed during adversarial training. The number of layers and dimensions of the target domain neural network need to be adjusted according to the dimensions of the preprocessed data. The dropout for the target domain encoder is set to 0.1, and the dropout for the discriminator is set to 0.2. This is because, during the initial stages of adversarial training, the discriminator learns quickly, which can lead to the target domain encoder failing to learn useful knowledge, thereby affecting the model's performance. An appropriate dropout value can prevent overfitting to some extent and ensure the model's generalization capability. The learning rate is set to 1×10^{-4}, which is a relatively small learning rate that helps the model to converge stably during training. Especially in the early stages of training, a smaller learning rate can prevent drastic fluctuations in model parameters, allowing the model to gradually learn the complex patterns and features in the data.

7   Conclusions and Outlook

In this study, we proposed a novel framework based on adversarial transfer learning for predicting single-cell drug sensitivity. One of the key innovations of our framework is the adversarial learning mechanism, which ensures that the feature distributions of the source and target domains are aligned. This alignment is crucial for the successful transfer of knowledge from the labeled cell line data to the unlabeled single-cell data. Through our experiments, we have demonstrated that our model outperforms existing methods in terms of prediction accuracy and robustness.

Despite the progress made in this study, there are several avenues for future work that we are actively exploring. One of the primary directions is the integration of multi-modal data. Currently, our framework relies primarily on single-cell RNA sequencing data; however, incorporating additional modalities such as proteomics, metabolomics and imaging data could provide a more comprehensive understanding of the cellular mechanisms underlying drug resistance. Multi-modal data integration has the potential to reveal hidden relationships between different biological layers and enhance the accuracy and reliability of causal gene identification.

Another promising direction is the extension of our framework to handle multi-class drug sensitivity predictions. Currently, our model focuses on binary classification of drug sensitivity. Expanding this to multi-class classification would allow for the prediction of varying levels of sensitivity to multiple drugs, providing a more detailed and nuanced understanding of drug responses at the single-cell level. This could lead to more effective and personalized treatment strategies in the future.

References

1. Wang, X., Zhang, H., Chen, X.: Drug resistance and combating drug resistance in cancer. Cancer Drug Resist. **2**(2), 141 (2019)
2. Siegfried, Z., Karni, R.: The role of alternative splicing in cancer drug resistance. Curr. Opin. Genet. Dev. **48**, 16–21 (2018)
3. Konieczkowski, D.J., Johannessen, C.M., Garraway, L.A.: A convergence-based framework for cancer drug resistance. Cancer Cell **33**(5), 801–815 (2018)
4. Panda, M., Biswal, B.K.: Cell signaling and cancer: a mechanistic insight into drug resistance. Mol. Biol. Rep. **46**(5), 5645–5659 (2019). https://doi.org/10.1007/s11033-019-04958-6
5. Ramsey, J., Glymour, M., Sanchez-Romero, R., Glymour, C.: A million variables and more: the fast greedy equivalence search algorithm for learning high-dimensional graphical causal models, with an application to functional magnetic resonance images. Int. J. Data Sci. Anal. **3**, 121–129 (2017)
6. Zhang, W., Deng, L., Zhang, L., Dongrui, W.: A survey on negative transfer. IEEE/CAA J. Automatica Sinica **10**(2), 305–329 (2022)
7. Gambardella, G., Viscido, G., Tumaini, B., Isacchi, A., Bosotti, R., Di Bernardo, D.: A single-cell analysis of breast cancer cell lines to study tumour heterogeneity and drug response. Nat. Commun. **13**(1), 1714 (2022)
8. Schmidt, F., Efferth, T.: Tumor heterogeneity, single-cell sequencing, and drug resistance. Pharmaceuticals **9**(2), 33 (2016)
9. Van de Sande, B., et al.: Applications of single-cell rna sequencing in drug discovery and development. Nat. Rev. Drug Disc. **22**(6), 496–520 (2023)
10. Shapiro, E., Biezuner, T., Linnarsson, S.: Single-cell sequencing-based technologies will revolutionize whole-organism science. Nat. Rev. Genet. **14**(9), 618–630 (2013)
11. Kim, K.-T., et al.: Single-cell mrna sequencing identifies subclonal heterogeneity in anti-cancer drug responses of lung adenocarcinoma cells. Genome Biol. **16**, 1–15 (2015)
12. Kempa, E.E., Hollywood, K.A., Smith, C.A., Barran, P.E.: High throughput screening of complex biological samples with mass spectrometry-from bulk measurements to single cell analysis. Analyst **144**(3), 872–891 (2019)
13. Goldie, J.H., Coldman, A.J.: The genetic origin of drug resistance in neoplasms: implications for systemic therapy. Cancer Res. **44**(9), 3643–3653 (1984)
14. Marine, J.-C., Dawson, S.-J., Dawson, M.A.: Non-genetic mechanisms of therapeutic resistance in cancer. Nat. Rev. Cancer **20**(12), 743–756 (2020)
15. Ganin, Y., et al.: Domain-adversarial training of neural networks. J. Mach. Learn. Res. **17**(59), 1–35 (2016)
16. Long, M., Cao, Y., Wang, J., Jordan, M.: Learning transferable features with deep adaptation networks. In: International Conference on Machine Learning, pp. 97–105. PMLR (2015)
17. Tzeng, E., Hoffman, J., Zhang, N., Saenko, K., Darrell, T.: Deep domain confusion: maximizing for domain invariance. arXiv preprint arXiv:1412.3474 (2014)
18. Ben-David, S., Blitzer, J., Crammer, K., Kulesza, A., Pereira, F., Vaughan, J.W.: A theory of learning from different domains. Mach. Learn. **79**, 151–175 (2010)
19. Sun, B., Feng, J., Saenko, K.: Return of frustratingly easy domain adaptation. In: Proceedings of the AAAI Conference on Artificial Intelligence, vol. 30 (2016)
20. Fang, Y., Yap, P.-T., Lin, W., Zhu, H., Liu, M.: Source-free unsupervised domain adaptation: a survey. Neural Netw. **174**, 106230 (2024)

21. Zhao, Z., Alzubaidi, L., Zhang, J., Duan, Y., Yuantong, G.: A comparison review of transfer learning and self-supervised learning: definitions, applications, advantages and limitations. Expert Syst. Appl. **242**, 122807 (2024)
22. Tzeng, E., Hoffman, J., Darrell, T., Saenko, K.: Simultaneous deep transfer across domains and tasks. In: Proceedings of the IEEE International Conference on Computer Vision, pp. 4068–4076 (2015)
23. Quiñonero-Candela, J., Sugiyama, M., Schwaighofer, A., Lawrence, N.: Covariate shift and local learning by distribution matching. In: Dataset Shift in Machine Learning, pp. 131–160 (2008)
24. Long, M., Cao, Y., Wang, J., Jordan, M.I.: Learning transferable features with deep adaptation networks. arXiv: Learning (2015)
25. Zhenyu, W., Zhang, H., Guo, J., Ji, Y., Pecht, M.: Imbalanced bearing fault diagnosis under variant working conditions using cost-sensitive deep domain adaptation network. Expert Syst. Appl. **193**, 116459 (2022)
26. Qian, Q., Qin, Y., Luo, J., Wang, Y., Fei, W.: Deep discriminative transfer learning network for cross-machine fault diagnosis. Mech. Syst. Signal Process. **186**, 109884 (2023)
27. Ganin, Y., Lempitsky, V.: Unsupervised domain adaptation by backpropagation. In: International Conference on Machine Learning, pp. 1180–1189. PMLR (2015)
28. Wang, J., Jiang, J.: Conditional coupled generative adversarial networks for zero-shot domain adaptation. In: Proceedings of the IEEE/CVF International Conference on Computer Vision, pp. 3375–3384 (2019)
29. Sun, Y., Yang, G., Ding,D., Cheng, G., Xu, J., Li, X.: A gan-based domain adaptation method for glaucoma diagnosis. In: 2020 International Joint Conference on Neural Networks (IJCNN), pp. 1–8. IEEE (2020)
30. Panigrahi, S., Nanda, A., Swarnkar, T.: A survey on transfer learning. In: Mishra, D., Buyya, R., Mohapatra, P., Patnaik, S. (eds.) Intelligent and Cloud Computing. SIST, vol. 194, pp. 781–789. Springer, Singapore (2021). https://doi.org/10.1007/978-981-15-5971-6_83
31. Goodfellow, I., et al.: Generative adversarial networks. Commun. ACM **63**(11), 139–144 (2020)
32. Chen, J., et al.: Deep transfer learning of cancer drug responses by integrating bulk and single-cell rna-seq data. Nat. Commun. **13**(1), 6494 (2022)
33. Wolf, F.A., Angerer, P., Theis, F.J.: SCANPY: large-scale single-cell gene expression data analysis. Genome Biol. **19**, 1–5 (2018)
34. Pedregosa, F., et al.: Scikit-learn: machine learning in python. J. Mach. Learn. Res. **12**, 2825–2830 (2011)
35. Fushiki, T.: Estimation of prediction error by using k-fold cross-validation. Stat. Comput. **21**, 137–146 (2011)

Pneumonia Detection in Chest X-Ray Images with Deep Learning

Savannah Hebert and Yan Zhang$^{(\boxtimes)}$

School of Computer Science and Engineering, California State University San Bernardino, San Bernardino, CA 92407, USA
007543322@coyote.csusb.edu, Yan.Zhang@csusb.edu

Abstract. Pneumonia is a severe lung infection that can be classified as bacterial and viral. Accurate and timely detection is essential for effective treatment planning and improved patient outcomes. Chest X-ray imaging is widely used for pneumonia diagnosis, offering detailed visual representations of lung structures. However, manual interpretation can be subjective and time-consuming, highlighting the need for automated diagnostic solutions. This study explores the application of deep learning models for pneumonia detection using a dataset of over 5,800 Chest X-ray images. We evaluate and compare three deep learning models: VGG-16, DenseNet-121, and EfficientNet-B3. Employing transfer learning with ImageNet pre-trained weights, we fine-tune each model for three-class classification and evaluate their performance. A comprehensive analysis is conducted to assess their strengths, limitations and performance for pneumonia detection. Experimental results show DenseNet-121 leading with 0.84 accuracy and balanced recalls (0.84 Normal, 0.97 Bacterial, 0.64 Viral), benefiting from its efficient 8 million parameters and dense connectivity. EfficientNet-B3 achieves 0.76 accuracy, excelling in Bacterial (0.97) and Viral (0.81) recalls, but lagging in Normal detection (0.52). VGG-16, with 0.71 accuracy, demonstrates high Bacterial (0.88) and Viral (0.80) recalls but poor Normal recall (0.47). The study highlights DenseNet-121's superiority for clinical use, particularly for bacterial pneumonia, and suggests future enhancements through ensemble approaches or expanded datasets.

Keywords: Pneumonia detection · X-ray image analysis · deep learning · Visual Geometry Group (VGG) · DenseNet · EfficientNet

1 Introduction

Pneumonia remains a significant global health challenge, ranking among the leading causes of death, particularly in children and the elderly [13]. This respiratory infection, caused by bacteria, viruses, or fungi, manifests in the lungs and can lead to severe complications if not diagnosed and treated promptly. Chest X-ray imaging is the gold standard for pneumonia diagnosis due to its accessibility and ability to reveal lung abnormalities. However, interpreting these

Q. Zhang et al. (Eds.): IJCRS 2025, LNAI 15710, pp. 257–271, 2025.
https://doi.org/10.1007/978-3-031-92741-6_19

images relies heavily on expert radiologists, whose assessments can be subjective, time-consuming, and prone to error. In resource-limited settings, where medical expertise and infrastructure are scarce, such challenges amplify the need for automated, reliable diagnostic tools.

Deep learning, particularly convolutional neural networks (CNN), has emerged as a transformative approach in medical imaging, offering the potential to enhance diagnostic accuracy and efficiency [18,21]. By learning hierarchical features directly from raw image data, CNNs can identify patterns indicative of pneumonia. Motivated by this promise, this study explores the application of three CNN architectures, that is, Visual Geometry Group (VGG-16), DenseNet-121, and EfficientNet-B3, to classify chest X-ray images into Normal, Bacterial, and Viral categories. Comparing these models allows us to assess their strengths in feature extraction, computational efficiency, and generalization across diverse image characteristics. This evaluation is crucial to determine the most effective architecture for real-world clinical deployment, balancing accuracy and resource demands. Using the Kermany dataset - a dataset of chest X-ray images from the Guangzhou Women and Children's clinic in Guangzhou [11], we aim to train, evaluate and compare the performance of three models, addressing the clinical imperative of rapid bacterial pneumonia detection for antibiotic treatment and the broader goal of supporting radiologists with robust tools.

The paper is structured as follows: Sect. 2 reviews related work on deep learning models in pneumonia detection; Sect. 3 describes the dataset and preprocessing; Sect. 4 details the methodologies of the selected models; Sect. 5 presents the experimental setup, training, evaluation, and comparative discussion; and Sect. 6 concludes with key findings and future directions. This study seeks to advance computer-aided diagnosis, improving patient outcomes through precise and automated pneumonia classification.

2   Related Work

The application of deep learning to medical imaging has revolutionized the field of pneumonia detection, particularly through the use of convolutional neural networks (CNNs), which excel at extracting meaningful features from complex datasets such as chest X-ray images. Many researchers and practitioners have explored a variety of CNN architectures, including the Visual Geometry Group (VGG) network, known for its depth and simplicity, as well as more advanced models like DenseNet and EfficientNet, which optimize feature reuse and computational efficiency, respectively. These studies highlight the potential of deep learning to enhance diagnostic accuracy, paving the way for pneumonia detection—as addressed in this study.

Rajpurkar et al. developed CheXNet, a 121-layer CNN model that was trained on the ChestX-ray14 dataset which contained over 100,000 chest X-ray images depicting 14 diseases including pneumonia [16]. The CheXNet model is an implementation of DenseNet and the weights of the model are preset with weights from a model trained on ImageNet, a large-scale collection of labeled images with hierarchical organization. The authors compared the performance of CheXNet with that of four radiologists. They found that CheXNet exceeded average radiologist performance on the F1 metric, achieving 0.435 compared to 0.387 [16].

Ayan and Unver evaluated VGG-16 and Xception CNN models for classifying chest X-ray images as normal or pneumonia cases, leveraging Kermany's dataset—the same dataset used in this study [3]. Employing transfer learning with fine-tuning, they improved model accuracy. Their findings showed VGG-16 outperforming Xception in accuracy, specificity, pneumonia precision, and pneumonia F1-score, while Xception outperformed VGG-16 in sensitivity, normal precision, and pneumonia recall [3]. Sharma et al. employed a VGG16 model with Neural Networks (NN) to classify chest X-ray images from two datasets. The first dataset achieves 0.92 accuracy. The second dataset (6,436 images: pneumonia, normal, COVID-19) yields 0.95 accuracy. VGG16-NN outperforms VGG16 with SVM, KNN, RF, and NB [17].

Jain et al. applied CNN models to detect pneumonia using X-ray images [10]. Six CNN models were trained to classify X-ray images into two classes, pneumonia and non-pneumonia, by changing various parameters, hyperparameters and number of convolutional layers. Two custom models with two and three convolutional layers achieved 0.85 and 0.92 accuracy, respectively. Pre-trained models—VGG-16, VGG-19, ResNet-50, and Inception-v3 —yielded accuracies of 0.87, 0.88, 0.78, and 0.71, respectively [10]. Hammoudi et al. assessed tailored deep learning models for pneumonia detection using the reorganized Kermany dataset, leveraging CNN backbones ResNet-34, ResNet-50, and DenseNet-169, with all exceeding 0.84 accuracy [6].

Ali et al. explored six deep learning models, CNN, InceptionResNetV2, Xception, VGG16, ResNet50, and EfficientNetV2L [1]. These models are implemented, trained on chest X-ray images, and optimized with the Adam optimizer to adjust epochs effectively. Performance results show accuracies of 0.88 (CNN), 0.89 (InceptionResNetV2), 0.91 (Xception), 0.92 (VGG16), 0.88 (ResNet50), and 0.94 (EfficientNetV2L), with EfficientNetV2L achieving the highest accuracy [1]. These findings underscore the robustness of deep learning in detecting pneumonia, supporting clinical decision-making and improving patient outcomes.

An et al. introduced a novel deep learning approach for pneumonia diagnosis via chest X-ray analysis, combining EfficientNetB0 and DenseNet121 with advanced attention mechanisms [2]. The method employs a deep convolutional neural network enhanced by multi-head self-attention modules, channel-attention-based feature fusion, residual blocks, and dynamic pooling strategies for precise feature extraction and efficient processing [2]. Trained on a dataset of chest X-rays from healthy and pneumonia-affected individuals, the model achieves an accuracy of 0.95.

Hashmi et al. utilized an ensemble approach to detect pneumonia in X-ray images [7]. Using the Kermany dataset, they created a weighted classifier module that took into account the weighted predictions of five pretrained models (ResNet-18, DenseNet-121, Inception, Xception, and MobileNetV2) to give a final weighted prediction. Each pretrained model was given a weight based on the confidence in that model. Using optimized weights in their prediction, they achieved a testing accuracy of 0.98 [7]. Kundu et al. developed a computer-aided diagnosis system for pneumonia detection using chest X-rays, addressing data scarcity with deep transfer learning and an ensemble of GoogLeNet, ResNet-18, and DenseNet-121. A novel weighted average technique improved performance, achieving 0.99 and 0.87 accuracy on the Kermany and RSNA datasets, respectively, outperforming state-of-the-art methods [12]. Robustness was validated via statistical tests, with code available online. Chouhan et al. applied deep learning framework with transfer learning to detect pneumonia from chest X-rays [5]. Five pretrained models—AlexNet, DenseNet-121, Inception-v3, GoogLeNet, and ResNet-18—extracted features from the Kermany dataset, followed by classification [5]. An ensemble of these models achieved a state-of-the-art accuracy of 0.96 and recall of 0.99, outperforming individual models.

Ma and Lv introduced the Swin Transformer to enhance pneumonia detection in chest X-ray images, addressing the performance plateau of traditional CNNs [14]. Optimized for X-ray characteristics, the model outperforms CNN-based backbones, boosting accuracy from 0.76 to 0.87 and 0.92 to 0.97 across two datasets [14]. Image enhancement tailored to X-ray features further improved accuracy compared to unenhanced results.

The reviewed studies underscore the evolution of deep learning in pneumonia detection, each offering unique strengths in accuracy, sensitivity, and lesion identification across diverse chest X-ray datasets. While ensemble methods and transfer learning have pushed performance boundaries, challenges remain in optimizing models for specific pneumonia types and generalizing across populations. Building on these insights, this study compares VGG, DenseNet, and EfficientNet, aiming to elucidate their comparative efficacy in classifying normal, bacterial, and viral pneumonia cases, leveraging the well-established Kermany dataset.

3   Dataset and Preprocessing

This study uses a dataset of chest X-ray images from the Guangzhou Women and Children's clinic in Guangzhou, P.R.China [11]. After excluding unreadable and low-quality scans, expert physicians labeled the remaining images as normal, bacterial pneumonia, or viral pneumonia. The dataset comprises 5,232 training images (1,349 Normal, 2,538 Bacterial, 1,345 Viral) and 624 testing images (234 Normal, 242 Bacterial, 148 Viral), as shown in Table 1. The training set was further split, with 90% used for training and 10% for validation. Images were assigned labels: '0' for Normal, '1' for Bacterial, and '2' for Viral.

The dataset, curated by Kermany et al., exhibits an intentional imbalance, with more bacterial than viral pneumonia images. This reflects clinical priorities, as bacterial pneumonia—treatable with antibiotics—demands rapid and accurate diagnosis, whereas viral pneumonia relies on supportive care and does not require antibiotic treatment [11]. This distinction between the two types of

Table 1. Dataset Distribution for Training and Testing Sets

Category	Training Set (No. of Images)	Testing Set (No. of Images)
Normal (0)	1349	234
Bacterial (1)	2538	242
Viral (2)	1345	148
Total	5232	624

pneumonia explains the greater number of bacterial instances to train a model that is better at detecting bacterial pneumonia.

For preprocessing, all images were resized to 224 × 224 pixels to match VGG-16's default input size. Although EfficientNet and DenseNet do not strictly require this resolution, we standardized all models to 224 × 224 pixels for consistency, despite the dataset's original varied image dimensions.

4   Methodologies

In this section, we outline the deep learning methodologies employed in this study to detect pneumonia from chest X-ray images, focusing on three convolutional neural network (CNN) architectures: Visual Geometry Group (VGG), DenseNet, and EfficientNet.

4.1   Convolutional Neural Networks (CNNs)

Convolutional Neural Networks (CNNs) are a class of deep learning models specifically used for image classification, segmentation, object detection, video processing, natural language processing, and speech recognition [15]. Unlike traditional neural networks, CNNs leverage spatial hierarchies in data through convolutional layers, which apply filters to extract local features—edges, textures, or patterns—while preserving spatial relationships [4]. These layers reduce the dimensionality of the input, enabling the network to focus on relevant features without requiring extensive manual feature engineering. Following convolution, pooling layers (e.g., max-pooling) further downsample the feature maps, enhancing computational efficiency and reducing overfitting by retaining only the most prominent features [15]. The CNN architecture typically includes stacked convolutional and pooling layers, followed by fully connected layers that integrate extracted features for classification. Activation functions, such as ReLU (Rectified Linear Unit), introduce non-linearity, allowing the network to model complex patterns. During training, the network optimizes its weights using backpropagation and gradient descent, minimizing a loss function (e.g., categorical cross-entropy) to improve prediction accuracy. CNNs' ability to learn hierarchical feature representations directly from raw pixel data makes them advantageous over traditional methods, which often rely on hand-crafted features [4].

This foundational mechanism underpins the three models evaluated in this research: VGG, DenseNet, and EfficientNet. Each builds upon the CNN framework, adapting it with unique structural innovations to balance accuracy, depth, and efficiency, as explored in the subsequent subsections.

4.2   Visual Geometry Group (VGG)

The Visual Geometry Group (VGG) network, proposed by Simonyan and Zisserman, is a deep convolutional neural network known for its simplicity, and straightforward uniform design [19]. These features make it a strong contender for classifying chest X-ray images in this study. We utilize VGG-16, featuring 16 layers—13 convolutional and 3 fully connected—characterized by small 3×3 filters stacked deeply to extract detailed features like edges and textures critical for distinguishing normal, bacterial, and viral pneumonia. Max-pooling layers are inserted throughout to reduce the size of volume, which makes the computation fast and also prevents overfitting. The 3 fully connected layers synthesize features before a softmax layer outputs class probabilities.

VGG-16's depth enhances its ability to capture hierarchical patterns, a key advantage for medical imaging tasks. With approximately 138 million parameters, VGG-16 demands many computational resources [19]. In this study, we used transfer learning by initializing VGG-16 with ImageNet pre-trained weights, harnessing its pre-learned feature extraction capabilities. We then fine-tuned the final layers to tailor the model to our dataset's three-class classification, optimizing it for pneumonia-specific patterns. This approach balances VGG's strengths with the need for efficiency, positioning it as a benchmark against DenseNet and EfficientNet in our comparative analysis.

4.3   DenseNet

DenseNet, introduced by Huang et al., is a convolutional neural network (CNN) architecture designed to address the vanishing gradient problem, where learned features diminish across deep layers, hindering training [8]. Traditional CNNs propagate features sequentially, but Huang et al. proposed a novel solution: connecting each layer directly to all subsequent layers to maintain gradient integrity [8]. In DenseNet, every layer receives inputs from all preceding layers and passes its outputs forward, creating dense connectivity that preserves gradient flow [8]. This approach mitigates vanishing gradients, enhances feature reuse, and reduces parameter count, making the model less prone to overfitting compared to deeper architectures like VGG.

For this study, we adopt DenseNet-121, a variant with 121 layers, to classify chest X-ray images into normal, bacterial, and viral pneumonia categories. Organized into dense blocks, DenseNet-121 concatenates features across layers, enabling the capture of intricate patterns—from basic edges to pneumonia-specific markers—while maintaining computational efficiency. Pre-trained on ImageNet, we apply transfer learning to leverage its generalized feature extraction, followed by fine-tuning to optimize performance on our dataset. This dense connectivity distinguishes DenseNet-121 from VGG's linear stacking, offering a sophisticated balance of depth and efficiency. Its reduced parameter complexity and robust gradient flow make it an ideal candidate for our comparative analysis.

4.4   EfficientNet

The advancements in CNN accuracy, as seen in models like VGG and DenseNet, were achieved by increasing the depth, width, or resolution of the network, often leading to parameter-heavy architectures. For instance, the GPipe model uses 557 million parameters, demanding extensive memory and parallel processing for training [9]. Tan and Le introduced EfficientNet, a family of CNNs that employs compound scaling, a method that uniformly balances depth, width, and resolution with a single scaling coefficient, to enhance accuracy without excessively inflating parameter counts [20]. This approach optimizes performance while maintaining computational efficiency.

In this study, we adopt EfficientNet-B3 for classifying chest X-ray images into normal, bacterial, and viral pneumonia categories. Pre-trained on ImageNet, EfficientNet-B3, with approximately 12 million parameters, builds on the baseline B0 model by moderately scaling its dimensions, incorporating inverted residual blocks and squeeze-and-excitation units to enhance feature extraction and computational efficiency. We fine-tune EfficientNet-B3 on our dataset, leveraging its balanced scaling to achieve high accuracy with fewer resources than VGG. Compared to DenseNet's dense connectivity or VGG's parameter-heavy structure, EfficientNet-B3 provides a refined compromise, optimizing performance for medical imaging.

5   Experimental Result and Discussion

5.1   Experiment Setup

This section details the experimental setup designed to evaluate the performance of VGG-16, DenseNet-121, and EfficientNet-B3 in pneumonia detection from chest X-rays. The implementation begins with defining hardware requirements tailored to each model's computational demands.

VGG-16, with its deep architecture and approximately 138 million parameters, requires significant resources. Effective training and inference demand a dual-core processor, at least 4 GB of RAM, and a GPU with a minimum of 4 GB VRAM, such as the NVIDIA GTX 1060 or better, reflecting its computationally intensive nature in CNN applications.

DenseNet-121, featuring 121 layers and dense connectivity, balances efficiency with depth, leveraging roughly 8 million parameters. To execute DenseNet-121, we utilize a dual-core processor with 4 GB of RAM and a GPU providing 4 GB VRAM, ensuring robust feature propagation without excessive resource strain.

EfficientNet-B3, optimized via compound scaling and employing about 12 million parameters, achieves high accuracy with fewer demands. It requires a dual-core processor, a minimum of 2 GB RAM, and a GPU with at least 2 GB VRAM, such as the NVIDIA GTX 1050 or higher.

5.2   VGG-16 Training and Evaluation

For our custom VGG-16 model implementation from the Keras library, we combined a pre-trained VGG-16 base model, which was originally trained on the
ImageNet dataset, with several custom layers to form a tailored classification
head for pneumonia detection. The architecture begins with VGG-16's convolutional and max-pooling layers, which extract hierarchical spatial features, such
as edges and textures, from chest X-ray images. These layers, leveraging their
pre-trained weights, provide a robust foundation for feature extraction. Next,
a two-dimensional Global Average Pooling layer reduces the feature maps into
a compact, lower-dimensional feature vector, simplifying subsequent processing.
This is followed by three Dense layers with a decreasing neuron count, designed
to refine high-level feature extraction and capture finer pneumonia-specific patterns. Finally, the model concludes with a classification output layer employing
a softmax activation function, assigning probabilities to the three target classes:
Normal, Bacterial, and Viral pneumonia.

Training used the Adam optimizer with a learning rate of 0.001, minimizing
categorical cross-entropy loss over 20 epochs, with a batch size of 32. We fine-
tuned the top layers while retaining the pre-trained convolutional base, adapting
it to our dataset. Figure 1a and 1b depict the training and validation loss and
accuracy, respectively, when training VGG-16 over 20 epochs. The declining
training loss depicted in Fig. 1a and increasing training accuracy in Fig. 1b indicate that the VGG-16 model was able to learn the distinct features of pneumonia
in the training set. However, VGG-16 overfits to the training set and generalized
features were not found in the validation set. Figure 1a shows the validation loss
increasing despite decreasing training loss, and the validation accuracy in Fig. 1b
stagnates between 0.75 and 0.77.

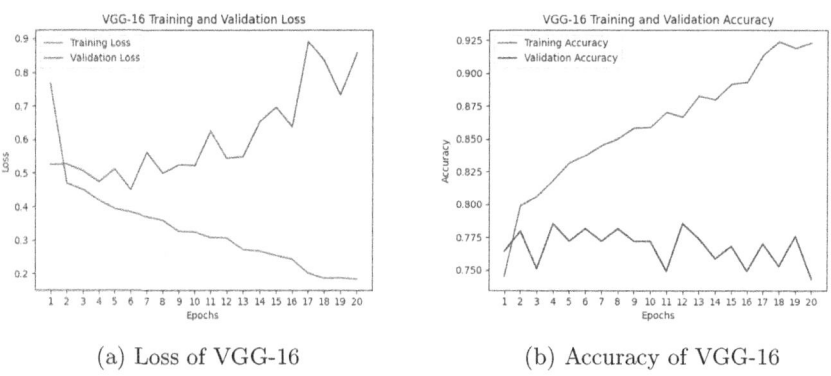

(a) Loss of VGG-16   (b) Accuracy of VGG-16

Fig. 1. The training and validation loss and accuracy of VGG-16

To evaluate VGG-16's performance, we analyzed its confusion matrix and
classification report on the test set of 624 images. The confusion matrix for

VGG-16 in Table 2 shows that the model correctly identified 110 of 234 Normal cases, but struggled the most with differentiating between Normal and Viral images with 93 Normal images incorrectly predicted as Viral. Bacterial pneumonia detection was stronger, with 212 of 242 cases correctly predicted.

Table 2. The confusion matrix of VGG-16.

True/Predicted	Normal	Bacterial	Viral
Normal	110	31	93
Bacterial	4	212	26
Viral	1	29	118

The classification report shown in Table 3 further details precision, recall, and F1-scores. Normal cases achieves a precision of 0.96 with a recall of 0.47, reflecting high accuracy when predicting Normal cases but low sensitivity due to missed Normal cases. Bacterial pneumonia scores a precision of 0.78 and recall of 0.88, indicating solid detection. Viral pneumonia has a precision of 0.50 but a recall of 0.80, revealing difficulty in identification. The F1-scores of 0.63 (Normal), 0.82 (Bacterial), and 0.61 (Viral) highlight VGG-16's uneven performance, excelling in classifying Bacterial cases yet faltering on Viral and Normal cases, possibly due to feature overlap in X-rays. Overall test accuracy is 0.71, suggesting moderate success constrained by Normal misclassification. VGG-16's depth aids Bacterial detection, but its 138 million parameters may contribute to overfitting or inefficiency on this task.

Table 3. The classification report of VGG-16.

Class	Precision	Recall	F1	Support
Normal	0.96	0.47	0.63	234
Bacterial	0.78	0.88	0.82	242
Viral	0.50	0.80	0.61	148
Accuracy	0.71			624

5.3   DenseNet-121 Training and Evaluation

For our DenseNet-121 implementation we utilized the Keras library and used an approach similar to VGG-16, integrating a pre-trained DenseNet-121 base model, trained on ImageNet, with custom layers to form a classification head tailored for pneumonia detection. The architecture is structured as follows: the base DenseNet-121 layers, with their dense connectivity, extract multi-scale features from chest X-ray images, leveraging the model's efficient gradient flow.

This is followed by a two-dimensional Global Average Pooling layer, which flattens the output of the base DenseNet layers. Then a 50% Dropout layer is introduced to reduce overfitting by randomly disabling half the units during training. Next, three fully connected Dense layers continue feature learning and extract high-level pneumonia-specific patterns. After that, a second 50% Dropout layer precedes the final classification layer to further reduce overfitting. Finally, the classification output layer employs a softmax activation function to output probabilities for the three classes: Normal, Bacterial, and Viral pneumonia.

Training is conducted using the Adamax optimizer with a learning rate of 0.001, minimizing categorical cross-entropy loss over 20 epochs, with a batch size of 32. We fine-tune the top layers while retaining the pre-trained base, optimizing for our dataset. Figure 2a and 2b show the training and validation loss and accuracy, respectively, when training DenseNet-121 over 20 epochs. Although the validation loss jumps and increases, DenseNet-121 is able to learn generalized features resulting in the validation accuracy increasing despite this.

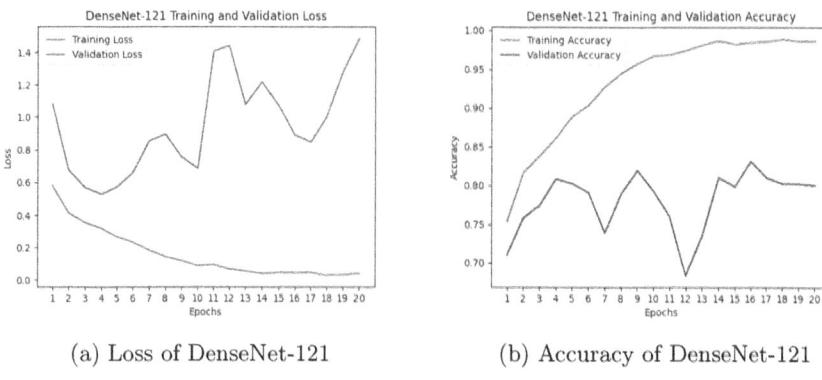

(a) Loss of DenseNet-121   (b) Accuracy of DenseNet-121

Fig. 2. The training and validation loss and accuracy of DenseNet-121

The confusion matrix for DenseNet-121 in Table 4 demonstrates its ability to classify pneumonia types from chest X-rays with notable precision. For Normal cases, DenseNet-121 correctly identifies 197 out of 234 instances, misclassifying 11 as Bacterial and 26 as Viral, suggesting high sensitivity to Normal patterns but some confusion with pneumonia features. Bacterial pneumonia detection is robust, with 234 of 242 cases correctly predicted, though five are mistaken for Normal and three for Viral, indicating minor overlap in feature recognition. Viral pneumonia shows 94 of 148 cases correctly classified, with seven mislabeled as Normal and 47 as Bacterial, reflecting challenges in distinguishing Viral-specific signatures.

The classification report further shows DenseNet-121's performance, reporting precision, recall, and F1-scores across classes, with an overall accuracy of 0.84. For Normal cases, precision reachs 0.94 and recall 0.84, yielding an F1-score of 0.89, indicating strong identification with minimal false positives. Bacterial class

Table 4. The confusion matrix of DenseNet-121.

True/Predicted	Normal	Bacterial	Viral
Normal	197	11	26
Bacterial	5	234	3
Viral	7	47	94

achieves a precision of 0.80 and recall of 0.97, resulting in an F1-score of 0.88, reflecting reliable detection due to the model's feature reuse. Viral class scores a precision of 0.76 and recall of 0.64, with an F1-score of 0.69, highlighting moderate performance, possibly from subtle X-ray feature overlap (Table 5).

Table 5. The classification report of DenseNet-121.

Class	Precision	Recall	F1	Support
Normal	0.94	0.84	0.89	234
Bacterial	0.80	0.97	0.88	242
Viral	0.76	0.64	0.69	148
Accuracy	0.84			624

5.4   EfficientNet-B3 Training and Evaluation

For our EfficientNet-B3 implementation we again utilize the Keras library, and a pre-trained base model, originally trained on ImageNet, augmented with custom layers to create a classification head tailored for pneumonia detection from chest X-rays. The architecture is structured as follows: the base EfficientNet-B3 layers extract hierarchical features—ranging from edges to complex patterns—from the input images. A Batch Normalization layer follows, stabilizing training by normalizing activations and emphasizing the prior batch's contributions. Next, a fully connected Dense layer refines these high-level features, enhancing the model's ability to discern pneumonia-specific signatures. To combat overfitting, a Dropout layer with a 50% rate deactivates neurons during training, promoting generalization. Finally, a classification output layer with a softmax activation assigns probabilities to the three target classes: Normal, Bacterial, and Viral pneumonia.

Training employs the Adamax optimizer with a learning rate of 0.001, minimizing categorical cross-entropy loss over 20 epochs, using a batch size of 32. We fine-tune the top layers while retaining the pre-trained base, optimizing for our dataset. Figure 3a and 3b show the changes in training and validation loss and accuracy, respectively, when training our EfficientNet-B3 model over 20 epochs. Both the training and validation loss in Fig. 3a show a significant decrease as

EfficientNet-B3 learns generalized features from the training data. The validation loss remains higher than the training loss at the 20th epoch. In Fig. 3b, the training accuracy climbs significantly indicating that the model overfits to the training data. Despite overfitting, the validation accuracy increases as well.

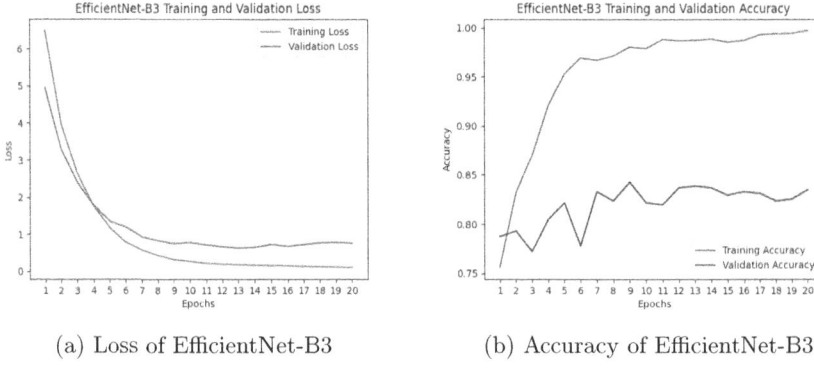

(a) Loss of EfficientNet-B3   (b) Accuracy of EfficientNet-B3

Fig. 3. The training and validation loss and accuracy of EfficientNet-B3

The confusion matrix for EfficientNet-B3 in Table 6 shows its performance in classifying chest X-rays into Normal, Bacterial, and Viral pneumonia. For Normal cases, it only correctly identifies 121 of 234 instances, with 100 misclassified as Viral, indicating difficulty distinguishing Normal-specific signatures from Viral X-rays. Bacterial pneumonia detection is robust, with 234 of 242 cases accurately predicted, though one is mistaken for Normal and seven for Viral, suggesting very minor feature overlap. Viral pneumonia proves more challenging, with 120 of 148 cases correctly classified, but 28 mislabeled as Bacterial, demonstrating an overlap in learned Viral and Bacterial features.

Table 6. The confusion matrix of EfficientNet-B3.

True/Predicted	Normal	Bacterial	Viral
Normal	121	13	100
Bacterial	1	234	7
Viral	0	28	120

The classification report in Table 7 provides deeper insight into EfficientNet-B3's performance, reporting precision, recall, and F1-scores, with an overall accuracy of 0.76. For Normal cases, the precision is 0.99 with a recall of 0.52, yielding an F1-score of 0.68, reflecting high accuracy but susceptibility to false positives from Viral misclassifications. Bacterial class achieves a precision of 0.85 and recall of 0.97, resulting in an F1-score of 0.91, showcasing strong and reliable

Table 7. The classification report of EfficientNet-B3.

Class	Precision	Recall	F1	Support
Normal	0.99	0.52	0.68	234
Bacterial	0.85	0.97	0.91	242
Viral	0.53	0.81	0.64	148
Accuracy	0.76			624

detection. Viral class scores a precision of 0.53 but a recall of 0.81, with an F1-score of 0.64, underscoring moderate performance limited by low precision.

5.5   Model Comparison and Discussion

The comparative analysis of VGG-16, DenseNet-121, and EfficientNet-B3 shows distinct strengths and trade-offs in pneumonia detection from chest X-ray images, as evidenced by their test accuracies: 0.71, 0.84, and 0.76, respectively. VGG-16 performs well in Bacterial (0.88) and Viral (0.80) case recalls but falters with Normal cases (recall 0.47), misclassifying 93 of 234 Normal cases as Viral, likely due to its parameter-heavy architecture overfitting to subtle feature overlaps. DenseNet-121, using 8 million parameters and dense connectivity, achieves the highest accuracy (0.84) and balanced performance across classes, with recalls of 0.84 (Normal), 0.97 (Bacterial), and 0.64 (Viral), misclassifying fewer instances (e.g., 26 Normal as Viral) than VGG-16. EfficientNet-B3, with 12 million parameters and compound scaling, achieves 0.76 accuracy, boasting near-perfect Bacterial recall (0.97) and strong Viral detection (recall 0.81), but struggles with Normal recall (0.52), mislabeling 100 of 234 Normal cases as Viral.

DenseNet-121's superior generalization and efficiency highlight its advantage in this task. These results reflect architectural impacts: VGG-16's depth aids sensitivity but sacrifices specificity, DenseNet-121's feature reuse enhances consistency, and EfficientNet-B3's balanced scaling optimizes resource use yet misses Normal nuances.

For clinical applications, DenseNet-121's higher accuracy and robust F1-scores (0.89 Normal, 0.88 Bacterial, 0.69 Viral) suggest it as the most reliable model, particularly for detecting bacterial pneumonia, which is critical for timely antibiotic treatment. EfficientNet-B3's efficiency and Bacterial detection suit screening scenarios, while VGG-16's limitations underscore the need for lighter models.

6   Conclusion

This study compared the performance of three CNN deep learning models, VGG-16, DenseNet-121, and EfficientNet-B3, in detecting pneumonia from chest X-ray images, classifying them as Normal, Bacterial, or Viral using the Kermany dataset. DenseNet-121 emerged as the top performer, achieving an accuracy of 0.84, with balanced recalls (0.84 Normal, 0.97 Bacterial, 0.64 Viral) and F1-scores (0.89, 0.88, 0.69), with its dense connectivity and efficient 8 million parameters. EfficientNet-B3 followed with 0.76 accuracy, excelling in Bacterial detection with near-perfect recall (0.97) and F1-score (0.91), but struggled with Normal recall (0.52), reflecting its 12 million-parameter design's trade-offs. VGG-16, despite a high Bacterial recall (0.88), lagged at 0.71 accuracy, hindered by Normal misclassifications (recall 0.47) and its 138 million parameters, suggesting overfitting risks. These findings highlight DenseNet-121's suitability for clinical pneumonia diagnosis, particularly for Bacterial cases requiring prompt treatment, while EfficientNet-B3 offers efficiency for resource-constrained settings. VGG-16's computational intensity limits its practicality. The intentional dataset imbalance favoring Bacterial instances enhanced model training for this critical class, aligning with clinical needs.

Future research could 1) refine Viral detection through ensemble techniques, 2) incorporate larger datasets and train more deep learning models, 3) apply data augmentation techniques to the dataset to improve model robustness, 4) conduct more comprehensive hyperparameter tuning for each model to optimize performance, 5) explore the combination of these models through ensemble methods for enhanced accuracy, and 6) perform feature importance analysis for each model to better understand their decision-making processes. Integrating these deep learning models with real-time diagnostic tools could enhance their utility, supporting radiologists in improving pneumonia management and patient outcomes.

Acknowledgments. We would like to express our gratitude to the High Performance Computing Center at CSUSB for providing computational resources.

References

1. Ali, M., et al.: Pneumonia detection using chest radiographs with novel efficient-netv2l model. IEEE Access (2024)
2. An, Q., Chen, W., Shao, W.: A deep convolutional neural network for pneumonia detection in x-ray images with attention ensemble. Diagnostics 14(4), 390 (2024)
3. Ayan, E., Ünver, H.M.: Diagnosis of pneumonia from chest x-ray images using deep learning. In: 2019 Scientific Meeting on Electrical-Electronics & Biomedical Engineering and Computer Science (EBBT), pp. 1–5. IEEE (2019)
4. Bhatt, D., et al.: Cnn variants for computer vision: history, architecture, application, challenges and future scope. Electronics 10(20), 2470 (2021)
5. Chouhan, V., et al.: A novel transfer learning based approach for pneumonia detection in chest x-ray images. Appl. Sci. 10(2), 559 (2020)
6. Hammoudi, K., et al.: Deep learning on chest x-ray images to detect and evaluate pneumonia cases at the era of covid-19. J. Med. Syst. 45(7), 75 (2021)
7. Hashmi, M.F., Katiyar, S., Keskar, A.G., Bokde, N.D., Geem, Z.W.: Efficient pneumonia detection in chest xray images using deep transfer learning. Diagnostics 10(6), 417 (2020)

8. Huang, G., Liu, Z., Van Der Maaten, L., Weinberger, K.Q.: Densely connected convolutional networks. In: Proceedings of the IEEE Conference on Computer Vision and Pattern Recognition, pp. 4700–4708 (2017)
9. Huang, Y., et al.: Gpipe: efficient training of giant neural networks using pipeline parallelism. Adv. Neural Inf. Process. Syst. **32** (2019)
10. Jain, R., Nagrath, P., Kataria, G., Kaushik, V.S., Hemanth, D.J.: Pneumonia detection in chest x-ray images using convolutional neural networks and transfer learning. Measurement **165**, 108046 (2020)
11. Kermany, D.S., et al.: Identifying medical diagnoses and treatable diseases by image-based deep learning. Cell **172**(5), 1122–1131 (2018)
12. Kundu, R., Das, R., Geem, Z.W., Han, G.T., Sarkar, R.: Pneumonia detection in chest x-ray images using an ensemble of deep learning models. PLoS ONE **16**(9), e0256630 (2021)
13. Lim, W.S.: Pneumonia—overview. In: Encyclopedia of Respiratory Medicine, p. 185 (2021)
14. Ma, Y., Lv, W.: Identification of pneumonia in chest x-ray image based on transformer. Int. J. Antennas Propagat. **2022**(1), 5072666 (2022)
15. Purwono, P., Ma'arif, A., Rahmaniar, W., Fathurrahman, H.I.K., Frisky, A.Z.K., ul Haq, Q.M.: Understanding of convolutional neural network (cnn): a review. Int. J. Rob. Control Syst. **2**(4), 739–748 (2022)
16. Rajpurkar, P., et al.: Chexnet: radiologist-level pneumonia detection on chest x-rays with deep learning (2017). https://arxiv.org/abs/1711.05225
17. Sharma, S., Guleria, K.: A deep learning based model for the detection of pneumonia from chest x-ray images using vgg-16 and neural networks. Procedia Comput. Sci. **218**, 357–366 (2023)
18. Sharma, S., Guleria, K.: A systematic literature review on deep learning approaches for pneumonia detection using chest x-ray images. Multimedia Tools Appl. **83**(8), 24101–24151 (2024)
19. Simonyan, K., Zisserman, A.: Very deep convolutional networks for large-scale image recognition. arXiv preprint arXiv:1409.1556 (2014)
20. Tan, M., Le, Q.: Efficientnet: rethinking model scaling for convolutional neural networks. In: International Conference on Machine Learning, pp. 6105–6114. PMLR (2019)
21. Zhong, Y., Liu, Y., Gao, E., Wei, C., Wang, Z., Yan, C.: Deep learning solutions for pneumonia detection: performance comparison of custom and transfer learning models. In: International Conference on Automation and Intelligent Technology (ICAIT 2024), vol. 13401, pp. 95–100. SPIE (2024)

Attention-Driven with Gaussian Processes for Weakly Supervised Hemorrhage Detection in Brain CT Scans

Si-Kha Huynh[1,2], Nhat-Minh Trieu[1,2], and Duy-Hoang Tran[1,2(✉)]

[1] Faculty of Information Technology, University of Science,
Ho Chi Minh City, Vietnam
{hskha21,tnminh21}@clc.fitus.edu.vn, tdhoang@fit.hcmus.edu.vn
[2] Vietnam National University, Ho Chi Minh City, Vietnam

Abstract. Intracranial hemorrhage (ICH), a life-threatening emergency, requires rapid diagnosis through analysis of computed tomography (CT) scans. Automating intracranial hemorrhage detection using artificial intelligence is crucial, but is hampered by the scarcity of labeled training data. To overcome this limitation, we present a novel weakly supervised deep learning framework. This framework integrates Attention-based Multiple Instance Learning (Att-MIL) with Sparse Variational Gaussian Processes (SVGP) to improve diagnostic accuracy and interpretability. Att-MIL mimics clinical radiology practice by identifying abnormal CT slices and modeling critical dependencies between slices. Meanwhile, SVGP provide probabilistic diagnostic results, explicitly highlighting the influence of training data size on decision confidence. Additionally, we introduce a data redundancy reduction technique that enhances learning efficiency by removing semantically redundant information across imaging modalities. Our proposed method demonstrates superior performance in binary ICH classification compared to existing Gaussian Processes-based MIL approaches. Furthermore, it exhibits robust performance in sub-label categorization, highlighting its effectiveness and generalizability across diverse classification tasks. The source code is available and maintained in the GitHub repository (https://github.com/nhatminhtrieu/Brain-Stroke-Diagnosis/).

Keywords: Multiple instance learning · CT hemorrhage detection · Attention mechanism · Gaussian process

1 Introduction

Stroke, particularly its subtype intracranial hemorrhage (ICH), is a life threatening medical emergency that requires rapid and accurate diagnosis to guide critical interventions. The global incidence rate in young adults was approximately 15.54 per 100,000 individuals [15]. Current research emphasizes supervised deep learning for ICH detection in computed tomography (CT) scan slices,

---

S.K. Huynh and N.M. Trieu—These authors contributed equally to this work.

assigning labels to individual slices. Wang et al. [16] integrated a 2D convolutional neural network (CNN) with sequence models, achieving expert-level accuracy in acute ICH detection and subtype classification on the Radiological Society of North America (RSNA) Brain CT Hemorrhage Challenge dataset [6]. Building upon this, Barhoumi et al. [3] introduced Scopeformer, a hybrid architecture that combines multiple CNNs for local feature extraction with Vision Transformers (ViTs) for global context modeling. In a separate approach, Wolf et al. [17] explored self-supervised learning, where the proposed masked autoencoder, SparK, demonstrated strong performance in downstream medical binary classification tasks. A significant limitation of current slice-level ICH detection methods is their reliance on extensive, detailed annotations, demanding large datasets. Slice-by-slice annotation of CT scans is a time-consuming and inefficient process, particularly given the volume of data involved. The need for large, annotated datasets is exemplified by the RSNA challenge, a multinational effort that required 60 radiologists to dedicate thousands of hours to curate 25,312 expert-validated CT scans [6]. This contrasts sharply with clinical practice, where diagnoses are typically made at the scan level, revealing a disconnect between annotation requirements and real-world diagnostic workflows. To address this, researchers are investigating methods that utilize scan-level labels for training, rather than slice-level annotations.

Several studies have employed 3D CNNs for scan-level ICH detection, using the entire 3D scan as input and the scan label for training. Early work by Titano et al. [14] explored 3D CNNs for acute neurological illness detection in head CT scans, adapting a ResNet-50 architecture, but highlighted the substantial computational demands of full 3D CNN training. To mitigate this, Burduja et al. [4] introduced a hybrid CNN-LSTM model, that processes individual CT slices with a 2D CNN, and then uses an Long short-term memory (LSTM) to integrate features across slices, capturing some 3D relationships while reducing computational load. More recently, Asif et al. [2] proposed a parallel deep learning architecture using ResNet101-V2 and Inception-V4 to extract features from windowed and adjacent CT slices. However, these studies suffer from high computational demands and a lack of interpretability, specifically the inability to segment ICH-containing slices. This scenario calls for Multiple Instance Learning (MIL), which uses scan-level annotations to diagnose ICH while handling slice correlations.

MIL, a form of weakly supervised learning, has gained significant traction in DL, particularly within pathology. MIL allows analysis of more datasets without exhaustive labeling. In this approach, we frame ICH diagnosis as an MIL problem. A complete CT scan becomes a "bag", and each individual slice represents an "instance". Crucially, a scan is classified as ICH if even a single slice exhibits ICH, while a scan is deemed normal only if all slices are normal. Notably, Gaussian Process-based MIL (GP-MIL) methods have demonstrated compelling results in ICH detection [5,12,18], attributed to their powerful representational capacity and ability to quantify diagnostic uncertainty. Wu et al. [18] introduced a two-stage MIL approach, combining an Attention-enhanced Convolutional

Neural Network (Att-CNN) for feature extraction with Variational GP for MIL (VGPMIL) for classification. One limitation of VGPMIL is the use of the Jaakkola bound approximation to handle logistic functions, which can degrade predictive performance due to approximation error, as pointed out in [5]. Building on this foundation, subsequent research has explored enhancements to the VGPMIL component. López Pérez et al. [12] replaced the original VGPMIL with Deep GP for MIL (DGPMIL), leveraging stacked GP layers and Evidence Lower Bound (ELBO) optimization to capture more intricate data relationships and richer representations. While DGPMIL demonstrated superior ability to capture complex patterns compared to VGPMIL, the framework's hierarchical structure introduces significant limitations, including high computational complexity due to kernel matrix inversion across multiple GP layers, error propagation caused by sequential inference schemes, and restricted scalability arising from limited parallelization opportunities. In a parallel development, Castro-Macías et al. [5] reformulated VGPMIL using Pólya-Gamma random variables. Notably, this reformulation yields equivalent variational posterior approximations to the original VGPMIL, a result stemming from the dual representations of the hyperbolic secant distribution. In contrast to the two-stage approach of Wu et al. [18], Pérez et al. [13] proposed end-to-end architectures, derived from VGPMIL, that integrate feature extraction, attention-based aggregation, and probabilistic inference into a single framework. However, the reliance on Kullback-Leibler (KL) divergence for loss function refinement can create a challenge in balancing model complexity and generalization, potentially leading to suboptimal posterior approximations.

Expanding upon the two-stage framework of Wu et al. [18], our study proposes an enhanced Att-MIL model incorporating GPs for binary ICH detection. Specifically, we integrate GPs into the Att-MIL classification stage, while preserving the Att-CNN for feature extraction. Our main contributions are:

1. An enhanced Att-MIL model with Sparse Variational GPs (SVGP) is introduced for classification. This integration allows for effective approximation of the true posterior distribution, mitigating the impact of limited datasets and enhancing model scalability and robustness.
2. Data heterogeneity is addressed through a data redundancy reduction technique. By eliminating semantically redundant information from similar or repetitive data across imaging modalities, this method streamlines the learning process, resulting in a more focused and efficient model.
3. The proposed method demonstrates superior performance compared to existing GP-MIL approaches in binary ICH classification tasks. Furthermore, it exhibits robust performance in sub-label categorization, underscoring its effectiveness and generalizability.

2   Methodology

We propose a framework based on the two-stage approach of Wu et al. [18]. The first stage employs the Att-CNN (from [18]) for feature extraction, which

combines CNN and Att-MIL. The second stage integrates probabilistic aggregation with attention-enhanced GPs, allowing for selective focus on relevant data
points. A visual overview of our pipeline is presented in Fig. 1.

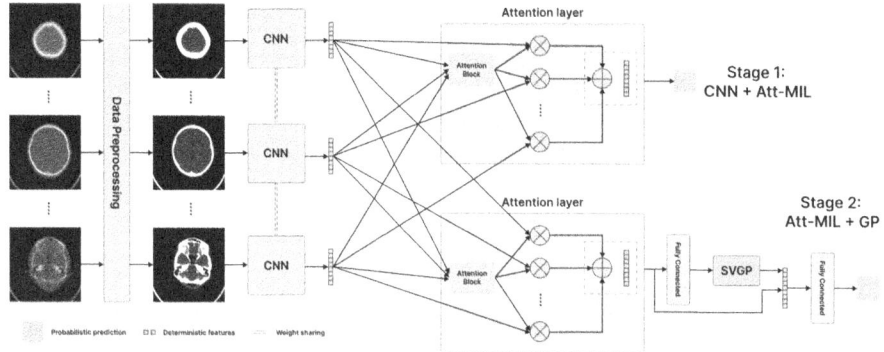

Fig. 1. Our AttMIL-SVGP architecture is a two-stage process. Stage 1 employs a CNN
and Att-MIL for feature extraction. Stage 2 distinguishes itself by integrating SVGP
with Att-MIL for probabilistic classification.

2.1 Stage 1: Feature Extraction (Att-CNN)

MIL is a machine learning framework designed for binary classification, where
each scan (referred to as a bag) consists of multiple CT slices (termed instances),
and the task is to predict labels for entire bags based on their constituent
instances. Formally, given a dataset of bags $\{B_1, B_2, \ldots, B_N\}$ with labels
$y_i \in \{0,1\}$, each bag B_i contains instances $\{x_{1i}, x_{2i}, \ldots, x_{Ni}\}$. The framework employs a CNN to process each instance x_{ji} independently, generating
a feature vector defined as $f(x_{ji}) = \text{CNN}(x_{ji})$, where $f(x_{ji}) \in \mathbb{R}^{M \times 1}$. As
detailed in the first paragraph of Sect. 2, our MIL framework employs a CNN
to extract features from individual CT slices and an adaptive attention mechanism to aggregate these into a discriminative bag-level representation. The CNN
architecture comprises six convolutional blocks, each containing a convolutional
layer and a max-pooling layer to extract discriminative features from CT slices,
followed by a flatten layer and a fully connected layer that reduces feature
dimensionality before passing them to an attention layer. To synthesize bag-
level representations, a MIL pooling function aggregates all instance features:
$g(B_i) = \text{MIL\_pooling}(\{f(x_{1i}), f(x_{2i}), \ldots, f(x_{Ni})\})$.

In traditional MIL approaches, max and mean pooling aggregate instance
embeddings into a bag-level representation. Max-pooling follows the MIL
assumption but ignores inter-slice dependencies, limiting contextual reasoning.
Mean pooling treats all slices equally, diluting rare ICH-positive signals with
noise. This trade-off sacrifices either contextual information or discriminative
features. MIL frameworks typically employ two distinct approaches for feature

aggregation: instance-level and embedding-level. On the one hand, the instance-level approach applies classifiers directly to instances and aggregates predictions to determine the bag label. This is the approach of Additive MIL (Javed et al. [10]), which allows for direct interpretation of each instance's contribution to the final prediction. On the other hand, the embedding-level approach first maps instances to feature vectors and then aggregates these representations into a bag-level embedding before classification, as implemented by Ilse et al. [9]. In our research, we implement the latter methodology, which more adeptly retains contextual relations among instances while enabling attention-based weighting to focus on key distinguishing features. The attention weight a_j for the j-th instance in bag B_i is computed as:

$$a_j = \frac{\exp\left\{\mathbf{w}^\top \tanh\left(\mathbf{V}f(\mathbf{x}_{ji})^\top\right)\right\}}{\sum_{m=1}^{N_i} \exp\left\{\mathbf{w}^\top \tanh\left(\mathbf{V}f(\mathbf{x}_{mi})^\top\right)\right\}} \tag{1}$$

where $f(\mathbf{x}_{ji}) \in \mathbb{R}^{M \times 1}$ is the M-dimensional feature vector of the j-th instance in bag B_i, $\mathbf{V} \in \mathbb{R}^{L \times M}$ and $\mathbf{w} \in \mathbb{R}^{L \times 1}$ are learnable parameters, and $\tanh(\cdot)$ introduces nonlinearity that generates both positive and negative values in the gradient flow. The attention mechanism projects each instance embedding $f(\mathbf{x}_{ji})$ into an L-dimensional space via $\mathbf{V}$, computes scalar relevance scores through $\mathbf{w}$, and applies softmax normalization to ensure $\sum_{j=1}^{N_i} a_j = 1$. The higher the attention weight a_j, the greater the probability that the corresponding CT slice contains features indicative of ICH for the patient.

2.2   Stage 2: Attention-Based MIL with GPs (AttMIL-SVGP)

In Stage 2, we retain the attention mechanism described in Stage 1, apply it once again to generate weighted bag-level representations $g(B_i)$ from instance features. These attention-weighted representations then serve as direct inputs to our SVGP framework. As introduced by Hensman et al. [8], the SVGP methodology provides a structured approach to variational approximation that substantially lowers computational demands while preserving the statistical integrity of the model. Standard GP inference requires $\mathcal{O}(N^3)$ time complexity, making it prohibitive for large datasets. The SVGP framework reduces this to $\mathcal{O}(NM^2 + M^3)$, where M is the number of inducing points and typically $M \ll N$, enabling scalable GP classification even for millions of data points. Our approach consists of the following steps:

Prior Distribution Over Inducing Variables. We begin by introducing a set of M inducing points $\mathbf{Z} = \{\mathbf{z}_1, \ldots, \mathbf{z}_M\} \in \mathbb{R}^{M \times D}$ and their corresponding function values $\mathbf{u} \in \mathbb{R}^M$. The prior distribution over these inducing variables is defined as:

$$p(\mathbf{u}) = \mathcal{N}(\mathbf{u} \mid \mathbf{0}, \mathbf{K}_{MM}) \tag{2}$$

where $\mathbf{K}_{MM} \in \mathbb{R}^{M \times M}$ is the kernel matrix computed between all pairs of inducing points.

Conditional Distribution of Function Values. Next, we establish how the function values $\mathbf{f} \in \mathbb{R}^N$ at observed data points depend on the inducing variables through the conditional distribution:

$$p(\mathbf{f} \mid \mathbf{u}) = \mathcal{N}(\mathbf{f} \mid \mathbf{K}_{NM}\mathbf{K}_{MM}^{-1}\mathbf{u}, \mathbf{K}_{NN} - \mathbf{Q}_{NN}) \tag{3}$$

where $\mathbf{K}_{NM} \in \mathbb{R}^{N \times M}$ is the cross-covariance matrix between observed data points and inducing points, $\mathbf{K}_{NN} \in \mathbb{R}^{N \times N}$ is the covariance matrix for observed points, and $\mathbf{Q}_{NN} = \mathbf{K}_{NM}\mathbf{K}_{MM}^{-1}\mathbf{K}_{MN}$ represents the Nyström approximation of $\mathbf{K}_{NN}$.

Variational Approximation to the Posterior. Instead of working with the true posterior $p(\mathbf{u} \mid \mathbf{y})$, which is intractable for classification problems, we approximate it with a variational distribution:

$$q(\mathbf{u}) = \mathcal{N}(\mathbf{u} \mid \mathbf{m}, \mathbf{S}) \tag{4}$$

where $\mathbf{m} \in \mathbb{R}^M$ and $\mathbf{S} \in \mathbb{R}^{M \times M}$ are learnable variational parameters that are optimized during training.

Approximate Posterior Over Function Values. Finally, we derive the approximate posterior over function values by integrating out the inducing variables:

$$q(\mathbf{f}) = \int p(\mathbf{f} \mid \mathbf{u})q(\mathbf{u})d\mathbf{u} = \mathcal{N}(\mathbf{f} \mid \boldsymbol{\mu}_f, \boldsymbol{\Sigma}_f) \tag{5}$$

where:

$$\boldsymbol{\mu}_f = \mathbf{K}_{NM}\mathbf{K}_{MM}^{-1}\mathbf{m} \tag{6}$$

$$\boldsymbol{\Sigma}_f = \mathbf{K}_{NN} - \mathbf{K}_{NM}\mathbf{K}_{MM}^{-1}(\mathbf{K}_{MM} - \mathbf{S})\mathbf{K}_{MM}^{-1}\mathbf{K}_{MN} \tag{7}$$

This integration is analytically tractable due to the conjugacy between the conditional distribution and the Gaussian variational distribution.

Handling Non-Gaussian Likelihoods with Pólya-Gamma augmentation. A fundamental challenge in GP classification is handling the non-Gaussian Bernoulli likelihood:

$$p(y_i = 1|\mathbf{f}_i) = \sigma(\mathbf{f}_i) = \frac{1}{1 + \exp(-\mathbf{f}_i)} \tag{8}$$

This non-conjugate likelihood makes posterior inference computationally challenging. Pólya-Gamma augmentation addresses this by introducing auxiliary variables that transform the problem into a conditionally Gaussian one. The key insight is that by augmenting our model with Pólya-Gamma random variables, we can represent the logistic likelihood in a form that enables efficient closed-form updates. Unlike methods that rely on numerical quadrature or sampling, this approach provides analytical expressions for optimization, significantly improving computational efficiency. In practical terms, the augmentation acts as a mathematical bridge that allows us to work with the non-Gaussian likelihood using familiar Gaussian inference tools, making the entire inference process more stable and efficient.

Combined Loss Function Approach. To enhance classification performance, we propose a combined loss function that integrates the ELBO with binary cross-entropy (BCE):

$$\mathcal{L} = \alpha \cdot \mathcal{L}_{\mathrm{ELBO}} + (1 - \alpha) \cdot \mathcal{L}_{\mathrm{BCE}} \tag{9}$$

where α is a weighting parameter and $\mathcal{L}_{\mathrm{BCE}}$ is the standard cross-entropy loss used in classification tasks:

$$\mathcal{L}_{\mathrm{BCE}} = \frac{1}{N} \sum_{i=1}^{N} \sum_{c=1}^{C} [y_{ic} \log \sigma(p_{ic}) + (1 - y_{ic}) \log(1 - \sigma(p_{ic}))] \tag{10}$$

$$\mathcal{L}_{\mathrm{ELBO}} = \mathbb{E}_{q(\mathbf{f})}[\log p(\mathbf{y}|\mathbf{f})] - \mathrm{KL}[q(\mathbf{u})\|p(\mathbf{u})] \tag{11}$$

where y_{ic} represents the label for sample i and class c, and f_{ic} is the corresponding model output.

2.3   Data Redundancy Removal

Previous methodologies as articulated by Wu et al. [18] and López Pérez et al. [12] have been conducted utilizing complete image slices derived from a scan. This approach is encumbered by issues stemming from superfluous data, which engenders noise and subsequently impairs the efficacy of the attention model. Fu et al. [7] demonstrated that redundancy removal significantly improved model performance (from 0.6702 to 0.9071 in F1 score), particularly for attention-based architectures where excessive slices disrupt the attention mechanism's ability to focus on diagnostically relevant information.

Building on their findings, we implemented a systematic approach to data reduction that maintains diagnostic integrity while eliminating redundancy. The 5 mm slice thickness standard was selected as it represents the clinical protocol commonly used for ICH diagnosis [7]. Our implementation follows these steps:

1. **Slice sorting:** DICOM images are initially arranged in ascending order based on their Z-axis coordinates to establish proper anatomical alignment across the volumetric sequence.
2. **Slice spacing calculation:** The original slice spacing is calculated from the differences in Z-coordinates of consecutive slices.

3. **Resampling:** Volumetric studies are systematically resampled to a uniform 5 mm slice spacing by retaining every n^{th} slice, where n is determined by the ratio of the target spacing to the mean original spacing.
4. **Slice number control:** To ensure dataset uniformity, studies are capped at 28 slices (RSNA) with thresholds empirically aligned to each multi-center cohort's imaging protocols and slice-distribution characteristics.

Our methodology confronts the issue of heterogeneity in multi-center data, characterized by variability in scanning protocols (ranging from 3 to 5 mm slice thickness), while concurrently mitigating the computational demands and possible degradation in performance that may arise from excessive padding in MIL frameworks. Through the standardization of both slice thickness and quantity, we preserve diagnostic integrity while enhancing computational efficiency and the efficacy of the model.

3   Experiments

3.1   Dataset and Preprocessing

Our study inherits the experimental framework from Stage 1 of the architecture proposed by Wu et al. [18], leveraging their preprocessed datasets and extending the analysis through Stage 2. The training and validation sets consist of 1,150 head CT scans sourced from the RSNA challenge dataset, comprising 411 hemorrhage-positive (P-scans) and 589 negative scans (N-scans). For testing, 150 RSNA scans (72 P-scans, 78 N-scans) were retained. The distribution of ICH types within the dataset is detailed in Table 1.

To enhance the visualization of relevant anatomical features in CT brain images, a windowing technique was applied, as described in the research of [12]. This approach involves applying three different window settings to each CT slice to highlight brain, blood, and soft tissue, respectively. The window settings used were [W:80, C:40], [W:200, C:80], and [W:380, C:40], where W represents the window width and C represents the window center in Hounsfield Units (HU) [12]. These settings allow for the optimization of image interpretation by mimicking the practice of radiologists adjusting window widths and centers to focus on specific structures. The resulting windowed images were concatenated into three channels and normalized to the range $[0, 1]$ to facilitate further processing [12].

Table 1. Distribution of ICH sub-labels in the RSNA challenge dataset

Sub-label of ICH	N-scans (0)	P-scans (1)
Any	667	483
Subdural	951	199
Epidural	1130	20
Intraparenchymal	863	287
Intraventricular	950	200
Subarachnoid	953	197

3.2   Experimental Design

Binary classification performance was measured using AUC-ROC, F1 score, Accuracy, and Cohen's Kappa. AUC-ROC served as the primary metric, and F1 score captured error balance. Adam optimizer was selected over NGD for faster convergence. Results were averaged over five independent runs for statistical robustness. Experiments were conducted on an NVIDIA GeForce RTX 4070 Super GPU, though the lightweight second-phase model allowed smooth execution on a CPU. Training a binary classification model took ~ 50 s per run, while sub-labels classification (six labels) averaged 290 s.

3.3   Results and Discussions

We proposed two models, AttMIL-SVGP and R-AttMIL-SVGP, where the latter incorporates a redundancy technique, as described in Sect. 2.3. The results demonstrate that R-AttMIL-SVGP outperforms all baseline methods on the RSNA challenge dataset of all metrics, see Table 3 for more details. Additionally, Table 4 compares our method with other approaches across different datasets. Despite using a smaller dataset (1,150 scans), R-AttMIL-SVGP achieves the highest Bag AUC, surpassing methods like Jnawali et al. [11] and Titano et al. [14], which rely on larger datasets and 3D CNNs.

Data Redundancy Improvement. Our proposed data redundancy method demonstrates significant performance improvements in ICH classification, with R-AttMIL-SVGP achieving superior metrics (AUC-PR: 0.975 ± 0.003) compared to baseline AttMIL-SVGP (Table 3). We hypothesize that standardizing variable-length multi-center scans through adaptive sequence alignment enables more effective attention mechanism operation, where the model selectively focuses on diagnostically relevant slices while attenuating protocol-specific noise. This approach not only enhances model accuracy, but also reduces computational memory

Table 2. Performance metrics for various hyperparameter settings of redundant data for AttMIL-SVGP model. The first column lists the hyperparameter combinations, defined by the scaling factor (α) in Eq. 9 and the number of inducing points in the GP

Configuration	Accuracy	F1	ROC-AUC	AUC-PR
0.1, 50	0.911 ± 0.013	0.910 ± 0.013	0.971 ± 0.004	0.972 ± 0.004
0.5, 50	0.913 ± 0.022	0.913 ± 0.022	0.972 ± 0.004	0.974 ± 0.003
0.9, 50	0.913 ± 0.022	0.913 ± 0.022	0.957 ± 0.019	0.967 ± 0.009
1.0, 50	0.904 ± 0.014	0.904 ± 0.014	0.965 ± 0.008	0.959 ± 0.016
0.1, 100	0.911 ± 0.013	0.910 ± 0.013	0.971 ± 0.004	0.972 ± 0.004
0.5, 100	0.915 ± 0.012	0.914 ± 0.012	0.972 ± 0.005	0.974 ± 0.004
0.9, 100	0.913 ± 0.013	0.913 ± 0.013	0.971 ± 0.008	0.969 ± 0.015
1.0, 100	0.899 ± 0.024	0.898 ± 0.024	0.965 ± 0.015	0.964 ± 0.019
0.1, 150	0.792 ± 0.240	0.787 ± 0.251	0.829 ± 0.286	0.846 ± 0.255
0.5, 150	0.913 ± 0.018	0.913 ± 0.018	0.973 ± 0.004	0.975 ± 0.004
0.9, 150	0.919 ± 0.015	0.918 ± 0.016	$\mathbf{0.974 \pm 0.004}$	0.975 ± 0.003
1.0, 150	$\mathbf{0.920 \pm 0.015}$	$\mathbf{0.920 \pm 0.015}$	0.974 ± 0.004	$\mathbf{0.976 \pm 0.003}$

Table 3. Binary classification results on RSNA challenge dataset

Model	Accuracy	F1 score	Cohen's kappa	AUC	AUC-PR
Att-MIL	0.781 ± 0.023	0.811 ± 0.017	0.569 ± 0.045	0.951 ± 0.011	0.841 ± 0.013
VGPMIL [18]	0.780 ± 0.089	0.814 ± 0.059	0.567 ± 0.172	0.964 ± 0.006	0.846 ± 0.043
DGPMIL [12]	0.825 ± 0.006	0.839 ± 0.006	0.654 ± 0.011	0.957 ± 0.011	0.961 ± 0.011
E2E-Att-GP [13]	0.876 ± 0.023	0.886 ± 0.011	N/A	0.965 ± 0.007	N/A
G-VGPMIL [5]	0.891 ± 0.017	0.886 ± 0.015	N/A	0.960 ± 0.008	N/A
AttMIL-SVGP	0.912 ± 0.014	0.904 ± 0.015	0.823 ± 0.027	0.960 ± 0.003	0.967 ± 0.003
R-AttMIL-SVGP	**0.919 ± 0.015**	**0.918 ± 0.016**	**0.836 ± 0.031**	**0.974 ± 0.004**	**0.975 ± 0.003**

Table 4. Comparisons with other methods with different dataset

Source	Dataset size	Labeling type	Method	Bag AUC
Jnawali [11]	40,357 scans	Scan	3D CNNs	0.87
Titano [14]	37,236 scans	Scan	3D CNNs	0.88
Arbabshirani [1]	45,583 scans	Scan	3D CNNs	0.85
VGPMIL [18]	1,150 scans	Scan	MIL	0.964
DGPMIL2 [12]	1,150 scans	Scan	MIL	0.957
E2E-Att-GP [13]	1,150 scans	Scan	MIL	0.965
G-VGPMIL [5]	1,150 scans	Scan	MIL	0.966
R-AttMIL-SVGP (Ours)	1,150 scans	Scan	MIL	**0.974**

Table 5. Results of redundant data for AttMIL-SVGP Model for ICH sub-labels

Hemorrhage type	Accuracy	Precision	Recall	F1 score
Any	0.912 ± 0.020	0.914 ± 0.020	0.912 ± 0.020	0.912 ± 0.020
Epidural	0.987 ± 0.000	0.987 ± 0.000	0.987 ± 0.000	0.980 ± 0.000
Intraparenchymal	0.847 ± 0.011	0.851 ± 0.010	0.847 ± 0.011	0.848 ± 0.011
Intraventricular	0.884 ± 0.006	0.892 ± 0.010	0.884 ± 0.006	0.887 ± 0.007
Subarachnoid	0.801 ± 0.003	0.814 ± 0.035	0.801 ± 0.003	0.739 ± 0.034
Subdural	0.800 ± 0.005	0.813 ± 0.035	0.800 ± 0.005	0.739 ± 0.036
Overall	0.872 ± 0.004	0.878 ± 0.011	0.872 ± 0.004	0.851 ± 0.007

requirements compared to conventional padding or truncation strategies, making it clinically practical for heterogeneous real-world imaging datasets while maintaining strict medical reliability standards.

Model Performance and Hyperparameter Trade-Offs. Our analysis of the scaling factor α (governing ELBO loss contribution) and inducing point count reveals critical performance relationships in Table 2. First, extreme α values demonstrate instability: low $\alpha = 0.1$ configurations suffer from insufficient GP regularization (Accuracy: 0.792 ± 0.240), while high $\alpha = 0.9$ setups exhibit over-regularization (AUC-PR: 0.967 ± 0.009 vs 0.976 optimal). Second, while $\alpha = 1.0$ achieves peak performance with 150 inducing points (AUC-PR: 0.976 ± 0.003), it shows significant variance at lower inducing counts (ΔAUC-PR:

−1.7% at 50 points vs $\alpha = 0.5$). These results confirm that maintaining cross-entropy influence ($\alpha < 1$) provides essential regularization, particularly under computational constraints, while sufficient inducing points (≥ 150) enable full exploitation of GP expressivity when emphasizing ELBO objectives.

Comparisons with the State of the Art. Our R-AttMIL-SVGP framework achieves state-of-the-art performance on RSNA challenge dataset (AUC-PR: 0.975 ± 0.003) in Table 3. While end-to-end architectures like E2E-Att-GP show marginal AUC advantages, our phased training offers superior clinical utility through explicit attention quantification and higher Cohen's kappa. Notably, our framework demonstrates remarkable data efficiency, achieving superior performance with 97% fewer training scans than 3D CNN baselines [14]. This efficiency is crucial for preserving inter-slice relationships, which are vital for accurate diagnosis in medical imaging. Our approach balances accuracy and interpretability, providing more confident and reliable predictions essential in medical classification.

Adaptive Architecture for Comprehensive ICH Assessment. Our modular architecture maintains high binary detection accuracy in hemorrhage sub-label analysis, shown by the "Any" category's strong performance (Accuracy: 0.912 ± 0.020, F1 score: 0.912 ± 0.020) in Table 5. This structure enables the core detector and sub-label classifiers to operate simultaneously without performance loss, which is crucial for reliable detection alongside detailed diagnostics. The design allows for selective activation of sub-label branches, ensuring computational efficiency while preserving diagnostic reliability. The stable metrics in the "Any" category indicate possibilities for future developments in multi-label classification through cross-branch attention mechanisms.

3.4   Visualization and Clinical Interpretability

Our framework employs Att-MIL to improve interpretability and diagnostic assistance. The attention weights emphasize high-risk slices, directing radiologists to essential areas. In Fig. 2, for negative instances, the weights are assigned a value of zero, whereas positive instances range from 0 to 100%, reflecting the likelihood of abnormalities in accordance with clinical reasoning.

Figures 3 shows that, in contrast to a few larger distributions of positive slices, an attention-based method could identify blank and negative slices with a lower weight distribution. By concentrating primarily on important or highly weighted slices of the attention mechanism, the bias issue with the model prediction produced by an imbalanced distribution of positive and negative slices in the context of ICH scans can be resolved.

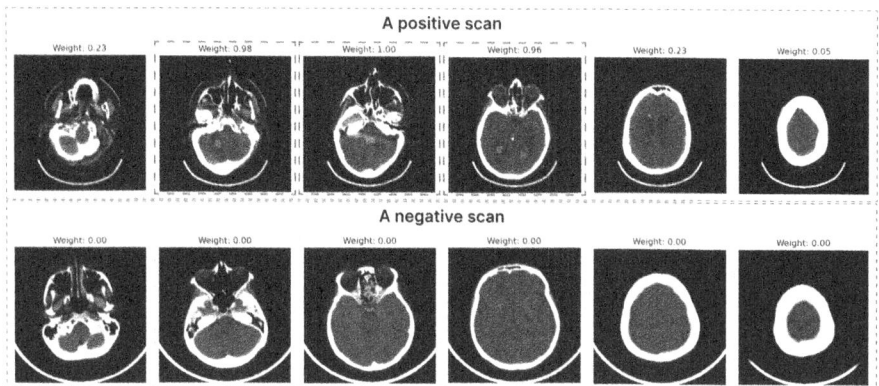

Fig. 2. Comparison between a positive and a negative scan. The stage 2 attention weights assigned by the model are displayed above each slice. Higher ones (highlighted with red borders) indicate where the model concentrates its attention, mimicking radiologists' emphasis on areas most likely to exhibit ICH features. (Color figure online)

In conjunction with elevating MIL performance, our attention mechanism presents a considerable benefit compared to traditional slice-level annotation methods, which require extensive manual labeling - an unfeasible undertaking in medical imaging due to time pressures and resource constraints. By enabling training with just a single scan-level label, our strategy significantly alleviates the challenges of dataset creation while preserving high diagnostic precision.

Moreover, our attention weights effectively capture the dependencies between adjacent slices. In practical radiology settings, specialists do not evaluate slices in isolation but analyze them within context, frequently revisiting surrounding slices to validate abnormalities. Our model inherently mimics this behavior by assigning related attention scores to neighboring slices, thereby reinforcing diagnostic reliability.

Figure 3 demonstrates the significant impact of our data redundancy reduction methods. By reducing the number of padding blank slices from 3,296 to just 13, and noisy real slices from 5,254 to 4,187, we achieve substantial computational efficiency. This efficiency comes at a minimal cost: only 3 out of 1,150 scan-level labels (approximately 0.3%), representing scans with 1–2 positive slices, were excluded. These exclusions occurred under complex, hard-to-detect conditions, requiring detailed expert review regardless. The trade-off is highly advantageous: a 99.6% reduction in blank slices and a 20.3% reduction in noisy real slices, with only a 0.3% loss of borderline cases. In essence, our data redundancy strategies offer a compelling balance, significantly enhancing computational efficiency while preserving crucial data integrity for practical medical applications.

(a) Original dataset

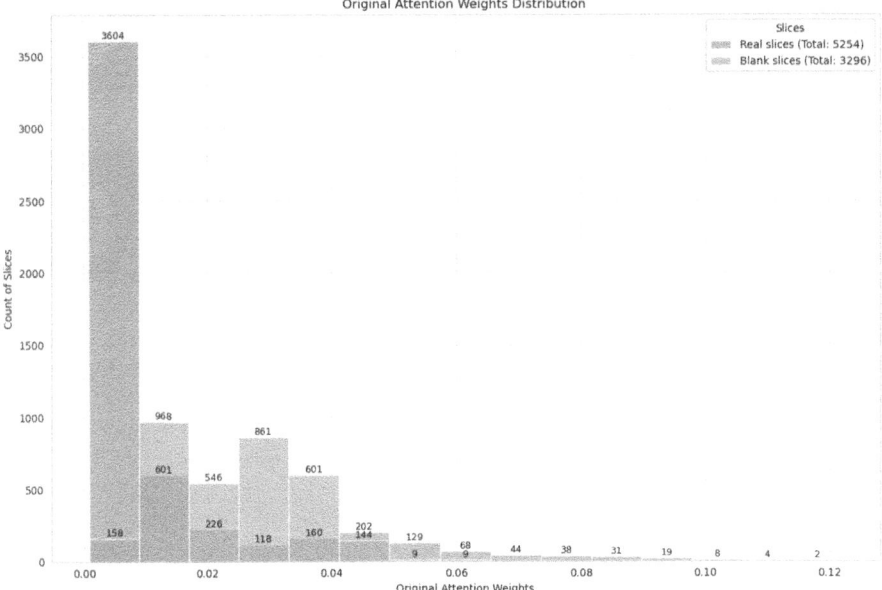

(b) Redundant dataset

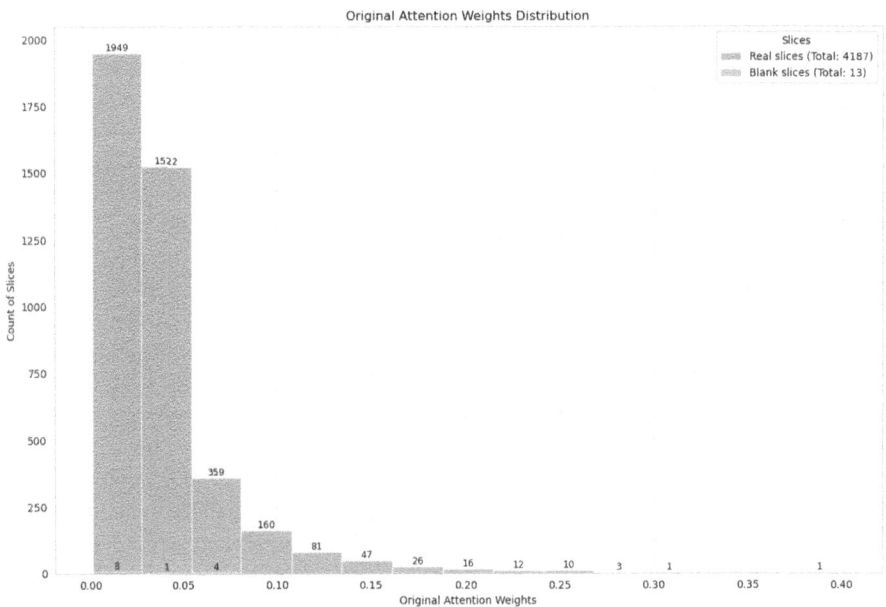

Fig. 3. Redundant dataset

4   Conclusions

This paper introduces a novel weakly supervised DL framework that effectively addresses the challenges of automated ICH detection. By integrating Att-MIL with GPs, our approach not only mimics clinical radiological practices but also provides probabilistic diagnostic results that enhance interpretability. We also implemented a data redundancy reduction technique to combat data heterogeneity. Our proposed method demonstrates superior performance compared to existing GP-MIL approaches in binary ICH classification tasks, and exhibits robust performance in sub-label categorization, underscoring its effectiveness and generalizability.

Future work will primarily focus on extending the capabilities of our proposed framework to address the more complex multi-label ICH classification problem. This involves developing a robust weakly supervised learning algorithm capable of accurately identifying and classifying multiple ICH sub-labels within a single CT scan, even with limited labeled data. Furthermore, we aim to evaluate the algorithm's performance on larger and more diverse datasets, ensuring its scalability and clinical applicability.

Acknowledgments. This research is funded by University of Science, VNU-HCM under grant number CNTT 2024-23.

References

1. Arbabshirani, M.R., et al.: Advanced machine learning in action: identification of intracranial hemorrhage on computed tomography scans of the head with clinical workflow integration. NPJ Dig. Med. **1**(1), 9 (2018)
2. Asif, M., et al.: Intracranial hemorrhage detection using parallel deep convolutional models and boosting mechanism. Diagnostics **13**(4), 652 (2023)
3. Barhoumi, Y., Bouaynaya, N.C., Rasool, G.: Efficient Scopeformer: toward scalable and rich feature extraction for intracranial hemorrhage detection. IEEE Access **11**, 81656–81671 (2023)
4. Burduja, M., Ionescu, R.T., Verga, N.: Accurate and efficient intracranial hemorrhage detection and subtype classification in 3D CT scans with convolutional and long short-term memory neural networks. Sensors **20**(19), 5611 (2020)
5. Castro-Macías, F.M., Morales-Álvarez, P., Wu, Y., Molina, R., Katsaggelos, A.K.: Hyperbolic Secant representation of the logistic function: application to probabilistic multiple instance learning for CT intracranial hemorrhage detection. Artif. Intell. **331**, 104115 (2024)
6. Flanders, A.E., et al.: Construction of a machine learning dataset through collaboration: the RSNA 2019 brain CT hemorrhage challenge. Radiol. Artif. Intell. **2**(3), e190211 (2020)
7. Fu, G., Li, J., Wang, R., Ma, Y., Chen, Y.: Attention-based full slice brain CT image diagnosis with explanations. Neurocomputing **452**, 263–274 (2021)
8. Hensman, J., Matthews, A., Ghahramani, Z.: Scalable variational Gaussian process classification. In: Artificial Intelligence and Statistics, pp. 351–360. PMLR (2015)

9. Ilse, M., Tomczak, J., Welling, M.: Attention-based deep multiple instance learning. In: International Conference on Machine Learning, pp. 2127–2136. PMLR (2018)

10. Javed, S.A., Juyal, D., Padigela, H., Taylor-Weiner, A., Yu, L., Prakash, A.: Additive MIL: intrinsically interpretable multiple instance learning for pathology. Adv. Neural. Inf. Process. Syst. **35**, 20689–20702 (2022)

11. Jnawali, K., Arbabshirani, M.R., Rao, N., Patel, A.A.: Deep 3D convolution neural network for CT brain hemorrhage classification. In: Medical Imaging 2018: Computer-Aided Diagnosis, vol. 10575, pp. 307–313. SPIE (2018)

12. López-Pérez, M., Schmidt, A., Wu, Y., Molina, R., Katsaggelos, A.K.: Deep Gaussian processes for multiple instance learning: application to CT intracranial hemorrhage detection. Comput. Methods Prog. Biomed. **219**, 106783 (2022)

13. Pérez-Cano, J., et al.: An end-to-end approach to combine attention feature extraction and Gaussian Process models for deep multiple instance learning in CT hemorrhage detection. Expert Syst. Appl. **240**, 122296 (2024)

14. Titano, J.J., et al.: Automated deep-neural-network surveillance of cranial images for acute neurologic events. Nat. Med. **24**(9), 1337–1341 (2018)

15. Vos, T., et al.: Global burden of 369 diseases and injuries in 204 countries and territories, 1990–2019: a systematic analysis for the Global Burden of Disease Study 2019. Lancet **396**(10258), 1204–1222 (2020)

16. Wang, X., et al.: A deep learning algorithm for automatic detection and classification of acute intracranial hemorrhages in head CT scans. NeuroImage: Clin. **32**, 102785 (2021)

17. Wolf, D., et al.: Self-supervised pre-training with contrastive and masked autoencoder methods for dealing with small datasets in deep learning for medical imaging. Sci. Rep. **13**(1), 20260 (2023)

18. Wu, Y., Schmidt, A., Hernández-Sánchez, E., Molina, R., Katsaggelos, A.K.: Combining attention-based multiple instance learning and gaussian processes for CT hemorrhage detection. In: de Bruijne, M., et al. (eds.) MICCAI 2021. LNCS, vol. 12902, pp. 582–591. Springer, Cham (2021). https://doi.org/10.1007/978-3-030-87196-3_54

Applications of Deep Learning and Soft Computing

A User-Oriented Perspective on Soft Clustering: Explainability and Uncertainty Quantification

Andrea Campagner[1][(✉)], Federico Cabitza[1,2], and Davide Ciucci[2]

[1] IRCCS Istituto Ortopedico Galeazzi, Milan, Italy
andrea.campagner@unimib.it
[2] MUDIlab, University of Milano-Bicocca, Milan, Italy

Abstract. Soft clustering refers to clustering analysis methods that not only assign instances to clusters, but also provide indication about the uncertainty in cluster assignments. In this article we focus on the interplay between soft clustering and explainable AI, by adopting a user-oriented perspective aimed at comparing different (soft) clustering approaches in terms of their differing effectiveness in conveying uncertainty to users. To this aim, we designed a simulated, but realistic, medical decision-making problem in which users had to take a clinically relevant decision with the support of a clustering algorithm, and analyzed differences in the ability of different methods to convey uncertainty as well as in how users perceived their usefulness and clarity. Our results and statistical analysis providing initial, but suggestive, empirical evidence towards the differing capabilities of soft clustering approaches to convey uncertainty.

Keywords: Clustering · soft clustering · eXplainable AI · uncertainty quantification

1 Introduction

Clustering analysis, that is the sub-field of data science aimed at discovering meaningful groups in data, remains to date one of the most important tasks in data analysis [41]. For this purpose, several algorithms have been developed, from classical k-means and hierarchical clustering to more modern approaches such as HDBSCAN [29] or DP-Means [15]), and applied to a variety of settings, from life sciences [46] to finance [27] and physics [19].

Soft clustering [18], in particular, has attracted particular interest in recent years, due to the possibility of supplementing traditional clustering analysis with an indication of the uncertainty in cluster assignment. Several approaches have been discussed, including generalization of traditional clustering methods as well as original ones, each of which adopts different ways to represent, quantify, and communicate uncertainty: examples include rough clustering [26], fuzzy clustering [40] and probabilistic clustering [21]. Furthermore, various articles have tried investigating the relative advantages of different uncertainty quantification

© The Author(s), under exclusive license to Springer Nature Switzerland AG 2025
Q. Zhang et al. (Eds.): IJCRS 2025, LNAI 15710, pp. 289–300, 2025.
https://doi.org/10.1007/978-3-031-92741-6_21

approaches [7,8,13,23], compared to state-of-the-art hard clustering methods, adopting a perspective centered on machine learning performance validation.

In this article, in contrast, we focus on a different perspective, more closely aligned with the aims and tenets of explainable AI [16] (XAI). XAI emerged in the recent years as a way to address the increasing complexity of state-of-the-art artificial intelligence methods (such as those based on deep learning) by emphasizing the need to make AI systems and their outputs interpretable and more easily understandable to human users [38]. Current research in the XAI field has mostly focused on the technical perspective, that is, the proposal of algorithms to improve the interpretability of AI methods [16], as well as metrics to evaluate interpretability [32]. Nonetheless, increasing interest has been devoted to the investigation of the user-oriented perspective which, grounding on human-computer interaction [37], aims at understanding the user perception of XAI, as well as the impact this latter may have on human decision-making [4,42,44]. Under this lens, uncertainty quantification and communication play a particularly important role as explainability approaches [43,48], as a way to help users calibrate their trust in explanations and AI support, and thus foster more risk-aware decision-making [9].

Despite recent increasing interest in research issues at the intersection of clustering analysis and XAI [1,35], only limited work has focused on the user-oriented perspective mentioned above [17] and, to our knowledge, no previous work has compared different (soft) clustering approaches in terms of their differing effectiveness in conveying uncertainty or perceived usefulness.

In this work, we provide a first step toward bridging this gap, by presenting the results of a pilot study aimed at understanding differences, from a user-centered perspective, among some of the most commonly studied soft clustering approaches. To this aim, we designed a simulated, but realistic, medical decision-making problem in which users had to take a clinically relevant decision with the support of a (soft) clustering algorithm, and analyzed differences in the ability of different methods to convey uncertainty, as well as in how users perceived their usefulness and clarity.

2   Background and Methods

2.1   Background on Clustering

Clustering (also called cluster analysis) refers to unsupervised machine learning methods that aim to automatically discover groups in data [46]. Let $\mathcal{X}$ be a set of objects or instances, and $\mathcal{C}$ a set of clusters. A *hard clustering* C is a function $C : \mathcal{X} \rightarrow \mathcal{C}$ that assigns objects to clusters. Note that in a hard clustering, cluster assignment is unique: that is, each object is assigned to exactly one cluster; formally, $\forall x \in \mathcal{X}, \exists! c \in \mathcal{C}$ s.t. $C(x) = c$. Several algorithms exist to automatically discover a clustering function from the data: as in this article we will be mainly concerned with the abstract definition of a clustering, with no reference to specific algorithms or techniques, we invite the interested reader to consult the recent review by Saxena et al. [41].

In contrast to hard clustering, *soft clustering* denotes a varied family of clustering methods and representation formalisms that relax the single-cluster assignment assumption as a way to represent uncertainty in the clustering assignment [18]. Several formalisms have been proposed to model such uncertain cluster assignments, among them *rough clustering* [26], probabilistic clustering (also called *model-based clustering* [21]), *fuzzy clustering* [40], *possibilistic clustering* [47] and evidential clustering [12]; each of these approaches relies on a different uncertainty representation theory to model uncertainty in clustering results. In this article, we focus on rough, fuzzy, and possibilistic clustering: we decided to exclude probabilistic clustering, as its representation is equivalent to the most commonly adopted representation for fuzzy clustering (see below), as well as evidential clustering, since this formalism is inherently more difficult to represent and interpret [12], as it requires specifying mass function values for all possible elements of the powerset of $\mathcal{C}$.

A rough clustering is a function $R : \mathcal{X} \rightarrow 2^{\mathcal{C}}$: that is, each object is assigned to a set of clusters. Given an object $x \in \mathcal{X}$, if $|R(x)| > 1$, this means that the cluster to which x belongs is unknown but likely to be among those reported in $R(x)$. An alternative, but equivalent, representation of rough clustering is based on associating with each cluster $c \in \mathcal{C}$ two sub-clusters, denoted as $Core(c)$ and $Fringe(c)$, under the following assumptions: (i) $\forall x \in \mathcal{X}, (\exists c \in \mathcal{C}, x \in Core(c) \implies \nexists c' \neq c \in \mathcal{C}, x \in Core(c') \cup Fringe(c'))$, (ii) $\forall x \in \mathcal{X}, (\exists c \in \mathcal{C}, x \in Fringe(c) \implies \exists c' \neq c \in \mathcal{C}, x \in Fringe(c'))$. Intuitively, $Core(c)$ contains all objects that are definitely known to belong to cluster c, while $Fringe(c)$ contains objects that may (or may not) belong to cluster c. Clearly, the two representations are equivalent to each other [6].

Rough clustering is strongly related to another soft clustering formalism called *three-way clustering* [45]: the main difference between rough clustering and three-way clustering regards the fact that the latter relaxes assumption (ii) above (that is, an object can belong to the Fringe region of a single cluster). Nonetheless, the two representations are isomorphic, in the sense that each rough clustering corresponds to a unique three-way clustering, while each three-way clustering can be associated with a canonical rough clustering [6]. Thus, in this article, we will not distinguish between rough and three-way clustering, and generally refer to these soft clustering formalisms as *rough clustering*.

A *fuzzy clustering* is a function $F : \mathcal{X} \rightarrow [0, 1]^{\mathcal{C}}$ s.t. $\forall x \in \mathcal{X}, \sum_{c \in \mathcal{C}} F(x)_c = 1$. That is, a fuzzy clustering assigns to each object x and cluster c pair a number in $[0, 1]$, which is interpreted as a *degree of membership* of x in c. Intuitively, this represents a degree of uncertainty with respect to the object belonging to the clusters: the higher the degree of membership of x in c, the more plausible it is that c is the true clustering to which x belongs. As the degree of memberships are required to sum to 1[1], fuzzy clustering can be given a probabilistic

---

[1] We refer here to the traditional definition of fuzzy clustering originating from the work on fuzzy c-means [3] and based on the definition of a Ruspini fuzzy partition [39]. Nevertheless, other approaches to fuzzy clustering that are based on other definitions of a fuzzy partition have been proposed in the literature: we refer the interested reader to [40].

semantics [7,31], where the degrees of membership are interpreted as probabilities of membership of objects to clusters.

A *possibilistic clustering* is a function $P : \mathcal{X} \to [0,1]^{\mathcal{C}}$, with no additional constraints. That is, a possibilistic clustering can be interpreted as a relaxation of a fuzzy clustering in which the degrees of membership are not required to sum to 1, and hence cannot be interpreted probabilistically. Instead, the degrees of memberships are usually interpreted as possibility distributions [34], in the sense of possibility theory, or also as a restricted form of a plausibility distribution [7], in the sense of evidence theory.

2.2 Experimental Design

As mentioned in the Introduction, our aim was to study differences (if any) in the ability of different soft clustering representation formalisms to convey uncertainty, as well as in their interpretability and comprehensibility.

To this end, we designed a questionnaire-based experiment in which we asked respondents to participate in a simulated decision-making scenario. The questionnaire was developed using the LimeSurvey questionnaire platform, and the respondents were involved by the authors and interviewed in person: we decided to administer the interviews in-person, rather than remotely, in order to minimize the risk of misunderstandings in the interpretation of the questionnaire. The questionnaire was implemented and administered in Italian and the responses were collected anonymously (that is, we did not record any identifying information about the respondents).

The questionnaire encompassed five different questions, presented to the respondents in three different pages. In the first page, we presented general information, including a disclaimer informing respondents that the data was collected anonynomously. In the second page, we presented participants with the decision-making task they were asked to address. Specifically, participants were asked to interpret the role of a patient who needs surgery for the treatment of disc herniation. We decided to focus on the specific medical case of disc herniation, as this latter is a common condition (with a prevalence of approximately 3% in the general population, which can increase to approximately 30% among those affected by low back pain) and can occur independently of age (being more prevalent among young and middle-aged adults) and sex [28].

Participants were informed that there were three different treatment options among which they should choose: a traditional treatment (treatment T), suitable for the general population, with moderate effectiveness and reduced risk of complications; and two personalized treatments (treatments A and B), each of which tailored for a particular patient profile (group A and group B, encompassing general health status, clinical history, and genetic characteristics), guaranteeing perfect recovery to patients in the group matching the intervention (i.e., treatment A and group A, or treatment B and group B), but having high risk of severe complications for patients in the opposite group (i.e., treatment A and group B, or treatment B and group A).

Participants were then told that their doctor employed a clustering algorithm to compare their characteristics with those of patients belonging to groups A and B, to help them make a more informed decision. The algorithm (which was simulated using a Wizard of Oz approach [10], in which the provided support was set beforehand by the authors) presented its information in one of four possible ways, detailed below:

– Hard clustering: The participant was told that their characteristics most closely matched those of patients in group A;
– Rough clustering: The participant was told that their characteristics partly matched those of both patients in group A and in group B;
– Fuzzy clustering: The participant was told that their characteristics were similar to those of patients in group A with degree 55% and those of patients in group B with degree 45%;
– Possibilistic clustering: The participant was told that their characteristics were similar to those of patients in group A and in group B, each with degree 55%;

Each of the participants was shown only one of the four settings and the assignment of participants to settings was performed randomly. Patients were then asked to select between the three treatment options. The presented problem represents a problem of decision-making under ignorance [36]: indeed, although the (simulated) algorithms provided the participants with some indication about their characteristics, this information was neither presented as fully certain nor supplemented with statistical evidence. Therefore, the tendency of participants to select one of the options, instead of the others, was understood as providing evidence of the ability of the clustering formalisms to influence the perception of uncertainty among the participants. In particular, treatment option T was understood to denote the highest level of perceived uncertainty.

In the last page, participants were asked to rate the comprehensibility of the advice provided by the algorithm, expressed on a 4-point Likert scale, ranging from 1 (minimum comprehensibility) to 4 (maximum comprehensibility). Participants were then asked to state their gender (male, female, other/unspecified), age range (18–28, 29–49, 50–75, >75 years), and whether they underwent surgery in the past.

2.3   Statistical Analysis

Our aim was to assess two hypotheses:

1. Whether there were differences in the perception of uncertainty between groups of patients assigned with the support of different clustering formalisms. As mentioned in the previous section, a participant selecting option T was considered to be an indication that the assigned clustering formalism was able to adequately convey the uncertainty in the decision-making task. The group of participants randomly assigned to the hard clustering support was considered as the control group (since this support only reported similarity to group A, with no uncertainty indication);

2. Whether there were any differences in perceived comprehensibility among the different types of support provided by clustering formalisms.

To evaluate these hypotheses, we adopted a statistical testing approach based on regression analysis [20]. To this end, first, we binarized the treatment selection response in two categories (treatment T, representing higher perceived uncertainty, and others, encompassing both treatment A and treatment B); then, we applied one-hot encoding to both the response indicating the type of clustering support (hard, fuzzy, rough, possibilistic), as well as the demographical variables (gender, age, prior interventions).

Then, for the first hypothesis, we trained a logistic regression model: the binarized treatment selection response was treated as the target variable to predict, the one-hot encoded variables representing the types of clustering support (hard, fuzzy, rough, possibilistic) were considered as the intervention, and the demographical variables (gender, age, prior interventions) were treated as confounders. For the second hypothesis, we instead trained an ordinal linear regression model: the comprehensibility was treated as the target variable to predict, the one-hot encoded variables representing the types of clustering support (hard, fuzzy, rough, possibilistic) were considered as the intervention, and the demographical variables (gender, age, prior interventions) were treated as confounders. In both cases, the coefficients of the one-hot encoded variables representing the types of clustering support were interpreted as the strength of association between the type of clustering support and the target of interest (either uncertainty perception or comprehensibility). Formally, the coefficient for a given clustering support type x is interpreted as the log-odds of the probability of the target variable, corrected for the confounding variables [22,33]. Therefore, positive coefficients denote a positive association between the clustering support types and the target variables.

To assess statistical differences among the different clustering support types, for both hypotheses, we employed an interval analysis approach [30] based on the bootstrap method. In detail, the above described training procedure was repeated 100 times (each of which based on a different bootstrap sample of the participants, randomly drawn with replacement from the set of all participants), and the obtained bootstrap distributions of regression coefficients, for the different clustering support types, were employed to construct 95% confidence intervals using the quantile method [14]. Two clustering support types were deemed to be significantly different if the corresponding confidence intervals did not overlap.

3   Results and Discussion

We collected the responses from 40 participants. 10 participants were assigned to the hard clustering group, 12 to the rough clustering group, 8 to the possibilistic clustering group, and 10 to the fuzzy clustering group. 22 (55%) participants identified as male, while the remaining 18 (45%) identified as female. 19 participants were in the 18–28 age group, 13 in the 50–75 group, and the remaining

8 in the 2-9-49 group: no participant reported having more than 75 years. A majority of participants did not undergo previous surgery (25 out of 40, 62.5%).

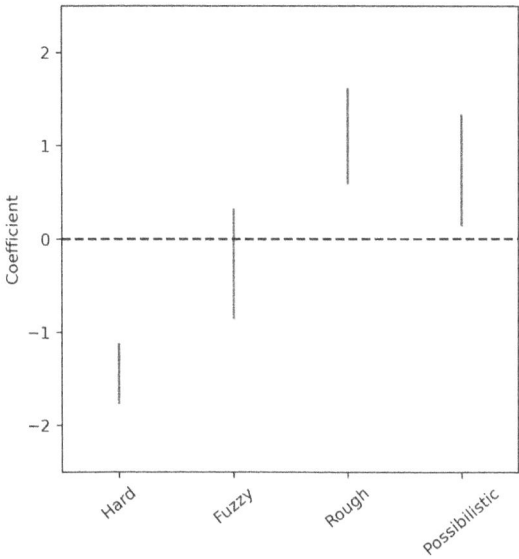

Fig. 1. Confidence intervals for the coefficients of the different clustering support types in the regression analysis of the uncertainty perception among the participants to the study.

The results in terms of perceived uncertainty are illustrated in Fig. 1, while the results on perceived comprehensibility are illustrated in Fig. 2.

All of the soft clustering approaches (fuzzy, rough and possibilistic clustering) were more strongly associated with higher perceived uncertainty than hard clustering, and significantly so. As hard clustering represented the control group, this result confirms the increased ability of soft clustering to convey decisional uncertainty compared to only providing clear-cut indications about assignments of objects to clusters.

More interestingly, we found significant differences among the three types of soft clustering support. In particular, fuzzy clustering was the method associated with the lowest perception of uncertainty. Indeed, there was no evidence of such a positive association, even though the coefficient for fuzzy clustering was not significantly smaller than that for possibilistic clustering. Among the participants in the fuzzy clustering group, 6 (60%) selected treatment A, while the remaining 4 (40%) selected instead treatment T. Replacing treatment B for treatment T, the reported percentages almost perfectly matched the similarity degrees displayed by the simulated fuzzy clustering algorithm (namely, 55% for treatment A, 45% for treatment B). Therefore, and also as a consequence of the fact that we represented memberships to clusters in terms of percentages, we conjecture that the results observed for fuzzy clustering derive from

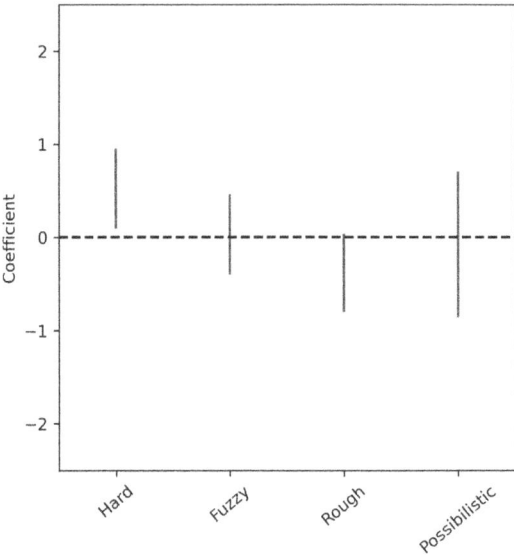

Fig. 2. Confidence intervals for the coefficients of the different clustering support types in the regression analysis of the perceived comprehensibility of the support.

the participants interpreting the reported membership degrees as probabilities. As additional evidence toward this conjecture, we note that the answer proportions for possibilistic clustering were much less balanced (6 participants out of 8 selected treatment T), despite both fuzzy and possibilistic clustering being based on the same underlying uncertainty representation formalism, as well as presenting close similarity degrees. While the original semantics of fuzzy clustering [40] is different from that of probabilistic clustering, in that its aim is not to recover a supposed generative model of the data but rather to make explicit uncertainty in clustering assignments[2], our result highlights how the representation formalism most commonly adopted in fuzzy clustering may implicitly influence users in interpreting membership degrees as probabilities [24], a perspective that seems to be justified also by recent theoretical analysis of the classical fuzzy c-means algorithm [31]. By contrast, the evidently non-probabilistic nature of possibilistic clustering was more effective in communicating higher uncertainty to users, an

---

[2] The main difference between fuzzy clustering and probabilistic clustering, from a theoretical point of view, is that in the latter we assume the existence of a probabilistic model that generated the data and that we aim to recover. As an example, in the commonly adopted Gaussian mixture model it is assumed that each cluster is sampled from a multivariate Gaussian distribution: the goal of clustering, then, is to recover the parameters of the model, namely the parameters (mean, standard deviation) of the Gaussian distributions and the mixing probability for the mixture. No such generative model, by contrast, is usually assumed in fuzzy clustering, though recent results have shown how an implicit generative model may underlie several classical fuzzy clustering methods [31].

observation that is in line with previous research on the use of possibility theory for risk communication [25]. Consequently, we believe it could be particularly fruitful to explore alternative representation formalisms for fuzzy clustering that do not necessarily rely on using percentages, such as using linguistic terms, or also graphical representations [49,50].

In contrast, rough clustering was the type of clustering support associated with the highest perception of uncertainty, although the difference with possibilistic clustering was not statistically significant. This result was not unexpected. Indeed, in binary settings, such as the one considered in our case study, rough clustering can only associate instances either to a single cluster (similarly to hard clustering) or to the full set of clusters (expressing vacuous knowledge, and hence full uncertainty). Interestingly, despite this potential limitation, rough clustering was not found to be significantly less useful or comprehensible than other types of clustering support, except for hard clustering. Indeed, the three soft clustering support types were not significantly different from each other in terms of perceived comprehensibility, when adjusting for the demographics' characteristics of the participants. Furthermore, neither fuzzy nor possibilistic clustering were found to be less comprehensible than hard clustering. Concluding, while these results highlight the capacity of rough clustering (as well as possibilistic clustering) to effectively convey uncertainty in decision-making settings, we believe that these findings should be investigated further, and especially so in multi-class settings. Indeed, even though set-valued representations have been shown in recent studies to be an effective uncertainty quantification formalism and alternative to more widely adopted probabilistic methods [2,5,11], no previous study has focused specifically on the clustering setting.

4   Conclusion

In this article, we studied soft clustering as an approach to convey and communicate uncertainty in decision-making settings by means of a user study, on a simulated but realistic medical decision-making scenario, through which we compared three of the most popular soft clustering formalisms (namely, rough, fuzzy and possibilistic clustering) in terms of both uncertainty communication capability and perceived comprehensibility.

Our results highlighted how, despite there not being any differences among soft clustering methods in terms of perceived comprehensibility, both rough and possibilistic clustering were much better in conveying uncertainty to the users than both hard and fuzzy clustering: in particular, for rough clustering, these differences were statistically significant.

We believe that our study and results provide the first analysis of the interplay between clustering analysis, XAI, and uncertainty communication, providing initial, but suggestive, empirical evidence towards the differing capabilities of soft clustering approaches to convey uncertainty effectively. Limitations of our study include focus on the binary setting, limited sample size as well as the simulated, Wizard-of-Oz design we adopted. As such, the generalizability of our findings to other settings, or fields (e.g., finance, education), should be further assessed by means of follow-up studies. More in general, future studies should be aimed at addressing the above mentioned limitations, by exploring the uncertainty quantification perspective on soft clustering in more complex settings (including multi-class ones), evaluating the effects of soft clustering on other psychometric (e.g., trust) or quantitative (e.g., decision accuracy) dimensions, as well as comparing practically used clustering algorithms, and decision-support systems based on them, in real-world decision-making scenarios.

References

1. Alvarez-Garcia, M., Ibar-Alonso, R., Arenas-Parra, M.: A comprehensive framework for explainable cluster analysis. Inf. Sci. **663**, 120282 (2024)
2. Babbar, V., Bhatt, U., Weller, A.: On the utility of prediction sets in human-ai teams. arXiv preprint arXiv:2205.01411 (2022)
3. Bezdek, J.C., Ehrlich, R., Full, W.: Fcm: the fuzzy c-means clustering algorithm. Comput. Geosci. **10**(2–3), 191–203 (1984)
4. Cabitza, F., Campagner, A., Ronzio, L., Cameli, M., Mandoli, G.E., Pastore, M.C., Sconfienza, L.M., Folgado, D., Barandas, M., Gamboa, H.: Rams, hounds and white boxes: Investigating human-ai collaboration protocols in medical diagnosis. Artif. Intell. Med. **138**, 102506 (2023)
5. Campagner, A., Cabitza, F., Berjano, P., Ciucci, D.: Three-way decision and conformal prediction: isomorphisms, differences and theoretical properties of cautious learning approaches. Inf. Sci. **579**, 347–367 (2021)
6. Campagner, A., Ciucci, D., Denœux, T.: Belief functions and rough sets: survey and new insights. Int. J. Approximate Reasoning **143**, 192–215 (2022)
7. Campagner, A., Ciucci, D., Denœux, T.: A distributional framework for evaluation, comparison and uncertainty quantification in soft clustering. Int. J. Approximate Reasoning **162**, 109008 (2023)
8. Campagner, A., Ciucci, D., Denœux, T.: A general framework for evaluating and comparing soft clusterings. Inf. Sci. **623**, 70–93 (2023)
9. Chiaburu, T., Haußer, F., Bießmann, F.: Uncertainty in xai: human perception and modeling approaches. Mach. Learn. Knowl. Extraction **6**(2), 1170–1192 (2024)
10. Dahlbäck, N., Jönsson, A., Ahrenberg, L.: Wizard of oz studies: why and how. In: Proceedings of the 1st International Conference on Intelligent User Interfaces, pp. 193–200 (1993)
11. De Toni, G., Okati, N., Thejaswi, S., Straitouri, E., Gomez-Rodriguez, M.: Towards human-ai complementarity with predictions sets. arXiv preprint arXiv:2405.17544 (2024)
12. Denoeux, T., Kanjanatarakul, O.: Evidential clustering: a review. In: International Symposium on Integrated Uncertainty in Knowledge Modelling and Decision Making, pp. 24–35. Springer (2016)
13. Denoeux, T., Li, S., Sriboonchitta, S.: Evaluating and comparing soft partitions: an approach based on dempster-shafer theory. IEEE Trans. Fuzzy Syst. **26**(3), 1231–1244 (2017)
14. Diciccio, T.J., Romano, J.P.: A review of bootstrap confidence intervals. J. R. Stat. Soc. Ser. B Stat Methodol. **50**(3), 338–354 (1988)

15. Dinari, O., Freifeld, O.: Revisiting dp-means: fast scalable algorithms via parallelism and delayed cluster creation. In: Uncertainty in Artificial Intelligence, pp. 579–588. PMLR (2022)

16. Dwivedi, R., et al.: Explainable ai (xai): core ideas, techniques, and solutions. ACM Comput. Surv. **55**(9), 1–33 (2023)

17. Feldman-Maggor, Y., Nazaretsky, T., Alexandron, G.: Explainable ai for unsupervised machine learning: a proposed scheme applied to a case study with science teachers. In: CSEDU (1), pp. 436–444 (2024)

18. Ferraro, M.B., Giordani, P.: Soft clustering. Wiley Interdisciplinary Reviews: Computational Statistics **12**(1), e1480 (2020)

19. Fotopoulou, S.: A review of unsupervised learning in astronomy. Astronomy and Computing, p. 100851 (2024)

20. Gelman, A., Hill, J., Vehtari, A.: Regression and other stories. Cambridge University Press (2020)

21. Gormley, I.C., Murphy, T.B., Raftery, A.E.: Model-based clustering. Ann. Rev. Stat. Appl. **10**(1), 573–595 (2023)

22. Harrell, Jr, F.E., Harrell, F.E.: Ordinal logistic regression. Regression modeling strategies: with applications to linear models, logistic and ordinal regression, and survival analysis, pp. 311–325 (2015)

23. Hullermeier, E., Rifqi, M., Henzgen, S., Senge, R.: Comparing fuzzy partitions: a generalization of the rand index and related measures. IEEE Trans. Fuzzy Syst. **20**(3), 546–556 (2011)

24. Laviolette, M., Seaman, J.W.: The efficacy of fuzzy representations of uncertainty. IEEE Trans. Fuzzy Syst. **2**(1), 4–15 (2002)

25. Lawson, J.R.: Communicating risk with possibility, not probability. arXiv preprint arXiv:2410.21664 (2024)

26. Lingras, P., Peters, G.: Rough clustering. Wiley Interdisciplinary Rev. Data Mining Knowl. Discovery **1**(1), 64–72 (2011)

27. Majumdar, S., Laha, A.K.: Clustering and classification of time series using topological data analysis with applications to finance. Expert Syst. Appl. **162**, 113868 (2020)

28. McGill, S.: Low back disorders: evidence-based prevention and rehabilitation. Human Kinetics (2015)

29. McInnes, L., Healy, J., Astels, S., et al.: hdbscan: hierarchical density based clustering. J. Open Source Softw. **2**(11), 205 (2017)

30. Meeker, W.Q., Hahn, G.J., Escobar, L.A.: Statistical intervals: a guide for practitioners and researchers, vol. 541. John Wiley & Sons (2017)

31. Mencar, C., Castiello, C.: A bayesian interpretation of fuzzy c-means. In: Conference of the European Society for Fuzzy Logic and Technology, pp. 443–454. Springer (2023)

32. Nauta, M., Trienes, J., Pathak, S., Nguyen, E., Peters, M., Schmitt, Y., Schlötterer, J., Van Keulen, M., Seifert, C.: From anecdotal evidence to quantitative evaluation methods: a systematic review on evaluating explainable ai. ACM Comput. Surv. **55**(13s), 1–42 (2023)

33. Norton, E.C., Dowd, B.E.: Log odds and the interpretation of logit models. Health Serv. Res. **53**(2), 859–878 (2018)

34. Pal, N.R., Pal, K., Keller, J.M., Bezdek, J.C.: A possibilistic fuzzy c-means clustering algorithm. IEEE Trans. Fuzzy Syst. **13**(4), 517–530 (2005)

35. Peng, X., Li, Y., Tsang, I.W., Zhu, H., Lv, J., Zhou, J.T.: Xai beyond classification: Interpretable neural clustering. J. Mach. Learn. Res. **23**(6), 1–28 (2022)

36. Peterson, M.: An introduction to decision theory. Cambridge University Press (2017)
37. Raees, M., Meijerink, I., Lykourentzou, I., Khan, V.J., Papangelis, K.: From explainable to interactive ai: a literature review on current trends in human-ai interaction. International Journal of Human-Computer Studies, p. 103301 (2024)
38. Rong, Y., et al.: Towards human-centered explainable ai: a survey of user studies for model explanations. IEEE Trans. Pattern Anal. Mach. Intell. (2023)
39. Ruspini, E.H.: A new approach to clustering. Inf. Control 15(1), 22–32 (1969)
40. Ruspini, E.H., Bezdek, J.C., Keller, J.M.: Fuzzy clustering: a historical perspective. IEEE Comput. Intell. Mag. 14(1), 45–55 (2019)
41. Saxena, A., et al.: A review of clustering techniques and developments. Neurocomputing 267, 664–681 (2017)
42. Schemmer, M., Hemmer, P., Nitsche, M., Kühl, N., Vössing, M.: A meta-analysis of the utility of explainable artificial intelligence in human-ai decision-making. In: Proceedings of the 2022 AAAI/ACM Conference on AI, Ethics, and Society, pp. 617–626 (2022)
43. Tomsett, R., et al.: Rapid trust calibration through interpretable and uncertainty-aware ai. Patterns 1(4) (2020)
44. Vasconcelos, H., et al.: When do xai methods work? a cost-benefit approach to human-ai collaboration. In: CHI Workshop on Trust and Reliance in AI-Human Teams, pp. 1–15 (2022)
45. Wang, P., Yang, X., Ding, W., Zhan, J., Yao, Y.: Three-way clustering: foundations, survey and challenges. Appl. Soft Comput. 151, 111131 (2024)
46. Xu, R., Wunsch, D.C.: Clustering algorithms in biomedical research: a review. IEEE Rev. Biomed. Eng. 3, 120–154 (2010)
47. Yang, M.S., Wu, K.L.: Unsupervised possibilistic clustering. Pattern Recogn. 39(1), 5–21 (2006)
48. Zhang, X., Chan, F.T., Mahadevan, S.: Explainable machine learning in image classification models: an uncertainty quantification perspective. Knowl.-Based Syst. 243, 108418 (2022)
49. Zhao, Y., Luo, F., Chen, M., Wang, Y., Xia, J., Zhou, F., Wang, Y., Chen, Y., Chen, W.: Evaluating multi-dimensional visualizations for understanding fuzzy clusters. IEEE Trans. Visual Comput. Graphics 25(1), 12–21 (2018)
50. Zhou, F., et al.: Fuzzyradar: visualization for understanding fuzzy clusters. J. Visualization 22, 913–926 (2019)

Generative Negative Sample Enhancement Based Few-Shot Named Entity Recognition

Zhen Duan[1,2,3], Shenghua Xiao[1,2,3], Jie Chen[1,2,3], Shu Zhao[1,2,3(✉)], and Yanping Zhang[1,2,3]

[1] Key Laboratory of Intelligent Computing and Signal Processing, Ministry of Education, Hefei 230601, Anhui, People's Republic of China
zhaoshuzs2002@hotmail.com
[2] School of Computer Science and Technology, Anhui University, Hefei 230601, People's Republic of China
[3] Information Materials and Intelligent Sensing Laboratory of Anhui Province, Hefei 230601, People's Republic of China

Abstract. Few-shot Named Entity Recognition (Few-shot NER) aims to identify and classify unseen named entity types with a limited labeled samples. In recent years, large language models have achieved remarkable results in various NLP tasks due to their excellent zero/few-shot generalization capabilities. Recent research has extensively focused on using large language models to study the named entity recognition task. Negative sample entities have been demonstrated to be effective when performing this task using in-context learning paradigm. However, the problem of low efficiency and poor quality in manually constructing negative sample entities has been exposed. To address this issue, in this paper, we propose a Few-shot NER method based on generative negative sample enhancement (GNSE). Specifically, GNSE first utilizes a large language model to pre-identify potential entity spans in the support set text paragraphs in a zero-shot manner and generate the semantics of the spans based on the paragraph context. Then, the pre-identified candidate spans are compared with the ground-truth labels to determine positive or negative sample entity spans. Finally, based on Chain-of-Thought, the span semantics, positive and negative sample entities, and ground-truth labels are inserted into the example template for the in-context learning paradigm, and the test text from the query set is concatenated to construct a complete prompt. GNSE was evaluated on three widely used datasets: CoNLL'03, WNUT'17 and GUM. Experiments show that GNSE has achieved better performance.

Keywords: Few-shot Named Entity Recognition · In-context Learning · Large Language Model · Prompt Engineering

1 Introduction

Named Entity Recognition (NER) stands as a cornerstone and pivotal task within the natural language processing (NLP) realm. Its objective is to precisely identify and classify predefined entity types, such as person, organization

© The Author(s), under exclusive license to Springer Nature Switzerland AG 2025
Q. Zhang et al. (Eds.): IJCRS 2025, LNAI 15710, pp. 301–313, 2025.
https://doi.org/10.1007/978-3-031-92741-6_22

and location, within unstructured text [1,12]. As a core language comprehension task, NER serves as an indispensable component in a lot of NLP applications. These applications span a wide spectrum, such as information extraction [2–4] and knowledge graph construction [5]. In scenarios where there is an ample supply of labeled data, deep-learning-based approaches [10] have demonstrated their prowess by attaining remarkable performance. However, in many real-world fields, labeled data is often scarce, leading to a significant decline in the performance of traditional NER models. Therefore, in recent years, Few-shot NER has become the focus of research, which involves recognizing previously unseen entity types based on only a handful of labeled instances (support set) per category.

For the few-shot named entity recognition task, the limited labeled samples lead to the core problem of knowledge deficiency. Thus, efficiently utilizing the limited samples is an effective and crucial approach. In the early stage, methods for the Few-shot NER task based on large language models (LLMs) merely used the text and ground-truth labels of the support set, without considering negative sample entities of the text to improve the utilization efficiency of the limited support set. The idea of considering both positive and negative samples has certain similarities to the contrastive learning method. Contrastive learning extracts meaningful features by comparing positive and negative examples, while the in-context Learning (ICL) paradigm does not require training. But we can add positive and negative samples to the demonstration part of ICL enables large language models to learn meaningful knowledge.

Recently, some related studies have demonstrated the effectiveness of using negative samples. For example, the PromptNER [18] method randomly selects several samples from the support set to manually construct demonstrations for the In-context learning paradigm. By adding "True" or "False" after the candidate entities in the demonstrations to indicate whether the candidate entity is a positive or negative sample entity, the demonstration used in this method are static. The RT [20] method also adds "True" or "False" after candidate entities to mark positive and negative sample entities. Differently, the RT method dynamically retrieves relevant demonstrations from the support set for the test text based on sentence-level semantic similarity. To this end, this method manually constructs negative sample entities for all sentences in the support set. Both of the above mentioned methods construct negative sample entities manually, which leads to problems of low efficiency and poor quality in constructing negative sample entities.

To address the above mentioned problems, we propose a few-shot named entity recognition method based on generative negative sample enhancement (GNSE), aiming to improve the efficiency and quality of constructing negative sample entities. The proposal of GNSE is based on the following key considerations: First, GNSE uses a large language model to pre-identify potential entity spans in the support set text in a zero-shot manner and compares them with ground-truth labels to determine positive and negative samples. This automated way improves the efficiency of constructing negative sample entities. Second, it generates higher-quality negative samples because GNSE enables the LLM to comprehend its own mistakes, namely the negative sample entities produced during the pre-identification stage, which is unattainable through manual methods. The contributions of this paper are as follows:

(1) We propose a generative negative sample entity generation method which can efficiently generate high-quality negative sample entities.
(2) Based on generative negative sample generation, we propose an Few-shot NER method based on generative negative sample enhancement (GNSE). This method demonstrates an effective prompt construction strategy, which improves the performance of LLMs on Few-shot NER task.
(3) Experimental results show that GNSE achieves competitive performance on three widely used English datasets, which demonstrates the effectiveness and progressiveness of GNSE.

2   Related Work

2.1   Few-Shot Named Entity Recognition

Fine-Tuning Based Methods. Few-shot Named Entity Recognition (Few-shot NER) aims to identify and classify entities when limited labeled data is available. In prior research, numerous few-shot NER methods have been proposed. ProtoBert [11] initially adapted the prototypical network model to make it suitable for the NER task, validating its efficacy in few-shot scenarios. NNShot [12] pre-train BERT using a traditional classification method during the source-domain training phase. During the target-domain inference phase, the class of each unlabeled token was determined by finding the nearest neighbors. CON-TaiNER [14] introduced a novel few-shot NER approach grounded in contrastive learning. This approach trains BERT using contrastive learning loss to enhance generalization capabilities and then employs the nearest neighbor method proposed by NNShot for inference. DecomposedNER [13] addresses the few-shot NER task by decomposing it into two phases: span detection and type classification. TadNER [15] discover the problems of over-detection of false spans and inaccurate and unstable prototype in the two-stage few-shot NER approach. To tackle these problems, TadNER proposed type-aware contrastive learning and type-aware span filtering strategies. These fine-tuning based methods founded upon deep learning neural networks exhibit commendable performance. Nevertheless, they typically entail intricate neural network design and training processes. In real-world scenarios, such complexity can make these approaches labor-intensive and resource-consuming.

Large Language Model Based Methods. Recently, with the remarkable progress of large language models (LLMs), their effectiveness in the application to NER has been demonstrated. GPT-NER [19] applies In-context Learning (ICL) to NER. It utilizes the K-Nearest Neighbors (KNN) method to retrieve the nearest-neighbor demonstrations based on sentence-level and entity-level features, thereby enhancing the performance of ICL. PromptNER [18] applies the chain-of-thought (CoT) [22] to NER for ICL by adding thought chains to fixed demonstrations. RT [20] fuses nearest-neighbor demonstrations with chain-of-thought and applies this approach in the biomedical domain. Double-Checker [26] believes that the span detector in the two-stage few-shot NER method based on fine-tuning cannot guarantee that only identify entity spans specific to the target domain and solve this problem by utilizing the collaboration between large language models and small models. These methods establish connections with closed-source LLMs via online APIs. Subsequently, they extract predictions by parsing the generated responses.

2.2   In-Context Learning

LLM employs a new paradigm called In-context Learning (ICL) [6,7,21] to formulate NLP tasks under the paradigm of language generation and make predictions by learning from a number of demonstrations. This approach has exhibited highly competitive performance across a spectrum of NLP tasks. In contrast to supervised learning approaches, which involve updates to model parameters during training or fine-tuning stages, ICL operates without any parameter update, generating responses directly from the provided prompt. Brown et al. [21] provide a comprehensive analysis of ICL, conducting extensive experiments across diverse tasks using their GPT-3 model. Additionally, research [8] has indicated that the dynamic selection of few-shot demonstrations for each test sample, as opposed to a static set, substantially enhances the ICL performance of LLMs. They also indicate that instances that are semantically closer to the context produce better results than instances that are further away. Furthermore, Wei et al. [22] revealed the reasoning ability of LLM by manually adding intermediate reasoning steps before giving an answer, which is called Chain-of-Thought (CoT). This prompt technique can be used in ICL.

3   Methodology

3.1   Problem Formulation

Named Entity Recognition is often defined as a token-level sequence labeling task. Given an input sentence with n tokens, $X = \{x_1, x_2, ..., x_n\}$, and a predefined label set T. NER model aims to assigns a label $y_i \in T$ to each token $x_i \in X$ and then output a target label sequence $Y = \{y_1, y_2, ..., y_n\}$. If a token does not belong to predefined classes, it will be labeled as O (Other).

In few-shot named entity recognition, model can only learn from a limited amount of labeled data. Given an N-way K-shot NER task, the goal is to solve this problem for each instance in the query set based on the support set. Specifically, the predefined label set T consists of N types of entity and the support set includes K instances for each label. The query set comprises the test input instance for each label. Given that the N and K are typically quite small, predicting the location and type of entities in query instances with limited labeled data represents a considerable challenge.

3.2   Model Architecture

As shown in Fig. 1, our GNSE model consists of two main components: the Negative Sample Generation Module and Inference Module.

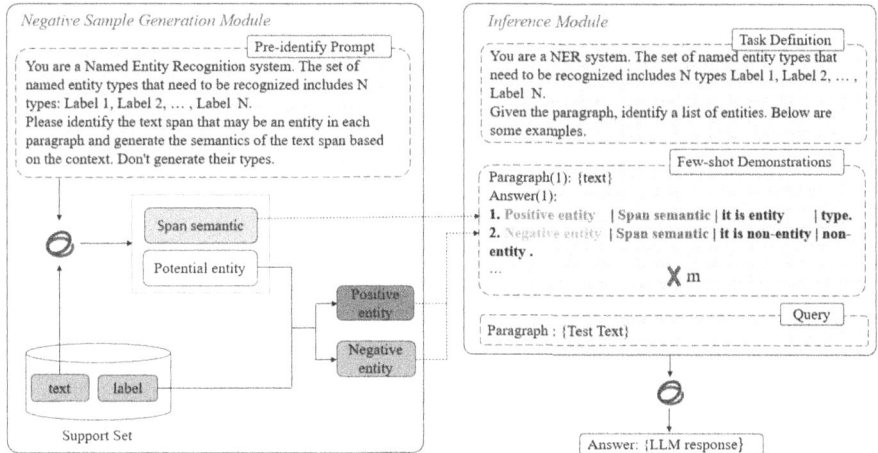

Fig. 1. The architecture of our proposed GNSE.

Negative Sample Generation Module utilizes large language model to pre-identify potential entity spans in the support set text in a zero-shot manner, and determines positive and negative sample entities

Inference Module describes how to construct an example template for the chain of thought approach and the formation of the final prompt.

3.3   Negative Sample Generation Module

First, we construct a simple prompt to utilize a large language model to pre-identify potential entity spans in the support set text in a zero-shot manner. This prompt consists of a task definition, support set text blocks, and format requirements. The task definition instructs the large language model to perform the named entity recognition task and provides predefined entity types. Then, it requires the large language model to identify text spans that might be entities in each text paragraph and generate the meaning of the span based on the paragraph context. The support set text blocks list all text paragraphs in a support set. For the sake of logicality and to facilitate the large language model's full understanding of our intentions, in this section, the support set text blocks are placed in the middle of the task definition. The format requirements specify the output format of the large language model, which is convenient for string parsing of the output. Note that during pre-identification, this paper only requires the large model to generate potential entity spans without generating types. This is because, on the one hand, there may be a large number of entity types in the dataset. On the other hand, the text paragraphs in the support set may be very long, resulting in a large number of candidate entities. These factors will make it very complex to determine positive and negative sample entities when comparing with true labels. At the same time, it may lead to a very long prompt in the final recognition stage, and may even exceed the word limit of the prompt.

By parsing the output of pre-identification, we can obtain the candidate entity spans recognized by the large language model for each text paragraph in the support set. At this point, for each text paragraph in the support set, there is a set of candidate entity spans C obtained through pre-identification and a set of real entity spans G derived from the ground-truth labels. Then traverse the set C, and if the span is not in the set G, the span is a negative sample entity. Therefore, the negative sample entity set $N = \{c_i \in C | c_i \notin G\}$. In addition, the real entity span set G of the paragraph is taken as the positive sample entity set P ($P = G$). By analogy, we can determine negative sample entities for the set of candidate entity spans of each text paragraph in the support set, and further obtain a positive sample entity set and a negative sample entity set for each text paragraph in the support set.

3.4   Inference Module

The negative sample generation module can yield the set of positive sample entity spans, the set of negative sample entity spans, and the meanings of entity spans for each text paragraph in the support set. To enhance the relevance between the demonstrations and the test text, in this section, we retrieve several of the most similar text paragraphs from the support set for the test text as few-shot demonstrations based on the SIMCSE [16] model and the KNN algorithm. Based on the Chain-of-Thought prompting idea, this section constructs a prompt template and reconstructs the few-shot demonstrations by inserting the corresponding content into this template. The Chain-of-Thought prompt template is "Positive/negative span | Span semantics | Is it entity/non-entity | type/non-entity". This template decomposes the named entity recognition task into a series of steps using the Chain-of-Thought prompt template, and obtains entity spans step by step and classifies them. After applying this template in the demonstrations of the ICL paradigm, the LLM will imitate its form during the generation process, thereby unleashing the reasoning potential of the LLM. The reasoning path is as follows: The LLM first determines the text spans that might be entities in the test text, then understands the meaning of the span based on the paragraph context, next judges whether the text span is an entity, and finally determines its entity type.

After reconstructing the few-shot demonstrations based on the Chain-of-Thought idea, the task definition, few-shot demonstrations, and test query are concatenated to form a complete prompt. When input into a large language model, the answer to the named entity recognition task for the test text can be parsed from its output.

4   Experiments

4.1   Experiment Setup

Datasets. Following the previous study, we used three English datasets to evaluate the performance of GNSE: CoNLL'03 [23], WNUT'17 [24] and GUM [25]. The statistics of these datasets are shown in Table 1. Due to the high cost of running our methods on the GLM-4 API, we evaluated all API based methods by sampling $100 \times N$ test queries from the test set for each dataset similar to other studies [19]. For example, the CoNLL dataset has 4 entity types, so we randomly sampled 400 test samples for this dataset.

Table 1. Dataset statistics.

Dataset	Domain	#Classes	#Sentences	#Entities
CoNLL'03	News	4	140.8k	29.2k
WNUT'17	Social	6	5.7k	3.9k
GUM	Wiki	11	20.7k	35.1k

Baselines. To validate the effectiveness of the proposed method, we conduct experiments comparing with the following baselines. We quote the performance reported in the papers corresponding to these baselines.

ProtoBERT [11] makes the prediction based on prototypical network [9]. The method characterizes each class by computing the average of token representations that share the same label. Subsequently, for each token within the query set, its label is determined according to its closeness to the relevant class prototype.

NNShot [12] pre-trains BERT [10] for token embedding by conventional classification training, and makes the prediction based on a token-level nearest neighbor method at testing.

CONTaiNER [14] trains BERT using token-level contrast learning, fine-tunes it on the support set to adequately fit the target domain, and employs a nearest neighbour approach during inference.

TadNER [15] uses a type-aware span filtering strategy to remove spans that are semantically distant from the type name in order to filter out false spans. Then, it use a type-aware contrastive learning strategy that combines support samples and type names as references to construct more accurate and stable prototypes for classification.

Double-Checker [26] leverages the collaboration between LLMs and small models, using LLMs to verify the candidate spans predicted by the small model, and eliminate any spans outside the scope of the target domain.

LLM Settings. We use GLM-4 [17] (glm-4-plus) as the LLM backbone for all experiments. For the main parameters of LLM, *temperature* is set to 0.95, and top_p is set to 0.05. The other parameters are set by default.

Tagging Scheme. To illustrate the setup of the tagging scheme, we used the BIO tagging scheme. In this scheme, "B" indicates that the tagged word is the beginning of the entity and "I" indicates inside of entity, while "O" refers to all other non-entity words.

Evaluation Metrics. We report the average Micro-F1 score in five random experiments with different seeds.

4.2   Experimental Results and Parameter Analysis

In this section, we present our main experiment results with previous methods in Table 2 and Table 3. The best results are in bold.

Table 2. Results of KAINER and baselines on CoNLL, WNUT and GUM datasets in 1-shot settings.

Models	CoNLL'03	WNUT17	GUM	Avg.
ProtoBERT (SAC 2019)	49.9	17.4	17.8	28.4
NNShot (EMNLP 2020)	61.2	22.7	10.5	31.5
CONTaiNER (ACL 2022)	61.2	27.5	18.5	35.7
TadNER (EMNLP 2023)	70.4	32.8	24.2	42.5
Double-Checker (EMNLP 2024)	72.4	39.9	25.4	45.9
GNSE (ours)	**77.8**	**50.9**	**32.2**	**53.6**

Table 3. Results of KAINER and baselines on CoNLL, WNUT and GUM datasets in 5-shot settings.

Models	CoNLL'03	WNUT17	GUM	Avg.
ProtoBERT (SAC 2019)	61.3	22.8	19.5	34.5
NNShot (EMNLP 2020)	74.1	27.3	15.9	39.1
CONTaiNER (ACL 2022)	75.8	32.5	25.2	44.5
TadNER (EMNLP 2023)	80.5	34.5	35.1	50.0
Double-Checker (EMNLP 2024)	80.8	40.4	38.0	53.1
GNSE (ours)	**84.3**	**51.0**	**40.1**	**58.5**

Experimental Results. From the two tables, we can observe that:

(1) Our method GNSE significantly outperforms other baseline methods. In the 1-shot setting, GNSE achieves an average result of 53.6% on three datasets, exceeding the optimal baseline by 7.5%. And in the 5-shot setting, GNSE achieves an average result of 58.5% on three datasets, exceeding the optimal baseline by 5.4%. This shows the effectiveness and progressiveness of GNSE.

(2) Compared with the fine-tuning based approach, the performance improvement is greater in 1-shot setting than in 5-shot setting. This suggests that even though the LLMs pre-trained with a large-scale corpus have an extremely strong zero/few shot generalization ability, this advantage weakens as the number of training samples increases.

(3) Double-Checker also employs LLMs, which it uses to verify the entity spans and their types predicted by a smaller model. However, Double-Checker was comprehensively outperformed by GNSE, indicating that while using LLMs to assist a small model is a feasible approach, its advancement is insufficient due to the performance being limited by the small model. In contrast, GNSE only uses LLMs to identify entities and perform classification, thereby not being subject to such limitations.

Parameter Analysis. In order to investigate the stability of GNSE when the GLM-4 parameters are varied, we conduct experiments with different values of temperature and top_p in 5-shot setting on CoNLL dataset. As shown in Fig. 2, we can observe that with the changes of temperature and top_p, the performance of GNSE only produces slight fluctuations, which shows that GNSE has good stability.

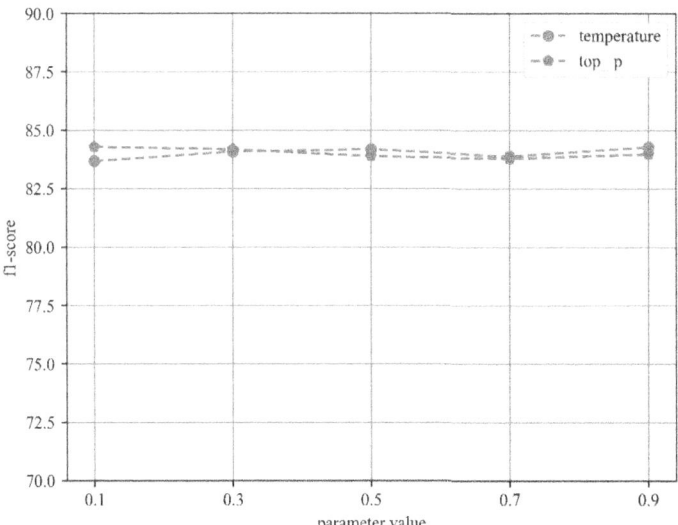

Fig. 2. Parameter Analysis about temperature and top-p on CoNLL dataset.

4.3 Ablation Study

In this subsection, we conduct an ablation study on CoNLL dataset in 1-shot and 5-shot settings in order to evaluate the effectiveness of negative sample generation module in our method. The results are listed in Table 4. When constructing the few-shot demonstration part of the ultimate prompt, we only use positive sample entities from the support set and do not use negative sample entities. This resulted in a 6.8% decrease of model performance in 1-shot setting and 5.0% in 5-shot setting. This demonstrates the advantages of negative sample generation module and the effectiveness of using negative sample entities in Few-shot NER task.

Table 4. Results of Ablation Study on CoNLL'03 dataset.

Model	1-shot	5-shot
GNSE	77.8	84.3
w/o negative sample	71.0	79.3

5   Conclusion

In this paper, we propose a novel In-context learning based Few-shot NER approach named GNSE. GNSE leverages large language models to pre-identify potential entity spans in the support set text in a zero-shot manner, and compares them with real labels to determine positive and negative samples. This approach enhances the efficiency and quality of constructing negative sample entities. Experimental results demonstrate that GNSE exhibits outstanding performance, validating its effectiveness and advancement in the few-shot NER task. This work provides a new reference for how future few-shot NER methods based on large language models can efficiently construct high-quality negative samples. Our future work will focus on further improving the quality of negative samples, while also running GNSE on more large model foundations to verify its generalizability.

Acknowledgments. Our work is supported by the National Natural Science Foundation of China (62476003), Anhui Province Excellent Scientific Research and Innovation Team (2024AH010004), Anhui Provincial Natural Science Foundation - Water Science Joint Fund (2408055US006), the University Synergy Innovation Program of Anhui Province (GXXT-2023-050), and SMP-Zhipu.AI Large Model Cross-Disciplinary Fund (SMP-Zhipu20240210). We also acknowledge the support from Zhipu AI-Anhui University Joint Research Center, and the High-Performance Computing Platform of Anhui University.

References

1. Li, J., Sun, A., Han, J., Li, C.: A survey on deep learning for named entity recognition. J. IEEE Trans. Knowl. Data Eng. **34**(1), 50–70 (2022). https://doi.org/10.1109/TKDE.2020.2981314

2. Li, X., Luo, X., Dong, C., Yang, D., Luan, B., He, Z.: TDEER: an efficient translating decoding schema for joint extraction of entities and relations. In: Moens, M.-F., Huang, X., Specia, L., Yih, S.W.-t. (eds.) Proceedings of the 2021 Conference on Empirical Methods in Natural Language Processing, pp. 8055–8064. Association for Computational Linguistics, Online and Punta Cana, Dominican Republic (2021). https://doi.org/10.18653/v1/2021.emnlp-main.635

3. Lu, Y., Liu, Q., Dai, D., Xiao, X., Lin, H., Han, X., Sun, L., Wu, H.: Unified structure generation for universal information extraction. In: Muresan, S., Nakov, P., Villavicencio, A. (eds.) Proceedings of the 60th Annual Meeting of the Association for Computational Linguistics (Volume 1: Long Papers), pp. 5755–5772. Association for Computational Linguistics, Dublin, Ireland (2022). https://doi.org/10.18653/v1/2022.acl-long.395

4. Sainz, O., García-Ferrero, I., Agerri, R., Lopez de Lacalle, O., Rigau, G., Agirre, E.: GoLLIE: annotation guidelines improve zero-shot information-extraction. In: The Twelfth International Conference on Learning Representations (2024)

5. He, K., Yao, L., Zhang, J., Li, Y., Li, C.: Construction of genealogical knowledge graphs from obituaries: Multitask neural network extraction system. J. Med. Internet Res. **23**(8), e25670 (2021). Publisher: JMIR Publications Inc

6. Dong, Q., et al.: A survey on in-context learning. In: Al-Onaizan, Y., Bansal, M., Chen, Y.-N. (eds.) Proceedings of the 2024 Conference on Empirical Methods in Natural Language Processing, pp. 1107–1128. Association for Computational Linguistics, Miami, Florida, USA (2024). https://doi.org/10.18653/v1/2024.emnlp-main.64

7. Min, S., Lyu, X., Holtzman, A., Artetxe, M., Lewis, M., Hajishirzi, H., Zettlemoyer, L.: Rethinking the role of demonstrations: what makes in-context learning work? In: Goldberg, Y., Kozareva, Z., Zhang, Y. (eds.) Proceedings of the 2022 Conference on Empirical Methods in Natural Language Processing, pp. 11048–11064. Association for Computational Linguistics, Abu Dhabi, United Arab Emirates (2022). https://doi.org/10.18653/v1/2022.emnlp-main.759

8. Liu, J., Shen, D., Zhang, Y., Dolan, B., Carin, L., Chen, W.: What makes good in-context examples for GPT -3?. In: Agirre, E., Apidianaki, M., Vulić, I. (eds.) Proceedings of Deep Learning Inside Out (DeeLIO 2022): The 3rd Workshop on Knowledge Extraction and Integration for Deep Learning Architectures, pp. 100–114. Association for Computational Linguistics, Dublin, Ireland and Online (2022). https://doi.org/10.18653/v1/2022.deelio-1.10

9. Snell, J., Swersky, K., Zemel, R.: Prototypical networks for few-shot learning. In: Proceedings of the 31st International Conference on Neural Information Processing Systems (NIPS'17), pp. 4080–4090. Curran Associates Inc., Red Hook (2017). Location: Long Beach, California, USA

10. Devlin, J., Chang, M.-W., Lee, K., Toutanova, K.: BERT: pre-training of deep bidirectional transformers for language understanding. In: Burstein, J., Doran, C., Solorio, T. (eds.) Proceedings of the 2019 Conference of the North American Chapter of the Association for Computational Linguistics: Human Language Technologies, Volume 1 (Long and Short Papers), pp. 4171–4186. Association for Computational Linguistics, Minneapolis, Minnesota (2019). https://doi.org/10.18653/v1/N19-1423

11. Fritzler, A., Logacheva, V., Kretov, M.: Few-shot classification in named entity recognition task. In: Proceedings of the 34th ACM/SIGAPP Symposium on Applied Computing (SAC '19), pp. 993–1000. Association for Computing Machinery, New York (2019). https://doi.org/10.1145/3297280.3297378

12. Yang, Y., Katiyar, A.: Simple and effective few-shot named entity recognition with structured nearest neighbor learning. In: Webber, B., Cohn, T., He, Y., Liu, Y. (eds.) Proceedings of the 2020 Conference on Empirical Methods in Natural Language Processing (EMNLP), pp. 6365–6375. Association for Computational Linguistics, Online (2020). https://doi.org/10.18653/v1/2020.emnlp-main.516

13. Ma, T., Jiang, H., Wu, Q., Zhao, T., Lin, C.-Y.: Decomposed meta-learning for few-shot named entity recognition. In: Muresan, S., Nakov, P., Villavicencio, A. (eds.) Findings of the Association for Computational Linguistics: ACL 2022, pp. 1584–1596. Association for Computational Linguistics, Dublin, Ireland (2022). https://doi.org/10.18653/v1/2022.findings-acl.124

14. Das, S.S.S., Katiyar, A., Passonneau, R., Zhang, R.: CONTaiNER: few-shot named entity recognition via contrastive learning. In: Muresan, S., Nakov, P., Villavicencio, A. (eds.) Proceedings of the 60th Annual Meeting of the Association for Computational Linguistics (Volume 1: Long Papers), pp. 6338–6353. Association for Computational Linguistics, Dublin, Ireland (2022). https://doi.org/10.18653/v1/2022.acl-long.439

15. Li, Y., Yu, Y., Qian, T.: Type-aware decomposed framework for few-shot named entity recognition. In: Bouamor, H., Pino, J., Bali, K. (eds.) Findings of the Association for Computational Linguistics: EMNLP 2023, pp. 8911–8927. Association for Computational Linguistics, Singapore (2023). https://doi.org/10.18653/v1/2023.findings-emnlp.598

16. Gao, T., Yao, X., Chen, D.: SimCSE: simple contrastive learning of sentence embeddings. In: Moens, M.-F., Huang, X., Specia, L., Yih, S.W.-t. (eds.) Proceedings of the 2021 Conference on Empirical Methods in Natural Language Processing, pp. 6894–6910. Association for Computational Linguistics, Online and Punta Cana, Dominican Republic (2021). https://doi.org/10.18653/v1/2021.emnlp-main.552

17. Team GLM, Zeng, A., et al.: ChatGLM: A Family of Large Language Models from GLM-130B to GLM-4 All Tools (2024). https://arxiv.org/abs/2406.12793

18. Ashok, D., Lipton, Z.C.: PromptNER: Prompting For Named Entity Recognition (2023). https://arxiv.org/abs/2305.15444

19. Wang, S., Sun, X., Li, X., Ouyang, R., Wu, F., Zhang, T., Li, J., Wang, G.: GPT-NER: Named Entity Recognition via Large Language Models (2023). https://arxiv.org/abs/2304.10428

20. Li, M., Zhou, H., Yang, H., Zhang, R.: RT: a Retrieving and Chain-of-Thought framework for few-shot medical named entity recognition. J. Am. Med. Inform. Assoc. **31**(9), 1929–1938 (2024). https://doi.org/10.1093/jamia/ocae095

21. Brown, T.B., et al.: Language models are few-shot learners. In: Proceedings of the 34th International Conference on Neural Information Processing Systems (NIPS '20), articleno = 159, pp. 25. Curran Associates Inc., Red Hook, NY, USA (2020). Location: Vancouver, BC, Canada

22. Wei, J., et al.: Chain-of-Thought Prompting Elicits Reasoning in Large Language Models. In: Advances in Neural Information Processing Systems, vol. 35, pp. 24824–24837. Inc, Curran Associates (2022)

23. Tjong Kim Sang, E.F., De Meulder, F.: Introduction to the CoNLL-2003 shared task: language-independent named entity recognition. In: Proceedings of the Seventh Conference on Natural Language Learning at HLT-NAACL 2003, pp. 142–147 (2003)

24. Derczynski, L., Nichols, E., van Erp, M., Limsopatham, N.: Results of the WNUT2017 shared task on novel and emerging entity recognition. In: Derczynski, L., Xu, W., Ritter, A., Baldwin, T. (eds.) Proceedings of the 3rd Workshop on Noisy User-generated Text, pp. 140–147. Association for Computational Linguistics, Copenhagen, Denmark (2017). https://doi.org/10.18653/v1/W17-4418
25. Zeldes, A.: The GUM corpus: creating multilayer resources in the classroom. Lang. Resour. Eval. **51**(3), 581–612 (2016). https://doi.org/10.1007/s10579-016-9343-x
26. Chen, W., Zhao, L., Zheng, Z., Xu, T., Wang, Y., Chen, E.: Double-checker: large language model as a checker for few-shot named entity recognition. In: Al-Onaizan, Y., Bansal, M., Chen, Y.-N. (eds.) Findings of the Association for Computational Linguistics: EMNLP 2024, pp. 3172–3181. Association for Computational Linguistics, Miami, Florida, USA (2024). https://doi.org/10.18653/v1/2024.findings-emnlp.180

Auxiliary Attributes-Guided Face Age Estimation

Fan You[1], Fujin Zhong[1,2,3]($\boxtimes$), Hong Yu[1,2,3], Jun Hu[1,2,3], Yan Yang[1], and Mengqi Liu[1]

[1] School of Computer Science and Technology, Chongqing University of Posts and Telecommunications, Chongqing 400065, China
zhongfj@cqupt.edu.cn
[2] Key Laboratory of Big Data Intelligent Computing, Chongqing University of Posts and Telecommunications, Chongqing 400065, China
[3] Key Laboratory of Cyberspace Big Data Intelligent Security, Ministry of Education, Chongqing 400065, China

Abstract. Facial age estimation is a critical task within computer vision, with wide-ranging applications in demographic analysis, video retrieval, and access control. Prior research has highlighted that facial aging is significantly influenced by attributes such as gender and ethnicity, which present substantial challenges to achieving accurate and robust age prediction. To tackle this challenge, we propose MSAN, an innovative end-to-end age estimation framework that integrates auxiliary information, including gender and ethnicity, during the prediction phase and dynamically models their interactions with age. Extensive evaluations on multiple public datasets demonstrate that our model outperforms several state-of-the-art approaches. This study provides an effective solution for enhancing the accuracy of facial age estimation.

Keywords: Age estimation · Auxiliary-guided · Multitask learning

1 Introduction

Facial age estimation represents a critical area of research within computer vision, demonstrating significant potential for applications in demographic analysis, intelligent security systems, and personalized healthcare services. However, existing studies indicate that age-related biological features are influenced by factors such as gender and ethnicity, which can compromise the accuracy of age estimation. Additionally, the imbalanced distribution of sample categories in current public datasets further exacerbates the challenges faced during the learning process.

To address these issues, we propose an attribute-guided dynamic weighting allocation mechanism aimed at more accurately modeling the interrelationships among gender, ethnicity, and age. By analyzing prediction discrepancies across specific attribute branches at various age ranges, we dynamically evaluate the intensity of influence exerted by these attributes and develop a two-stage optimization strategy accordingly. For attribute-sensitive age ranges (e.g., middle-aged individuals), we implement differential reinforcement training

© The Author(s), under exclusive license to Springer Nature Switzerland AG 2025
Q. Zhang et al. (Eds.): IJCRS 2025, LNAI 15710, pp. 314–320, 2025.
https://doi.org/10.1007/978-3-031-92741-6_23

to optimize particular branches; conversely, for attribute-insensitive age ranges (e.g., infancy), we enhance model performance through collaborative learning across multiple branches. This approach effectively alleviates problems arising from biases in sample distribution.

2   Related Works

With the great success of deep learning in various computer vision tasks, deep learning methods have also been widely used for facial age estimation in recent years. Most deep learning studies on facial age estimation focused solely on age prediction, without considering the influence of other facial attributes. Deng et al. [2] introduced Progressive Margin Loss (PML) for unconstrained facial age classification, which adaptively improves the age label mode by enforcing pair-wise margins. Deep Regression Forests (DRFs) [11] combine random forests and CNNs for age regression, implicitly encoding the order relationship between adjacent age groups. Deep Label Distribution Learning Forest (DLDLF) [11] extends DRFs by introducing label distribution learning [3] at leaf nodes to model data more effectively.

Additionally, a study [7] has shown that ignoring auxiliary attributes such as gender and race in age estimation can increase estimation errors. To address this, Xie et al. [16] proposed a multi-stage model. This model uses a weighted summation of estimated probabilities under different preset conditions to obtain age estimates, effectively incorporating gender, race, and age attributes. However, dataset bias will significantly affect the model's learning, as relying on auxiliary classifiers to control the learning of sub-classifiers makes it difficult for small-sample classes to receive effective learning. This issue is especially pronounced given the current prevalence of bias in facial datasets.

3   Method

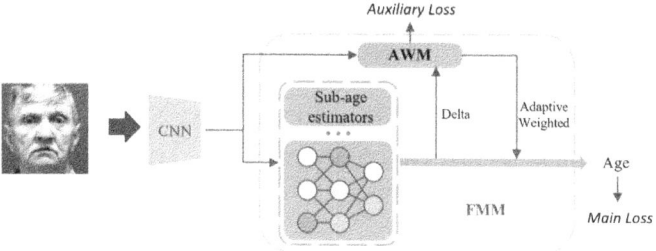

Fig. 1. Overall framework diagram

The MSAN framework operates end-to-end, directly mapping raw facial images to age estimates via multi-task learning (Fig. 1). This design eliminates manual feature engineering and ensures holistic optimization of cross-task dependencies, dynamically balancing contributions from auxiliary attributes (e.g., gender, ethnicity) during training. The end-to-end architecture simplifies the pipeline while enhancing discriminative representation learning for demographic variations.

3.1   Auxiliary Attributes Guidance

Adaptive Weight Module. Attribute differences influence the aging process, reflected in the prediction discrepancy δ_{age}. Here, we let s denote the parameter quantifying this influence. A larger δ_{age} indicates reduced robustness of one age estimation branch for a given sample, while a smaller δ_{age} suggests similar aging processes among different genders. Hence, $s \propto \delta_{age}$, and this parameter is learnable.

We map δ_{age} to parameter s using two fully connected layers (the first with 512 nodes and the second with 1 node). Concurrently, feature F is processed through two fully connected layers (the first with 512 nodes and the second with 2 nodes) to obtain gender classification probabilities $P = [p_{male}, p_{female}]$, supervised by gender labels, as illustrated in Fig. 2.

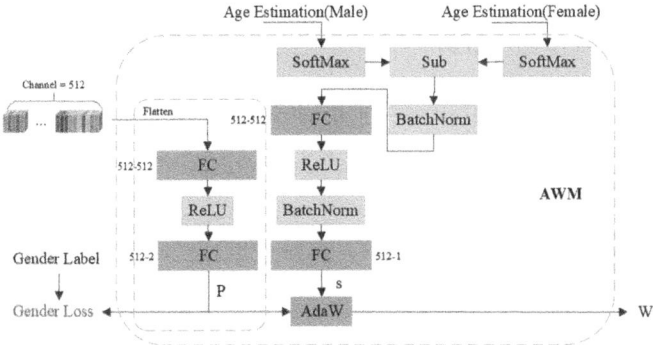

Fig. 2. AWM guided by gender

Subsequently, we can calculate the weights $W = [w_{male}, w_{female}]$ corresponding to the age estimation branches based on the parameter s and P. We refer to this process as the AdaW operation. The calculation of W can be given as follows:

$$W = \text{AdaW}(s, P) = (P - 1/N) \cdot s + 1/N, \tag{1}$$

where N denotes the number of categories for the auxiliary attribute. As s approaches negative infinity, both w_{male} and w_{female} approach $1/N = 1/2$. Thus, smaller s values lead to uniform weights across age estimation branches, indicating that predictions are independent of auxiliary information. Conversely, larger s values result in greater reliance on the output from one specific branch. This principle holds for any $N \geq 2$.

Feature Mapping Module. The Feature Mapping Module (FMM) integrates gender and ethnicity during prediction via attribute-specific sub-estimators (Fig. 1). Each sub-estimator is a 512-512-101 MLP designed to capture demographic-specific aging patterns (e.g., Male/Female, Black/White/Other). The Adaptive Weighting Module (AWM) dynamically adjusts weights for all attributes based on prediction discrepancies, ensuring age estimation accounts for demographic variations. For datasets with additional attributes, FMM can append sub-estimators with the same architecture, maintaining consistency while expanding coverage.

4 Experiments

4.1 Datasets

To validate the proposed method, experiments were conducted on three facial age datasets: MORPH Album2 [10], FG-NET [9], and AgeDB [8].

MORPH Album2 dataset contains over 50k facial images of 13,618 individuals aged 16–77 in controlled settings, annotated with age, gender, and race attributes. Two evaluation protocols were used on MORPH: the RS protocol (80% training set, 20% testing set) and the S1-S2-S3 protocol (accounting for imbalanced gender and race distributions [14]). FG-NET contains 1,002 images from 82 individuals aged 0 to 69, using the leave-one-person-out (LOPO) evaluation protocol [16]. AgeDB includes 16,488 images from 568 individuals aged 1 to 101, employing the RS protocol.

4.2 Evaluation Metrics

Mean Absolute Error (MAE) is used to evaluate the performance of our facial age estimation methods [15].

4.3 Experimental Settings

Each image is preprocessed using the MTCNN face detector [4] and aligned based on facial landmarks. All images are then resized to $256 \times 256 \times 3$ pixels. The training set is augmented through three steps: (a) random cropping to $224 \times 224 \times 3$ pixels, (b) random horizontal flipping, and (c) random rotation within $[-5°, 5°]$.

Table 1. MAEs of the proposed approach and several SOTA methods on MORPH Album2 dataset (RS Protocol). (* indicates that the model was pre-trained on the IMDB-WIKI dataset.)

Methods	Backbone	Params	MAE
Xie et al. [16]	VGG-16	41M	2.22
DCDL	VGG-16-BN	-	2.26
DLDLF [12]	VGG-16	-	2.19
DRF [12]	VGG-16	-	2.14
PML [2]	ResNet-34	16M	2.15
FP-Age [6]	RTNet	27M	**2.04**
DAA [1]	ResNet-18	11M	2.25/2.06*
MSAN(Ours)	ResNet-18	16M	2.06

Table 2. MAEs of the proposed approach and several SOTA methods on MORPH Album2 dataset (S1-S2-S3 Protocol)

Methods	Train	Test	MAE	Avg
Xie et al. [16]	S1	S2+S3	2.80	2.81
	S2	S1+S3	2.81	
DLDLF [12]	S1	S2+S3	-	2.99
	S2	S1+S3	-	
DCDL	S1	S2+S3	2.83	2.86
	S2	S1+S3	2.89	
DRF [12]	S1	S2+S3	-	2.90
	S2	S1+S3	-	
DAA	S1	S2+S3	3.43	3.04
	S2	S1+S3	3.41	
MSAN(Ours)	S1	S2+S3	**2.71**	**2.72**
	S2	S1+S3	**2.76**	

Table 3. Experience on FG-NET dataset (LOPO Protocol)

Methods	Backbone	MAE
Xie et al. [16]	VGG-16	3.58
DOEL [17]	-	3.44
DCDL	VGG-16-BN	4.48
DLDLF [12]	VGG-16	3.71
DRF [12]	VGG-16	3.47
AMR [19]	ResNet-50	3.61
DAA [1]	ResNet-18	3.41/**2.19***
MSAN (Ours)	ResNet-18	**3.39**

Table 4. Experience on AgeDB dataset (RS Protocol)

Methods	MAE
AGEn [5]	6.22
CMT [18]	6.19
SAF [7]	6.87
Xie et al. [16]	5.74
DOEL [17]	5.69
DCDL	6.13
DAA	6.60
MSAN(Ours)	**5.63**

Experiments are conducted on two NVIDIA RTX 2080Ti GPUs. We employ the SGD optimization algorithm for network training, with a weight decay of 0.0005, momentum of 0.9, and a batch size of 32. Training involves 120 epochs, starting with an initial learning rate of 0.001, which is reduced by a factor of 10 at the 50th, 70th, and 90th epochs. The hyperparameters α and β in the joint loss function are both set to 0.01.

4.4   Experiments Results and Analysis

Tables 1 and 2 present MORPH Album2 results under two protocols. MSAN outperforms auxiliary-attribute methods like SAF [7], Xie et al. [16], and DCDL [13] under the RS protocol. While MSAN's MAE is 0.2 higher than FP-Age [6], the latter requires significantly more parameters for comparable performance. On FG-Net (Table 3), MSAN achieves the lowest MAE of 3.39 without IMDB-WIKI pre-training, surpassing DAA [1] by 0.19. This highlights MSAN's superior generalization to limited-data scenarios. In AgeDB (Table 4), MSAN reduces MAE by 0.09 compared to Xie et al.'s VGG-16 approach, validating its robustness in real-world conditions. MSAN's dynamic attribute weighting and multi-branch architecture enable better generalization across datasets with varying distributions.

5   Conclusion

In this study, we enhance facial age estimation by using auxiliary attribute guidance. We introduce the FMM to model the relationships between attributes such as gender, ethnicity, and age, while AWM adjusts the weight of auxiliary information to mitigate data bias. Experimental results demonstrate the effectiveness of our approach. Future work will focus on developing loss functions for multi-task learning and exploring more efficient feature extraction techniques tailored for facial images.

Acknowledgments. This work is supported in part by the National Natural Science Foundation of China under Grant 62136002 and Grant 62221005; in part by the National Natural Science Foundation of Chongqing under Grant cstc2022ycjh-bgzxm0004 and cstb2022nscq-msx0578; in part by the Key Cooperation Project of Chongqing Municipal Education Commission under Grant HZ2021008; and in part by the Science and Technology Research Program of Chongqing Municipal Education Commission under Grant KJQN202300608.

Disclosure of Interests. The authors have no competing interests to declare that are relevant to the content of this article.

References

1. Chen, P., Zhang, X., Li, Y., Tao, J., Xiao, B., Wang, B., Jiang, Z.: DAA: A delta age adain operation for age estimation via binary code transformer. In: IEEE/CVF Conference on Computer Vision and Pattern Recognition, CVPR 2023, Vancouver, BC, Canada, June 17–24, 2023, pp. 15836–15845. IEEE (2023). https://doi.org/10.1109/CVPR52729.2023.01520

2. Deng, Z., Liu, H., Wang, Y., Wang, C., Yu, Z., Sun, X.: Pml: progressive margin loss for long-tailed age classification. In: 2021 IEEE/CVF Conference on Computer Vision and Pattern Recognition (CVPR), pp. 10498–10507 (2021). https://doi.org/10.1109/CVPR46437.2021.01036

3. Geng, X.: Label distribution learning. IEEE Trans. Knowl. Data Eng. **28**(7), 1734–1748 (2016). https://doi.org/10.1109/TKDE.2016.2545658

4. Geng, X., Wang, Q., Xia, Y.: Facial age estimation by adaptive label distribution learning. In: 2014 22nd International Conference on Pattern Recognition, pp. 4465–4470 (2014). https://doi.org/10.1109/ICPR.2014.764

5. Girshick, R.: Fast r-cnn. In: 2015 IEEE International Conference on Computer Vision (ICCV), pp. 1440–1448 (2015). https://doi.org/10.1109/ICCV.2015.169

6. Lin, Y., Shen, J., Wang, Y., Pantic, M.: Fp-age: leveraging face parsing attention for facial age estimation in the wild. CoRR abs/2106.11145 (2021)

7. Liu, K.H., Liu, T.J.: A structure-based human facial age estimation framework under a constrained condition. IEEE Trans. Image Process. **28**(10), 5187–5200 (2019). https://doi.org/10.1109/TIP.2019.2916768

8. Moschoglou, S., Papaioannou, A., Sagonas, C., Deng, J., Kotsia, I., Zafeiriou, S.: Agedb: the first manually collected, in-the-wild age database. In: 2017 IEEE Conference on Computer Vision and Pattern Recognition Workshops (CVPRW), pp. 1997–2005 (2017). https://doi.org/10.1109/CVPRW.2017.250

9. Panis, G., Lanitis, A.: An overview of research activities in facial age estimation using the FG-NET aging database. In: Agapito, L., Bronstein, M.M., Rother, C. (eds.) Computer Vision - ECCV 2014 Workshops - Zurich, Switzerland, September 6–7 and 12, 2014, Proceedings, Part II. LNCS, vol. 8926, pp. 737–750. Springer (2014). https://doi.org/10.1007/978-3-319-16181-5_56

10. Ricanek, K., Tesafaye, T.: Morph: a longitudinal image database of normal adult age-progression. In: 7th International Conference on Automatic Face and Gesture Recognition (FGR06), pp. 341–345 (2006). https://doi.org/10.1109/FGR.2006.78

11. Shen, W., Guo, Y., Wang, Y., Zhao, K., Wang, B., Yuille, A.: Deep regression forests for age estimation. In: 2018 IEEE/CVF Conference on Computer Vision and Pattern Recognition, pp. 2304–2313 (2018). https://doi.org/10.1109/CVPR.2018.00245

12. Shen, W., Guo, Y., Wang, Y., Zhao, K., Wang, B., Yuille, A.: Deep differentiable random forests for age estimation. IEEE Trans. Pattern Anal. Mach. Intell. **43**(2), 404–419 (2021). https://doi.org/10.1109/TPAMI.2019.2937294

13. Sun, H., Pan, H., Han, H., Shan, S.: Deep conditional distribution learning for age estimation. IEEE Trans. Inf. Forensics Secur. **16**, 4679–4690 (2021). https://doi.org/10.1109/TIFS.2021.3114066

14. Szasz, T., Harrison, E., Liu, P.J., Lin, P.C., Birali Runesha, H., Adukia, A.: Measuring representation of race, gender, and age in children's books: Face detection and feature classification in illustrated images. In: 2022 IEEE/CVF Winter Conference on Applications of Computer Vision (WACV), pp. 3371–3380 (2022). https://doi.org/10.1109/WACV51458.2022.00343

15. Tan, Z., Yang, Y., Wan, J., Guo, G., Li, S.Z.: Deeply-learned hybrid representations for facial age estimation. In: Kraus, S. (ed.) Proceedings of the Twenty-Eighth International Joint Conference on Artificial Intelligence, IJCAI 2019, Macao, China, August 10-16, 2019, pp. 3548–3554. ijcai.org (2019). https://doi.org/10.24963/IJCAI.2019/492

16. Xie, J.C., Pun, C.M.: Chronological age estimation under the guidance of age-related facial attributes. IEEE Trans. Inf. Forensics Secur. **14**(9), 2500–2511 (2019). https://doi.org/10.1109/TIFS.2019.2902823

17. Xie, J.C., Pun, C.M.: Deep and ordinal ensemble learning for human age estimation from facial images. IEEE Trans. Inf. Forensics Secur. **15**, 2361–2374 (2020). https://doi.org/10.1109/TIFS.2020.2965298

18. Yoo, B., Kwak, Y., Kim, Y., Choi, C., Kim, J.: Deep facial age estimation using conditional multitask learning with weak label expansion. IEEE Signal Process. Lett. **25**(6), 808–812 (2018). https://doi.org/10.1109/LSP.2018.2822241

19. Zhao, Z., Qian, P., Hou, Y., Zeng, Z.: Adaptive mean-residue loss for robust facial age estimation. In: 2022 IEEE International Conference on Multimedia and Expo (ICME), pp. 1–6 (2022). https://doi.org/10.1109/ICME52920.2022.9859703

Steering Angle Prediction Based on Travelable Regions and Bio-Inspired Neural Circuit Policy

Yali Wang, Ye Wang, Hong Yu, and Ke Liu$^{(\boxtimes)}$ ⑩

School of Computer Science and Technology, Chongqing University of Posts and
Telecommunications, Chongqing 400065, China
liuke@cqupt.edu.cn

Abstract. Steering angle control is crucial for the safe operation of
autonomous vehicles, as even a minor error can lead to serious traffic acci-
dents. However, many end-to-end steering angle prediction systems overlook the extraction of task-relevant features. Including irrelevant features
such as trees, buildings, and the sky can significantly reduce the performance of steering angle prediction. To address this issue, we propose
TLNBm-NCP, a new steering angle prediction network that utilizes drivable region segmentation and bio-inspired neural circuit policy. Specifically, TLNBm-NCP employs TwinLiteNet to Generate images containing
both drivable and non-drivable areas. Then, a binary mask and a multi-
scale spatio-temporal feature extraction module (Multc) are introduced
to extract the features of the lane regions. Finally, the bio-inspired neural circuit policies (NCPs) is employed to learn temporal dependencies
between image frames and provide the steering angle. After extensive
experiments on Udacity, Carla, and SullyChen datasets, TLNBm-NCP
demonstrates superior performance compared to existing models. This
model shows promise for autonomous driving steering angle prediction.
The code and supplementary materials are available at https://github.
com/wyl121/TLNBm-NCP.

Keywords: Autonomous Vehicles · Steering Angle Prediction · Neural
Circuit Policies (NCPs) · Travelable Area

1 Introduction

Over the past decade, significant advancements have been made in automated
driving technology, attracting considerable interest from various industries and
academic institutions. This progress is largely attributed to ongoing research
and advancements in deep neural networks and high-performance computing
hardware [1]. A critical aspect driving this progress is the emphasis on steering
angle prediction, a vital component in automated driving systems (ADS). This
predictive capability is crucial as it directly influences the vehicle's safe and
efficient navigation [2].

The traditional approach is to divide driving tasks into standard modules
such as target detection, localization, path planning, etc. and then build rule-
based methods to connect the different modules [3]. The modular architecture

© The Author(s), under exclusive license to Springer Nature Switzerland AG 2025
Q. Zhang et al. (Eds.): IJCRS 2025, LNAI 15710, pp. 321–334, 2025.
https://doi.org/10.1007/978-3-031-92741-6_24

enables the development of each module in an independent fashion facilitating the collaboration between all elements of the engineering team. One of the great advantages of these modular approaches is the ease of interpretation. However, the major disadvantage of modular systems is the arduous task of developing and maintaining the interconnection between all modules in the system. For example, different scenarios may require different connections between modules [4], which compromises the modularity paradigm. The modular architecture is also prone to error propagation, in which a minor error in one module can produce catastrophic results in another. Additionally, as the modules are task-specialized, they may fail to generalize to unusual conditions and unexpected situations.

Deep learning has recently advanced significantly in areas like image classification and speech recognition, leading to its increased use in autonomous driving applications, which involve planning, perception, mapping, and localization. Vehicle control in autonomous driving heavily depends on perception, prompting many techniques to utilize convolutional neural networks (CNNs) for predicting control actions directly from scene images. However, CNNs primarily capture local information, often overlooking temporal data, which can hinder performance in complex scenarios. In contrast, recurrent neural networks (RNNs) are trained on finite-length labeled sequences using backpropagation on their unfolded feedforward representation. However, training RNNs has been difficult due to issues like exploding or vanishing gradients. The introduction of advanced gated RNNs, such as long short-term memory (LSTM) networks, has mitigated these challenges by maintaining a constant error flow through fixed recurrent weights and removing nonlinearities in the feedback path. However, in end-to-end autonomous driving, learning long-term dependencies can be counterproductive, as the task often relies on short-term causality. For instance, when driving, humans typically do not recall road images from more than a few seconds prior. Furthermore, while vehicle camera images provide crucial driving information, they also include irrelevant features like the sky and trees. Many researchers have overlooked the impact of these irrelevant features on autonomous driving models, opting instead to process road images directly with CNNs—either mapping them to steering commands or generating feature maps for RNNs. This approach forces RNNs to incorporate irrelevant information, complicating the extraction of necessary features for decision-making. Consequently, while current end-to-end methods may perform well in training, they can struggle in diverse scenarios and incur higher computational costs, compounded by the black-box nature of deep neural networks. This complexity complicates the explanation of their operations, a critical aspect for security in end-to-end control systems.

In this work, we proposed a new model for steering angle prediction, which combines a specialized CNN to extract features from drivable areas and a Neural Circuit Policy (NCP) inspired by the neural circuits of the brain of C. elegans. This fusion aims to develop an interpretable and lane-centric steering angle prediction model. The main contributions are outlined as follows:

(1) We present a bio-inspired neural circuit framework that concentrates on drivable regions, extracting only a subset of decision-relevant features, enhancing

the model's performance for steering prediction and its generalizability across various scenarios, validated experimentally on three prominent datasets.
(2) We introduce binary mask and multi-scale spatio-temporal feature extraction module (Multc) to extract multi-scale texture features and dynamic temporal information of the drivable area, which avoids the misleading of interfering information to the model, and further acquires rich spatial and temporal information to guide the forecasting.

2   Related Work

With the advancement in hardware computing power and the rapid progress of deep learning methods, end-to-end driving models have gained increasing attention from researchers in recent years [5]. An early effort to directly link camera images to control commands through a simple network is ALVINN (Autonomous Land Vehicles in Neural Networks). Even in relatively simple driving scenarios with few obstacles, ALVINN demonstrated the significant potential of neural networks for autonomous navigation. In 2016, the study by [6] introduced a system named PilotNet, which uses CNNs to extract features from input images through a five-layer CNN and then outputs steering angle information directly through a fully connected layer. This method showed promising prediction results on highways. Subsequently, Xu et al. [7] expanded this framework by including normalization layers, convolutional layers, and fully connected layers, leveraging a large-scale video dataset to predict vehicle self-motion, thereby enhancing robustness across various scenarios.

While CNN-based models are adept at extracting local details, they often miss out on the temporal data essential for accurate steering angle predictions. Chi and Mu utilized Long Short-Term Memory (LSTM) networks to incorporate this temporal aspect, treating steering angles as continuous variables and training the networks to minimize the discrepancy between predicted and expert angles. However, from a time-series analysis perspective, maintaining a consistent error flow is crucial, as data sequences may exhibit long-term correlations. Yet, for end-to-end autonomous driving, focusing too much on long-term dependencies may backfire due to the task's inherent short-term focus. Addressing this, Lechner Mathias and colleagues introduced a biologically inspired model known as neural circuit policies (NCPs), featuring a streamlined algorithm with 19 control neurons that link 32 input features to outputs through 253 synapses. At the heart of NCPs are Liquid Time Constant (LTC) networks, which process outputs using numerical differential equation solvers. This approach, highlighted in neural ordinary differential equation frameworks, significantly boosts the predictive accuracy of time series tasks. NCPs, with their compact and sparse structure built on expressive neural mechanisms (LTC model), offer a marked improvement in lane-keeping performance over larger, more opaque learning systems.

Another challenge for deep neural networks in steering angle prediction is the model interpretability. To enhance the interpretability, the study in [8] aimed to pinpoint the most prominent targets or areas in the input image to determine the steering angle of the network. Furthermore [9,10], employed the attention mechanisms to reveal the specific regions of the road image that the model prioritizes. However, attention may be directed towards irrelevant features, leading to potentially misleading interpretations and increased computational overhead.

To minimize the impact of irrelevant data, the research in [11] utilized a tracking algorithm to predict steering angles. This approach directs the vehicle to follow the centerline of the lane by considering image-level labels. It effectively filters out unnecessary image features for decision-making while preserving essential features for lane-keeping. Recently, Wang Mohammadi and colleagues [12] introduced a cost-effective method to decrease input size without sacrificing accuracy. They employed the Feature Density Metric (FDM) to identify and filter areas of the input image with insufficient features. This optimization decreases the computational resources required for training and inference in Deep Convolutional Neural Networks (DCNN), eliminating unnecessary details such as sky portions and roadside buildings.

3   Method

This section outlines our proposed model for bio-inspired neural circuits focusing on drivable regions. As shown in Fig. 1. The proposed model consists of three modules, i.e., the travelable area extraction module, the lane feature extraction module, and the temporal information extraction module.

3.1   Travelable Area Extraction Module

The module utilizes a lane segmentation backbone network [13], featuring an Efficient Spatial Pyramid Network (ESPNet) encoder that balances low computational cost with high accuracy. Dual attention modules are included to capture global dependencies across spatial and channel dimensions, enhancing contextual understanding. The feature mapping $A \in \mathbb{R}^{32 \times C \times \frac{H}{8} \times \frac{W}{8}}$ obtained from ESPNet-C is processed through a double attention block. The decoder, designed for efficiency, employs a ConvTranspose layer followed by batch normalization and pReLU activation. We modified the decoder to focus solely on lane segmentation results, further reducing computational costs and improving model efficiency.

3.2   Lane Feature Extraction Module

This module employs a binary mask to accurately extract lane areas and avoid interference from non-lane areas. Particularly, in this paper, we propose a multiscale spatiotemporal feature extraction module (Multc) that simultaneously captures local detailed texture features and global information as well as local temporal correlations of the drivable region to avoid interference from non-lane areas. This mask is generated by assigning a value of 1 to pixels representing lane categories and 0 to non-lane pixels. Represented as a 2D array, each element corresponds to a pixel in the image, taking on values of 0 or 1. A value of

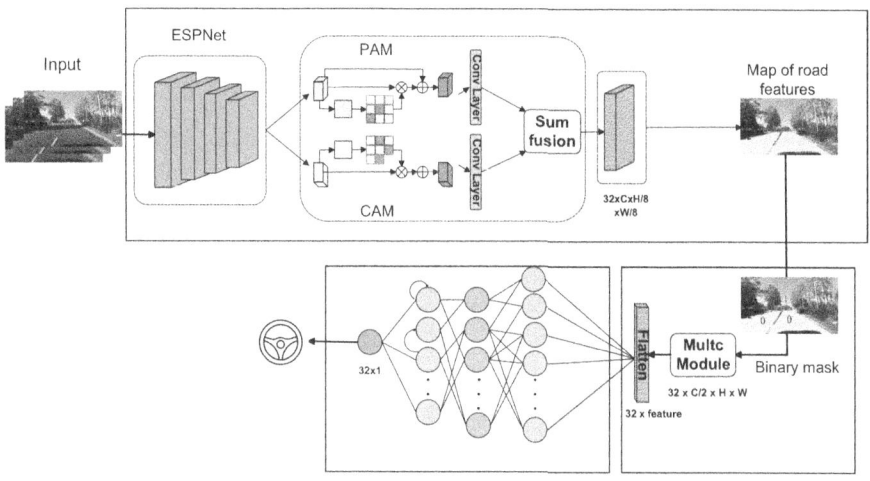

Fig. 1. Architecture overview of the proposed model. (a) Travelable area extraction module, which consists of an encoder, a decoder, and a dual attention module; (b) Lane feature extraction module, which employs a binary mask and a multi-scale spatio-temporal feature extraction module which employs a binary mask and a multi-scale spatio-temporal feature extraction module to focus only on the lane region; (c) Temporal information extraction module uses neural circuit policies NCPs.

0 indicates that the pixel belongs to the lane area. The binary mask, denoted as $\boldsymbol{B}$, has dimensions $M \times N$, where M is the height of the image and N is the width of the image. Each element $B_{i,j}$ in the binary mask represents the value of the pixel in row i and column j.

$$
B_{i,j} = \begin{cases} 0, & \text{if the pixel at } (i,j) \text{ belongs to the lane region} \\ 1, & \text{otherwise} \end{cases} \tag{1}
$$

Here, i varies from 1 to M for image height, and j varies from 1 to N for image width. The resulting binary mask indicates the drivable area by extracting features from the essential lane region for the task.

To capture multiscale texture features and retain complete temporal information, we introduce the multiscale spatiotemporal feature extraction module (Multc). As illustrated in Fig. 2, it consists of a small kernel convolution for local information, followed by parallel deep convolutions for contextual information across multiple scales, and concludes with temporal convolution to enhance spatiotemporal consistency. Mathematically, the Multc Module is represented as

$$
L_{in} = \text{Conv}_{k_s \times k_s} \left(X_{in} \right),
$$
$$
D^{(m)} = \text{DWConv}_{k^{(m)} \times k^{(m)}} \left(L_{in} \right), m = 1, \cdots, 4. \tag{2}
$$

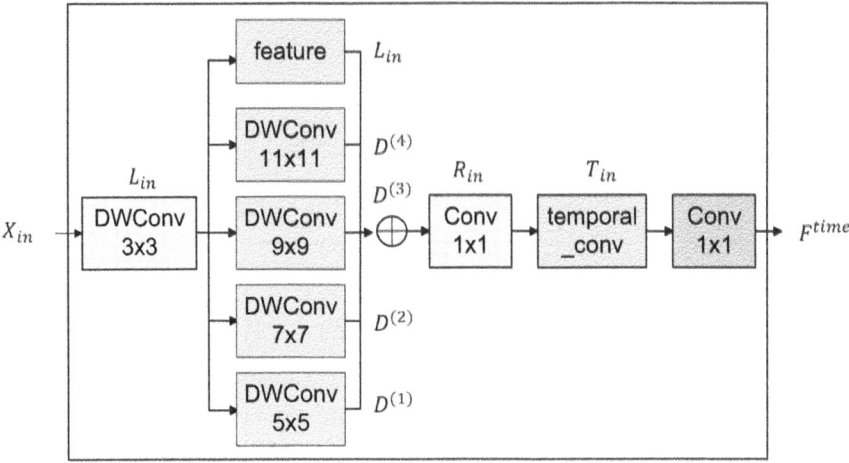

Fig. 2. Multc Module

Here Multc Module takes in $X_{in} \in \mathbb{R}^{T \times \frac{1}{2}C_l \times H_l \times W_l}$ and then output F^{time}, $\boldsymbol{L_{in}} \in \mathbb{R}^{T \times \frac{1}{2}C_l \times H_l \times W_l}$ is the local feature extracted by the $ks \times ks$ convolution, and $D^{(m)} \in \mathbb{R}^{T \times \frac{1}{2}C_l \times H_l \times W_l}$ is the context feature extracted by the m-th $k^{(m)} \times k^{(m)}$ depth-wise convolution (DWConv). Local texture information in the input data is initially extracted by a small convolutional kernel 3×3. This step ensures that fine-grained, high-resolution features are not lost and helps to capture subtle texture variations in the image. Fusion of local features and multi-scale context:

$$R_{in} = \text{Conv}_{1 \times 1} \left(L_{in} + \sum_{m=1}^{4} D^{(m)} \right) \qquad (3)$$

where $\boldsymbol{R_{in}} \in \mathbb{R}^{T \times \frac{1}{2}C_l \times H_l \times W_l}$ represents the rich spatial features. Different convolutional kernel sizes can extract contextual information from different scales. Larger convolutional kernels capture a larger range of contextual information, while smaller convolutional kernels preserve details. This multi-scale design makes the model more robust to changes in object size or texture. After the first channel fusion, temporal convolution is performed on the fused features along the time dimension

$$T_{in} = \text{Conv}_{2D} \left(R_{in}, 3 \right) \qquad (4)$$

here T_{in} is the output feature after applying temporal convolution with a convolution kernel of 3 contains information on dynamic changes across time steps. Such a design can learn the motion and dynamic relationships between consecutive frames in the sequence data, ensuring the complete retention and utilization of temporal information. Finally, the obtained spatiotemporal features are integrated and downscaled by convolution of size 1×1:

$$F^{time} = \text{Conv}_{1 \times 1} \left(T_{in} \right) \qquad (5)$$

where $F^{time} \in \mathbb{R}^{T \times \frac{1}{2}C_l \times H_l \times W_l}$ represents the output feature. In this way, our Multc module captures a wide range of contextual and temporal dynamic information without compromising the integrity of local texture features.

3.3   Temporal Information Extraction Module

In this module, we employ NCPs inspired by neural computation in the nematode brain [14]. Neural circuits in the nematode nervous system consist of various four-layer hierarchical network topologies. NCPs feature a nonlinear time-varying synaptic transmission mechanism that enhances their expressive power in time series modeling compared to traditional deep learning models. The neural components at the core of NCP are known as Liquid Time Constant (LTC) networks. LTC is a continuous-time recurrent neural network model, with outputs computed using numerical differential equation solvers. These networks demonstrate stable and bounded behavior, offer enhanced expressivity among neural ordinary differential equations, and lead to improved performance in time-series prediction tasks [15].

$$\frac{dx(t)}{dt} = -x(t)\left[\frac{1}{\tau} + f(x(t), I(t), \theta)\right] + Af(x(t), I(t), t, \theta) \qquad (6)$$

where f is the neural network, $I(t)$ is the input, τ is the time constant, $x(t)$ is the hidden layer, and θ and A are parameters. The neural network f not only determines the derivative of the hidden $x(t)$ state but also acts as a variable time constant with respect to the input.

3.4   Performance Metric and Loss Function

Performance Metric. In this work, we employed the root mean square error (RMSE) to assess the performance of the proposed model.

$$\text{RMSE} = \sqrt{\frac{1}{N}\sum_{i=1}^{N}(y_g^i - y_p^i)^2} \qquad (7)$$

where y_g^i is the actual value. y_p^i denotes the predicted value of sample i. N represents the overall sample size. A lower RMSE value indicates better predictive performance. For statistical significance assessment, we performed Wilcoxon rank sum tests within the proposed model against each benchmark algorithm, adjusting for the false discovery rate (FDR).

Loss Function. The parameters of the proposed model are updated by minimizing the following cost function

$$\mathcal{L}_{steer} = \sqrt{\frac{1}{n}\sum_{i=1}^{n}\left(\mathcal{Y}_i - \hat{\mathcal{Y}}_i\right)^2} \qquad (8)$$

where $\mathcal{L}_{steer}$ represents the loss function, $\mathcal{Y}_i$ denotes the true values. $\hat{\mathcal{Y}}_i$ represents the model's predictions, n is the number of samples.

4   Experiments and Implementation Details

4.1   Dataset

SullyChen Dataset [16]: The data was collected using a 2014 Honda Civic in Rancho Palos Verdes and San Pedro, California. The dataset comprises approximately 45,500 frontal road images captured at a time resolution of 0.05 s. Qian et al. [17] utilized an initial version of this dataset in their research. It includes straight roads, mixed roads, intersections, main roads, and secondary roads. Over 40,000 data points were collected in the experiment, covering various terrains and weather conditions. In this experiment, we selected 90% of the data from each road category and encompassed various weather conditions. We discarded turns exceeding 180° to eliminate some unreasonable maneuvers.

CARLA Simulator Dataset [18]: This dataset features two towns with two-way roads, turns, intersections, buildings, vegetation, urban elements, traffic signs, lights, and dynamic objects like vehicles and pedestrians. Town 1 spans 2.9 km with 11 intersections, while Town 2 covers 1.4 km with 8 intersections. These towns offer exploration under six weather conditions: 'clear noon', 'clear after rain', 'heavy rain noon', 'clear sunset', 'wet cloudy noon', and 'soft rainy sunset'. In this experiment, we select 90% of the data for each town, including various weather conditions.

Udacity Simulator Dataset [19]: This dataset originates from Udacity's open-source autonomous driving simulation platform. It consists of 24,108 images captured by the simulator, corresponding to 8,036 steering wheel angle records (each record is generated by three simultaneous pictures from left, middle, and right cameras). The original image size is 160×320.

4.2   Benchmark Methods

In order to fully evaluate the performance of the proposed method, the following test algorithms are selected in this paper, They cover a wide range of strategies from traditional CNN methods to the fusion of temporal information and bio-inspired mechanisms, ensuring the diversity and breadth of comparison experiments, which allows for a more comprehensive validation of the robustness and generalization ability of our methods under different scenarios and network structures:

(1) PiloNet [6]: As a classical end-to-end convolutional neural network, its structure is relatively simple and provides us with a basic benchmark for comparison. The network comprises 9 layers, including a normalization layer, 5 convolutional layers, and 3 fully connected layers. The convolutional layers were designed for feature extraction, with strided convolutions in the first three layers using a 2×2 stride and a 5×5 kernel, and a non-strided convolution in the last two layers using a 3×3 kernel. These are followed by three fully connected layers that lead to an output control value (the steering angle).

(2) CNN-NCP [14]: Incorporating a bio-inspired neural circuit strategy that emphasizes the capture of temporal dynamic information, it is able to demonstrate the advantages of neural circuit strategy-based approaches in processing dynamic information. The convolutional head consists of five layers that extract features for the Neural Circuit Policies (NCP) to generate control commands. The first three convolutional layers use 5×5 kernels the last three layers use 3×3 kernels. Passive testing was conducted using around five hours of driving data from the Boston metropolitan area, followed by active testing on private roads, resulting in satisfactory outcomes.

(3) BO-ST-LSTM [20]: Tuning the hyperparameters of long temporal memory networks using Bayesian optimization demonstrates another effective solution for temporal-dependent learning. The network comprises ConvLSTM layers and a 3DConv layer. BO is used to optimize the hyperparameters of a long-term spatial memory (ST-LSTM) network, which yields good performance on the SullyChen dataset.

(4) LaksNet [21]: represents the latest prediction networks based on multi-scale convolutional design, which can reflect the recent advances in deep learning in dealing with complex scenarios. The model includes four convolutional layers, four max-pooling layers, two dropout layers, and two fully connected layers, with one serving as an output layer. The first three convolutional layers use 3×3 kernels, while the last layer uses 5×5 kernels, resulting in extended travel times within the simulator.

4.3   Implementation Details

Experiments were carried out using PyTorch with two NVIDIA RTX 3090ti GPUs. Pixel values of image frames were normalized to $[-1, 1]$, and inputted as RGB images. The epoch was set to 50, the batch size to 4 for our network, and 32 for other models. All time series lengths are 32. The initial learning rate was 0.001, gradually reduced by a factor of 0.9 every 5 iterations. Adam served as the optimizer. Data were split into ten equally sized non-overlapping blocks for cross-testing. Each model was trained in nine sets and evaluated on a separate test set. Training periods were optimized based on a validation set, distinct from the training data. All methods followed a consistent approach to training and validation. The source code of the proposed model is available at https://github.com/wyl121/TLNBm-NCP

5   Results

5.1   Compared with the State-of-the-Art Methods

We compared our proposed method with baseline approaches on three public datasets (SullyChen, Udacity, and Carla), summarizing the results in Table 1. All methods underwent ten-fold cross-validation. Our method consistently achieved the lowest RMSE values across all datasets, indicating superior performance in predicting steering angles, particularly on the real-world dataset SullyChen.

Notably, each dataset features diverse road environments with varying weather conditions and scenes. However, the influence of these changes on our model is minimal. This suggests that focusing on relevant features as well as the Multc module's extraction of multi-scale texture special and dynamic temporal information for relevant features can enhance the model's ability to generalize the model across different weather conditions and scenarios in different scenarios.

Table 1. RMSE values (mean $\pm$ std) of TLNBm-NCP and other methods for tenfold cross-validation on three datasets. The ** indicates that the prediction accuracy of TLNBm-NCP was significantly better than the compared methods with $p < 0.01$, where p is calculated by rank sum test corrected by FDR.

Method	SullyChen	Carla	Udacity
PiloNet (2017)	$18.81 \pm 0.7^{**}$	$0.13 \pm 0.007^{**}$	$0.37 \pm 0.01^{**}$
CNN-NCP (2020)	$3.40 \pm 0.5^{**}$	$0.14 \pm 0.005^{**}$	$0.45 \pm 0.007^{**}$
BO-ST-LSTM (2022)	$2.45 \pm 0.48^{**}$	$0.19 \pm 0.008^{**}$	$0.41 \pm 0.5^{**}$
LaksNet (2023)	$13.70 \pm 1.42^{**}$	$0.15 \pm 0.006^{**}$	$0.38 \pm 0.008^{**}$
TLNBm-NCP	0.90 ± 0.74	0.01 ± 0.0007	0.11 ± 0.0032

Figure 3 provides a qualitative comparison between our proposed method and the baseline on different tracks in the SullyChen dataset, the Udacity dataset, and the Carla dataset, all with the same number of frames. The environments of the tracks vary significantly: the SullyChen dataset features multiple sections with tree shading, the Udacity dataset includes multiple curved road segments, and the Carla dataset represents road segments with water puddles after rain, all of which pose significant challenges. In Fig. 3(a), our method displays smooth and consistent prediction curves closely aligning with the actual angle values in the initial frames, unlike other methods that show greater fluctuations and instability, an advantage that persists in subsequent frames. In Fig. 3(b), Fig. 3(b) shows that our model trends closer to the actual values, indicating higher accuracy compared to others. In Fig. 3(c), the predicted values from our approach align closely with the actual angles, demonstrating its superior performance. Taken together, these observations confirm that our method exhibits the strongest generalization ability as well as robustness across different orbits. This is mainly due to the fact that our method discards lane-independent interference regions, especially the Multc module proposed in this paper, which is crucial for capturing detailed texture and dynamic timing information by combining the design of small convolutional kernel localized feature extraction, parallel multiscale depth-separable convolution, and subsequent temporal convolution. This feature integration not only leads to smoother prediction results in the initial frame but also maintains a high level of consistency in subsequent frames.

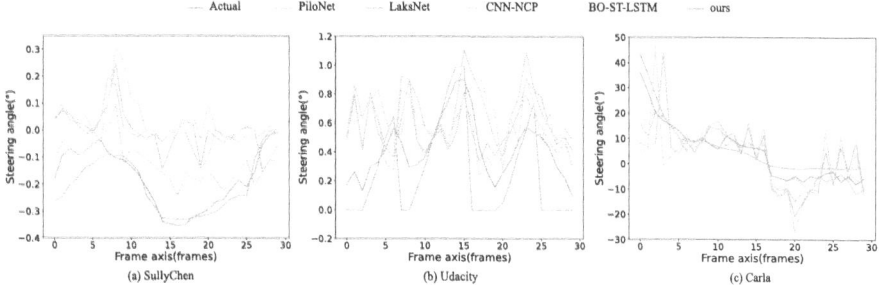

Fig. 3. Qualitative comparison of the proposed and baseline methods on three datasets. (a)Data taken from multi-shade sections in the SullyChen dataset. (b) Data taken from the Udacity simulator dataset of multi-curved road segments. (c) Data taken from wetland sections after rainfall in the Carla simulator dataset.

5.2   The Effect of Binary Masking and Multc Module

To evaluate the impact of the binary mask and the Multc module, we divided the model into four versions: our proposed model, TLNBm-NCP; a baseline model (without the binary mask and Multc module); a model incorporating only the binary mask module; and a model incorporating only the Multc module. These versions were compared across three datasets. As shown in Table 2, the removal of the Multc module from TLNBm-NCP results in a significant increase in RMSE and a greater fluctuation in prediction error. Conversely, while retaining the Multc module and removing the binary mask, we observe a significant increase in error fluctuation, although the RMSE value remains low. The baseline model, which lacks both the binary mask and the Multc module, shows notably higher RMSE values and larger error fluctuations. In contrast, the model that retains both the binary mask and the Multc module maintains a low mean RMSE value and minimal error fluctuation.

These findings strongly suggest that the binary masking module helps the model to select the most informative lane features, which are typically distorted by noise, causing the model to learn irrelevant information or misleading patterns. By using the Binary Masking Module, these distorted features are masked, thus reducing the interference with the model. In addition, it is fully verified that the design of the Multc module is a key factor in improving the model performance, and its ability to capture multi-scale spatial information and dynamic temporal features plays a decisive role in the generalization performance of the model in a variable environment. The local convolution kernel size k_s of the Multc module determines the ability to extract fine-grained texture information, while the multi-scale depth-separable convolution kernel size $k^{(m)}$ (3×3, 5×5, 7×7, 9×9, 11×11) directly affects the effectiveness of the model in capturing different ranges of contextual information. A temporal convolution with a convolution kernel of 3, on the other hand, can capture enough dynamic temporal information, which in turn helps the NCPs to better handle temporal relationships.

Table 2. Performance Comparison with and without Binary Masks and Multc Module. T: Travelable area segmentation module; B: Binary mask; M: Multc module.

Method	SullyChen	Carla	Udacity
T+NCP	$2.62 \pm 1.62^{**}$	$0.15 \pm 0.09^{**}$	$0.43 \pm 0.012^{**}$
T+B+NCP	$2.05 \pm 0.81^{**}$	$0.11 \pm 0.0038^{**}$	$0.37 \pm 0.0013^{**}$
T+M+NCP	$1.87 \pm 1.32^{**}$	$0.09 \pm 0.02^{**}$	$0.28 \pm 0.009^{**}$
TLNBm-NCP	0.90 ± 0.74	0.01 ± 0.0007	0.11 ± 0.0032

5.3   Visualization of Attention in Convolutional Layers

We calculated CAM heatmaps by applying CAM computation to various network convolutional feature maps. We set thresholds for the heatmaps (as the peak values of the heatmaps vary, requiring different suitable thresholds for each method, which we determined through multiple experiments), identifying regions representing the network's focus of attention. Subsequently, we adopted ROI selection boxes of consistent size (height: 20 pixels, width: 43 pixels) to enhance the visibility of visual results. As shown in Fig. 4, our method primarily focuses on the lane, akin to human attention during driving, emphasizing the lane while disregarding surrounding trees and buildings. In contrast, the PiloNet network seems to prioritize roadside features while neglecting the road itself. In LaksNet, the CNN mainly focuses on roadside trees to facilitate driving decisions, whereas NCP directs attention towards the horizon of the road. However, LSTM exhibits a tendency to focus more on roadside features.

Fig. 4. The selected ROI for each network. The ROI selection box displays the locations each network learns to focus on during the driving process.

6 Conclusion

The extraction of highly task-relevant feature subsets has been a primary focus in the realm of steering angle prediction using visual images. We proposed a neural circuit policy method based on drivable areas to predict vehicle steering angle. The proposed model employs a binary mask to distinguish drivable and non-drivable regions. Subsequently, we introduce a multiscale spatio-temporal feature extraction module (Multc) to extract the multiscale texture features and dynamic temporal information of the lane region, which is extracted to a subset of features that are closely related to the task at hand, and integrate the temporal information through a neural circuit policies (NCPs). Extensive experiments have been conducted on three distinct datasets. The experimental results indicate that our model effectively identifies lane markings and demonstrates superior predictive performance across diverse road conditions. In comparison to existing methods, our drivable area-based neural circuit policy method has exhibited enhanced performance in steering angle prediction under varying driving conditions.

Acknowledgement. This work was supported in part by the National Natural Science Foundation of China under Grants 62476034, and the Natural Science Foundation of Chongqing under Grant CSTB2022NSCQ-MSX0291.

References

1. Gidado, U.M., Chiroma, H., Aljojo, N., Abubakar, S., Popoola, S.I., Al-Garadi, M.A.: A survey on deep learning for steering angle prediction in autonomous vehicles. IEEE Access **8**, 163797–163817 (2020)
2. Saleem, H., Riaz, F., Mostarda, L., Niazi, M.A., Rafiq, A., Saeed, S.: Steering angle prediction techniques for autonomous ground vehicles: a review. IEEE Access **9**, 78567–78585 (2021)
3. Liu, F., Zihao, L., Lin, X.: Vision-based environmental perception for autonomous driving. Proc. Inst. Mech. Eng. Part D J. Automobile Eng. **239**(1), 39–69 (2025)
4. Bi, J., et al.: Lane detection for autonomous driving: Comprehensive reviews, current challenges, and future predictions. IEEE Trans. Intell. Transp. Syst. (2025)
5. Li, G., Ji, Z., Li, S., Luo, X., Xingda, Q.: Driver behavioral cloning for route following in autonomous vehicles using task knowledge distillation. IEEE Trans. Intell. Veh. **8**(2), 1025–1033 (2022)
6. Bojarski, M., et al.: End to end learning for self-driving cars. Computing Research Repository (2016)
7. Xu, H., Gao, Y., Yu, F., Darrell, T.: End-to-end learning of driving models from large-scale video datasets. In: Proceedings of the IEEE Conference on Computer Vision and Pattern Recognition, pp. 2174–2182 (2017)
8. Bojarski, M., et al.: Explaining how a deep neural network trained with end-to-end learning steers a car. arXiv preprint arXiv:1704.07911 (2017)
9. Kim, J., Canny, J.: Interpretable learning for self-driving cars by visualizing causal attention. In: Proceedings of the IEEE International Conference on Computer Vision, pp. 2942–2950 (2017)

10. Kim, J., Rohrbach, A., Darrell, T., Canny, J., Akata, Z.: Textual explanations for self-driving vehicles. In: Proceedings of the European Conference on Computer Vision (ECCV), September 2018
11. Wang, T., Luo, Y., Liu, J., Chen, R., Li, K.: End-to-end self-driving approach independent of irrelevant roadside objects with auto-encoder. IEEE Trans. Intell. Transp. Syst. **23**(1), 641–650 (2020)
12. Mohammadi, A., Jamshidi, K., Shahbazi, H., Rezaei, M.: Efficient deep steering control method for self-driving cars through feature density metric. Neurocomputing **515**, 107–120 (2023)
13. Che, Q.-H., Nguyen, D.-P., Pham, M.-Q., Lam, D.-K.: Twinlitenet: an efficient and lightweight model for driveable area and lane segmentation in self-driving cars. In: 2023 International Conference on Multimedia Analysis and Pattern Recognition (MAPR), pp. 1–6. IEEE (2023)
14. Lechner, M., Hasani, R., Amini, A., Henzinger, T.A., Rus, D., Grosu, R.: Neural circuit policies enabling auditable autonomy. Nature Mach. Intell. **2**(10), 642–652 (2020)
15. Hasani, R., Lechner, M., Amini, A., Rus, D., Grosu, R.: Liquid time-constant networks. In: Proceedings of the AAAI Conference on Artificial Intelligence **35**, 7657–7666 (2021)
16. Chen, S.: A collection of labeled car driving datasets. Collection of labeled car driving datasets (2018)
17. Qian, D., et al.: End-to-end learning driver policy using moments deep neural network. In: 2018 IEEE International Conference on Robotics and Biomimetics (ROBIO), pp. 1533–1538. IEEE (2018)
18. Dosovitskiy, A., Ros, G., Codevilla, F., Lopez, A., Koltun, V.: Carla: an open urban driving simulator. In: Conference on Robot Learning, pp. 1–16. PMLR (2017)
19. Manchekar, S., Parsi, B., Thakur, N., Bielaski, K.: Simulation of self driving car
20. Riboni, A., Ghioldi, N., Candelieri, A., Borrotti, M.: Bayesian optimization and deep learning for steering wheel angle prediction. Sci. Rep. **12**(1), 8739 (2022)
21. Polamreddy, L.R., Zhang, Y.: Laksnet: an end-to-end deep learning model for self-driving cars in udacity simulator. In: Proceedings of the Future Technologies Conference, pp. 1–13. Springer (2023)

Multi-exposure Correction via Feature Transfer and Calibration

Jinchi Li, Di Wang, and Xiuyi Jia[✉]

School of Computer Science and Engineering, Nanjing University of Science and Technology, Nanjing 210094, China
{jinchili,wd,jiaxy}@njust.edu.cn

Abstract. Images captured by cameras with incorrect exposure settings typically suffer from degradations in multiple aspects, including brightness, color, and structure. Existing methods often obtain a normally exposed image by fusing images with varying exposures. However, these methods need to calculate the exposure representations of different images, which increases the computational burden. Some other models work on correcting the underexposures of a single image. However, the large differences between different exposures may cause the poor generalization. In this paper, we propose a unified exposure correction network (MECNet), the core of which is an exposure normalization and adaptive transfer (ENAT) block. Specifically, ENAT block employs normalization to align different exposures, and transfers the information unaffected by abnormal exposures to mitigate the normalization effects. We also introduce the shuffle attention to establish communication between multiple-exposures, and make the model more focused on areas with abnormal exposures. In addition, we also design a residual-guided feature calibration (RGFC) block for the up and down sampling layers, which can control the flow of information at different scales. Extensive experiments demonstrate that our method achieves competitive performance while maintaining a low amount of parameters, and can be applied to other enhancement tasks, proving the potential and effectiveness in visual tasks.

Keywords: Multi-exposure correction · Exposure normalization · Adaptive transfer · Shuffle attention · Residual-guided feature calibration

1 Introduction

As camera equipment becomes more advanced, people can capture images in a wide range of scenarios. However, due to different lighting conditions in various scenes, the captured images often suffer from underexposure or overexposure, resulting in undesirable visual effects, which can even affect some downstream tasks such as object detection and image classification [3,12,39].

In recent years, many approaches have been proposed to solve the problem of images with incorrect exposures, including conventional and deep learning-based methods [1,7,11,30]. Some of these methods obtain a normally exposed image

© The Author(s), under exclusive license to Springer Nature Switzerland AG 2025
Q. Zhang et al. (Eds.): IJCRS 2025, LNAI 15710, pp. 335–349, 2025.
https://doi.org/10.1007/978-3-031-92741-6_25

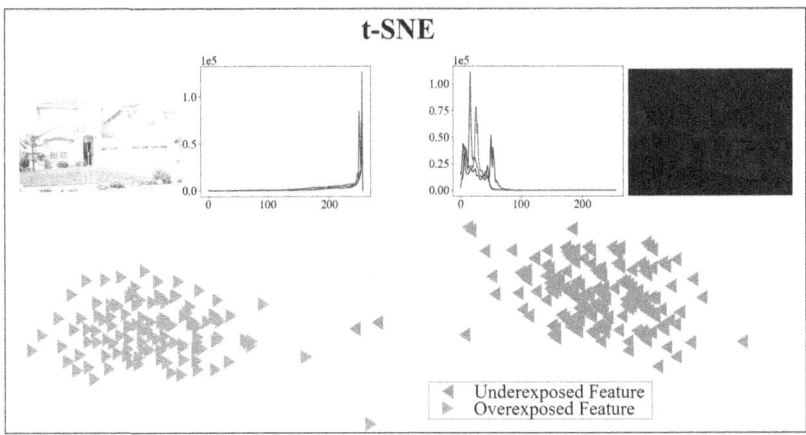

Fig. 1. t-SNE visualization of different exposure features.

by fusing an image sequence with different exposures. However, these models lack robustness in changing scenarios and have a heavy computational burden. Some models are designed to restore underexposed images, but overexposed feature representations and underexposed feature representations exhibit significant differences in their distributions, as shown in Fig. 1. Moreover, according to the statistical analysis of pixel values, overexposed images tend to lose more information (with a majority of pixels approaching 255), making their restoration more challenging. Directly applying an underexposure restoration model to recover overexposed images results in suboptimal performance. One feasible approach is fine-tuning the model on overexposed data, but this method significantly increases training time and parameter complexity. Additionally, the model needs to handle the diverse brightness and color variations during multiple training iterations, leading to a bias towards ignoring unfavorable data [37] and causing the imbalance performance across different exposure levels.

In this paper, we propose a unified exposure correction network (MECNet), the core of which is an exposure normalization and adaptive transfer block. The ENAT block aligns different exposure features to reduce the distribution variances. In order to address the issue of discriminative information loss [19,29] during the normalization process, we utilize Adaptive Instance Normalization (AdaIN) [18] to transfer the normal exposure information to the aligned features, and ensure the integrity of the information. To model dependencies between varying degrees of exposures, we also introduce the shuffle attention [25,44,45] to establish communication between different exposures, and make the model more focused on the areas not exposed properly. Our model is based on the multi-scale architecture, which can capture pattern information across scales. A residual-guided feature calibration block is proposed to build the interaction between up and down sampling layers, and control the flow of information by calculating the differences in feature distributions. Extensive experiments

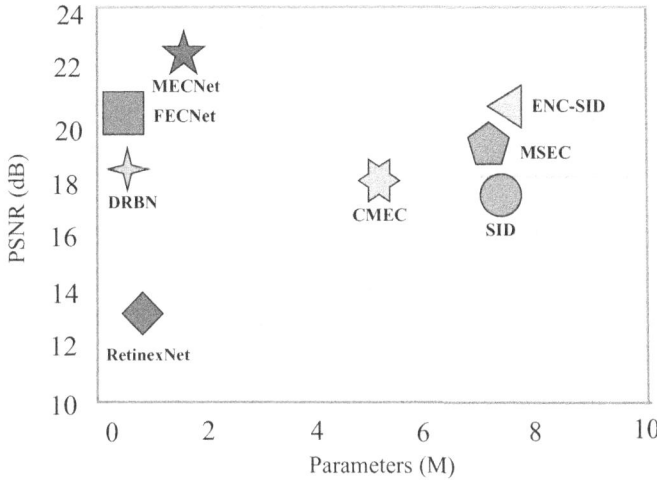

Fig. 2. The best PSNR-parameter trade-off of the SOTA models.

demonstrate that MECNet achieves competitive performance while maintaining a low amount of parameters, and can be applied to other enhancement tasks, proving the potential and effectiveness in visual tasks. The best PSNR-parameter trade-off of the state-of-the-art (SOTA) models is presented in Fig. 2.

The contributions of this paper are summarized as following:

– We propose the MECNet that can be trained on the multi-exposure datasets. MECNet can uniformly correct both the overexposed and underexposed images.
– We propose the exposure normalization and adaptive transfer (ENAT) block to align multi-exposure features, and use shuffle attention to make the model more attentive to areas with abnormal exposures. In addition, the normal exposure information is compensated to the normalized features by the Adaptive Instance Normalization (AdaIN).
– We propose the residual-guided feature calibration (RGFC) block to control the interaction between up and down sampling layers, and restore the lightness, color and other information by calibrating the feature distributions.

2   Related Work

Multi-exposure Image Fusion Methods. Extensive studies have been proposed to restore image exposures. Some conventional methods obtain a normally exposed result by fusing several images with different exposures [21, 22, 24]. These methods have a similar two-step fusion process, they first calculate the weight of each abnormally exposed image, and then fuse them by weights to get the final result. However, these methods rely on the hand-craft features, and they are less robust to the changing scenes. The deep learning-based methods fuse image pairs with extreme exposures [7,8,38]. Specifically, Deng et al. [8] propose a sparse coding network that can distinguish the common and unique features between underexposed and overexposed image pairs to achieve the optimal fusion effect. However, these models need to explore the information in both the underexposed and overexposed images, and process these information through multiple layers of convolution, which severely delays the inference speed.

Single Image Exposure Correction Networks. The conventional exposure correction methods tend to correct the lightness and contrast by histogram adjustment [1], and they can get the enhanced results with less contrast and brightness distortion. There are also some CNN-based models that are dedicated to restore the underexposed images. RetinexNet [34] decomposes the image into illumination and reflection, and then restore them by applying various constraints. MSEC [2] corrects the multiple-exposures with a pyramid structure, they enhance the components at different scales in a coarse-to-fine manner. FEC-Net [17] jointly recovers exposures in the frequency and spatial domains, which can restore the amplitude and phase representations that correspond to improving lightness and refining structures progressively. ENC-SID [16] develops an exposure normalization and compensation module, which is simple yet effective, but it only can be used as a plug-and-play module for existing exposure correction architectures.

Deep Learning-Based Image Restoration Methods. In recent years, CNN models have been shown to outperform conventional restoration approaches in visual tasks [9,13,27,36]. Among them, the encoder-decoder based architectures are widely used because of their ability to mine multi-scale features [6,31,42]. Some approaches introduce skip connections to pay special attention to the residual information and achieve state-of-the-art performance [10,14,23]. These architectures have great potential and deserve to be further investigated.

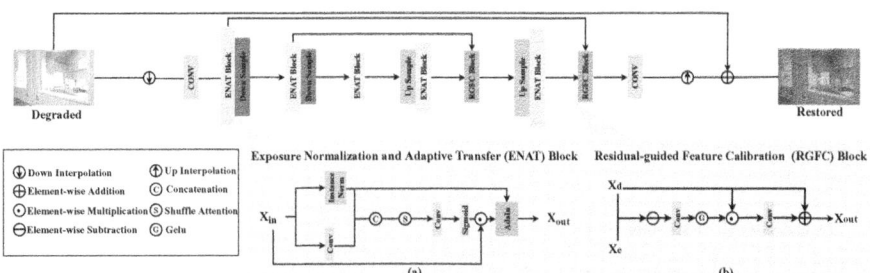

Fig. 3. Network structure of MECNet. Our model is composed of (a) the exposure normalization and adaptive transfer (ENAT) block and (b) the residual-guided feature calibration (RGFC) block.

3   Method

Images captured in various scenes are often overexposed or underexposed, which results in unsatisfactory visual effects. We aim to correct the multiple-exposures using a unified framework. The overall architecture of our MECNet is shown in Fig. 3, the core components of the proposed model are: (a) the exposure normalization and adaptive transfer (ENAT) block and (b) the residual-guided feature calibration (RGFC) block.

Given a degraded input $I \in \mathbb{R}^{H \times W \times 3}$, our network first applies a convolution to obtain the shallow features $F_s \in \mathbb{R}^{H \times W \times C}$, where $H \times W$ is the spatial dimension and C is the number of channels. Next, the features are processed by a multi-scale encoder-decoder architecture, where each level of the encoder and decoder contains ENAT blocks. The encoder gradually decreases the resolution of F_s while expanding the channels. The third level of the encoder outputs the high-level features $F_h \in \mathbb{R}^{\frac{H}{4} \times \frac{W}{4} \times 4C}$. The decoder then gradually restores F_h to the original size. In addition, there is an RGEC block at each scale to control the flow of information, and establish interaction between the up and down sampling layers to calibrate the intermediate features. Finally, a convolution layer is utilized to refine F_h, and then the original input is added to F_h to obtain the enhanced result. We describe our network in detail below.

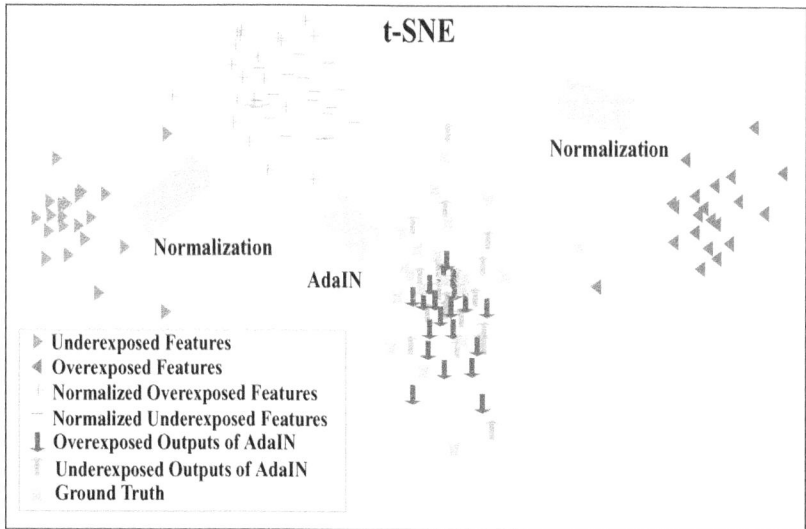

Fig. 4. t-SNE visualization of the features. As can be seen, after normalization, the multi-exposure features tend to be intersected together, and AdaIN can make them closer to the ground truth.

3.1   Exposure Normalization and Adaptive Transfer Block

There are large differences in the distributions of overexposed and underexposed features, which may negatively affect the training process and lead to performance degradation. We therefore design the ENAT block, as shown in Fig. 3(a). ENAT block starts by aligning the input multi-exposure features through normalization:

$$F_n = \gamma \frac{F - \mu(F)}{\sigma(F)} + \beta, \tag{1}$$

where γ and β are parameters learned from data, $\mu(\cdot)$ and $\sigma(\cdot)$ represent the mean and standard deviation of the input features F, respectively. The normalization operation can reduce the representation discrepancies of multiple-exposures and attenuate the exposure effect. But it can also cause the loss of information in the normally exposed areas that may benefit the subsequent reconstruction process. We further utilize AdaIN to transfer the information not affected by abnormal exposures to the aligned features, and ensure the integrity of information. This process can be described as:

$$AdaIN(F_s, F_c) = \sigma(F_s) * F_c + \mu(F_s), \tag{2}$$

where F_s is the style input and F_c is the content input. Our goal is to transfer the exposure style of F_s to the normalized content input F_c. We present the t-SNE [26] visualization of the above features in Fig. 4. It can be seen that normalization can narrow the gap of different exposure representations, and AdaIN can further restore the features closer to the ground truth.

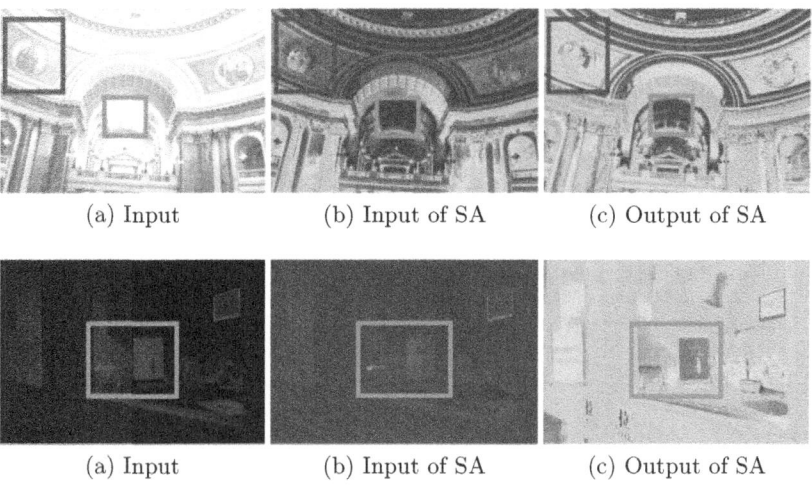

(a) Input   (b) Input of SA   (c) Output of SA

(a) Input   (b) Input of SA   (c) Output of SA

Fig. 5. Heatmaps of the features at different stages, where "SA" is the shuffle attention. The shuffle attention can give different weights to areas with various exposure levels.

The previous work [16] mitigates the impact of different exposure levels on features by incorporating supervised information from normally exposed images. We expect the model to have the ability to adaptively focus on areas that are more affected by abnormal exposures. We thus introduce the shuffle attention. We connect the aligned and original features, and then mix their channels to build communication between different exposure representations. Next, we calculate the attention map and add weights to the original features, and input them with the normalized features into the AdaIN module. The overall process can be written as:

$$F_{attn} = Sigmoid(Conv(Shuffle([X_n, X_{in}]))), \tag{3}$$

$$\widetilde{X_{in}} = F_{attn} * X_{in}, \tag{4}$$

$$X_{out} = AdaIN(\widetilde{X_{in}}, X_n), \tag{5}$$

where X_{in} denotes the input features, and X_n denotes the normalized features. $[\cdot]$ denotes the connection operation, $Shuffle$ is the channel shuffle operation, and the attention map F_{attn} is calculated by the Sigmoid function. Figure 5 shows the corresponding heatmaps. As can be seen, most areas of the original features have almost the same weights, which is not conducive to restoring the severely degraded parts. The shuffle attention gives higher weights to the areas with abnormal exposures, thus the model can pay more attention to the recovery of lightness, color, structure and other information in these areas.

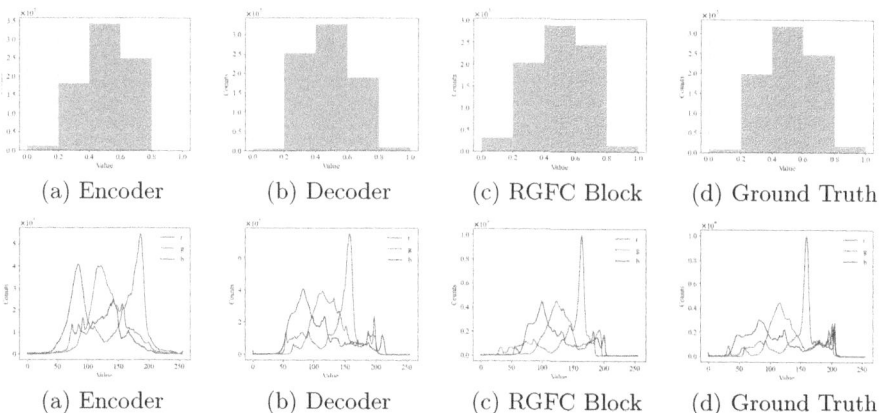

(a) Encoder   (b) Decoder   (c) RGFC Block   (d) Ground Truth

(a) Encoder   (b) Decoder   (c) RGFC Block   (d) Ground Truth

Fig. 6. Visualizations of feature distributions and the statistical results for the color channels.

3.2   Residual-Guided Feature Calibration Block

The shallow layers in the neural networks can capture the low-level features, such as lightness, color, outline and other information [43]. However, with an increase

Table 1. Quantitative evaluations on the two datasets in terms of PSNR, SSIM and the amount of parameters. The higher values of PSNR and SSIM indicate the better performance of the models.

Method	ME						SICE						Param
	Over		Under		Average		Over		Under		Average		
	PSNR	SSIM	PSNR	SSIM	PSNR	SSIM	PSNR	SSIM	PSNR	SSIM	PSNR	SSIM	
CLAHE	14.92	0.5910	16.83	0.6462	15.88	0.6186	12.27	0.5147	13.10	0.5044	12.69	0.5096	-
RetinexNet	10.43	0.5851	12.35	0.6113	11.39	0.5982	12.65	0.5369	13.96	0.5489	13.31	0.5429	0.84M
SID	18.75	0.7996	19.41	0.8122	19.08	0.8059	16.87	0.6517	18.15	0.6664	17.51	0.6591	7.40M
DRBN	19.64	0.8339	19.71	0.8423	19.68	0.8381	18.35	0.7005	18.47	0.7103	18.41	0.7054	0.53M
CMEC	22.58	0.8436	22.37	0.8149	22.48	0.8293	18.54	0.6918	17.96	0.6645	18.25	0.6782	5.40M
MSEC	19.81	0.8227	20.44	0.8295	20.13	0.8261	17.60	0.6673	19.21	0.6952	18.41	0.6813	7.04M
FECNet	23.22	0.8748	22.96	0.8598	23.09	0.8673	19.91	0.6961	22.01	0.6737	20.96	0.6849	**0.15M**
ENC-SID	22.36	0.8519	22.59	0.8423	22.48	0.8471	19.63	0.6941	**21.30**	0.6645	20.47	0.6793	7.45M
Ours	**24.24**	**0.9079**	**22.80**	**0.8852**	**23.52**	**0.8966**	**22.01**	**0.8943**	20.35	**0.8274**	**21.18**	**0.8609**	1.73M

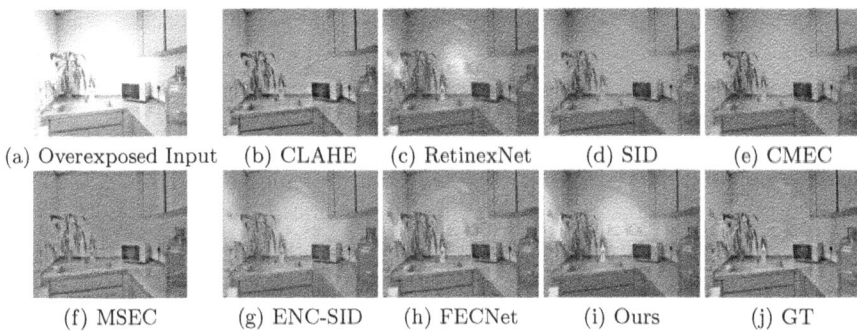

(a) Overexposed Input   (b) CLAHE   (c) RetinexNet   (d) SID   (e) CMEC

(f) MSEC   (g) ENC-SID   (h) FECNet   (i) Ours   (j) GT

Fig. 7. Visualization results on the SICE dataset of overexposure correction.

(a) Underexposed Input   (b) CLAHE   (c) RetinexNet   (d) SID   (e) CMEC

(f) MSEC   (g) ENC-SID   (h) FECNet   (i) Ours   (j) GT

Fig. 8. Visualization results on the ME dataset of underexposure correction.

of the network's depth, gradient vanishing and explosions may occur [14], and the low-level features degrade gradually. To deal with this issue, several previous works [14,31] add skip connections to integrate the shallow and deep features, and ensure the reusability. Similarly, in the exposure correction network, the lack of connections between the up and down sampling layers can lead to information bias, which in turn affects the quality of the images. We thus propose the RGFC block to guide the interaction of information, and model the feature differences of the up and down sampling layers.

The structure of RGEC block is shown in Fig. 3(b), the processing in which can be described as:

$$X_{res} = X_d - X_e, \tag{6}$$

$$X_{out} = Conv(X_d * GELU(Conv(X_{res}))) + X_d, \tag{7}$$

where X_d and X_e are the outputs of decoder and encoder, respectively. $Conv$ denotes the convolution operation, and $GELU$ represents the GELU activation function [41]. We also visualize the input and output features of RGEC block, and the statistical results for the color channels are shown in Fig. 6. As can be seen, when our model contains only encoder and decoder, there is a large deviation between the enhanced results and the ground truth. The differences are not only reflected in the overall pixel intensity distributions, but also in the inconsistency of color channels. After the processing of RGFC, these features are clearly moving closer to ground truth. These visualization results show that our RGEC block can effectively calibrate these features.

3.3   Loss Function

We use the L_1 loss as the reconstruction loss on the training set to penalize small errors:

$$L_{rec} = \frac{1}{N} \sum_{i=1}^{N} \left\| \widetilde{X}_i - X_i \right\|, \tag{8}$$

where N is the number of the training images, $\widetilde{X}_i$ is the output of our model, and X_i is the corresponding ground truth. We also introduce the perceptual loss [20] to constrain the recovery results at a deep feature level, and preserving the high-level semantic information of the original input image:

$$L_{per} = \frac{1}{C_j W_j H_j} \left\| \phi_j(\tilde{X}) - \phi_j(X) \right\|_2^2, \tag{9}$$

where VGG-19 [32] is selected as the feature extraction network $\phi(\cdot)$. C_j, W_j and H_j denote the feature size of the jth layer in the network. Therefore, the overall loss function can be further formulated as:

$$L_{total} = L_{rec} + \alpha L_{per}, \tag{10}$$

where α is set to 0.1 in this paper.

4   Experiment

4.1   Settings

We train our network on two public datasets, including the multiple exposure (ME) dataset, which is proposed in MSEC [2] and SICE dataset [4]. The images in the ME dataset contain five exposure levels, including 17675 images for training, 750 images for validation, and 5905 images for testing. The exposure levels in the SICE dataset are not uniform, thus we derive the middle-level exposure subset as the ground truth, and randomly select the corresponding overexposed and underexposed images as inputs. We adopt 300 images for training, 40 images for validation and 60 images for testing respectively.

4.2   Implementation Details

During training, we use the Adam optimizer with β_1 set to 0.9 and β_2 set to 0.999. We train our model with the patch size of 512×512 and batch size of 8. The initial learning rate is set to 1×10^{-5}, which decays by a factor value of 0.5 every 40 epochs. For the ME and SICE datasets, the total number of epochs is set as 80 and 160, respectively.

4.3   Evaluation

The network is compared with several state-of-the-art exposure correction methods. Among them, CLAHE [30] is the conventional method, RetinexNet [34], SID [5] and DRBN [40] are designed for restoring the low-light images, CMEC [28], MSEC [2], FECNet [17] and ENC-SID [16] can uniformly correct multiple-exposures. We use PSNR to measure the pixel-wise distance and SSIM to measure the structural similarity.

Quantitative Evaluation. The quantitative results are shown in Table 1. As can be observed, our model outperforms the other models in terms of PSNR and SSIM. The single image exposure correction algorithms CMEC, MSEC, FECNet and ENC-SID can achieve the similar performance to ours. It can also be seen that when trained on the multi-exposure datasets, there is a significant degradation in the performance of low-light enhancement networks, and these models are weak in their ability to restore the overexposed features. In addition, Our model contains relatively fewer parameters while maintaining the performance.

Qualitative Evaluation. Figure 7 and 8 present the correction results for multiple-exposures on the SICE and ME datasets. It can be seen that our model can restore the lightness and color, and produces the results colser to the ground truth.

Table 2. Quantitative results of the ablation studies. 'SA' denotes the shuffle attention, and 'RGFCB' is the RGFC block.

BaseLine	AdaIN	SA	RGFCB	PSNR
✓				18.03
✓	✓			19.46
✓	✓	✓		20.36
✓			✓	18.54
✓	✓		✓	19.97
✓	✓	✓	✓	**20.65**

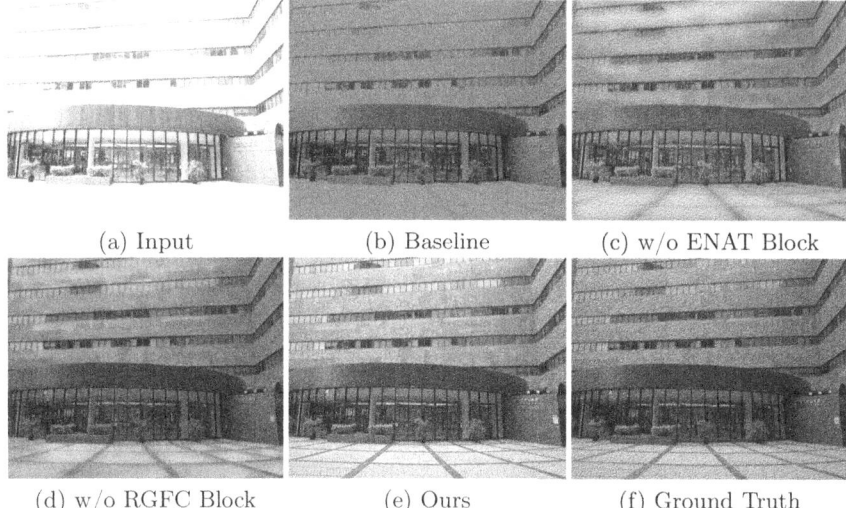

(a) Input   (b) Baseline   (c) w/o ENAT Block

(d) w/o RGFC Block   (e) Ours   (f) Ground Truth

Fig. 9. Visualization results of the ablation studies.

4.4 Ablation Studies

In this section, we conduct the ablation studies to demonstrate the effectiveness of our model. We remove the ENAT block and the RGFC block, and add skip connections between the encoder and the decoder, the simplified model is set as baseline. We also replace the shuffle attention with the channel-spatial attention that are widely used in visual tasks [15,33,35]. The experimental results for different modules are shown in Table 2, and the visualization results are presented in Fig. 9. The network performance drops significantly without AdaIN, illustrating the necessity of integrating the original and normalized features. The introduction of shuffle attention allows the model to focus more on areas with abnormal exposure, while improving the recovery performance. RGFC block enables the improved interaction between encoder and decoder, and calibrates the data distributions of features.

Table 3. Quantitative evaluations on the LOL dataset.

Method	CLAHE	RetinexNet	SID	DRBN	CMEC	MSEC	FECNet	ENC-SID	Ours
PSNR	18.43	18.94	19.75	19.77	21.03	21.57	23.10	22.76	**23.59**
SSIM	0.7176	0.7644	0.7839	0.7919	0.8214	0.8334	0.8404	0.8371	**0.9225**

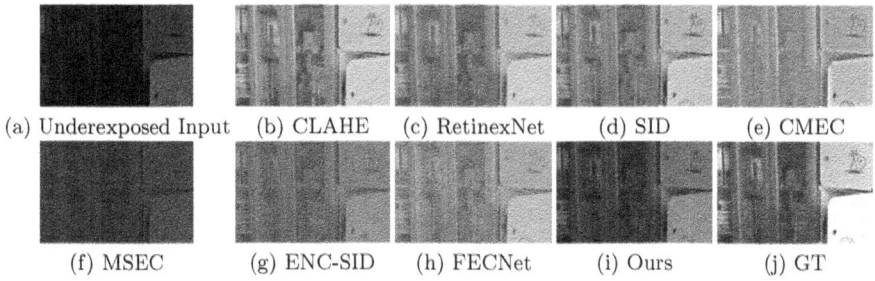

(a) Underexposed Input   (b) CLAHE   (c) RetinexNet   (d) SID   (e) CMEC

(f) MSEC   (g) ENC-SID   (h) FECNet   (i) Ours   (j) GT

Fig. 10. Visualization results on the LOL dataset.

4.5 Effectiveness for Low-Light Image Enhancement

Low-light image enhancement models mainly focus on increasing the lightness of the images and denoising them. We evaluate all the models on the LOL dataset [34]. The dataset consists of 458 images for training and 15 images for testing. The results are shown in Table 3 and the corresponding visualization results are shown in Fig. 10. It can be seen that our model can effectively enhance the lightness, even in the case of significant loss of luminance information. Our model achieves the best performance both quantitatively and qualitatively.

5   Limitation

Our model narrows the gap of different exposure representations by normalization, and transfers the normal features to compensate for the information loss, thus the network can effectively weaken the effects of abnormal exposures and improve the image quality. However, under the extreme exposure conditions, the raw sensor data has been physically truncated (e.g., pixel values are saturated or close to the noise floor), resulting in a complete loss of luminance gradients, texture details, and color information. Existing feature migration-based models rely on local statistical laws, making it difficult to generate semantically sound content from zero-information regions, and are prone to artifacts or excessive smoothing. The further researches need to be proposed to address the extreme multi-exposure correction problems, for example, a Differentiable Imaging Pipeline (DIP) can be designed to jointly estimate extreme exposure degradation parameters to provide physical constraints on the restoration network.

6 Conclusion

In this paper, we propose the MECNet for multi-exposure correction. MECNet consists of the exposure normalization and adaptive transfer (ENAT) block and the residual-guided feature calibration (RGFC) block. ENAT block uses normalization to align different exposures, and transfers the information not affected by abnormal exposures to eliminate the effects of normalization. The shuffle attention is also introduced to help the model distinguish areas that are affected by varying degrees of exposures, and focus on the severely degraded parts. The RGFC block is proposed for building connection between the up and down sampling layers. The block can control the flow of information at different scales, and allow the useful information to go further. Extensive experiments have demonstrated that our method achieves competitive performance while maintaining a low amount of parameters, and can be applied to other visual tasks.

Acknowledgments. This work is supported by the National Natural Science Foundation of China under Grants (62176123, 62476130), and the Natural Science Foundation of Jiangsu Province of China under Grant (BK20242045).

References

1. Abdullah-Al-Wadud, M., Kabir, M.H., Dewan, M., Chae, O.: A dynamic histogram equalization for image contrast enhancement. IEEE Trans. Consum. Electron. **53**(2), 593–600 (2007)
2. Afifi, M., Derpanis, K.G., Ommer, B., Brown, M.S.: Learning multi-scale photo exposure correction. In: IEEE/CVF Conference on Computer Vision and Pattern Recognition, pp. 9157–9167 (2021)
3. Afrasiyabi, A., Larochelle, H., Lalonde, J.F., Gagné, C.: Matching feature sets for few-shot image classification. In: IEEE/CVF Conference on Computer Vision and Pattern Recognition, pp. 9014–9024 (2022)
4. Cai, J., Gu, S., Zhang, L.: Learning a deep single image contrast enhancer from multi-exposure images. IEEE Trans. Image Process. **27**(4), 2049–2062 (2018)
5. Chen, C., Chen, Q., Xu, J., Koltun, V.: Learning to see in the dark. In: IEEE/CVF Conference on Computer Vision and Pattern Recognition, pp. 3291–3300 (2018)
6. Cho, S., Ji, S., Hong, J., Jung, S., Ko, S.: Rethinking coarse-to-fine approach in single image deblurring. In: International Conference on Computer Vision, pp. 4621–4630 (2021)
7. Cui, R., Niu, L., Hu, G.: Unsupervised exposure correction. In: European Conference on Computer Vision, pp. 252–268 (2025)
8. Deng, X., Dragotti, P.L.: Deep convolutional neural network for multi-modal image restoration and fusion. IEEE Trans. Pattern Anal. Mach. Intell. **43**(10), 3333–3348 (2021)
9. Feng, R., Li, C., Chen, H., Li, S., Loy, C.C., Gu, J.: Removing diffraction image artifacts in under-display camera via dynamic skip connection network. In: IEEE/CVF Conference on Computer Vision and Pattern Recognition, pp. 662–671 (2021)
10. Gu, S., Li, Y., Gool, L.V., Timofte, R.: Self-guided network for fast image denoising. In: International Conference on Computer Vision, pp. 2511–2520 (2019)

11. Guo, C., et al.: Zero-reference deep curve estimation for low-light image enhancement. In: IEEE/CVF Conference on Computer Vision and Pattern Recognition, pp. 1777–1786 (2020)

12. Han, J., Ren, Y., Ding, J., Pan, X., Yan, K., Xia, G.S.: Expanding low-density latent regions for open-set object detection. In: IEEE/CVF Conference on Computer Vision and Pattern Recognition, pp. 9591–9600 (2022)

13. He, K., Sun, J., Tang, X.: Single image haze removal using dark channel prior. IEEE Trans. Pattern Anal. Mach. Intell. **33**(12), 2341–2353 (2011)

14. He, K., Zhang, X., Ren, S., Sun, J.: Deep residual learning for image recognition. In: IEEE/CVF Conference on Computer Vision and Pattern Recognition, pp. 770–778 (2016)

15. Hu, J., Shen, L., Sun, G.: Squeeze-and-excitation networks. In: IEEE/CVF Conference on Computer Vision and Pattern Recognition, pp. 7132–7141 (2018)

16. Huang, J., et al.: Exposure normalization and compensation for multiple-exposure correction. In: IEEE/CVF Conference on Computer Vision and Pattern Recognition, pp. 6033–6042 (2022)

17. Huang, J., et al.: Deep Fourier-based exposure correction network with spatial-frequency interaction. In: European Conference on Computer Vision, pp. 163–180 (2022)

18. Huang, X., Belongie, S.J.: Arbitrary style transfer in real-time with adaptive instance normalization. In: International Conference on Computer Vision, pp. 1510–1519 (2017)

19. Jin, X., Lan, C., Zeng, W., Chen, Z., Zhang, L.: Style normalization and restitution for generalizable person re-identification. In: IEEE/CVF Conference on Computer Vision and Pattern Recognition, pp. 3140–3149 (2020)

20. Johnson, J., Alahi, A., Fei-Fei, L.: Perceptual losses for real-time style transfer and super-resolution. In: European Conference on Computer Vision, pp. 694–711 (2016)

21. Li, H., Ma, K., Yong, H., Zhang, L.: Fast multi-scale structural patch decomposition for multi-exposure image fusion. IEEE Trans. Image Process. **29**, 5805–5816 (2020)

22. Li, S., Kang, X.: Fast multi-exposure image fusion with median filter and recursive filter. IEEE Trans. Consum. Electron. **58**(2), 626–632 (2012)

23. Liu, X., Suganuma, M., Sun, Z., Okatani, T.: Dual residual networks leveraging the potential of paired operations for image restoration. In: IEEE/CVF Conference on Computer Vision and Pattern Recognition, pp. 7007–7016 (2019)

24. Ma, K., Li, H., Yong, H., Wang, Z., Meng, D., Zhang, L.: Robust multi-exposure image fusion: a structural patch decomposition approach. IEEE Trans. Image Process. **26**(5), 2519–2532 (2017)

25. Ma, N., Zhang, X., Zheng, H., Sun, J.: Shufflenet V2: practical guidelines for efficient CNN architecture design. In: European Conference on Computer Vision, pp. 122–138 (2018)

26. Van der Maaten, L., Hinton, G.: Visualizing data using t-SNE. Mach. Learn. Res. **9**(11) (2008)

27. Mo, H., Jiang, J., Wang, Q., Yin, D., Dong, P., Tian, J.: Frequency attention network: blind noise removal for real images. In: Asian Conference on Computer Vision, vol. 12623, pp. 168–184 (2020)

28. Nsampi, N.E., Hu, Z., Wang, Q.: Learning exposure correction via consistency modeling. In: British Machine Vision Conference, p. 12 (2021)

29. Pan, X., Luo, P., Shi, J., Tang, X.: Two at once: enhancing learning and generalization capacities via IBN-net. In: European Conference on Computer Vision, pp. 484–500 (2018)

30. Reza, A.M.: Realization of the contrast limited adaptive histogram equalization (CLAHE) for real-time image enhancement. Signal Process. Syst. Signal Image Video Technol. **38**, 35–44 (2004)

31. Ronneberger, O., Fischer, P., Brox, T.: U-Net: convolutional networks for biomedical image segmentation. In: Medical Image Computing and Computer Assisted Intervention, pp. 234–241 (2015)

32. Simonyan, K., Zisserman, A.: Very deep convolutional networks for large-scale image recognition. In: International Conference on Learning Representations (2015)

33. Wang, X., Girshick, R., Gupta, A., He, K.: Non-local neural networks. In: IEEE/CVF Conference on Computer Vision and Pattern Recognition, pp. 7794–7803 (2018)

34. Wei, C., Wang, W., Yang, W., Liu, J.: Deep retinex decomposition for low-light enhancement. In: British Machine Vision Conference, p. 155 (2018)

35. Woo, S., Park, J., Lee, J., Kweon, I.S.: CBAM: convolutional block attention module. In: European Conference on Computer Vision, pp. 3–19 (2018)

36. Wu, H., et al.: Contrastive learning for compact single image dehazing. In: IEEE/CVF Conference on Computer Vision and Pattern Recognition, pp. 10551–10560 (2021)

37. Xiao, J., Zhou, M., Fu, X., Liu, A., Zha, Z.: Improving de-raining generalization via neural reorganization. In: International Conference on Computer Vision, pp. 4967–4976 (2021)

38. Xu, H., Ma, J., Zhang, X.S.: MEF-GAN: multi-exposure image fusion via generative adversarial networks. IEEE Trans. Image Process. **29**, 7203–7216 (2020)

39. Yang, H., et al.: Balanced and hierarchical relation learning for one-shot object detection. In: IEEE/CVF Conference on Computer Vision and Pattern Recognition, pp. 7591–7600 (2022)

40. Yang, W., Wang, S., Fang, Y., Wang, Y., Liu, J.: From fidelity to perceptual quality: a semi-supervised approach for low-light image enhancement. In: IEEE/CVF Conference on Computer Vision and Pattern Recognition, pp. 3060–3069 (2020)

41. Yu, C., Su, Z.: Symmetrical gaussian error linear units (SGELUs). CoRR abs/1911.03925 (2019)

42. Zamir, S.W., et al.: Multi-stage progressive image restoration. In: IEEE/CVF Conference on Computer Vision and Pattern Recognition, pp. 14821–14831 (2021)

43. Zeiler, M.D., Fergus, R.: Visualizing and understanding convolutional networks. In: European Conference on Computer Vision, pp. 818–833 (2014)

44. Zhang, Q.L., Yang, Y.B.: SA-Net: shuffle attention for deep convolutional neural networks. In: IEEE International Conference on Acoustics, Speech, and Signal Processing, pp. 2235–2239 (2021)

45. Zhang, X., Zhou, X., Lin, M., Sun, J.: Shufflenet: an extremely efficient convolutional neural network for mobile devices. In: IEEE/CVF Conference on Computer Vision and Pattern Recognition, pp. 6848–6856 (2018)

Collaborative Optimization of Truck-Drone Based on Multi-agent Reinforcement Learning

Wenhao Zhang[1], Guoyin Wang[1,2(✉)], and Qun Liu[1]

[1] Key Laboratory of Cyberspace Big Data Intelligent Security, Ministry of Education, Chongqing University of Posts and Telecommunications, Chongqing 400065, China
s230233058@stu.cqupt.cn, {wanggy,liuqun}@cqupt.edu.cn
[2] National Center for Applied Mathematics in Chongqing, Chongqing Normal University, Chongqing 401331, China

Abstract. This study proposes a novel Multi-Head Attention-enhanced Proximal Policy Optimization (MHA-PPO) algorithm for collaborative truck-drone delivery systems in dynamic logistics scenarios. To address the limitations of traditional multi-agent reinforcement learning in handling complex spatiotemporal dependencies and real-time decision-making, we integrate a multi-head attention mechanism into the PPO framework, enabling adaptive feature fusion of heterogeneous agent observations (e.g., drone positions, truck routes, and package urgency levels). A key innovation lies in the design of a variable emergency mechanism that models a random urgency generation module and dynamically adjusts task priorities based on the urgency of customer demand for packages. The system is implemented in a high-fidelity Unity simulation environment incorporating 3D urban terrain, realistic drone aerodynamics, and traffic flow patterns. Comparative experiments against baseline algorithms (PPO, Multi-Agent Proximal Policy Optimization (MAPPO), Multi-Head Attention Deep Deterministic Policy Gradient (MHA-DDPG)) demonstrate that our MHA-PPO achieves superior performance.

Keywords: Reinforcement learning · Truck-drone cooperative delivery · Multi-agent problem · Task allocation

1 Introduction

As a novel intelligent logistics solution, the truck-drone collaborative delivery model utilizes trucks as mobile transportation bases and combines drones as the primary agents for end-of-line delivery, forming a composite material transportation network. In recent years, driven by the rapid expansion of e-commerce and the acceleration of urbanization, the logistics industry has faced increasing labor costs and pressure to upgrade its technology, necessitating the construction of a new intelligent transportation system. Drones [9], with their agile response characteristics, rapid deployment capabilities, and efficient operation features, have gradually become a key technological carrier for optimizing the "last mile"

Q. Zhang et al. (Eds.): IJCRS 2025, LNAI 15710, pp. 350–362, 2025.
https://doi.org/10.1007/978-3-031-92741-6_26

delivery efficiency. It is worth noting that companies such as FedEx and Meituan have already launched pilot applications, verifying the potential of this model in optimizing delivery timeliness [4,15].

Current research on truck-drone delivery systems focuses on two approaches: (1) Heuristic algorithms [1] follow four steps: problem modeling (defining collaboration mechanisms and constraints), initial solution generation via greedy strategies, iterative optimization using adaptive neighborhood search with destruction-repair operations, and parameter tuning based on problem scale. (2) Deep reinforcement learning (DRL) [7] models logistics as a Markov Decision Process (MDP), defining states (truck/drone status, customer urgency), actions (routing decisions), and reward functions. Agents iteratively optimize actions through environmental feedback to achieve delivery objectives.

Current path-planning methods face limitations. Heuristics are fast but sensitive to initial conditions, leading to variability [21]. At the same time, Deep Reinforcement Learning (DRL) optimizes paths and reduces costs but struggles with data dependency, resource consumption, and stability in dynamic environments. Both approaches lack effective multi-agent coordination, making static solutions unsuitable for dynamic logistics [23]. DRL, however, demonstrates stronger adaptability and the ability to handle complex state spaces, optimizing long-term rewards through continuous learning and reducing reliance on manual rules. In contrast, heuristic algorithms rely on fixed rules, limiting their flexibility in multi-objective and complex scenarios. This paper adopts DRL to address collaborative delivery, leveraging the attention mechanism [20] to integrate agent information and facilitate strategy adjustment. By using self- or cross-attention for state encoding and channel compression for feature fusion, the approach reduces complexity and enables efficient collaboration.

We use multi-head attention to enrich agent observations without increasing input dimensions. It fuses agent observations before policy network input. The network outputs decisions for all agents, treating trucks and drones as two agent types with separate policy and value networks. This differs from traditional multi-agent algorithms, solving a coordination problem and implementing multi-agent reinforcement learning.

In summary, the specific contributions of this paper are as follows: (1) We propose a multi-head attention-driven policy optimization framework, Multi-head Attention-Proximal Policy Optimization (MHA-PPO), which improves the shortcomings of traditional PPO in multi-agent collaborative decision-making regarding insufficient information fusion. (2) We develop a digital twin logistics system that integrates virtual and physical environments based on the Unity engine, providing a more accurate simulation environment for subsequent research. (3) We investigate the truck-drone hybrid delivery problem based on reinforcement learning, deepening the research in the field of truck-drone delivery models.

This environment provides a wealth of case studies for future research on truck-drone delivery. The structure of the paper is as follows: The second section introduces related work on solving truck-drone problems based on heuristic and reinforcement learning methods; the third section describes the scenario studied in this paper; the fourth section presents the design of the MHA-PPO architecture; finally, the fifth and sixth sections present the simulation experiments and conclusions of the proposed algorithm, respectively.

2 Related Work

In the field of mixed integer programming, Moshref-Javadi et al. [11] innovatively extended the Traveling Salesman Problem (TRP) by integrating Tabu Search with Simulated Annealing to achieve dynamic path planning for drones launched from multiple truck stops. Building on this, Jeong et al. [8] further incorporated load-bearing and energy consumption constraints along with no-fly zone parameters, and their developed two-stage heuristic algorithm reduced the response delay to weather/policy constraints to within 5 s. More recently, Freitas et al. [5] pioneered the construction of Mixed Integer Programming (MIP) models combined with Generalized Variable Neighborhood Tabu Search algorithms, successfully enhancing the solution efficiency for large-scale instances.

The Relax-and-Fix with Re-couple-Refine-and-Optimize(RF-RRO) heuristic algorithm proposed by Thomas's team [18] optimizes both delivery time and cost through Mixed Integer Linear Programming, enhancing drone utilization. Corresponding to this approach, Najy et al. [14] designed the Inventory-Routing Problem with Drone (IRP-D) model, which integrates unmanned aerial vehicles (UAVs) into traditional truck delivery systems to address the inventory routing problem (IRP). The model is solved using mixed-integer linear programming (MILP) and branch-and-bound methods, and a heuristic approach based on the non-drone IRP solution is proposed to improve computational efficiency. The main advantage of this model lies in its ability to effectively handle scenarios with significant differences in customer demand. Specifically, customers with small demands can be served through low-cost drone deliveries without the need for truck visits, thereby optimizing overall delivery efficiency and cost. A series of studies continue to expand the application boundaries of Mixed Integer Linear Programming(MILP): Vasquez et al. [19] constructed a dual-time constraint model that achieved over 9% precision in solving the VRP-D problem, and Liu et al. [12] pioneered an integrated transportation cost evaluation system.

For specific delivery scenarios, Schermer et al. [16] proposed an improved branch-and-cut algorithm that reduced the benchmark solution time for the TSP-D base case. Dell'Amico et al. [3] more innovatively defined dual-mode waiting strategies for drones, combined with a three-index verification mechanism to significantly lower the path planning error rate. When complex return demands are involved, Ham's multi-depot TCDCP model [6] successfully broke through the efficiency bottleneck of traditional single-path generation. It is worth noting that Liu's team [13] was the first to introduce reinforcement learning technology, enhancing service efficiency by 23.8% under random travel times using DQN and A2C algorithms. In the latest developments, Bi et al. [2] built a multi-drone collaborative environment based on the MPE library, achieving a 98.6% truck recovery success rate in a single-delivery task scenario with the Appo algorithm.

3   Problem Description

This paper introduces a hybrid truck-drone delivery model with a dynamic urgency mechanism. The system comprises q zones $\mathbb{P} = \{P_1, P_2, \ldots, P_q\}$, each with a distribution center and p customer nodes $\mathbb{C} = \{C_1, C_2, \ldots, C_p\}$. Customers are exclusively served by their zone's distribution center. We deploy n trucks ($\mathbb{T} = \{T_1, T_2, \ldots, T_n\}$) and m drones ($\mathbb{U} = \{u_1, u_2, \ldots, u_m\}$), with each truck carrying m drones. They depart the warehouse to: (1) select truck routes by zone priority, (2) release drones for last-mile delivery upon arrival at distribution centers, and (3) return to the warehouse after completing all tasks.

To enhance dynamism, we introduce a dual urgency mechanism: at the customer level, a package urgency parameter $h_i \in [0.1, 1]$ is defined, indicating the customer's urgency to receive the package. This value is randomly reset after each delivery to simulate demand fluctuations, prioritizing higher urgency deliveries. At the regional level, an aggregated index $H_j = \frac{1}{m} \sum_{i=1}^{m} h_i(j)$ represents the comprehensive urgency of the j-th partition. The system adopts a two-layer decision architecture: the truck scheduling layer visits in descending order of $\{H_j\}$, and the drone operation layer executes delivery in descending order of $\{h_i\}$ within its partition, forming a dynamic response mechanism to maximize time-sensitive delivery benefits.

The agent's path is represented as $\{d_1, d_2, d_3, \ldots\}$, where d_i denotes the distance to reach the i-th target point. d_i is calculated using a distance formula, where x_i, y_i are the current position coordinates. Drones must ensure enough endurance to return to the distribution center. Thus, delivery decisions must satisfy $d_i > d_k + d_{kj}$, where d_k is the distance to the next customer, and d_{kj} is the distance to the j-th distribution center. If $d_i > d_k + d_{kj}$ cannot be met, the drone returns directly to the distribution center. This paper assumes drones can serve multiple customers at once, differing from traditional one-customer-at-a-time service.

We hope that the agent can complete all delivery tasks within the shortest possible distance, prioritizing customers with higher urgency.

In summary, we can model this problem as follows:

$$\operatorname*{argmin}_{\tau} \quad L(\tau) + \rho \sum_{i \in \mathbb{C}} h_i T_i$$

s.t.

$$x_i = 1, \quad \forall i \in \mathbb{C} \tag{1}$$
$$E_i > d_k + d_{kj}, \quad \forall i \in \mathbb{U}, k \in \mathbb{C}$$
$$R_{2i} > 0, \quad \forall i \in \mathbb{C}$$

Here, $L(\tau)$ is the distance traveled by the agent along its trajectory, h_i denotes the urgency level, T_i is the decision cycle value, and x_i is a binary variable that equals 1 if the agent visits the customer, and 0 otherwise. R_{2i} stands for the delivery reward value, the specific meaning of which will be explained later in the text.

4   Method

4.1   Algorithm

Proximal Policy Optimization: PPO is an efficient reinforcement learning algorithm based on policy gradients [17]. It maximizes cumulative returns by optimizing the policy network π_θ while using the value network v_φ to estimate state values and guide policy updates. Compared to TRPO, which relies on complex second-order optimization, PPO employs a clipped surrogate objective function to limit the policy update ratio (e.g., within $[0.8, 1.2]$), combined with first-order gradient methods to simplify computations, ensuring stability while improving efficiency. The clipped surrogate objective function of PPO is formalized as follows:

$$L^{\mathrm{CLIP}}(s, a, \theta) = \mathbb{E}_t \left[\min(r(\theta), \mathrm{clip}(\theta, \varepsilon)) \hat{A}(t) \right], \tag{2}$$

where

$$\mathrm{clip}(\theta, \varepsilon) = \begin{cases} 1 - \varepsilon, & r(\theta) < 1 - \varepsilon \\ r(\theta), & 1 - \varepsilon \le r(\theta) < 1 + \varepsilon \\ 1 + \varepsilon, & 1 + \varepsilon \le r(\theta) \end{cases} \tag{3}$$

and

$$r(\theta) = \frac{\pi_\theta(a|s)}{\pi_{\theta_{\mathrm{old}}}(a|s)}, \tag{4}$$

where π_{old} is the old policy, and the update ratio is the step size, which limits the maximum difference between the new and old policies. This method ensures that updates are not too aggressive. When calculating the update loss of the policy network, it is necessary to calculate the dominance function $A(t)$ and we use the generalized advantage function:

$$\hat{A}(t) = A^{\mathrm{GAE}}(t) = \sum_{k=0}^{\infty} (\gamma\lambda)^k A(t + k). \tag{5}$$

The critic network is characterized by parameters φ, and its loss function is formulated as:

$$L^V(s, a, \varphi) = \mathbb{E}_t \left[(v_\varphi(s_t, a_t) - y(t))^2 \right], \tag{6}$$

where $y(t) = R_t + \gamma v_\varphi(s_{t+1}, a_{t+1})$ represents the temporal difference target value, and R_t denotes the reward obtained in the current environment. The parameters of the actor-network are updated using the following optimization:

$$\varphi_{k+1} = \arg\min_\varphi \mathbb{E} \left[L^V(s, a, \varphi_k) \right]. \tag{7}$$

To enhance the exploration of the policy and prevent overfitting, this paper also adds action entropy [22] to the training.

$$H(a) = -\sum_a \pi(a|s) \log \pi(a|s), \tag{8}$$

where $\pi(a|s)$ is the probability of choosing action a in state s. Action entropy measures the dispersion of the policy's action distribution: high entropy indicates uniform exploration, while low entropy suggests concentrated exploitation. Incorporating entropy regularization in PPO prevents local optima, enhances exploration efficiency, and mitigates policy overconfidence. The modified objective function becomes:

$$L^{\text{CLIP+ENT}}(\theta) = \mathbb{E}_{(s,a)\sim\pi_{\theta_{\text{old}}}}\left[L^{\text{CLIP}} + \beta \cdot \mathcal{H}(\pi_\theta(\cdot|s))\right], \tag{9}$$

where β is the entropy coefficient, a larger β encourages policy randomness, and a smaller β tends towards a deterministic policy.

After incorporating action entropy, the iterative objective of the policy network's parameters θ is updated as follows:

$$\theta_{k+1} = \arg\max_\theta \mathbb{E}_{(s,a)\sim\pi_{\theta_{\text{old}}}}\left[L^{\text{CLIP+ENT}}(\theta)\right], \tag{10}$$

The gradient update for the policy network can be expressed as:

$$\nabla_\theta J(\theta) = \mathbb{E}_{(s,a)\sim\pi_{\theta_{\text{old}}}}\left[\nabla_\theta \log \pi_\theta(a|s)\left(L^{\text{CLIP+ENT}}(\theta)\right)\right], \tag{11}$$

Attention: The multi-head attention mechanism is built upon the self-attention mechanism, which allows each position in the input sequence to attend to every other position. It dynamically computes a weighted sum as the output to capture the dependencies within the sequence. The core of the attention mechanism lies in utilizing the query (Query), key (Key), and value (Value) triplet to compute the dependency strength between elements, which is formally represented as:

$$\text{Attention}(Q, K, V) = \text{softmax}\left(\frac{QK^T}{\sqrt{d_k}}\right)V, \tag{12}$$

where d_k is the dimensionality of the key vectors. The scaling factor $\sqrt{d_k}$ is used to balance the dot product computation magnitude, preventing gradient anomalies.

The multi-head attention mechanism employs multiple parallel self-attention layers ("heads") to capture diverse feature representations. Each head independently processes input embeddings using distinct learnable parameter matrices W_i^Q, W_i^K, W_i^V, generating query (Q_i), key (K_i), and value (V_i) vectors. The outputs of all heads are concatenated and linearly transformed to produce the final result: $\text{MultiHead}(X) = \text{Concat}(\text{head}_1, ..., \text{head}_h)W^O$ Key components:

- **Parallel heads**: h independent self-attention computations
- **Feature diversity**: Each head learns different attention patterns
- **Fusion**: Linear projection (W^O) integrates multi-head features

This design enhances model expressiveness through cross-head interaction while maintaining computational efficiency [20].

Algorithm Procedure: As shown in Fig. 1, this study categorizes trucks and drones into two distinct classes, each equipped with its policy network and value network. The observations $s = \{s_1, s_2, \ldots, s_i\}$ obtained from the environment, where s_i represents the local observation of the i-th agent, are fed into a multi-head attention mechanism to achieve dynamic weighted fusion of multi-source observation information. Specifically, the observations are first preprocessed by the multi-head attention mechanism: $S_{\text{mixed}} = \text{multihead}(s)$. The resulting S_{mixed} contains global information, which is then input into the policy network to generate the actions of all agents: $A = actor(S_{\text{mixed}})$, where $A = \{a_1, a_2, \ldots, a_i\}$.

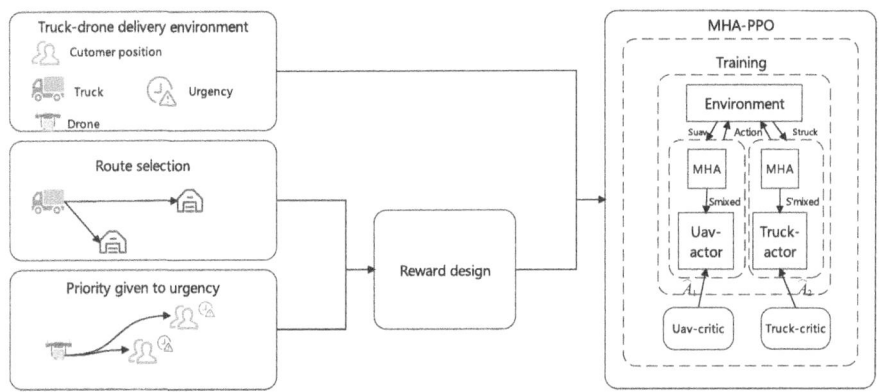

Fig. 1. MHA-PPO based Truck-drone training framework.

In this framework, homogeneous agents (e.g., drones) share identical observation dimensions: individual observation $s_i \in \mathbb{R}^{1 \times d_{\text{obs}}}$ aggregates to $S \in \mathbb{R}^{n \times d_{\text{obs}}}$ for n agents. The multi-head attention mechanism processes S through parallel FC layers generating $Q/K/V$ matrices while maintaining dimensional consistency ($S_{\text{mixed}} \in \mathbb{R}^{n \times d_{\text{obs}}}$). Policy networks then map S_{mixed} to action matrix $A \in \mathbb{R}^{n \times d_{\text{act}}}$. During training, both policy and value networks receive attention-processed S_{mixed}, with the attention mechanism jointly optimized through policy gradient backpropagation via $L^{\text{CLIP+ENT}}(\theta)$ (see Algorithm 1). This integrated approach enables parameter updates for both attention layers and policy networks while preserving dimensional stability throughout the architecture.

5   Experiment and Results

In this section, we conducted tests on the constructed Unity environment. Firstly, we evaluated the task completion rates of different algorithms under the same environment. Secondly, we performed multiple tests within this model to assess the impact of various parameters on the results. Finally, we evaluated the effectiveness of the environment and algorithms used, and compared the performance of the Multi-Head Attention-PPO algorithm with different parameter settings.

Algorithm 1. Training Algorithm of MHA-PPO

1: Initialize π_θ (with MHA), V_φ, buffer D, hyperparams $\varepsilon, \rho, \alpha$
2: **for** episode $= 1$ to N **do**
3:     Collect trajectories $\{(s_t, a_t, r_t)\}$ via π_θ and store in D
4:     Compute GAE advantages $\hat{A}^t$ and returns G^t
5:     **for** epoch $= 1$ to K **do**
6:         Sample mini-batches from shuffled D
7:         **for** each batch B **do**
8:             old_log_probs $\leftarrow$ detach$(\pi_\theta(a_t|s_t))$
9:             $S_{\text{mixed}} \leftarrow$ MultiHeadAttention(s_t)
10:            new_log_probs, entropy $\leftarrow \pi_\theta(a_t|S_{\text{mixed}})$
11:            ratio$_t \leftarrow$ exp(new_log_probs $-$ old_log_probs)
12:            $\mathcal{L}_{\text{clip}} \leftarrow -\mathbb{E}[\min(\text{ratio}_t A^t, \text{clip}(\text{ratio}_t, 1 \pm \varepsilon)\hat{A}^t)]$
13:            $\mathcal{L}_{\text{value}} \leftarrow \frac{1}{2}\|V_\varphi(s_t) - G^t\|^2$
14:            $\theta \leftarrow \theta - \alpha\nabla(\mathcal{L}_{\text{clip}} - \rho * \text{entropy} + \mathcal{L}_{\text{value}})$
15:            $\varphi \leftarrow \varphi - \alpha\nabla\mathcal{L}_{\text{value}}$
16:        **end for**
17:    **end for**
18: **end for**

5.1 Simulation Setting

All experiments in this paper are conducted using consistent environmental settings. The hardware environment includes a 12th Gen Intel(R) Core(TM) i7-12700 2.10 GHz CPU, 32GB RAM, and an NVIDIA RTX 1660 8 GB graphics card. The software environment includes Python 3.9.18, PyTorch 1.5.0, and Unity 2022.3.13.

State: This paper defines the state of the truck-drone collaborative delivery problem as:

$$S = ((x, y), (x', y'), (x_i, y_i), h_i, a_i, t, E), \tag{13}$$

where (x, y) is the position of the current agent, (x', y') is the positions of other agents, (x_i, y_i) is the position of the target point, h_i is the urgency of the customer, a_i is a binary variable, where $a_i=1$ if the agent has visited the i-th target point, and $a_i=0$ otherwise. t is the number of steps in which the current agent makes a decision and E is the remaining endurance of the drone.

Action: The actions of an agent are defined as the agent's destinations. The policy network outputs a target point, and the agent proceeds to that target point to complete the associated tasks. For trucks, the set of actions is as follows:

$$A_t = \{P_0, P_1, P_2, \ldots, P_q\}, \tag{14}$$

where P_i is the label of the distribution center for the region i , and p_0 represents the warehouse. For uav, the set of actions is as follows:

$$A_u = \{P_i, C_{i1}, C_{i2}, \ldots, C_{ij}, i \neq 0\}, \tag{15}$$

where c_{ij} is the customer j in region i. Since drones follow trucks to different regions for delivery, the action set available to drones also changes with the delivery area. The policy network outputs P_i to indicate that the drone returns to the truck's parking point to swap batteries and replenish goods.

Reward Function: The purpose of the reward function is to encourage all agents to proceed to their destinations as quickly as possible while maintaining a safe and reasonable state. It will determine the action patterns based on different observations and is defined as follows:

$$R = -R_1 + R_2. \tag{16}$$

Among them, R_1 represents the travel distance reward, and R_2 represents the delivery reward upon reaching the target point. R_1 is the effective distance traveled by the agent within the decision period, i.e., $R_1 = \sqrt{(x' - x_i)^2 + (y' - y_i)^2}$, where (x', y') is the position at the previous step, and (x_i, y_i) is the current position. R_2 is defined as $R_2 = A - t \cdot i$, where A is a base constant, t is the number of decision steps, and i is the urgency level of the target point.

In the Unity simulation, 12 customer points are evenly divided into 3 regions, each with 4 randomly distributed customers. The distance between customers and the distribution center is limited to half the drone's range to ensure return capability. For realism, the truck's weight is set to 5000 kg (medium-sized truck) and the drone's to 45 kg (medium-sized transport drone), based on real logistics data. Other parameters are listed in Table 1.

Table 1. Training Parameters Setup

Parameters	Values
max steps	50K
buffer capacity	256
mini batch size	64
Reaching Reward A	35
actor hidden size	[256, 128]
critic hidden size	[128, 128]
activation function	PReLU
optimizer	Adam
actor lr, critic lr	[5e−4, 5e−4]
PPO epsilon clip	0.25
γ in GAE	0.85

5.2   Convergence Performance

This study optimizes dynamic task allocation for truck-drone fleets to minimize delivery costs (obstacle-free assumption). Using PPO with 1 truck and 2 drones, we train discrete-action agents in episodic environments reset at maximum exploration steps. Actor-critic learning rates critically affect convergence, demonstrated by per-episode reward averages in Fig. 2. Initial exploration phases show volatile rewards, stabilizing as agents learn optimal actions. Episode-specific urgency levels introduce reward variance for identical routes, mitigated through triple-episode reward summation. Optimal convergence occurred at LR=5e-4, adopted for subsequent comparative experiments. It is worth noting that although the curves in the graph generally tend to stabilize, significant fluctuations remain. This is because the urgency level is randomly reassigned at the start of each episode, leading to varying optimal rewards in each round. As a result, the fluctuations observed in the graph are a normal phenomenon.

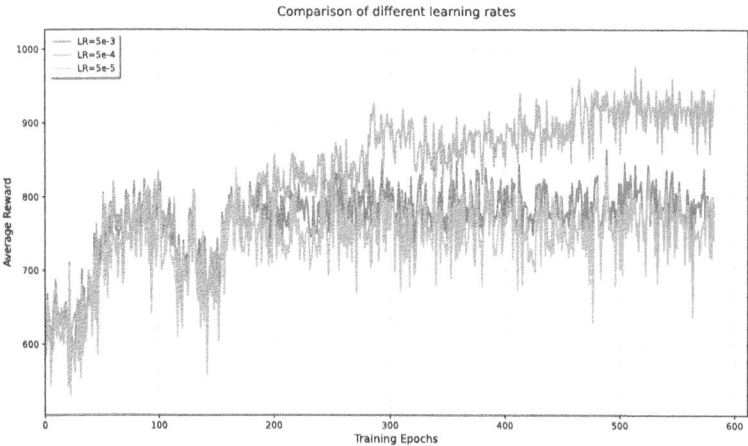

Fig. 2. Convergence of MHA-PPO under different learning rates.

5.3   Compare with Other Reinforcement Learning Methods

In this study, we focus on the application of reinforcement learning algorithms to the truck-drone collaborative delivery problem, particularly the autonomous optimization and multi-agent collaboration capabilities of MHA-PPO in dynamic environments. Since heuristic algorithms rely on predefined rules and struggle to adapt to complex, dynamic scenarios, and their optimization mechanisms fundamentally differ from those of reinforcement learning, a direct comparison would lack fairness. Therefore, this study chooses to compare with other reinforcement learning algorithms of the same category.

Our experiments in dynamic logistics scenarios (Fig. 3) revealed distinct algorithmic behaviors: MHA-DDPG initially accelerated reward gains through deterministic policies but later showed instability and lower convergence than PPO due to limited adaptability to multimodal state distributions. While MAPPO supported multi-agent coordination via centralized critics, gradient conflicts from shared value estimation delayed policy optimization. In contrast, PPO achieved superior performance through entropy-regulated exploration and distributed advantage computation, effectively addressing dynamic path planning challenges. These results highlight the critical alignment between environmental dynamics and algorithmic exploration mechanisms in multi-agent systems.

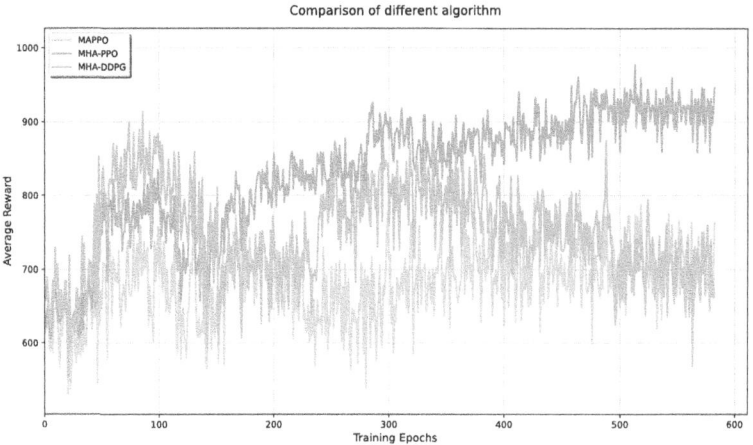

Fig. 3. Convergence of MHA-PPO under different algorithm.

6   Conclusions and Prospects

This paper investigates the optimization problem of truck-drone cooperative delivery based on reinforcement learning. To address this issue, we employed a multi-head attention mechanism and the PPO algorithm and conducted simulation training and validation in a Unity environment. Considering the characteristics of truck-drone delivery, techniques such as the multi-head attention mechanism and reward normalization were utilized to accelerate training speed and enhance structural convergence. It not only improved the probability of successful solutions but also expanded the methods for solving truck-drone cooperative delivery problems. Our experiments confirmed that the incorporation of the multi-head attention mechanism enhanced the decision-making quality of agents. Comparisons with other reinforcement learning algorithms demonstrated the effectiveness of this approach. Moreover, the Unity reinforcement learning environment used in this study, which integrates ML-Agents [10] and Python for communication, is more complex and closer to real-world scenarios compared to traditional environments built with Gym and MPE. This makes the training results more credible. However, this complexity also brings about higher computational costs.

In reinforcement learning, agents learn optimal strategies through continuous exploration. When conducting multiple experiments with the same parameters, the initial exploration process of the agents may not be entirely consistent. Additionally, reinforcement learning algorithms are highly sensitive to hyperparameter values, which can lead to significantly different training outcomes. Therefore, we aim to further improve the stability of training results based on this work.

Acknowledgments. This work is supported by The National Natural Science Foundations of China (62376045), Natural Science Foundations of Chongqing (cstc2021ycjh-bgzxm0013), The Key Cooperation Project of Chongqing Municipal Education Commission(HZ2021008). We would like to thank editor and reviewers for insightful comments and advice.

References

1. Ammouriova, M., Bertolini, M., Castaneda, J., Juan, A.A., Neroni, M.: A heuristic-based simulation for an education process to learn about optimization applications in logistics and transportation. Mathematics **10**(5), 830 (2022)
2. Bi, Z., Guo, X., Wang, J., Qin, S., Liu, G.: Truck-drone delivery optimization based on multi-agent reinforcement learning. Drones **8**(1), 27 (2024)
3. Chen, L., Xin, B., Chen, J.: Interactive multiobjective evolutionary algorithm based on decomposition and compression. Sci. China Inf. Sci. **64**(10), 1–16 (2021). https://doi.org/10.1007/s11432-020-3092-y
4. Erdelj, M., Natalizio, E.: UAV-assisted disaster management: applications and open issues. In: 2016 International Conference on Computing, Networking and Communications (ICNC). IEEE (2016)
5. Freitas, J.C., Penna, P., Toffolo, T.A.: Exact and heuristic approaches to truck-drone delivery problems. EURO J. Transport. Logist. **12**, 100094 (2023)
6. Ham, A.M.: Integrated scheduling of m-truck, m-drone, and m-depot constrained by time-window, drop-pickup, and m-visit using constraint programming. Transport. Res. Part C: Emerg. Technol. **91**, 1–14 (2018)
7. Intelligence, C., et al.: Retracted: Application of deep reinforcement learning algorithm in uncertain logistics transportation scheduling. Comput. Intell. Neurosci. **2023**, 9838603 (2023)
8. Jeong, H.Y., Song, B.D., Lee, S.: Truck-drone hybrid delivery routing: Payload-energy dependency and no-fly zones. Int. J. Prod. Econ. **214**, 220–233 (2019)
9. Jiandong, Z., Qiming, Y., Guoqing, S., Yi, L., Yong, W.: UAV cooperative air combat maneuver decision based on multi-agent reinforcement learning. J. Syst. Eng. Electron. **32**(6), 1421–1438 (2021)
10. Juliani, A., et al.: Unity: a general platform for intelligent agents. arXiv preprint arXiv:1809.02627 (2020). https://arxiv.org/pdf/1809.02627.pdf
11. Kitjacharoenchai, P., Ventresca, M., Moshref-Javadi, M., Lee, S., Tanchoco, J.M., Brunese, P.A.: Multiple traveling salesman problem with drones: mathematical model and heuristic approach. Comput. Ind. Eng. **129**, 14–30 (2019)
12. Liu, J., Guan, Z., Xie, X.: Truck and drone in tandem route scheduling under sparse demand distribution. In: 2018 8th International Conference on Logistics, Informatics and Service Sciences (LISS). IEEE (2018)

13. Liu, Z., Li, X., Khojandi, A.: The flying sidekick traveling salesman problem with stochastic travel time: a reinforcement learning approach. Transport. Res. Part E: Logist. Transport. Rev. **164**, 102816 (2022)
14. Najy, W., Archetti, C., Diabat, A.: Collaborative truck-and-drone delivery for inventory-routing problems. Transport. Res. Part C: Emerg. Technol. **146**, 103791 (2023)
15. Sajid, M., Mittal, H., Pare, S., Prasad, M.: Routing and scheduling optimization for UAV assisted delivery system: a hybrid approach. Appl. Soft Comput. **126**, 109225 (2022)
16. Schermer, D., Moeini, M., Wendt, O.: A hybrid VNS/Tabu search algorithm for solving the vehicle routing problem with drones and EN route operations. Comput. Oper. Res. **109**, 134–158 (2019)
17. Schulman, J., Wolski, F., Dhariwal, P., Radford, A., Klimov, O.: Proximal policy optimization algorithms. arXiv preprint arXiv:1707.06347 (2017)
18. Thomas, T., Srinivas, S., Rajendran, C.: Collaborative truck multi-drone delivery system considering drone scheduling and EN route operations. Ann. Oper. Res. **339**(1), 693–739 (2024)
19. Vásquez, S.A., Angulo, G., Klapp, M.A.: An exact solution method for the tsp with drone based on decomposition. Comput. Oper. Res. **127**, 105127 (2021)
20. Vaswani, A., et al.: Attention is all you need. In: Advances in Neural Information Processing Systems, vol. 30 (2017)
21. Wang, K., Yuan, B., Zhao, M., Lu, Y.: Cooperative route planning for the drone and truck in delivery services: a bi-objective optimisation approach. J. Oper. Res. Soc. **71**(10), 1657–1674 (2020)
22. Wulfmeier, M., Ondruska, P., Posner, I.: Maximum entropy deep inverse reinforcement learning. arXiv preprint arXiv:1507.04888 (2015)
23. Zhang, R., Dou, L., Xin, B., Chen, C., Deng, F., Chen, J.: A review on the truck and drone cooperative delivery problem. Unmanned Syst. **12**(05), 823–847 (2024)

Two-Stage Multi-modal Multi-objective Algorithm Based on Dynamic Niche Updating

Haonian Ji[1,2,3,4], Hongmei Chen[1,2,3,4(✉)], and Yong Mi[1,2,3,4]

[1] Institute of Artificial Intelligence, School of Computing and Artificial Intelligence,
Southwest Jiaotong University, Chengdu 611756, China
{2022111346,miyong}@my.swjtu.edu.cn, {hmchen,trli}@swjtu.edu.cn
[2] National Engering Laboratory of Integrated Transportation Big Data Application
Technology, Southwest Jiaotong University, Chengdu 611756, China
[3] Engering Research Center of Sustainable Urban Intelligent Transportation,
Ministry of Education, Southwest Jiaotong University, Chengdu 611756, China
[4] Manufacturing Industry Chains Collaboration and Information Support Technology
Key Laboratory of Sichuan Province, Southwest Jiaotong University,
Chengdu 611756, China

Abstract. In solving multi-modal multi-objective optimization problems (MMOPs), objective and decision space changes need to be considered. However, many multi-modal multi-objective evolutionary algorithms (MMOEAs) tend to prioritize diversity in the objective space during the optimization process, often leading to suboptimal diversity and distribution in the decision space. This study proposes the Dynamic-Niche-Based Two-Stage Evolution (DNTE) approach to address this issue. DNTE employs differential evolutionary strategies in two stages, utilizing dynamic niche to enhance solution viability and optimize decision space distribution at different phases. In the first stage, the dynamic niche is used to restrict the dominance range of Pareto domination. In contrast, in the second stage, it is integrated with differential evolution to refine the distribution of the decision space. Combining DNTE with DN-NSGA-II enhances the capability of solving MMOPs and demonstrates superior performance compared to other evolutionary algorithms on MMF and MMMOP benchmark functions.

Keywords: Multi-modal multi-objective optimization problem ·
Differential evolution · Dynamic niche

1 Introduction

The multi-objective optimization problems (MOPs) involve optimizing multiple conflicting objectives simultaneously. During the optimization process, improving one objective may lead to the degradation of others. Such problems are commonly encountered in real-world scenarios, such as rocket engine design [3], path planning [15], feature selection [18], and protein design [4]. These types of problems can be formally described as

$$\min \mathbf{F}(\mathbf{x}) = (f_1(\mathbf{x}), f_2(\mathbf{x}), \dots, f_m(\mathbf{x}))^T$$
$$\text{s.t.} \quad \mathbf{x} = (x_1, x_2, \dots, x_n), \quad \mathbf{x} \in S. \tag{1}$$

Q. Zhang et al. (Eds.): IJCRS 2025, LNAI 15710, pp. 363–382, 2025.
https://doi.org/10.1007/978-3-031-92741-6_27

where $\mathbf{x}$ is an n-dimensional decision vector, S is the feasible region of the decision space, and m is the total number of optimization objectives. The objective function space is defined by the mapping $f : S \to \mathbb{R}^m$, where $\mathbb{R}^m$ represents the objective function space.

In multi-objective optimization, a solution $\mathbf{x_a}$ is said to *Pareto dominate* another solution $\mathbf{x_b}$ if and only if the following conditions hold.

$$\forall i = 1, 2, \ldots, m, \quad f_i(\mathbf{x_a}) \leq f_i(\mathbf{x_b}), \tag{2}$$
$$\text{and} \quad \exists j = 1, 2, \ldots, m, \quad f_j(\mathbf{x_a}) < f_j(\mathbf{x_b}). \tag{3}$$

In other words, solution $\mathbf{x_a}$ is no worse than solution $\mathbf{x_b}$ in all objectives and strictly better than $\mathbf{x_b}$ in at least one objective. The set of Pareto optimal solutions is called a Pareto optimal set (PS). The image of the PS in objective space is known as the Pareto optimal front (PF).

There are already many methods that can effectively solve multi-objective optimization problems (MOPs), e.g., NSGA-II [1], MOEA/D [19], which can achieve good convergence and maintain the diversity of solutions in the benchmark tests for MOPs. However, a specific class of MOPs presents a significant challenge to the existing solution approaches. Multi-modal optimization problems (MMOPs) are problems that have multiple optimal solutions with very similar objective values, often resulting in different Pareto subsets corresponding to the same Pareto front in the objective space.

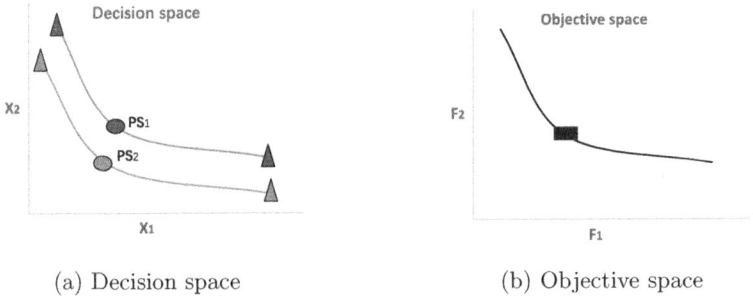

(a) Decision space   (b) Objective space

Fig. 1. Illustration of two distinct Pareto subsets.

Figure 1 illustrates that in a two-dimensional decision space composed of X_1 and X_2, there exist multiple distinct Pareto subsets (Fig. 1(a)). These subsets are mapped through $f : S \to \mathbb{R}^m$ to another two-dimensional objective space composed of F_1 and F_2 (Fig. 1 (b)). There are two distinct Pareto subsets, represented by blue circles (PS_1) and orange circles (PS_2), which correspond to the same Pareto front in the objective space, denoted as PF, and are shown as a black rectangle. Since both Pareto subsets exhibit good performance, the algorithm must retain both Pareto subsets corresponding to the same Pareto front. This means that in MMOPs, the algorithm must ensure convergence and diversity in the objective space and maintain a good distribution in the decision space [20] [11].

Zou et al. [20] proposed the dynamic-niching-based Pareto domination method (DNPN), which integrates the distribution information of individuals in the decision space into traditional Pareto domination, thereby constraining the range of domination for individuals. It avoids the early-stage domination of individuals distant from others in the decision space. Liu et al. [9] proposed a method with double niches to utilize the distribution of individuals in both the decision space and the objective space, thereby optimizing the decision space.

The niche method constrains the individual optimization process, effectively improving the diversity of solutions in the decision space during optimization. The DNPN method also improves the survival capacity of the solutions, making early convergence difficult in the optimization process. However, it does not consider the distribution of solutions in the decision space. The Dynamic-Niche-Based Two-Stage Evolution approach (DNTE) is proposed to address the above issues. The main contributions of this paper are as follows.

(i) A two-stage MMOEA framework is designed, which utilizes different evolutionary strategies to make the distribution in the decision space more uniform and incorporate decision space information into the evolutionary process, helping to ensure diversity in the decision space.

(ii) A decision space density optimization strategy is introduced, enabling the dynamic niche to optimize the population distribution. The density of the decision space is adaptively optimized at each stage through differential evolution to adjust the population distribution.

(iii) A hybridization of differential evolution and niche evolutionary algorithms is proposed, enhancing the population distribution capability.

The structure of this paper is as follows: Sect. 1 provides an introduction to the topic. Section 2 reviews related works and motivations. Section 3 details the design and process of the DNPN-DE algorithm. Section 4 presents experimental comparisons and result analysis. Finally, Sect. 5 concludes the paper and discusses future research directions.

2   Related Works and Motivations

The characteristics of MMOPs solutions dictate that the convergence of solutions and the uniform distribution of solutions in both decision and objective spaces must be considered when designing Multi-objective Metaheuristic Optimization Algorithms (MMOEAs). In recent years, numerous MMOEAs have been proposed, broadly categorized into three types: Pareto-dominance-based algorithms, decomposition-based algorithms, and indicator-based algorithms. These approaches generally perform well in terms of convergence and distribution in the objective space; however, issues such as the continuous deterioration of the decision space and the uneven distribution between the decision and objective spaces may arise during the optimization process. The DNTE approach is proposed to optimize the decision space.

2.1   Related Works

The diversity of the solution space and its convergence ability have always been key goals in the research on MMOEAs.

In Pareto-dominance-based algorithms, convergence is achieved through Pareto dominance. However, for space diversity, different algorithms adopt different strategies. In the Omni-optimizer [2], an alternative crowding distance is used in non-dominated sorting methods, considering both the crowding distance in the objective and decision space. In DN-NSGA-II [8], a niche selection mechanism is integrated into NSGA-II to improve diversity in the decision space. In CPDEA [10], the density of the decision space is explicitly considered during the evolutionary process.

Decomposition-based algorithms, on the other hand, differ in that they decompose the MMOP, exploring subspaces of each subproblem and combining them to find the optimal solution. The diversity of the space is optimized during the process of decomposing, re-integrating, and optimizing reference points or vectors. In MOEA/D-AD [12], niche technology is introduced to maintain distribution in the decision space.

In Indicator-Based Algorithms, an indicator is assigned to represent both the solutions' space diversity and convergence ability. When seeking the optimal solution, one only needs to continuously optimize the indicator, simultaneously optimizing the space diversity and solution representation. In MMEA-WI [7], a single indicator represents both the diversity in the decision space and the convergence information in the objective space, thereby maintaining diversity in the decision space.

In the research mentioned above, parameters are used to characterize both space diversity and solution convergence, aiming to achieve diversity in both the objective and decision spaces while ensuring good solution convergence.

In the optimal solution set, the distribution of the objective space on the Pareto front is uniform and regular. Therefore, most MMOEAs prioritize the diversity of the objective space, which may lead to a continuous deterioration of the diversity and distribution in the decision space, with offspring becoming increasingly unevenly distributed in the decision space. DNPN [20] introduces decision space information into Pareto dominance, restricting the dominance range of early individuals, which helps mitigate the ongoing deterioration of diversity in the decision space.

2.2   Motivations

The niche method has been proven effective in optimizing the distribution of the decision space during the evolutionary process [8] [9]. The niche size adapts accordingly as the population evolves, altering individuals' dominance range. During the evolutionary process, the niche radius serves as a tool to characterize the dominance range of individual solutions and reflects the overall population distribution. By adjusting the niche radius, the overall sparsity distribution of the population can be modified. This paper will elaborate on its motivations from two perspectives: modifying individuals' dominance capability and adjusting the overall sparsity distribution.

The Dominance Capability of Individuals. The dominance range of an individual in the population refers to the maximum distance at which it can participate in non-dominated sorting with other individuals. Suppose the dominance range is not constrained during evolution. In that case, many well-distributed individuals may be prematurely dominated by others with strong convergence properties in the non-dominated sorting process. This affects the survival duration of individuals, as well-distributed individuals may be dominated too early due to non-dominated sorting.

Zou et al. [20] proposed the DNPN method, which significantly enhances the survival capability of solutions. During each selection of parents through non-dominated sorting, decision space information is incorporated into the selection criteria. As the number of evolutionary iterations increases, the niche radius gradually expands. In the decision space, the dominance range of individuals is constrained within the niche radius, meaning that only individuals within this range undergo non-dominated sorting.

Figure 2 illustrates the domination process of DNPN, where concentric spheres of different colors represent the expanding domination range of a point during evolution. Figure 2(a) describes the domination process in DNPN, where the expanding domination range of a point prevents premature domination. From the yellow sphere to the blue sphere and then to the white sphere, the radius of the spheres gradually increases, demonstrating the expanding range of DNPN. The concentric spheres in the upper left corner intersect with the adjacent purple sphere. However, the yellow sphere does not overlap with the purple sphere, indicating that the domination range is relatively small at that stage, preventing the two solutions from being compared. As the domination range increases, an overlapping region emerges between the white and purple spheres. At this stage, non-dominated sorting occurs between the two, which means that when another point enters the domination range, non-dominated sorting is performed. This mechanism ensures that two individuals within a certain distance do not dominate or get dominated prematurely, thereby improving the survival ability of solutions.

The Overall Sparsity Distribution of the Population. The Pareto optimal solution set in the decision space has different positions and shapes. During the

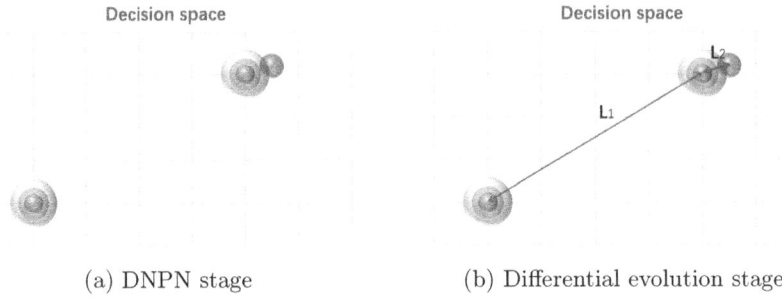

(a) DNPN stage   (b) Differential evolution stage

Fig. 2. The illustration of the domination process of DNPN

evolutionary process, individuals with a better distribution in the objective space may replace those with a slightly better distribution in the decision space, leading to a change in the sparsity of the decision space. This results in an uneven distribution of solutions in the decision space. Figure 2(b) shows a distribution with varying sparsity, with unequal distances between solutions. Three independently distributed points in the space represent three solutions during the evolutionary process. These points are not evenly distributed, with the L_1 distance significantly greater than the L_2 distance. As evolution progresses, this uneven sparsity distribution tends to worsen. After a non-dominated sorting process, the niche radius of the selected parent individuals represents the dominance range of locally optimal solutions. The decision space contains numerous locally well-performing individuals whose dominance range gradually expands to cover the entire decision space as the niche radius increases. However, their ability to explore the space remains insufficient when the niche radius is relatively small in the early stages.

Due to differences in sparsity distribution, the extent of space exploration varies, leading to an uneven distribution of offspring. As evolution progresses, this uneven distribution tends to deteriorate further. Therefore, it is necessary to generate new points to alter the distribution in the decision space.

3   The Proposed Method

This section presents the details of the proposed DNTE method.

3.1   Two-Stage Multi-modal Multi-objective Algorithm Based on Dynamic Niche Updating

This study proposes a two-stage, multi-modal, multi-objective algorithm based on dynamic niche updating. The dynamic niche is related to the number of evolutionary generations and represents the "niche radius". Its calculation formula is as

$$R_d = \alpha \times \left[1 + \left(\frac{Gen}{MaxGen} \right)^{\beta} \right]. \tag{4}$$

Based on dynamic niche, differential evolution is incorporated to adaptively generate new offspring based on the sparsity of the decision space. In each evolutionary process, offspring generation is divided into two stages. In the first stage, the dynamic niche incorporates decision space information into the non-dominated sorting process. This ensures that parents separated by a distance greater than the dynamic niche size cannot dominate each other.

In the second stage, the dynamic niche is again applied to optimize the decision space's sparsity. Parents selected from the best individuals in the first stage undergo further evolution, allowing new offspring to be generated between parents at least twice the dynamic niche size apart, thereby refining the distribution in the decision space. As the dynamic niche size gradually increases, the dominance ability of individuals in the first stage strengthens. In contrast, the number of new offspring generated in the second stage decreases, leading the algorithm toward convergence.

To formalize the offspring generation in the second stage, consider the following equation:

$$x_{\text{new}} = x_i + \gamma \left(x_j - x_i \right), \quad \text{if} \quad \| x_i - x_j \| \geq cR_d, \tag{5}$$

where x_i and x_j represent two parent individuals selected from the first stage, γ is a random number in the range $[0, 1]$ that controls the interpolation between the two parents, and c is a variable coefficient. This equation ensures that a new offspring is generated between two parent individuals only if their Euclidean distance is greater than or equal to cR_d, thereby improving the distribution in the decision space.

Additionally, the generation of new offspring between parents separated by at least cR_d allows for adaptive adjustment of decision space sparsity. In the late stages of evolution, as the dynamic niche size continues to grow, the dominance range of parents in the first stage eventually covers almost the entire decision space, making the second stage nearly inactive, further facilitating convergence. Figure 3 illustrates this process.

(a) No new offspring are generated   (b) Offspring generated in the second stage

Fig. 3. Schematic of offspring generation in the second stage

Figure 3(a) shows that when the distance L_2 is less than twice the dynamic niche size, no new offspring are generated. Figure 3(b) demonstrates the scenario where new points are generated when the distance L_1 is more significant than twice the dynamic niche size. The small green balls represent the offspring generated in the second stage.

In the first stage, solutions are randomly selected in the space, and non-dominated sorting is performed within a radius of R_d around the solution. This process ultimately produces a set of parent individuals that will be used to generate new offspring. The distribution of these parent individuals is sparse and uneven, so in the second stage, a differential algorithm is used to adjust the sparsity.

Specifically, two-parent individuals selected randomly from the first stage are examined for the distance between them. If their distance is greater than twice the R_d radius, meaning their dominance regions do not overlap, a new offspring is generated between them. This helps introduce a new offspring between two sparsely distributed solutions, optimizing the sparsity of the distribution.

At the same time, if the newly generated offspring is very close to another solution, i.e., the distance is less than R_d, it will undergo non-dominated sorting in the first stage of the subsequent evolution, ensuring that the sparsity is not worsened.

In the later stages of evolution, as R_d increases, twice the R_d radius will exceed the size of the entire solution space. At this point, the second stage will no longer generate new offspring, which helps the algorithm converge.

3.2   Framework

The general framework of DNTE is given in Algorithm 1.

4   Experimental Results and Analysis

A comparative experiment is conducted on PlatEMO [14], selecting 22 test problems to evaluate the performance of 8 algorithms. The algorithms are implemented in MATLAB R2024b and executed on a computer with an Intel Core i7-12700 CPU and 40 GB of memory. The operating system is Microsoft Windows 11.

4.1   Experiment Design

DNNSGA-II [8] is used as the base algorithm, with DNTE incorporated into it. Six advanced multi-objective evolutionary algorithms (MMOEAs) are selected for comparison to assess the performance of the proposed algorithm, including TriMOEATAR [11], CoMMEA [5], DNNSGA-II [8], MMEA-WI [7], HREA [6], MO_Ring_PSO_SCD [16], as well as DNNSGA-II_DNPN [20] for demonstrating the effectiveness of the proposed algorithm.

Algorithm 1. General framework of DNTE

Input: Maximum generations $MaxGen$, population size N, current algorithm iteration number Gen, population P

Output: A set of optimal solutions

1: $P \leftarrow Initialization(N)$ // Initialize the population with size N.
2: $R_d \leftarrow \alpha \times \left[1 + \left(\frac{Gen}{MaxGen}\right)^{\beta}\right]$ // Calculate R_d based on Gen and $MaxGen$.
3: **while** $NotTerminated$ **do**
4:     $ParentA \leftarrow TournamentSel(P, R_d, N/2)$
5:     $OffspringA \leftarrow Variation(ParentA)$ // Generate offspring.
6:     $ParentB \leftarrow TournamentSel(ParentA, R_d, N/2)$ // Select more sparse solutions from ParentA.
7:     $OffspringB \leftarrow VariationDE(ParentB)$ // Generate offspring using differential evolution.
8:     **if** $No individual 1 in ParentB$ **then**
9:         $[Population, R_d] \leftarrow EnvSel([Population, OffspringA], R_d, N)$ // Select a new population when there is no individual in ParentB.
10:    **else**
11:        $[Population, R_d] \leftarrow EnvSel([Population, OffspringA, OffspringB], R_d, N)$ // Select new population when individuals are present in ParentB.
12:    **end if**
13:    $R_d \leftarrow \alpha \times \left[1 + \left(\frac{Gen}{MaxGen}\right)^{\beta}\right]$ // Update R_d based on Gen and $MaxGen$.
14: **end while**

The population size and maximum function evaluations (max) for all algorithms are set to 200 and 10,000, respectively, to ensure an objective evaluation of the algorithms' performance. The optimal parameters of DNPN [20] are used as the default parameters for both DNNSGA-II_DNPN [20] and DNNSGA-II_DNTE. For a fair comparison, all algorithms are independently run 30 times on each of the selected test problems.

A total of 22 test problems are chosen to evaluate the performance of the algorithms, including the recently proposed MMF1-9 [17], MMF1_e, MMF1_z, MMF14, MMF14_a, MMMOP1-6 [11], as well as SYM-PART [13] and Omni-test [2]. For performance evaluation, the Inverted Generational Distance (IGD) metric is used to assess the algorithms' convergence and diversity. The IGDf [8] is employed to evaluate the performance in the objective space, where a smaller IGDf value indicates better diversity and convergence in the objective space. The expression for IGDf is as

$$\text{IGDf} = \frac{\left(\sum_{i=1}^{|\text{PF}|} d_i^2\right)^{\frac{1}{2}}}{|\text{PF}|}, \tag{6}$$

where PF represents the Pareto Front, and d_i is the Euclidean distance between the obtained solution and the closest solution on the Pareto Front.

On the other hand, IGDx [8] reflects the convergence and diversity in the decision space. Its mathematical expression is as

$$IGD_X = \frac{\left(\sum_{j=1}^{|PS|} d_j^2\right)^{\frac{1}{2}}}{|PS|}. \tag{7}$$

In both expressions, d_j represents the Euclidean distance between the (objective) decision vector and the closest (objective) decision vector obtained, where $|PS|$ is the size of the solution set.

4.2 Result Analysis

The experimental results are presented in Tables 1, 2, 3, 4, 5 and 6.

Table 1. IGDf of DNNSGAII, DNNSGAII_DNPN, and DNNSGAII_DNTE

Problem	DNNSGAII	DNNSGAII_DNPN	DNNSGAII_DNTE
MMF1	6.3488e-3 (2.82e-4) −	6.6444e-3 (3.42e-4) −	**3.3601e-3 (1.84e-4)**
MMF2	4.5229e-2 (3.52e-2) −	**1.7927e-2 (7.40e-3)** ≈	1.6217e-2 (4.47e-3)
MMF3	2.4470e-2 (1.91e-2) −	**1.2729e-2 (2.58e-3)** ≈	1.4141e-2 (5.54e-3)
MMF4	6.2251e-3 (3.67e-4) −	6.9500e-3 (4.06e-4) −	**3.4206e-3 (2.33e-4)**
MMF5	6.2953e-3 (3.22e-4) −	7.4190e-3 (4.71e-4) −	**3.8275e-3 (1.73e-4)**
MMF6	6.2463e-3 (3.14e-4) −	6.7959e-3 (4.32e-4) −	**3.4726e-3 (1.53e-4)**
MMF7	6.9866e-3 (5.15e-4) −	7.2665e-3 (5.30e-4) −	**3.4833e-3 (1.60e-4)**
MMF8	8.6119e-3 (1.18e-3) −	8.7031e-3 (6.44e-4) −	**4.8903e-3 (3.88e-4)**
MMF9	2.7609e-2 (1.86e-3) −	2.8278e-2 (2.10e-3) −	**1.4385e-2 (1.47e-3)**
Omni-test	**2.3470e-2 (1.64e-3)** +	4.1184e-2 (5.67e-3) −	2.5047e-2 (2.18e-3)
SYM-PART-rotated	3.2530e-2 (3.23e-3) −	3.5089e-2 (4.51e-3) −	**1.9461e-2 (2.07e-3)**
SYM_PART-simple	2.6692e-2 (2.99e-3) −	2.7698e-2 (2.85e-3) −	**1.4080e-2 (1.74e-3)**
MMMOP1	**5.0661e-3 (3.43e-4)** +	2.2296e-2 (2.19e-2) ≈	1.9030e-2 (2.65e-2)
MMMOP2	1.3109e-1 (2.78e-1) −	1.4695e-2 (4.49e-3) −	**7.1215e-3 (1.48e-3)**
MMMOP3	6.5924e-3 (3.43e-4) −	9.1376e-3 (8.50e-4) −	**4.6784e-3 (2.87e-4)**
MMMOP4	**6.5672e-3 (4.74e-4)** +	1.9268e-2 (1.63e-2) ≈	1.8561e-2 (1.26e-2)
MMMOP5	**6.7349e-3 (5.98e-4)** +	1.7278e-2 (1.08e-2) ≈	1.8375e-2 (1.68e-2)
MMMOP6	**2.7846e-2 (2.69e-2)** +	2.3922e-2 (6.87e-3) +	3.1125e-2 (8.56e-3)
MMF1_e	4.1996e+0 (3.84e-1) −	**3.6582e+0 (3.91e-1)** ≈	3.7047e+0 (4.64e-1)
MMF1_z	1.2585e+0 (2.19e-3) −	1.2585e+0 (3.15e-3) −	**1.2573e+0 (1.16e-4)**
MMF14	**9.2089e-2 (2.04e-3)** +	9.5385e-2 (2.38e-3) ≈	9.4996e-2 (3.06e-3)
MMF14_a	**1.3181e-1 (4.50e-3)** +	1.3621e-1 (4.89e-3) ≈	1.3464e-1 (5.72e-3)
+/ − / ≈	7/15/0	1/13/8	

From Tables 1, 2, 3, 4, 5 and 6, the following observations are obtained.

(i) DNNSGAII_DNTE performs well overall on IGDf and shows decent results on MMF1-9, MMF1_z, MMF1_e, and MMMOP2-3. However, it does not perform as well as the current state-of-the-art algorithms on MMMOP1, 4, 5, 6, and MMF14_a. MMMOP2 is biased in the decision space, and the spacing between the Pareto optimal solutions is highly varied. This further

Table 2. IGDf of CoMMEA HREA and DNNSGAII_DNTE

Problem	CoMMEA	HREA	DNNSGAII_DNTE
MMF1	7.9213e-3 (8.37e-4) −	6.6517e-3 (4.50e-4) −	**3.3601e-3 (1.84e-4)**
MMF2	**1.7930e-2 (5.23e-3)** ≈	1.9356e-2 (9.59e-3) ≈	**1.6217e-2 (4.47e-3)**
MMF3	**1.3106e-2 (1.71e-3)** ≈	1.3910e-2 (3.57e-3) ≈	**1.4141e-2 (5.54e-3)**
MMF4	8.3820e-3 (6.99e-4) −	6.8716e-3 (4.09e-4) −	**3.4206e-3 (2.33e-4)**
MMF5	8.1945e-3 (5.94e-4) −	7.2652e-3 (7.45e-4) −	**3.8275e-3 (1.73e-4)**
MMF6	8.5077e-3 (9.03e-4) −	6.8350e-3 (5.35e-4) −	**3.4726e-3 (1.53e-4)**
MMF7	8.5681e-3 (6.28e-4) −	7.5578e-3 (7.61e-4) −	**3.4833e-3 (1.60e-4)**
MMF8	1.0826e-2 (1.37e-3) −	7.9010e-3 (8.84e-4) −	**4.8903e-3 (3.88e-4)**
MMF9	2.4137e-2 (1.51e-3) −	7.7073e-3 (1.10e-2) −	**1.4385e-2 (1.47e-3)**
Omni-test	**1.5508e-2 (1.15e-3)** +	2.5397e-2 (2.28e-3) ≈	2.5047e-2 (2.18e-3)
SYM-PART-rotated	4.1839e-2 (3.96e-3) −	2.9727e-2 (3.64e-3) −	**1.9461e-2 (2.07e-3)**
SYM-PART-simple	3.2455e-2 (2.77e-3) −	2.9509e-2 (3.96e-3) −	**1.4080e-2 (1.74e-3)**
MMMOP1	**6.2554e-3 (8.92e-4)** +	9.6339e-3 (3.66e-3) ≈	1.9030e-2 (2.65e-2)
MMMOP2	2.3597e-2 (2.84e-3) −	2.0971e-2 (7.24e-3) −	**7.1215e-3 (1.48e-3)**
MMMOP3	1.8118e-2 (1.50e-3) −	1.3876e-2 (2.17e-3) −	**4.6784e-3 (2.87e-4)**
MMMOP4	**6.1630e-3 (7.81e-4)** +	8.5824e-3 (2.87e-3) +	1.8561e-2 (1.26e-2)
MMMOP5	**6.3357e-3 (6.09e-4)** +	4.8268e-2 (1.82e-1) ≈	1.8375e-2 (1.68e-2)
MMMOP6	**1.8210e-2 (3.97e-3)** +	4.4339e-2 (2.56e-2)	3.1125e-2 (8.56e-3)
MMF1_e	4.0488e+0 (4.04e-1) −	3.9972e+0 (4.83e-1) −	**3.7047e+0 (4.64e-1)**
MMF1_z	1.2576e+0 (4.91e-4) −	1.2584e+0 (3.90e-3) −	**1.2573e+0 (1.16e-4)**
MMF14	**7.5374e-2 (1.27e-3)** +	1.0544e-1 (4.49e-3) −	9.4996e-2 (3.06e-3)
MMF14_a	**8.5891e-2 (2.24e-3)** +	8.8026e-2 (2.51e-3) +	1.3464e-1 (5.72e-3)
+/−/≈	7/13/2	2/15/5	

Table 3. IGDf of MMEAWI, MO_Ring_PSO_SCD, and TriMOEATAR

Problem	MMEAWI	MO_Ring_PSO_SCD	TriMOEATAR
MMF1	6.4141e-3 (5.90e-4) −	8.5483e-3 (1.14e-3) −	5.9084e-3 (6.81e-4) −
MMF2	2.4467e-2 (1.47e-2) −	5.2339e-2 (2.05e-2) −	4.9011e-2 (6.72e-2) −
MMF3	**1.8458e-2 (1.36e-2)** ≈	4.6683e-2 (2.40e-2) −	2.9874e-2 (1.49e-2) −
MMF4	7.1533e-3 (7.03e-4) −	7.0367e-3 (4.50e-4) −	5.4686e-3 (8.53e-2) −
MMF5	6.2455e-3 (4.91e-4) −	7.3959e-3 (7.11e-4) −	6.5181e-3 (2.05e-3) −
MMF6	**3.5091e-3 (2.18e-4)** ≈	7.4610e-3 (5.89e-4) −	5.5972e-3 (1.49e-3) −
MMF7	3.9264e-3 (2.21e-4) −	7.2658e-3 (7.94e-4) −	5.1210e-3 (1.32e-3) −
MMF8	**3.7546e-3 (1.86e-4)** +	7.6641e-3 (6.36e-4) −	4.4743e-3 (9.54e-5) +
MMF9	1.5833e-2 (1.72e-3) −	3.2875e-2 (6.48e-3) −	7.5087e-2 (8.03e-4) −
Omni-test	**1.5072e-2 (1.87e-3)** −	4.0079e-2 (2.97e-3) −	2.8720e-2 (4.74e-3) −
SYM-PART-rotated	1.7865e-2 (1.95e-3) +	4.1662e-2 (4.73e-3) −	4.8361e-2 (6.10e-3) −
SYM-PART-simple	1.5287e-2 (2.08e-3) −	3.7118e-2 (3.88e-3) −	4.8693e-2 (2.03e-3) −
MMMOP1	7.6457e-2 (6.16e-2) −	5.1886e-1 (2.74e-1) −	**5.0232e-3 (1.78e-3)** +
MMMOP2	**4.3094e-3 (3.98e-4)** +	4.1965e-1 (2.38e-1) −	1.2714e-1 (2.80e-1) −
MMMOP3	**3.8101e-3 (2.37e-4)** +	1.5628e-2 (1.88e-3) −	4.0495e-3 (5.31e-5) +
MMMOP4	5.4207e-2 (3.42e-2) −	4.3526e-2 (1.81e-1) −	**6.5922e-3 (3.46e-3)** +
MMMOP5	3.3027e-2 (3.94e-2) −	4.4174e-1 (3.02e-1) −	**5.1275e-3 (8.01e-4)** +
MMMOP6	4.9334e-2 (2.22e-2) −	4.1130e-2 (7.63e-3) −	**3.2018e-2 (1.96e-2)** ≈
MMF1_e	4.1494e+0 (4.41e-1) −	**3.7395e+0 (4.37e-1)** ≈	4.2470e+0 (2.53e-1) −
MMF1_z	**1.2576e+0 (8.82e-4)** ≈	1.2601e+0 (2.34e-3) −	1.2628e+0 (8.17e-3) −
MMF14	**8.6579e-2 (1.84e-3)** +	1.0383e-1 (4.48e-3) −	9.0111e-2 (1.31e-3) +
MMF14_a	9.3294e-2 (2.40e-3) +	9.9910e-2 (3.84e-3) +	**8.9980e-2 (2.35e-3)** +
+/−/≈	7/12/3	1/20/1	7/14/1

Table 4. IGDx of DNNSGAII, DNNSGAII_DNPN, and DNNSGAII_DNTE

Problem	DNNSGAII	DNNSGAII_DNPN	DNNSGAII_DNTE
MMF1	7.3366e-2 (5.60e-3) −	7.0084e-2 (2.49e-3) −	**4.1930e-2 (1.35e-3)**
MMF2	1.2740e-1 (7.06e-2) −	**3.2015e-2 (1.15e-2)** ≈	2.8657e-2 (5.98e-3)
MMF3	7.4698e-2 (3.37e-2) −	2.8915e-2 (5.86e-3) −	**2.6261e-2 (7.80e-3)**
MMF4	5.0903e-2 (4.77e-3) −	5.5253e-2 (7.00e-3) −	**2.7679e-2 (1.89e-3)**
MMF5	1.3494e-1 (1.26e-2) −	1.2503e-1 (7.48e-3) −	**7.1943e-2 (2.91e-3)**
MMF6	1.1467e-1 (9.83e-3) −	1.1202e-1 (5.99e-3) −	**6.7651e-2 (2.40e-3)**
MMF7	3.9696e-2 (3.03e-3) −	3.8961e-2 (2.84e-3) −	**2.2463e-2 (1.47e-3)**
MMF8	1.7034e-1 (6.23e-2) −	1.2903e-1 (2.37e-2) −	**5.9756e-2 (3.87e-3)**
MMF9	1.1874e-2 (2.02e-3) −	1.1451e-2 (9.55e-4) −	**6.0527e-3 (3.58e-4)**
Omni-test	5.3088e-1 (2.76e-1) −	2.9490e-1 (7.10e-2) −	**1.3160e-1 (2.37e-2)**
SYM-PART-rotated	2.0778e+0 (1.17e+0) −	1.8013e+0 (1.32e+0) −	**8.9871e-2 (1.02e-2)**
SYM-PART-simple	2.2606e-1 (5.37e-1) −	2.2921e-1 (5.92e-1) −	**6.8060e-2 (8.69e-3)**
MMMOP1	2.1431e-1 (8.45e-2) −	1.2474e-1 (3.61e-2) −	**8.1985e-2 (2.65e-2)**
MMMOP2	2.8748e-1 (3.50e-1) −	3.6007e-2 (2.09e-2) −	**1.5571e-2 (7.43e-3)**
MMMOP3	2.9458e-2 (1.30e-2) −	2.6579e-2 (2.36e-3) −	**1.6328e-2 (1.52e-3)**
MMMOP4	1.6276e-1 (1.00e-1) −	4.0728e-2 (2.77e-2) −	**1.8479e-2 (1.03e-2)**
MMMOP5	1.8948e-1 (9.31e-2) −	3.5689e-2 (3.00e-2) −	**2.0274e-2 (1.53e-2)**
MMMOP6	4.2485e-1 (4.02e-2) −	1.4315e-1 (2.78e-2) −	**1.0824e-1 (1.88e-2)**
MMF1_e	1.4718e+0 (5.98e-1) −	**2.6171e-1 (1.60e-2)** ≈	2.5610e-1 (1.77e-2)
MMF1_z	6.9114e-2 (5.18e-2) −	6.2079e-2 (2.51e-3) −	**5.9431e-2 (1.61e-3)**
MMF14	**6.0451e-2 (1.55e-3)** +	6.1764e-2 (1.47e-3) ≈	6.1911e-2 (2.00e-3)
MMF14_a	**1.0361e-1 (5.14e-3)** ≈	1.0611e-1 (3.89e-3) ≈	1.0571e-1 (4.89e-3)
+/ − / ≈	1/20/1	0/18/4	

Table 5. IGDx of CoMMEA, HREA, and DNNSGAII_DNTE

Problem	CoMMEA	HREA	DNNSGAII_DNTE
MMF1	5.8706e-2 (9.91e-4) −	5.6558e-2 (1.89e-3) −	**4.1930e-2 (1.35e-3)**
MMF2	**3.1275e-2 (6.61e-3)** ≈	3.9002e-2 (1.18e-2) −	2.8657e-2 (5.98e-3)
MMF3	**2.3497e-2 (3.82e-3)** ≈	3.1630e-2 (8.42e-3) −	2.6261e-2 (7.80e-3)
MMF4	4.0182e-2 (1.54e-3) −	3.2884e-2 (1.46e-3) −	**2.7679e-2 (1.89e-3)**
MMF5	9.5720e-2 (2.08e-3) −	9.5292e-2 (3.52e-3) −	**7.1943e-2 (2.91e-3)**
MMF6	8.6179e-2 (2.13e-3) −	8.3418e-2 (2.14e-3) −	**6.7651e-2 (2.40e-3)**
MMF7	4.2812e-2 (3.04e-3) −	3.3820e-2 (1.96e-3) −	**2.2463e-2 (1.47e-3)**
MMF8	8.0598e-2 (3.15e-3) −	7.2782e-2 (7.22e-3) −	**5.9756e-2 (3.87e-3)**
MMF9	1.0968e-2 (1.17e-3) −	1.7834e-2 (1.06e-3) −	**6.0527e-3 (3.58e-4)**
Omni-test	3.3508e-1 (1.79e-1) −	1.4426e-1 (2.60e-2) −	**1.3160e-1 (2.37e-2)**
SYM-PART-rotated	1.6382e-1 (1.78e-1) −	9.5681e-2 (6.95e-3) −	**8.9871e-2 (1.02e-2)**
SYM-PART-simple	9.8898e-2 (5.00e-3) −	7.2660e-2 (3.79e-3) −	**6.8060e-2 (8.69e-3)**
MMMOP1	**4.9543e-2 (3.13e-2)** +	5.7383e-2 (2.03e-2) +	8.1985e-2 (2.65e-2)
MMMOP2	2.0602e-2 (4.89e-3) −	**1.7192e-2 (1.33e-2)** ≈	1.5571e-2 (7.43e-3)
MMMOP3	4.4384e-2 (3.29e-3) −	2.9628e-2 (4.13e-3) −	**1.6328e-2 (1.52e-3)**
MMMOP4	**5.5234e-3 (1.83e-4)** +	1.9552e-2 (1.20e-2) ≈	1.8479e-2 (1.03e-2)
MMMOP5	**5.5615e-3 (1.16e-4)** +	3.3023e-2 (5.10e-2) ≈	2.0274e-2 (1.53e-2)
MMMOP6	**9.2338e-2 (4.23e-2)** +	1.7896e-2 (6.05e-2) −	1.0824e-1 (1.88e-2)
MMF1_e	3.1551e-1 (9.41e-2) −	3.1046e-1 (3.95e-2) −	**2.5610e-1 (1.77e-2)**
MMF1_z	**5.5094e-2 (1.47e-3)** +	5.7874e-2 (2.16e-3) +	5.9431e-2 (1.61e-3)
MMF14	**5.0151e-2 (9.80e-4)** +	5.9520e-2 (2.17e-3) +	6.1911e-2 (2.00e-3)
MMF14_a	6.6904e-2 (1.24e-3) +	**6.1263e-2 (1.93e-3)** +	1.0571e-1 (4.89e-3)
+/ − / ≈	7/13/2	4/15/3	

Table 6. IGDx of MMEAWI, MO_Ring_PSO_SCD, and TriMOEATAR

Problem	MMEAWI	MO_Ring_PSO_SCD	TriMOEATAR
MMF1	6.5624e-2 (2.81e-3) −	9.9103e-2 (1.24e-2) −	9.7493e-2 (1.91e-2) −
MMF2	7.1100e-2 (4.58e-2) −	1.1130e-1 (4.53e-2) −	1.2679e-1 (5.52e-2) −
MMF3	4.6917e-2 (2.40e-2) −	8.9596e-2 (4.30e-2) −	9.2930e-2 (3.22e-2) −
MMF4	4.2244e-2 (2.58e-3) −	7.2131e-2 (1.44e-2) −	1.8342e-1 (1.46e-1) −
MMF5	1.0862e-1 (3.71e-3) −	1.6282e-1 (2.07e-2) −	1.7410e-1 (3.27e-2) −
MMF6	**6.5811e-2 (2.91e-3)** +	1.3915e-1 (1.22e-2) −	1.5319e-1 (3.76e-2) −
MMF7	2.4546e-2 (1.81e-3) −	6.7045e-2 (1.57e-2) −	6.8379e-2 (1.59e-2) −
MMF8	8.6851e-2 (2.07e-2) −	2.1622e-1 (1.00e-1) −	6.5015e-1 (2.57e-1) −
MMF9	6.3024e-3 (2.07e-4) −	3.1556e-2 (1.57e-2) −	**3.4342e-3 (3.09e-5)** +
Omni-test	5.4293e-1 (1.69e-1) −	8.0300e-1 (1.57e-1) −	1.0570e+0 (2.94e-1) −
SYM-PART-rotated	1.9755e-1 (3.95e-1) −	3.6641e-1 (3.49e-1) −	4.1291e+0 (1.80e+0) −
SYM-PART-simple	**5.9284e-2 (4.45e-3)** +	6.2484e-1 (5.60e-1) −	3.8975e-1 (9.14e-1) −
MMMOP1	2.5820e-1 (5.85e-2) −	2.7068e-1 (5.00e-2) −	**1.7984e-2 (2.83e-2)** +
MMMOP2	**1.3793e-2 (1.09e-2)** +	4.4114e-1 (2.93e-1) −	1.3041e-1 (2.61e-1) −
MMMOP3	1.2277e-2 (8.89e-4) +	6.5467e-2 (1.66e-2) −	**6.2722e-3 (2.01e-3)** +
MMMOP4	9.7587e-2 (4.23e-2) −	2.1168e-1 (7.00e-2) −	5.6537e-2 (1.01e-1) −
MMMOP5	1.2182e-1 (6.76e-2) −	2.0893e-1 (6.44e-2) −	**1.5613e-2 (1.55e-2)** +
MMMOP6	3.4640e-1 (6.80e-2) −	3.0210e-1 (4.16e-2) −	3.9398e-1 (9.93e-2) −
MMF1_e	7.7462e-1 (3.09e-1) −	8.2128e-1 (3.95e-1) −	1.8022e+0 (7.26e-1) −
MMF1_z	6.2110e-2 (2.94e-3) −	8.1900e-2 (1.20e-2) −	8.5211e-2 (7.24e-3) −
MMF14	5.2505e-2 (1.03e-3) +	7.4155e-2 (3.81e-3) −	**3.8173e-2 (5.07e-4)** +
MMF14_a	7.2936e-2 (2.22e-3) +	8.4665e-2 (4.82e-3) +	**6.9702e-2 (4.11e-3)** +
+/ − / ≈	6/16/0	1/21/0	6/16/0

increases the difficulty of retaining all of these solutions. In MMMOP3, a point on the Pareto front has multiple Pareto optimal solutions, which have different diversity and/or convergence-related decision variable values. These features challenge MMOEAs regarding diversity in the decision space, while DNNSGAII_DNTE performs well in both tests. However, in algorithms with many local optima, DNNSGAII_DNTE does not perform as well.

(ii) DNNSGAII_DNTE performs very well on IGDx, showing excellent results in almost all test metrics. It is weaker than advanced algorithms such as CoMMA in MMMOP4-6, indicating that DNNSGAII_DNTE has indeed optimized the distribution in the decision space but faces some difficulties in exploring local optima.

(iii) Compared to DNNSGAII_DNPN, DNNSGAII_DNTE has advantages on both IGDx and IGDf, suggesting that including the differential algorithm plays a significant role in solving MMOPs problems.

Table 7. IGDf on different γ

Problem	$\gamma = 0.1$	$\gamma = 0.2$	$\gamma = 0.3$
MMF1	**3.2956e-3 (1.74e-4)** +	3.3487e-3 (1.89e-4) +	3.4344e-3 (1.69e-4) +
MMF2	**1.1781e-2 (2.89e-3)** +	1.3464e-2 (2.39e-3) +	1.4027e-2 (2.86e-3) +
MMF3	1.4761e-2 (5.66e-3) $\approx$	1.3237e-2 (5.39e-3) $\approx$	1.3626e-2 (6.33e-3)
MMF4	**3.4376e-3 (2.77e-4)** +	3.5235e-3 (2.52e-4) +	3.4753e-3 (2.53e-4) +
MMF5	3.7936e-3 (2.09e-4) +	3.7780e-3 (2.21e-4) +	3.7949e-3 (1.53e-4) +
MMF6	**3.4808e-3 (1.41e-4)** +	3.5219e-3 (1.69e-4) +	3.5385e-3 (1.63e-4) +
MMF7	3.5756e-3 (1.26e-4) +	3.4515e-3 (1.53e-4) +	**3.4144e-3 (2.32e-4)** +
MMF8	**4.4229e-3 (3.62e-4)** +	4.7678e-3 (3.33e-4) +	4.7563e-3 (4.33e-4) +
MMF9	1.4455e-2 (1.03e-3) +	1.4476e-2 (1.21e-3) +	1.4365e-2 (1.18e-3) +
Omni-test	2.5705e-2 (2.04e-3) +	2.5647e-2 (2.44e-3) +	2.5450e-2 (2.59e-3) +
SYM-PART-rotated	**1.8923e-2 (2.36e-3)** +	1.9369e-2 (2.13e-3) +	1.9999e-2 (3.33e-3) +
SYM-PART-simple	**1.3424e-2 (1.75e-3)** +	1.4549e-2 (1.69e-3) +	1.4095e-2 (1.51e-3) +
MMMOP1	3.4183e-2 (6.78e-2) $\approx$	4.2656e-2 (1.39e-1) $\approx$	**1.9323e-2 (1.20e-2)** $\approx$
MMMOP2	**6.5758e-3 (9.54e-4)** +	7.0390e-3 (1.67e-3) +	7.1471e-3 (1.53e-3) +
MMMOP3	4.8639e-3 (3.89e-4) +	**4.6614e-3 (3.58e-4)** +	4.7936e-3 (4.08e-4) +
MMMOP4	2.2296e-2 (2.16e-2) $\approx$	2.3330e-2 (2.79e-2) $\approx$	**1.6254e-2 (1.53e-2)** $\approx$
MMMOP5	1.5063e-2 (1.68e-2) +	2.1042e-2 (3.49e-2) $\approx$	**1.1439e-2 (5.80e-3)** +
MMMOP6	4.0115e-2 (1.06e-2) −	4.6078e-2 (1.53e-2) −	4.9821e-2 (1.30e-2) −
MMF14	9.5255e-2 (2.83e-3) $\approx$	9.5391e-2 (1.59e-3) $\approx$	9.5156e-2 (3.09e-3) $\approx$
MMF14_a	9.6664e-2 (3.56e-3) +	9.3967e-2 (2.59e-3) +	9.4212e-2 (2.17e-3) +
MMF1_e	3.7552e+0 (4.51e-1) $\approx$	**3.6393e+0 (3.95e-1)** $\approx$	3.6496e+0 (3.98e-1) $\approx$
MMF1_z	1.2582e+0 (3.30e-3) +	1.2573e+0 (1.04e-4) +	1.2573e+0 (1.33e-4) +
+/ − / $\approx$	16/1/5	15/1/6	16/1/5
Problem	$\gamma = 0.4$	$\gamma = 0.5$	DNNSGAII_DNPN
MMF1	3.3547e-3 (1.65e-4) +	3.3739e-3 (1.47e-4) +	6.6444e-3 (3.42e-4)
MMF2	1.4872e-2 (2.51e-3) $\approx$	1.5216e-2 (2.11e-3) $\approx$	1.7927e-2 (7.40e-3)
MMF3	1.3225e-2 (2.99e-3) $\approx$	1.2980e-2 (2.49e-3) $\approx$	**1.2729e-2 (2.58e-3)**
MMF4	3.5001e-3 (2.40e-4) +	3.4999e-3 (1.93e-4) +	6.9500e-3 (4.06e-4)
MMF5	**3.7524e-3 (1.74e-4)** +	3.7944e-3 (1.58e-4) +	7.4190e-3 (4.71e-4)
MMF6	3.5265e-3 (1.93e-4) +	3.5551e-3 (1.12e-4) +	6.7959e-3 (4.32e-4)
MMF7	3.5187e-3 (1.82e-4) +	3.4866e-3 (2.59e-4) +	7.2665e-3 (5.30e-4)
MMF8	4.8470e-3 (3.56e-4) +	4.9114e-3 (4.19e-4) +	8.7031e-3 (6.44e-4)
MMF9	1.4248e-2 (1.28e-3) +	**1.4232e-2 (1.37e-3)** +	2.8278e-2 (2.10e-3)
Omni-test	2.5388e-2 (2.06e-3) +	**2.4801e-2 (1.80e-3)** +	4.1184e-2 (5.67e-3)
SYM-PART-rotated	2.0565e-2 (3.09e-3) +	1.9383e-2 (2.79e-3) +	3.5089e-2 (4.51e-3)
SYM-PART-simple	1.3792e-2 (1.33e-3) +	1.4629e-2 (1.89e-3) +	2.7698e-2 (2.85e-3)
MMMOP1	3.1429e-2 (5.09e-2) $\approx$	3.1741e-2 (4.02e-2) $\approx$	2.2296e-2 (2.19e-2)
MMMOP2	7.3129e-3 (1.62e-3) +	6.7252e-3 (1.34e-3) +	1.4695e-2 (4.49e-3)
MMMOP3	4.8390e-3 (5.37e-4) +	4.7945e-3 (4.18e-4) +	9.1376e-3 (8.50e-4)
MMMOP4	2.0656e-2 (1.29e-2) $\approx$	2.8001e-2 (3.69e-2) $\approx$	1.9268e-2 (1.63e-2)
MMMOP5	1.3802e-2 (2.50e-2) +	1.7752e-2 (2.39e-2) $\approx$	1.7278e-2 (1.08e-2)
MMMOP6	7.1663e-2 (2.69e-2) −	7.2816e-2 (2.09e-2) −	**2.3922e-2 (6.87e-3)**
MMF14	**9.4474e-2 (2.51e-3)** $\approx$	9.4678e-2 (2.51e-3) $\approx$	9.5385e-2 (2.38e-3)
MMF14_a	9.4279e-2 (3.19e-3) +	**9.3192e-2 (1.73e-3)** +	1.3621e-1 (4.89e-3)
MMF1_e	3.6606e+0 (3.64e-1) $\approx$	3.6928e+0 (4.48e-1) $\approx$	3.6582e+0 (3.91e-1)
MMF1_z	**1.2573e+0 (8.20e-5)** +	1.2573e+0 (1.56e-4) +	1.2585e+0 (3.15e-3)
+/ − / $\approx$	15/1/6	14/1/7	

Table 8. IGDx on different γ

Problem	$\gamma = 0.1$	$\gamma = 0.2$	$\gamma = 0.3$
MMF1	4.2886e-2 (1.87e-3) +	4.2228e-2 (1.47e-3) +	4.1477e-2 (1.28e-3) +
MMF2	2.2039e-2 (6.87e-3) +	**2.1801e-2 (3.87e-3)** +	2.2678e-2 (5.31e-3) +
MMF3	2.8524e-2 (8.15e-3) ≈	**2.4772e-2 (5.42e-3)** +	2.6052e-2 (6.70e-3) +
MMF4	2.9826e-2 (2.49e-3) +	2.8193e-2 (2.07e-3) +	2.7852e-2 (2.55e-3) +
MMF5	7.7683e-2 (4.27e-3) +	7.4018e-2 (2.98e-3) +	**7.2638e-2 (3.40e-3)** +
MMF6	7.3106e-2 (5.15e-3) +	6.9540e-2 (3.28e-3) +	6.7766e-2 (3.08e-3) +
MMF7	2.2671e-2 (1.14e-3) +	2.2596e-2 (1.83e-3) +	2.1908e-2 (1.25e-3) +
MMF8	6.2598e-2 (5.14e-3) +	**6.0265e-2 (6.76e-3)** +	6.1472e-2 (5.58e-3) +
MMF9	6.1710e-3 (4.01e-4) +	**6.0900e-3 (4.85e-4)** +	6.1904e-3 (4.09e-4) +
Omni-test	1.4488e-1 (3.62e-2) +	1.3762e-1 (2.73e-2) +	1.4902e-1 (3.91e-2) +
SYM-PART-rotated	6.3483e-1 (7.11e-1) +	3.3508e-1 (4.00e-1) +	4.0016e-1 (5.76e-1) +
SYM-PART-simple	7.5686e-2 (1.48e-2) +	7.0130e-2 (8.54e-3) +	7.2765e-2 (1.09e-2) +
MMMOP1	8.7592e-2 (3.39e-2) +	9.6468e-2 (2.80e-2) +	8.2563e-2 (2.59e-2) +
MMMOP2	2.2009e-2 (1.86e-2) +	**1.6711e-2 (6.21e-3)** +	1.7476e-2 (7.08e-3) +
MMMOP3	1.6013e-2 (1.28e-3) +	1.5843e-2 (1.57e-3) +	1.5859e-2 (1.31e-3) +
MMMOP4	2.0453e-2 (1.37e-2) +	1.7031e-2 (8.84e-3) +	**1.6426e-2 (1.14e-2)** +
MMMOP5	**1.5449e-2 (8.90e-3)** +	2.0093e-2 (1.43e-2) +	1.6444e-2 (9.54e-3) +
MMMOP6	1.0439e-1 (1.63e-2) +	1.0221e-1 (1.09e-2) +	1.0599e-1 (1.52e-2) +
MMF14	6.2576e-2 (1.83e-3) ≈	6.2491e-2 (1.59e-3) ≈	6.2207e-2 (2.23e-3) ≈
MMF14_a	7.7208e-2 (3.77e-3) +	**7.4544e-2 (2.10e-3)** +	7.4919e-2 (2.21e-3) +
MMF1_e	2.5788e-1 (1.66e-2) ≈	2.5843e-1 (1.37e-2) ≈	**2.5244e-1 (1.52e-2)** +
MMF1_z	6.0684e-2 (2.00e-3) +	5.9112e-2 (2.03e-3) +	5.9257e-2 (1.91e-3) +
+/ − /≈	19/0/3	20/0/2	21/0/1
Problem	$\gamma = 0.4$	$\gamma = 0.5$	DNNSGAII_DNPN
MMF1	**4.0947e-2 (1.30e-3)** +	4.1646e-2 (1.33e-3) +	7.0084e-2 (2.49e-3)
MMF2	2.4599e-2 (3.86e-3) +	2.4334e-2 (2.99e-3) +	3.2015e-2 (1.15e-2)
MMF3	2.8365e-2 (8.10e-3) ≈	2.6612e-2 (5.02e-3) ≈	2.8915e-2 (5.86e-3)
MMF4	2.8117e-2 (2.56e-3) +	**2.7338e-2 (2.02e-3)** +	5.5253e-2 (7.00e-3)
MMF5	7.2783e-2 (2.89e-3) +	7.3516e-2 (3.19e-3) +	1.2503e-1 (7.48e-3)
MMF6	**6.6606e-2 (2.91e-3)** +	6.7947e-2 (3.24e-3) +	1.1202e-1 (5.99e-3)
MMF7	**2.1876e-2 (9.51e-4)** +	2.2008e-2 (9.63e-4) +	3.8961e-2 (2.84e-3)
MMF8	6.1302e-2 (4.87e-3) +	6.3308e-2 (4.32e-3) +	1.2903e-1 (2.37e-2)
MMF9	6.1579e-3 (3.51e-4) +	6.1599e-3 (4.20e-4) +	1.1451e-2 (9.55e-4)
Omni-test	1.4844e-1 (2.86e-2) +	**1.3160e-1 (2.80e-2)** +	2.9490e-1 (7.10e-2)
SYM-PART-rotated	2.2328e-1 (2.98e-1) +	**8.9132e-2 (1.44e-2)** +	1.8013e+0 (1.32e+0)
SYM-PART-simple	6.8833e-2 (5.92e-3) +	**6.6037e-2 (6.01e-3)** +	2.2921e-1 (5.92e-1)
MMMOP1	**8.0022e-2 (2.49e-2)** +	8.9486e-2 (2.43e-2) +	1.2474e-1 (3.61e-2)
MMMOP2	1.8645e-2 (1.05e-2) +	2.6048e-2 (2.75e-2) +	3.6007e-2 (2.09e-2)
MMMOP3	**1.5835e-2 (1.57e-3)** +	1.5883e-2 (1.68e-3) +	2.6579e-2 (2.36e-3)
MMMOP4	1.7393e-2 (8.87e-3) +	2.0099e-2 (1.77e-2) +	4.0728e-2 (2.77e-2)
MMMOP5	1.8709e-2 (1.36e-2) +	1.9410e-2 (1.60e-2) +	3.5689e-2 (3.00e-2)
MMMOP6	**1.0067e-1 (1.06e-2)** +	1.0468e-1 (1.40e-2) +	1.4315e-1 (2.78e-2)
MMF14	6.1880e-2 (1.85e-3) ≈	6.2411e-2 (1.88e-3) ≈	**6.1764e-2 (1.47e-3)**
MMF14_a	7.5068e-2 (2.93e-3) +	7.4921e-2 (2.87e-3) +	1.0611e-1 (3.89e-3)
MMF1_e	2.5745e-1 (1.46e-2) ≈	2.5848e-1 (1.53e-2) ≈	2.6171e-1 (1.60e-2)
MMF1_z	5.9239e-2 (1.47e-3) +	**5.8873e-2 (1.91e-3)** +	6.2079e-2 (2.51e-3)
+/ − /≈	19/0/3	19/0/3	

Table 9. IGDf on different c

Problem	$c = 1$	$c = 1.5$	$c = 2$	$c = 2.5$
MMF1	3.3754e-3 (2.00e-4) +	**3.3496e-3 (1.69e-4)** +	6.6013e-3 (3.60e-4) ≈	3.4180e-3 (1.60e-4) +
MMF2	1.5123e-2 (2.37e-3) ≈	1.5657e-2 (2.98e-3) ≈	1.7000e-2 (5.29e-3) ≈	1.4966e-2 (3.19e-3) ≈
MMF3	1.3098e-2 (2.37e-3) ≈	1.3647e-2 (3.41e-3) ≈	1.2931e-2 (2.13e-3) ≈	1.3300e-2 (2.55e-3) ≈
MMF4	3.4764e-3 (2.02e-4) +	3.4608e-3 (2.54e-4) +	**3.4189e-3 (2.19e-4)** +	3.4894e-3 (2.55e-4) +
MMF5	3.7809e-3 (2.00e-4) +	3.7909e-3 (1.20e-4) +	3.7835e-3 (1.74e-4) +	3.7708e-3 (1.66e-4) +
MMF6	3.5080e-3 (1.43e-4) +	3.5364e-3 (1.60e-4) +	3.5269e-3 (1.61e-4) +	3.5321e-3 (1.49e-4) +
MMF7	**3.4592e-3 (1.90e-4)** +	3.4730e-3 (1.73e-4) +	3.5225e-3 (2.31e-4) +	3.5633e-3 (1.90e-4) +
MMF8	4.8976e-3 (4.41e-4) +	**4.7476e-3 (3.54e-4)** +	4.7806e-3 (3.05e-4) +	4.7625e-3 (3.75e-4) +
MMF9	1.4481e-2 (1.48e-3) +	1.4256e-2 (1.24e-3) +	1.4470e-2 (1.24e-3) +	1.4525e-2 (1.12e-3) +
Omni-test	2.4511e-2 (2.00e-3) +	2.4538e-2 (2.07e-3) +	2.4720e-2 (2.19e-3) +	2.4805e-2 (2.50e-3) +
SYM-PART-rotated	1.9172e-2 (2.32e-3) +	1.8603e-2 (2.78e-3) +	**1.8472e-2 (2.58e-3)** +	1.9232e-2 (2.23e-3) +
SYM-PART-simple	1.3605e-2 (1.49e-3) +	1.3873e-2 (1.72e-3) +	1.3595e-2 (1.61e-3) +	1.4313e-2 (1.87e-3) +
MMMOP1	7.6494e-2 (6.30e-2) −	2.4655e-2 (2.77e-2) ≈	5.3012e-2 (1.34e-1) ≈	2.0526e-2 (1.82e-2) ≈
MMMOP2	1.6615e-2 (5.37e-3) +	6.9391e-3 (1.25e-3) +	6.9643e-3 (1.55e-3) +	1.4695e-2 (4.49e-3) +
MMMOP3	5.2276e-3 (5.34e-4) +	4.7786e-3 (4.07e-4) +	**4.5780e-3 (2.96e-4)** +	4.8595e-3 (4.54e-4) +
MMMOP4	9.2036e-2 (6.01e-2) −	1.7021e-2 (1.95e-2) ≈	3.4686e-2 (9.31e-2) ≈	3.1658e-2 (8.01e-2) ≈
MMMOP5	5.8629e-2 (2.84e-2) −	**1.2269e-2 (7.56e-3)** +	1.2466e-2 (6.92e-3) +	1.5318e-2 (1.02e-2) ≈
MMMOP6	9.2025e-2 (2.49e-2) −	7.0389e-2 (2.24e-2) −	3.2596e-2 (1.21e-2) −	3.0133e-2 (7.89e-3) −
MMF14	**9.3196e-2 (2.54e-3)** +	9.4341e-2 (2.12e-3) ≈	9.4939e-2 (4.08e-3) ≈	9.5026e-2 (2.52e-3) ≈
MMF14_a	**9.3253e-2 (2.13e-3)** +	9.4308e-2 (2.94e-3) +	9.5661e-2 (2.69e-3) +	9.4981e-2 (2.95e-3) +
MMF1_e	3.6739e+0 (3.55e-1) ≈	3.6399e+0 (4.14e-1) ≈	3.7767e+0 (3.67e-1) ≈	**3.5489e+0 (3.90e-1)** ≈
MMF1_z	**1.2573e+0 (1.58e-4)** +	1.2573e+0 (1.02e-4) ≈	1.2573e+0 (1.48e-4) +	1.2587e+0 (3.23e-3) ≈
+/ − / ≈	14/4/4	14/1/7	14/1/7	13/1/8
Problem	$c = 3$	$c = 3.5$	$c = 4$	DNNSGAII_DNPN
MMF1	3.3884e-3 (1.64e-4) +	3.3928e-3 (2.06e-4) +	3.3788e-3 (1.88e-4) +	6.6444e-3 (3.42e-4)
MMF2	**1.4716e-2 (3.01e-3)** ≈	1.7121e-2 (6.80e-3) ≈	1.5883e-2 (5.36e-3) ≈	1.7927e-2 (7.40e-3)
MMF3	1.3295e-2 (3.59e-3) ≈	1.3487e-2 (4.24e-3) ≈	1.3906e-2 (3.96e-3) ≈	**1.2729e-2 (2.58e-3)**
MMF4	3.4513e-3 (2.27e-4) +	3.4788e-3 (2.55e-4) +	3.5133e-3 (2.35e-4) +	6.9500e-3 (4.06e-4)
MMF5	3.8033e-3 (1.92e-4) +	3.7630e-3 (1.64e-4) +	**3.7313e-3 (1.33e-4)** +	7.4190e-3 (4.71e-4)
MMF6	3.5247e-3 (1.54e-4) +	**3.4958e-3 (1.38e-4)** +	3.5191e-3 (1.16e-4) +	6.7959e-3 (4.32e-4)
MMF7	3.5400e-3 (1.89e-4) +	3.4831e-3 (1.67e-4) +	3.5107e-3 (1.45e-4) +	7.2665e-3 (5.30e-4)
MMF8	4.9752e-3 (3.83e-4) +	4.9492e-3 (2.98e-4) +	4.7607e-3 (2.68e-4) +	8.7031e-3 (6.44e-4)
MMF9	1.4435e-2 (1.41e-3) +	**1.4235e-2 (1.22e-3)** +	1.4410e-2 (1.29e-3) +	2.8278e-2 (2.10e-3)
Omni-test	2.4984e-2 (2.05e-3) +	**2.3806e-2 (1.83e-3)** +	2.4908e-2 (2.35e-3) +	4.1184e-2 (5.67e-3)
SYM-PART-rotated	1.8664e-2 (2.68e-3) +	1.9134e-2 (2.64e-3) +	1.9018e-2 (2.48e-3) +	3.5089e-2 (4.51e-3)
SYM-PART-simple	1.3674e-2 (1.66e-3) +	**1.3257e-2 (1.24e-3)** +	1.4380e-2 (1.34e-3) +	2.7698e-2 (2.85e-3)
MMMOP1	1.7130e-2 (1.48e-2) ≈	1.9046e-2 (2.59e-2) ≈	**1.3947e-2 (8.46e-3)** ≈	2.2296e-2 (2.19e-2)
MMMOP2	**6.6064e-3 (1.26e-3)** +	7.4141e-3 (2.30e-3) +	6.8744e-3 (1.41e-3) +	1.4695e-2 (4.49e-3)
MMMOP3	4.7904e-3 (5.45e-4) +	4.5956e-3 (3.50e-4) +	4.7718e-3 (5.16e-4) +	9.1376e-3 (8.50e-4)
MMMOP4	1.9178e-2 (1.81e-2) ≈	**1.6111e-2 (1.23e-2)** ≈	1.7980e-2 (1.60e-2) ≈	1.9268e-2 (1.63e-2)
MMMOP5	1.6613e-2 (1.55e-2) ≈	2.7515e-2 (6.33e-2) ≈	1.4663e-2 (2.00e-2) +	1.7278e-2 (1.08e-2)
MMMOP6	2.9928e-2 (8.26e-3) −	2.9953e-2 (5.65e-3) −	2.8700e-2 (8.44e-3) −	**2.3922e-2 (6.87e-3)**
MMF14	9.4569e-2 (2.63e-3) ≈	9.5325e-2 (2.79e-3) ≈	9.4738e-2 (3.02e-3) ≈	9.5385e-2 (2.38e-3)
MMF14_a	9.5676e-2 (3.06e-3) +	9.5279e-2 (2.93e-3) +	1.3621e-1 (4.89e-3)	
MMF1_e	3.6813e+0 (3.91e-1) ≈	3.7669e+0 (4.73e-1) ≈	3.7692e+0 (4.09e-1) ≈	3.6582e+0 (3.91e-1)
MMF1_z	1.2586e+0 (2.52e-3) ≈	1.2583e+0 (2.66e-3) ≈	1.2584e+0 (2.71e-3) ≈	1.2585e+0 (3.15e-3)
+/ − / ≈	13/1/8	13/1/8	14/1/7	

4.3 Parameter Sensitive Analysis

In the previous comparative experiments, we set Eq. 5, with $\gamma = 0.5$ and $c = 2$. Meanwhile, the parameters in Eq. 4 were set to the optimal values obtained by DNPN [20]. The experiments showed that setting parameter β to 30 and α to $\frac{1}{3}$ yielded better performance.

Table 10. IGDx on different c

Problem	$c = 1$	$c = 1.5$	$c = 2$	$c = 2.5$
MMF1	**4.0641e-2 (1.18e-3)** +	4.1338e-2 (1.28e-3) +	6.7982e-2 (3.21e-3) +	4.4941e-2 (1.51e-3) +
MMF2	**2.4276e-2 (3.34e-3)** +	2.5445e-2 (4.82e-3) +	2.9391e-2 (7.09e-3) ≈	2.6491e-2 (3.64e-3) ≈
MMF3	2.6566e-2 (4.18e-3) ≈	2.7904e-2 (5.51e-3) ≈	2.6818e-2 (4.79e-3) ≈	**2.5673e-2 (4.96e-3)** +
MMF4	2.7316e-2 (1.60e-3) +	2.7304e-2 (1.93e-3) +	**2.7130e-2 (2.18e-3)** +	2.9519e-2 (2.90e-3) +
MMF5	**7.1624e-2 (2.24e-3)** +	7.3031e-2 (3.23e-3) +	7.2130e-2 (2.10e-3) +	7.2447e-2 (4.00e-3) +
MMF6	**6.7111e-2 (2.69e-3)** +	6.7400e-2 (2.90e-3) +	6.7624e-2 (3.75e-3) +	7.0500e-2 (2.63e-3) +
MMF7	2.2340e-2 (1.20e-3) +	**2.2050e-2 (9.41e-4)** +	2.2373e-2 (1.30e-3) +	2.2546e-2 (2.33e-3) +
MMF8	6.0105e-2 (4.13e-3) +	**5.9805e-2 (4.61e-3)** +	5.9938e-2 (4.21e-3) +	6.0268e-2 (5.25e-3) +
MMF9	6.0888e-3 (2.30e-4) +	6.0955e-3 (3.37e-4) +	6.1014e-3 (3.67e-4) +	6.1624e-3 (3.16e-4) +
Omni-test	1.2592e-1 (1.60e-2) +	1.2987e-1 (2.41e-2) +	1.3704e-1 (2.74e-2) +	1.2432e-1 (1.68e-2) +
SYM-PART-rotated	9.2349e-2 (1.37e-2) +	1.2075e-1 (1.81e-1) +	8.7218e-2 (9.21e-3) +	**8.6573e-2 (1.17e-2)** +
SYM-PART-simple	**6.5032e-2 (6.76e-3)** +	6.5171e-2 (6.41e-3) +	6.5370e-2 (5.48e-3) +	6.7408e-2 (1.07e-2) +
MMMOP1	**6.2743e-2 (1.34e-2)** +	9.0793e-2 (2.26e-2) +	9.3177e-2 (3.21e-2) +	8.3260e-2 (2.38e-2) +
MMMOP2	2.4896e-2 (1.14e-2) +	1.7661e-2 (1.31e-2) +	1.9490e-2 (1.31e-2) +	2.0770e-2 (1.26e-2) +
MMMOP3	1.6987e-2 (1.43e-3) +	1.5973e-2 (1.52e-3) +	**1.5416e-2 (1.34e-3)** +	1.5844e-2 (1.53e-3) +
MMMOP4	3.4011e-2 (1.01e-2) ≈	1.8621e-2 (1.25e-2) +	2.5290e-2 (4.28e-2) +	2.3154e-2 (2.11e-2) +
MMMOP5	2.8030e-2 (9.16e-3) ≈	**1.4543e-2 (6.29e-3)** +	1.6489e-2 (8.84e-3) +	1.6672e-2 (8.71e-3) +
MMMOP6	1.1067e-1 (1.11e-2) +	1.0558e-1 (1.06e-2) +	1.0716e-1 (2.23e-2) +	1.0553e-1 (1.80e-2) +
MMF14	6.2002e-2 (2.13e-3) ≈	**6.1708e-2 (1.70e-3)** ≈	6.2026e-2 (2.41e-3) ≈	6.2068e-2 (1.92e-3) ≈
MMF14_a	**7.3074e-2 (2.04e-3)** +	7.4527e-2 (2.16e-3) +	7.5267e-2 (3.01e-3) +	7.5306e-2 (2.38e-3) +
MMF1_e	**2.5766e-1 (1.65e-2)** +	2.6166e-1 (1.47e-2) ≈	2.5773e-1 (2.07e-2) ≈	2.6414e-1 (1.79e-2) ≈
MMF1_z	5.8958e-2 (1.81e-3) +	**5.8289e-2 (1.69e-3)** +	5.9764e-2 (2.00e-3) +	6.2149e-2 (2.42e-3) ≈
+/ - / ≈	17/0/5	19/0/3	18/0/4	18/0/4

Problem	$c = 3$	$c = 3.5$	$c = 4$	DNNSGAII_DNPN
MMF1	4.4745e-2 (1.72e-3) +	4.4076e-2 (1.41e-3) +	4.4823e-2 (1.65e-3) +	7.0084e-2 (2.49e-3)
MMF2	2.7480e-2 (4.38e-3) ≈	2.8647e-2 (8.25e-3) ≈	2.9255e-2 (7.83e-3) ≈	3.2015e-2 (1.15e-2)
MMF3	2.7564e-2 (8.03e-3) ≈	2.5865e-2 (6.04e-3) ≈	2.8006e-2 (6.39e-3) ≈	2.8915e-2 (5.86e-3)
MMF4	2.8320e-2 (2.05e-3) +	2.9011e-2 (1.79e-3) +	2.9199e-2 (2.43e-3) +	5.5253e-2 (7.00e-3)
MMF5	7.4646e-2 (3.26e-3) +	7.5390e-2 (3.66e-3) +	7.7965e-2 (4.06e-3) +	1.2503e-1 (7.48e-3)
MMF6	7.1771e-2 (4.11e-3) +	7.0150e-2 (3.13e-3) +	7.1816e-2 (3.48e-3) +	1.1202e-1 (5.99e-3)
MMF7	2.2578e-2 (1.28e-3) +	2.2890e-2 (1.68e-3) +	2.2208e-2 (1.42e-3) +	3.8961e-2 (2.84e-3)
MMF8	6.2352e-2 (3.94e-3) +	6.0178e-2 (4.54e-3) +	6.0963e-2 (5.87e-3) +	1.2903e-1 (2.37e-2)
MMF9	6.2751e-3 (4.68e-4) +	6.1315e-3 (4.43e-4) +	**6.0573e-3 (3.50e-4)** +	1.1451e-2 (9.55e-4)
Omni-test	**1.2323e-1 (1.32e-2)** +	1.3247e-1 (2.71e-2) +	1.3424e-1 (2.85e-2) +	2.9490e-1 (7.10e-2)
SYM-PART-rotated	1.2374e-1 (1.81e-1) +	8.9277e-2 (9.44e-3) +	8.8713e-2 (1.02e-2) +	1.8013e+0 (1.32e+0)
SYM-PART-simple	6.6515e-2 (5.97e-3) +	6.6677e-2 (7.69e-3) +	6.7197e-2 (7.49e-3) +	2.2921e-1 (5.92e-1)
MMMOP1	8.5609e-2 (2.59e-2) +	8.9233e-2 (3.22e-2) +	9.0802e-2 (3.26e-2) +	1.2474e-1 (3.61e-2)
MMMOP2	**1.7019e-2 (8.25e-3)** +	1.8218e-2 (7.99e-3) +	2.2396e-2 (1.65e-2) +	3.6007e-2 (2.09e-2)
MMMOP3	1.6177e-2 (1.91e-3) +	1.5890e-2 (1.60e-3) +	1.5789e-2 (1.34e-3) +	2.6579e-2 (2.36e-3)
MMMOP4	1.9826e-2 (1.59e-2) +	**1.8246e-2 (1.98e-2)** +	2.0258e-2 (1.76e-2) +	4.0728e-2 (2.77e-2)
MMMOP5	1.6199e-2 (6.85e-3) +	1.6724e-2 (1.08e-2) +	1.6228e-2 (8.49e-3) +	3.5689e-2 (3.00e-2)
MMMOP6	**1.0124e-1 (1.97e-2)** +	1.0156e-1 (1.58e-2) +	1.0781e-1 (2.08e-2) +	1.4315e-1 (2.78e-2)
MMF14	6.1836e-2 (1.83e-3) ≈	6.2162e-2 (1.80e-3) ≈	6.2016e-2 (1.62e-3) ≈	6.1764e-2 (1.47e-3)
MMF14_a	7.5603e-2 (2.43e-3) +	7.6249e-2 (3.04e-3) +	7.5387e-2 (1.68e-3) +	1.0611e-1 (3.89e-3)
MMF1_e	2.5952e-1 (1.83e-2) ≈	2.6063e-1 (1.82e-2) ≈	2.6150e-1 (1.90e-2) ≈	2.6171e-1 (1.60e-2)
MMF1_z	6.2044e-2 (3.12e-3) ≈	6.1515e-2 (2.57e-3) ≈	6.0700e-2 (2.75e-3) +	6.2079e-2 (2.51e-3)
+/ - / ≈	17/0/5	17/0/5	18/0/4	

Effect of Parameter γ. The parameter γ, as a scaling factor in the differential algorithm, is used in the second phase to control the distance between the newly generated particles and the existing particles, playing a role in adjusting the particle distribution in the decision space. A total of 22 test problems are chosen to find the optimal value of γ, including the recently proposed MMF1-9 [17], MMF1_e, MMF1_z, MMF14, MMF14_a, MMMOP1-6 [11], as well as SYM-PART [13] and Omni-test [2].

Table 7 and Table 8 show the IGDX results of DNNSGAII_DNTE on the 22 test functions. The values of γ are selected from $\{0.1, 0.2, 0.3, 0.4, 0.5\}$, and c is set to 1.5. It can be clearly seen that when $\gamma = 0.3$, there is an advantage in both IGDX and IGDF performance.

Effect of Parameter c. c serves as the distance threshold to control the interaction range of particles, constraining the minimum effective domain of generated particles in the second stage. The value of c is chosen from 1, 1.5, 2, 2.5, 3, 3.5, 4. It can be clearly observed in Table 9 and Table 10 that when c is set to 1.5, the performance in both IGDf and IGDx is relatively better.

5   Conclusions

This paper proposes a multi-modal, multi-objective evolutionary algorithm, termed DNTE, which uses dynamic niche and differential evolutionary strategies to enhance the algorithm's ability to maintain diversity and distribution in the decision space. In DNTE, the evolution process is divided into two stages. In each stage, the dynamic niche is used to provide information about the decision space to non-dominated sorting, which is then applied to select parents. In the first stage, the influence range of solutions is restricted, and only individuals within the niche have a dominant relationship. The second stage continues the selection process based on the parents obtained in the first stage, selecting sub-parents who are located more than twice the niche distance away. During the offspring generation, both parents and subparents will generate corresponding offspring.

Additionally, the size of the niche changes dynamically. In the early stages, the niche value is small, which not only helps increase the algorithm's diversity in the first stage but allows for the dynamic generation of sub-parents in the second stage based on the sparsity of the decision space, leading to a richer population. In the later stages, the niche becomes larger, allowing for faster convergence in the first stage. However, because the niche is larger, obtaining new sub-parents in the second stage becomes difficult, which optimizes the convergence speed.

However, experiments show that DNTE still has the following shortcomings: its performance is suboptimal when solving problems with many local optima. Furthermore, although the population distribution in the decision space has improved, there is some performance loss in the objective space. Since DNNS-GAII performs well in the objective space, DNNSGAII_DNTE still maintains good performance in the objective space.

Acknowledgments. This work is supported by the National Natural Science Foundation of China (Nos. 62376230, 62076171, 62476182, 62306196, and 62306054).

References

1. Deb, K., Pratap, A., Agarwal, S., Meyarivan, T.: A fast and elitist multiobjective genetic algorithm: NSGA-II. IEEE Trans. Evol. Comput. **6**(2), 182–197 (2002)
2. Deb, K., Tiwari, S.: Omni-optimizer: a generic evolutionary algorithm for single and multi-objective optimization. Eur. J. Oper. Res. **185**(3), 1062–1087 (2008)
3. Kudo, F., Yoshikawa, T., Furuhashi, T.: A study on analysis of design variables in pareto solutions for conceptual design optimization problem of hybrid rocket engine. In: 2011 IEEE Congress of Evolutionary Computation (CEC), pp. 2558–2562 (2011)
4. Lapizco-Encinas, G., Kingsford, C., Reggia, J.: Particle swarm optimization for multimodal combinatorial problems and its application to protein design. In: IEEE Congress on Evolutionary Computation, pp. 1–8. IEEE (2010)
5. Li, W., Yao, X., Li, K., Wang, R., Zhang, T., Wang, L.: Coevolutionary framework for generalized multimodal multi-objective optimization. IEEE/CAA J. Automatica Sinica **10**(7), 1544–1556 (2023)
6. Li, W., Yao, X., Zhang, T., Wang, R., Wang, L.: Hierarchy ranking method for multimodal multiobjective optimization with local pareto fronts. IEEE Trans. Evol. Comput. **27**(1), 98–110 (2023)
7. Li, W., Zhang, T., Wang, R., Ishibuchi, H.: Weighted indicator-based evolutionary algorithm for multimodal multiobjective optimization. IEEE Trans. Evol. Comput. **25**(6), 1064–1078 (2021)
8. Liang, J.J., Yue, C.T., Qu, B.Y.: Multimodal multi-objective optimization: a preliminary study. In: 2016 IEEE Congress on Evolutionary Computation (CEC), pp. 2454–2461 (2016)
9. Liu, Y., Ishibuchi, H., Nojima, Y., Masuyama, N., Han, Y.: Searching for local pareto optimal solutions: a case study on polygon-based problems. In: 2019 IEEE Congress on Evolutionary Computation (CEC), pp. 896–903 (2019)
10. Liu, Y., Ishibuchi, H., Yen, G.G., Nojima, Y., Masuyama, N.: Handling imbalance between convergence and diversity in the decision space in evolutionary multimodal multiobjective optimization. IEEE Trans. Evol. Comput. **24**(3), 551–565 (2020)
11. Liu, Y., Yen, G.G., Gong, D.: A multimodal multiobjective evolutionary algorithm using two-archive and recombination strategies. IEEE Trans. Evol. Comput. **23**(4), 660–674 (2019)
12. Peng, Y., Ishibuchi, H.: A decomposition-based hybrid evolutionary algorithm for multi-modal multi-objective optimization. In: 2021 IEEE International Conference on Systems, Man, and Cybernetics (SMC), pp. 160–167 (2021)
13. Rudolph, G., Naujoks, B., Preuss, M.: Capabilities of EMOA to detect and preserve equivalent pareto subsets. In: Obayashi, S., Deb, K., Poloni, C., Hiroyasu, T., Murata, T. (eds.) Evolutionary Multi-Criterion Optimization, pp. 36–50. Springer, Heidelberg (2007)
14. Tian, Y., Cheng, R., Zhang, X., Jin, Y.: Platemo: a matlab platform for evolutionary multi-objective optimization [educational forum]. IEEE Comput. Intell. Mag. **12**(4), 73–87 (2017)
15. Yao, X., Li, W., Pan, X., Wang, R.: Multimodal multi-objective evolutionary algorithm for multiple path planning. Comput. Ind. Eng. **169**, 108145 (2022)
16. Yue, C., Qu, B., Liang, J.: A multiobjective particle swarm optimizer using ring topology for solving multimodal multiobjective problems. IEEE Trans. Evol. Comput. **22**(5), 805–817 (2018)

17. Yue, C., Qu, B., Yu, K., Liang, J., Li, X.: A novel scalable test problem suite for multimodal multiobjective optimization. Swarm Evol. Comput. **48**, 62–71 (2019)
18. Yue, C., Liang, J.J., Qu, B.Y., Yu, K., Song, H.: Multimodal multiobjective optimization in feature selection. In: 2019 IEEE Congress on Evolutionary Computation (CEC), pp. 302–309. IEEE (2019)
19. Zhang, Q., Li, H.: MOEA/D: a multiobjective evolutionary algorithm based on decomposition. IEEE Trans. Evol. Comput. **11**(6), 712–731 (2007)
20. Zou, J., Deng, Q., Liu, Y., Yang, X., Yang, S., Zheng, J.: A dynamic-niching-based pareto domination for multimodal multiobjective optimization. IEEE Trans. Evol. Comput. **28**(5), 1529–1543 (2024)

Predicting Medium-Term Trends of Stock Market Indexes with Deep Learning

Yan Zhang$^{(\boxtimes)}$ and David Vargas-Monroy

School of Computer Science and Engineering, California State University San Bernardino, San Bernardino, CA 92407, USA
Yan.Zhang@csusb.edu, 007060554@coyote.csusb.edu

Abstract. Stock market indexes, such as the S&P 500, Dow Jones Industrial Average, and Nasdaq, represent the performance of a specific market by holding a diversified portfolio of underlying assets. The movement of an index is influenced by the performance of its constituent assets, making trend prediction a challenging task due to the complexity, randomness, and inherent unpredictability of financial markets. While short-term price movements are erratic, medium-term trends (beyond 20 trading days) are critical for investors and analysts relying on historical data and technical indicators to make stratigic decisions. This study employs deep learning to forecast medium-term trends (Down (-1), Stable (0), Up (1)) of these indexes, using three 10-year datasets, each comprising 2524 records from June 23, 2010, to June 30, 2020. Five models, single-layer Long Short-Term Memory (LSTM), single-layer Gated Recurrent Unit (GRU), stacked LSTM, stacked GRU, and a hybrid LSTM-GRU, are trained and evaluated on the three datasets. The hybrid LSTM+GRU model outperforms all others, achieving the highest accuracies across 20-, 40-, and 60-day horizons for all three indexes. Comparative analysis underscores its superior ability to capture medium-term patterns, advancing reliable predictive tools for financial decision-making and highlighting deep learning's promise in stock market forecasting.

Keywords: Stock Market Prediction · Medium-term Trends Prediction · Gated Recurrent Units (GRU) · Long Short-Term Memory (LSTM) · Deep Learning

1 Introduction

Stock market trend prediction remains a cornerstone of financial analysis, offering important insights for investors, traders, and policymakers aiming to navigate the complexities of economic markets. Accurate forecasting of medium-term trends—spanning periods such as 20, 40, and 60 trading days—enables stakeholders to anticipate market movements, optimize investment strategies, and mitigate risks. However, the inherent volatility, non-linearity, and noise in stock market data pose significant challenges to traditional statistical methods.

Q. Zhang et al. (Eds.): IJCRS 2025, LNAI 15710, pp. 383–404, 2025.
https://doi.org/10.1007/978-3-031-92741-6_28

Recent advancements in deep learning have improved time series forecasting by providing useful tools to model complex patterns in sequential data. Recurrent Neural Networks (RNNs), particularly Long Short-Term Memory (LSTM) units and Gated Recurrent Units (GRUs), have showed capability in capturing long-term dependencies, making them suitable for financial applications. Unlike traditional approaches, deep learning models can deal with large datasets and diverse features to find hidden trends without relying on assumptions about data distributions. The potential of this capability has driven many attempts to apply deep learning to stock market prediction. However, current deep learning-based stock prediction frameworks still face challenges, including overfitting to noisy financial data, sensitivity to hyperparameter tuning, and difficulty in generalizing across diverse market conditions. Moreover, the comparative performance of different architectures for medium-term forecasting remains underexplored.

Motivated by this gap, this study investigates the performance of five deep learning models, single-layer LSTM, single-layer GRU, stacked LSTM, stacked GRU, and a hybrid LSTM+GRU model, in predicting medium-term trends of the S&P 500, Dow Jones Industrial Average, and Nasdaq indexes. These indexes, widely recognized benchmarks of U.S. equity market performance, are selected for their comprehensive representation of large-cap stocks and extensive historical data, spanning June 23, 2010, to June 30, 2020, comprising 2524 records. Medium-term trend horizons (20-, 40-, and 60-trading days) are chosen to balance short-term volatility with longer-term investment perspectives, classifying trends as Down (-1), Stable (0), or Up (1) based on predefined thresholds. The primary objective of this research is to compare the predictive performance of these models using a 60-day look-back period and a 16-feature input set, including price data and technical indicators. By evaluating precision, recall, F1-score, and accuracy on a test set, this study aims to identify the most effective architecture for medium-term trend prediction and discuss the impact of model complexity on prediction accuracy.

This paper is structured as follows: Sect. 2 reviews related work on stock market prediction and deep learning; Sect. 3 describes the dataset and preprocessing steps; Sect. 4 outlines the methodology; Sect. 5 presents model architectures, experimental results and discussion; and Sect. 6 concludes with findings and future directions.

2   Related Work

Deep learning models have been extensively applied to stock market prediction, addressing both short-term price movements and medium-term trends. Traditional statistical and machine learning approaches often struggle to capture the complex temporal dependencies and non-linear patterns present in financial data. Recurrent Neural Networks (RNNs) and their advanced variants, Long Short-Term Memory (LSTM) and Gated Recurrent Units (GRU), have emerged as powerful tools for analyzing sequential stock market data. This section reviews existing research on stock market prediction using deep learning models, with a particular focus on LSTM and GRU architectures and their applications in financial forecasting.

Nelson et al. examined LSTM models to predict stock price trends using historical data and technical indicators [10]. The authors compared LSTM performance with traditional machine learning methods through multiple experiments. Their model achieved an average accuracy of 55.9% in predicting price increases, indicating LSTM's potential for stock market forecasting while highlighting the need for further improvements [10]. Nabipour et al. examined stock market trend prediction using machine learning and deep learning models, focusing on four stock groups from the Tehran Stock Exchange [9]. It compared 9 machine learning models with RNN and LSTM based on 10 years of historical stock data. The results showed that RNN and LSTM outperform traditional models [9]. These findings reinforced the effectiveness of RNNs and LSTMs for stock trend prediction. However, challenges in current deep learning frameworks, such as overfitting to volatile market noise and the computational burden of training complex models, often limit their practical deployment.

Bhandari et al. explored LSTM-based models for predicting the next-day closing price of the S&P 500 index using a combination of fundamental market data and technical indicators [3]. It compared single-layer and multi-layer LSTM models, evaluating their performance with RMSE, MAPE, and correlation coefficient. The results indicated that single-layer LSTM outperformed multi-layer LSTM, offering a higher prediction accuracy [3]. This highlights LSTM's effectiveness in capturing short-term market trends.

Gupta et al. introduced StockNet, a GRU-based model for stock index prediction, incorporating a data augmentation approach to prevent overfitting [6]. The model consisted of an injection module for overfitting prevention and an investigation module for forecasting. Validated on the CNX-Nifty index, the improved StockNet-c model reduced test loss compared to TargetNet, achieving 65.59% lower RMSE, 27.30% lower MAE, and 14.89% lower MAPE [6]. The results showed the effectiveness of GRU models with data augmentation in enhancing stock market prediction accuracy.

Trivedi and Patel examined LSTM, GRU, and a hybrid LSTM+GRU model for predicting HDFCBANK stock prices, addressing the challenges of noise, nonlinearity, and chaotic fluctuations in stock data [14]. The proposed hybrid model consisted of a GRU first layer followed by three LSTM layers. Evaluated using MSE, RMSE, and MAE, the hybrid model outperformed standalone LSTM and GRU models. Patra et al. employed a hybrid LSTM+GRU model to predict the adjusted closing price of the S&P 500 index, addressing stock market nonlinearity and volatility [11]. Evaluated using Return ratio, R^2, MSE, and Optimism/Pessimism ratios, the hybrid model outperformed standalone LSTM, GRU, and MLP models. Liu et al. introduced a Regularized GRU-LSTM model for short-term stock price forecasting, addressing the complexities of stock market fluctuations [7]. The model was applied to predict the closing prices of two stocks and was compared with standalone GRU and LSTM models. Experimental results demonstrated that the hybrid model achieved higher accuracy in time-series stock prediction [7]. These novel idea led us to consider the option of combining LSTM and GRU for medium-term trend prediction of stock market indexes.

The reviewed studies highlight the effectiveness of deep learning models, particularly LSTM, GRU, and hybrid approaches, in stock market prediction. These models consistently outperform traditional machine learning techniques due to their ability to capture sequential dependencies and nonlinear patterns in financial time series data. Data augmentation and feature expansion techniques, such as incorporating technical indicators, improve model performance. Nevertheless, prior research is often constrained by limitations such as a focus on short-term horizons, neglecting medium-term trends critical for strategic investment, and reliance on single-index datasets, which may not generalize across diverse markets. This motivates further research into optimizing deep learning architectures for robust stock market forecasting, particularly for medium-term trend prediction, as examined in this study.

3   Data Collection and Processing

3.1   Data Collection

The three datasets used in this study comprise historical stock market data for the S&P 500 (^GSPC), Dow Jones Industrial Average (^DJI), and NASDAQ 100 (^NDX), collected from Yahoo Finance [2] over the period from June 23, 2010, to September 24, 2020. Data retrieval was facilitated by the open-source Python library yfinance, which provides an efficient and reliable interface for accessing and downloading historical financial records. A total of 2,584 trading day records were collected in each dataset, covering the index's performance across over 10 years. Each dataset encompasses several key financial attributes critical for predicting market trends. Table 1 summarizes the dataset attributes and their meanings. These attributes collectively provide insights into market behavior, enabling the analysis of both short-term fluctuations and medium-term trends.

Table 1. Features and descriptions

Features	Description
Date	The date of market day
Open	The stock price at the market's opening
High	The highest price reached during the market day
Low	The lowest price recorded during the market day
Close	The stock price at the market close
Volume	The total number of shares traded on that market day

3.2   Technical Indicator Features

To enhance the predictive capability of the datasets, several technical indicators were calculated and added using TA-LIB, an open-source library that derives technical indicators from historical market data [1]. These indicators, detailed in Table 2, were selected to capture essential market trends, momentum, and volatility.

Table 2. The technical indicators added to datasets

Indicator	Description
MA20/MA40/MA60	Simple Moving Averages over 20, 40, and 60 days, used to smooth price fluctuations and identify trends
MACD_HIST	Moving Average Convergence/Divergence Histogram measures the difference between the MACD and its signal line, indicating momentum shifts
ATR	Average True Range quantifies market volatility by evaluating price range variations
RSI	Relative Strength Index, a momentum oscillator that measures the speed and change of price movements. The RSI oscillates between 0 and 100. RSI is considered overbought when above 70 and oversold when below 30
MFI	Money Flow Index, a momentum indicator that measures the flow of money into and out of a security over a specified period of time
WILLR	Williams' %R, is a momentum indicator that is the inverse of the Fast Stochastic Oscillator, it measures overbought or oversold conditions
STDDEV	Standard Deviation is a statistical measure of market volatility, measuring how widely prices are dispersed from the average
SLOWK	Slow Stochastic Oscillator (%K) is a momentum indicator that shows the location of the close relative to the high-low range over a set number of periods, and it is used to identify the strength of trends in price movements
SLOWD	Slow %D is a 3-day simple moving average of %K

These indicators enrich the datasets, providing insights into stock index movements and improving the potential for accurate trend prediction.

3.3   Adding Trend Labels

In the data preprocessing phase, three new output columns—trend20, trend40, and trend60—were added to represent the medium-term trends of the three indexes over horizons of 20, 40, and 60 trading days, respectively. These labels categorize the directional movement of the index into three distinct classes:

- **1**: Indicates an upward trend (significant increase).
- **0**: Indicates a stable trend (minimal or no significant change).
- **−1**: Indicates a downward trend (decrease).

Trend labels are determined by calculating the relative change in the index closing price between the current trading day and a future trading day, based on the specified horizon. Each trend horizon has its own set of thresholds to define the boundaries between downward, stable, and upward trends.

Determining Trend Labels. Let:

- P_t = the closing price on the current trading day t,
- P_{t+k} = the closing price k trading days later, where $k = 20, 40, 60$.

The difference in price, ΔP_k, is calculated as:

$$\Delta P_k = P_{t+k} - P_t.$$

The trend labels are assigned based on the following conditions for each trend horizon:

20-Day Trend (**trend20** *)* For the 20-day trend, the thresholds are set at 0 and $P_t \times 0.025$ (i.e., 2.5% of the current price):

- If $\Delta P_{20} < 0$: `trend20` = `-1` (downward trend).
- If $0 \leq \Delta P_{20} \leq P_t \times 0.025$: `trend20` = `0` (stable trend).
- If $\Delta P_{20} > P_t \times 0.025$: `trend20` = `1` (upward trend).

40-Day Trend (**trend40** *)* For the 40-day trend, the thresholds are set at $P_t \times 0.005$ (i.e., 0.5% of the current price) and $P_t \times 0.040$ (i.e., 4% of the current price):

- If $\Delta P_{40} < P_t \times 0.005$: `trend40` = `-1` (downward trend).
- If $P_t \times 0.005 \leq \Delta P_{40} \leq P_t \times 0.040$: `trend40` = `0` (stable trend).
- If $\Delta P_{40} > P_t \times 0.040$: `trend40` = `1` (upward trend).

60-Day Trend (**trend60** *)* For the 60-day trend, the thresholds are set at $P_t \times 0.01$ (i.e., 1% of the current price) and $P_t \times 0.055$ (i.e., 5.5% of the current price):

- If $\Delta P_{60} < P_t \times 0.01$: `trend60` = `-1` (downward trend).
- If $P_t \times 0.01 \leq \Delta P_{60} \leq P_t \times 0.055$: `trend60` = `0` (stable trend).
- If $\Delta P_{60} > P_t \times 0.055$: `trend60` = `1` (upward trend).

Distribution of Data on Trend Labels. Each dataset originally contained 2,584 trading day records, but the last 60 records were removed because they lacked complete trend labels for the 60-day trend. This resulted in the final datasets of 2,524 records, each with trend labels for the 20-day, 40-day, and 60-day trends. These labels categorize the market movement as Down (−1), Stable (0), or Up (1) based on the specified thresholds for each trend.

Table 3 presents the distribution of data on trend labels for each trend in S&P 500 dataset. Tables 4 and 5 present the distribution of data on trend labels for each trend in Dow Jones and Nasdaq datasets. Each column corresponds to a specific trend (20-, 40-, or 60-day trends), and each row indicates the number of records classified as Down (−1), Stable (0), or Up (1) for that trend.

Table 3. Distribution of data on trend labels in S&P 500 dataset

Label	20-day trend	40-day trend	60-day trend
Down (−1)	806	808	751
Stable (0)	849	846	961
Up (1)	869	870	812

Table 4. Distribution of data on trend labels in Dow Jones dataset

Label	20-day trend	40-day trend	60-day trend
Down (−1)	848	766	686
Stable (0)	824	567	446
Up (1)	852	1191	1392

Table 5. Distribution of data on trend labels in Nasdaq dataset

Label	20-day trend	40-day trend	60-day trend
Down (−1)	793	669	558
Stable (0)	612	407	342
Up (1)	1119	1448	1624

Purpose of Trend Labels. These trend labels serve as target variables for predictive modeling, capturing the directional shifts in the stock indexes over medium-term periods of 20-, 40-, and 60-trading days. The thresholds are tailored to each trend to reflect the varying magnitudes of price changes, ensuring the labels effectively distinguish between meaningful trends and noise.

3.4   Feature Selection and Removal

The DATE column, which establishes the temporal ordering of the data, does not provide numerical or categorical information for market trend prediction. So, it was excluded from the dataset to focus on features with immediate relevance to trend prediction.

The moving average features (MA20, MA40, MA60) were retained selectively based on their correspondence to the prediction trends, while other features were consistently included:

- For the 20-day trend prediction (trend20), MA20 was retained as a relevant moving average, while MA40 and MA60 were excluded. The remaining 13 features—OPEN, HIGH, LOW, CLOSE, VOL, MACD_HIST, ATR, RSI, MFI, WILLR, STDDEV, SLOWK and SLOWD—were included.
- For the 40-day trend prediction (trend40), MA40 was retained, while MA20 and MA60 were excluded. The same set of additional 13 features was retained.
- For the 60-day trend prediction (trend60), MA60 was retained, while MA20 and MA40 were excluded, with the additional 13 features included.

3.5   Normalization

To improve model performance and training stability, all 16 feature values were normalized to the range of 0 to 1 using the Min-Max normalizer. Normalization ensures that features are on a similar scale, preventing certain variables from dominating the learning process.

The preprocessed and refined dataset serves as the foundation for training and evaluating deep learning models to predict medium-term stock market trends.

4   Methodologies

The experiments use five deep learning models, all rooted in the Recurrent Neural Network (RNN) framework, to predict medium-term trend of three indexes: single-layer LSTM, single-layer GRU, stacked LSTM, stacked GRU, and hybrid LSTM+GRU.

4.1   Recurrent Neural Networks (RNNs)

Recurrent Neural Networks (RNNs) are designed to process sequential data by maintaining a hidden state that evolves over time [8]. Unlike traditional feedforward networks, RNNs share parameters across time steps, making them efficient for modeling temporal dependencies [13]. They learn through Backpropagation Through Time, allowing them to capture short-term patterns in data.

However, RNNs suffer from the vanishing and exploding gradient problem, which limits their ability to learn long-term dependencies [13]. As gradients diminish, earlier time steps have little influence on learning, making it difficult for the network to retain information over long sequences. Additionally, RNNs process data sequentially, increasing computational costs.

Due to these limitations, advanced architectures such as Long Short-Term Memory (LSTM) and Gated Recurrent Units (GRU) were developed. These models introduce gating mechanisms to improve memory retention and mitigate gradient-related issues, making them better suited for tasks like stock market prediction.

4.2   Long Short-Term Memory (LSTM)

Long Short-Term Memory (LSTM) networks improve standard RNNs by introducing memory cells and gating mechanisms to retain long-term dependencies [15]. LSTM uses input, forget, and output gates to control the flow of information, preventing the vanishing gradient problem and allowing the network to learn patterns over extended sequences [16].

This architecture makes LSTM highly effective in time-series forecasting, as it can capture both short-term fluctuations and medium-term trends in stock market data. By selectively retaining relevant information, LSTM models can identify patterns that influence future price movements.

Despite their advantages, LSTMs require more computational resources due to their complex structure, making training slower compared to simpler models [16]. Hyperparameter tuning, such as selecting the number of layers and hidden units, is crucial for optimizing performance. LSTM's ability to handle long-term dependencies makes it a strong candidate for stock market prediction, but alternative architectures like GRUs offer a more efficient approach.

4.3   Gated Recurrent Units (GRUs)

Gated Recurrent Units (GRUs) are a simplified alternative to LSTMs that reduce computational complexity while preserving the ability to model long-term dependencies [12]. Unlike LSTMs, which use three gates, GRUs employ only two: the update gate, which determines how much past information is retained, and the reset gate, which controls how much previous data influences the current state [4].

GRUs require fewer parameters, making them faster to train and more efficient in handling sequential data. They have been widely used in time-series forecasting, including stock market prediction, due to their ability to capture essential patterns in financial data [4].

While GRUs often perform similarly to LSTMs, they may struggle with extremely long-term dependencies due to the absence of a dedicated memory cell. However, their reduced computational cost makes them an attractive option for stock market trend forecasting, where balancing accuracy and efficiency is crucial.

4.4 Hybrid LSTM+GRU

The hybrid LSTM+GRU model integrates both LSTM and GRU layers to leverage their strengths [14].LSTMs effectively retain long-term dependencies through memory cells, while GRUs offer computational efficiency by simplifying the gating mechanism. By combining these architectures, the hybrid model enhances sequential data processing while reducing training time [14].

In stock market prediction, LSTM layers capture medium-term trends, and GRU layers refine information flow for improved efficiency [11]. This combination allows the model to learn complex temporal dependencies while mitigating computational overhead. Hybrid LSTM+GRU models have shown promising results in financial forecasting, balancing accuracy and efficiency [11]. Designing an optimal hybrid model requires careful selection of hyperparameters and network architecture. The added complexity may partially offset computational benefits, requiring experimentation to achieve the best performance. Despite this, the hybrid model remains a powerful approach for predicting stock market trends.

5 Experimental Results and Discussion

5.1 Parameter Configurations

The parameter configurations for the five deep learning models utilized in this study are explained in detail in this section. The experiments use the Keras neural network architecture [5] to build prediction models, leveraging its flexibility and robust implementation of recurrent neural network layers. These models were trained and evaluated on the dataset of 2524 records, divided into 90% training data and 10% testing data. Separate models were constructed for each trend prediction—20-day trend (trend20), 40-day trend (trend40), and 60-day trend (trend60). The parameter settings, encompassing data shape, neural unit count, batch size, epochs, loss function, activation function, optimizer, model architecture, and dropout, are described below.

Data Shape. The input data was structured as a three-dimensional array with the shape (samples, timesteps, features). Here, "samples" refers to the number of sequences (2217 for training, 247 for testing); "timesteps" indicates the look-back period, and "features" represents the number of input features per timestep. A look-back period of 60 trading days was chosen to capture extended historical patterns. For each trend, the feature set included 14 features. Thus, the input shape was (2217, 60, 14) for training and (247, 60, 14) for testing across all trends.

Neural Unit Count. The number of neural units per layer was uniformly set to 128 across all layers in each model.

Batch Size. A batch size of 32 was applied uniformly across all models and trends.

Epochs. Each model was trained for up to 100 epochs, with early stopping employed based on validation loss (using a 10% validation split from the training data, i.e., 227 records). The patience parameter for early stopping was set to 10 epochs to prevent overfitting.

Loss Function. As the trend prediction task involves multi-class classification (labels: -1, 0, 1), categorical cross-entropy was selected as the loss function. Target labels were one-hot encoded into three-dimensional vectors (e.g., -1 as $[1, 0, 0]$, 0 as $[0, 1, 0]$, 1 as $[0, 0, 1]$) to align with this loss function.

Activation Functions. Recurrent layers (LSTM and GRU) utilized their default activation functions: `tanh` for the recurrent state and `sigmoid` for the gates. The output layer employed a `softmax` activation to generate probability distributions over the three trend classes $(-1, 0, 1)$, consistent with the classification objective.

Optimizer. The Adam optimizer was used across all models, with a default learning rate of 0.001. This optimizer was selected for its adaptive learning rate properties, facilitating efficient convergence across the diverse architectures.

Model Architecture. The architectures for the five models were defined as follows:

- Single-layer LSTM: One LSTM layer (128 units) followed by a dense output layer (3 units).
- Single-layer GRU: One GRU layer (128 units) followed by a dense output layer (3 units).
- Stacked LSTM: Two LSTM layers (128 units each) followed by a dense output layer (3 units).
- Stacked GRU: Two GRU layers (128 units each) followed by a dense output layer (3 units).
- Hybrid LSTM+GRU: Two LSTM layers (128 units each), followed by two GRU layers (128 units each), and a dense output layer (3 units).

The `return_sequences=True` parameter enabled sequence propagation between recurrent layers in multi-layer models, with the final recurrent layer outputting a single vector to the dense layer.

Dropout. Dropout was applied after each recurrent layer across all models at a rate of 0.2 (20% of units dropped during training) to mitigate overfitting.

These parameter settings were designed to leverage the feature set for each trend (`trend20`, `trend40`, `trend60`), enabling the models to learn and predict medium-term trends in the S&P 500, Dow Jones, and Nasdaq indexes while maintaining consistency for comparative evaluation.

5.2   Single-Layer LSTM Evaluation

In this section, we evaluate the performance of the single-layer LSTM model in predicting medium-term trends of the S&P 500 index across the 20-day

(trend20), 40-day (trend40), and 60-day (trend60) trends. The model, configured with 128 units as detailed in Sect. 5.1, was trained and tested on a dataset
of 2524 records, with a test set of 247 samples after applying a 60-day look-back
period. Performance is assessed using precision, recall, and F1-score for each
class (Down (-1), Stable (0), Up (1)), along with overall accuracy.

The single-layer LSTM model was executed five times for each trend horizon,
and the average test accuracy was computed for each. The classification reports
presented in Tables 6, 7, and 8 correspond to the runs with accuracies closest to
the average for each trend horizon.

Table 6. Classification report for single-layer LSTM on trend20

Class	Precision	Recall	F1-Score	Support
Down (-1)	0.86	0.92	0.89	87
Stable (0)	0.71	0.66	0.68	83
Up (1)	0.79	0.79	0.79	77
Accuracy	0.79			247

Table 7. Classification report for single-layer LSTM on trend40

Class	Precision	Recall	F1-Score	Support
Down (-1)	0.84	0.78	0.81	82
Stable (0)	0.65	0.63	0.64	80
Up (1)	0.76	0.83	0.79	85
Accuracy	0.75			247

Table 8. Classification report for single-layer LSTM on trend60

Class	Precision	Recall	F1-Score	Support
Down (-1)	0.85	0.80	0.82	73
Stable (0)	0.73	0.82	0.77	87
Up (1)	0.88	0.80	0.84	87
Accuracy	0.81			247

Analysis. The single-layer LSTM model exhibits varying performance across the three prediction trend horizons, with overall accuracies of 0.79, 0.75, and 0.81 for `trend20`, `trend40`, and `trend60`, respectively. These results indicate that the model performs best on the longest trend horizon (`trend60`) and weakest on the intermediate horizon (`trend40`), with the shortest horizon (`trend20`) achieving an intermediate accuracy.

For `trend20`, the model demonstrates strong capability in predicting downward trends, with a recall of 0.92 and an F1-score of 0.89 for the Down (-1) class, reflecting effective detection of declining market movements. On `trend40`, the overall accuracy drops to 0.75, with the Stable (0) class again performing weakest (F1-score of 0.64). The lower precision (0.65) and recall (0.63) indicate persistent challenges in identifying stable trends, potentially due to the wider threshold range (0.5% to 4%) overlapping with subtle directional movements. For `trend60`, the model achieves its highest accuracy (0.81) and a more balanced performance across classes. The Up (1) class excels with a precision of 0.88 and an F1-score of 0.84, indicating high confidence in predicting upward trends over this horizon. The superior performance on `trend60` likely arises from the alignment of the 60-day look-back period with the prediction horizon, enabling the model to capture longer-term dependencies and smoother trends in the S&P 500 data.

5.3 Stacked LSTM Evaluation

Now we evaluate the performance of the stacked LSTM model, featuring two layers of 128 units each as described in Sect. 5.1, in predicting medium-term trends of the S&P 500 index across the 20-day (`trend20`), 40-day (`trend40`), and 60-day (`trend60`) horizons. The stacked LSTM model was executed five times for each trend horizon and the average test accuracy was calculated for each. The classification reports presented in Tables 9, 10, and 11 correspond to the runs with accuracies closest to the average for each trend horizon.

Table 9. Classification report for stacked LSTM on `trend20`

Class	Precision	Recall	F1-Score	Support
Down (-1)	0.85	0.87	0.86	84
Stable (0)	0.72	0.74	0.73	87
Up (1)	0.85	0.81	0.83	76
Accuracy	0.81			247

Analysis. The stacked LSTM model demonstrates good performance across the three prediction horizons, achieving overall accuracies of 0.81, 0.83, and 0.86 for `trend20`, `trend40`, and `trend60`, respectively. This upward trend in accuracy

Table 10. Classification report for stacked LSTM on `trend40`

Class	Precision	Recall	F1-Score	Support
Down (−1)	0.91	0.85	0.88	88
Stable (0)	0.70	0.78	0.74	76
Up (1)	0.87	0.84	0.86	83
Accuracy	0.83			247

Table 11. Classification report for stacked LSTM on `trend60`

Class	Precision	Recall	F1-Score	Support
Down (−1)	0.90	0.85	0.87	72
Stable (0)	0.79	0.86	0.82	92
Up (1)	0.91	0.87	0.89	83
Accuracy	0.86			247

with increasing horizon length suggests that the model benefits significantly from longer-term data patterns, outperforming the single-layer LSTM (accuracies of 0.79, 0.75, and 0.81) across all horizons.

For `trend20`, the model achieves balanced performance across classes, with F1-scores of 0.86 (Down (−1)), 0.73 (Stable (0)), and 0.83 (Up (1)). On `trend40`, the model's accuracy rises to 0.83, with exceptional performance on the Down (−1) class (F1-score of 0.88, precision of 0.91), reflecting high confidence in predicting downward trends. For `trend60`, the model reaches its peak accuracy of 0.86, with F1-scores of 0.87 (Down (−1)), 0.82 (Stable (0)), and 0.89 (Up (1)), reflecting robust performance across all classes.

The stacked LSTM's superior performance, particularly on `trend40` and `trend60`, can be attributed to its two-layer architecture, which enhances its capacity to model complex temporal relationships within the 16-feature input set over a 60-day look-back.

5.4   Single-Layer GRU Evaluation

The performance of the single-layer GRU model in predicting medium-term trends of the S&P 500 index across the 20-day (`trend20`), 40-day (`trend40`), and 60-day (`trend60`) trend horizons is assessed in this section. The model, configured with 128 units as detailed in Sect. 5.1, was trained and tested on a dataset. The single-layer GRU model was executed five times for each trend horizon, and the average test accuracy was computed for each. The classification reports presented in Tables 12, 13, and 14 correspond to the runs with accuracies closest to the average for each trend horizon.

Table 12. Classification report for single-layer GRU on `trend20`

Class	Precision	Recall	F1-Score	Support
Down (−1)	0.79	0.89	0.84	84
Stable (0)	0.68	0.67	0.67	83
Up (1)	0.85	0.74	0.79	80
Accuracy	0.77			247

Table 13. Classification report for single-layer GRU on `trend40`

Class	Precision	Recall	F1-Score	Support
Down (−1)	0.91	0.86	0.89	89
Stable (0)	0.72	0.80	0.75	78
Up (1)	0.88	0.84	0.86	80
Accuracy	0.83			247

Analysis. The single-layer GRU model exhibits varying performance across the three prediction horizons, with overall accuracies of 0.77, 0.83, and 0.81 for `trend20`, `trend40`, and `trend60`, respectively. For `trend20`, the Down (−1) class performs strongly with a recall of 0.89 and an F1-score of 0.84, indicating effective detection of downward trends. The Up (1) class shows high precision (0.85) but lower recall (0.74), indicating conservative predictions of upward trends. On `trend40`, the model reaches its accuracy of 0.83. The Down (−1) class excels with a precision of 0.91 and an F1-score of 0.89, reflecting high reliability in predicting declines. The Up (1) class achieves a balanced F1-score of 0.86, with strong precision (0.88) and recall (0.84). For `trend60`, the Down (−1) class shows high precision (0.92) and an F1-score of 0.85, though recall (0.80) indicates some missed declines.

The single-layer GRU's strong performance on `trend40` may stem from its simpler gating mechanism (update and reset gates). compared to LSTM's three-gate structure, potentially allowing faster adaptation to the intermediate horizon's patterns within the 60-day look-back.

Table 14. Classification report for single-layer GRU on `trend60`

Class	Precision	Recall	F1-Score	Support
Down (−1)	0.92	0.80	0.85	68
Stable (0)	0.71	0.79	0.75	91
Up (1)	0.84	0.83	0.83	88
Accuracy	0.81			247

5.5   Stacked GRU Evaluation

We execute the stacked GRU model, featuring two layers of 128 units each as described in Sect. 5.1, to predict medium-term trends of the S&P 500 index across the 20-day (trend20), 40-day (trend40), and 60-day (trend60) horizons. The model was trained and tested on the dataset. The stacked GRU model was executed five times for each trend horizon, and the average test accuracy was computed for each. The classification reports presented in Tables 15, 16, and 17 correspond to the runs with accuracies closest to the average for each trend horizon.

Table 15. Classification report for stacked GRU on trend20

Class	Precision	Recall	F1-Score	Support
Down (−1)	0.84	0.86	0.85	85
Stable (0)	0.70	0.66	0.68	87
Up (1)	0.81	0.84	0.83	75
Accuracy	0.79			247

Table 16. Classification report for stacked GRU on trend40

Class	Precision	Recall	F1-Score	Support
Down (−1)	0.85	0.92	0.88	83
Stable (0)	0.77	0.68	0.72	79
Up (1)	0.85	0.87	0.86	85
Accuracy	0.83			247

Table 17. Classification report for stacked GRU on trend60

Class	Precision	Recall	F1-Score	Support
Down (−1)	0.94	0.86	0.90	68
Stable (0)	0.76	0.87	0.81	92
Up (1)	0.90	0.83	0.86	87
Accuracy	0.85			247

Analysis. The stacked GRU model achieved overall accuracies of 0.79, 0.83, and 0.85 for `trend20`, `trend40`, and `trend60`, respectively. For `trend20`, the Down (−1) class performs well with an F1-score of 0.85, driven by balanced precision (0.84) and recall (0.86). On `trend40`, the accuracy rises to 0.83. The Down (−1) class excels with an F1-score of 0.88, driven by a high recall of 0.92. The Up (1) class maintains a strong F1-score of 0.86, reflecting reliable prediction of upward trends. For `trend60`, the model achieves its highest accuracy of 0.85. The Down (−1) class stands out with a precision of 0.94 and an F1-score of 0.90, indicating high confidence in predicting declines. The Up (1) class achieves an F1-score of 0.86, with balanced precision (0.90) and recall (0.83). The stacked GRU's performance benefits from its two-layer architecture, enhancing its ability to model temporal dependencies, particularly on `trend60`, where the 60-day look-back aligns with the horizon.

5.6 Hybrid LSTM+GRU Evaluation

This section evaluates the performance of the hybrid LSTM+GRU model, featuring two LSTM layers followed by two GRU layers, each with 128 units as described in Sect. 5.1, in predicting medium-term trends of the S&P 500 index across the 20-day (`trend20`), 40-day (`trend40`), and 60-day (`trend60`) horizons. The hybrid LSTM+GRU model was executed five times for each trend horizon, and the average test accuracy was computed for each. The classification reports presented in Tables 18, 19, and 20, correspond to the runs with accuracies closest to the average for each trend horizon.

Table 18. Classification report for hybrid LSTM+GRU on `trend20`

Class	Precision	Recall	F1-Score	Support
Down (−1)	0.91	0.87	0.89	85
Stable (0)	0.73	0.73	0.73	85
Up (1)	0.80	0.84	0.82	77
Accuracy	0.81			247

Table 19. Classification report for hybrid LSTM+GRU on `trend40`

Class	Precision	Recall	F1-Score	Support
Down (−1)	0.88	0.93	0.90	86
Stable (0)	0.77	0.73	0.75	79
Up (1)	0.88	0.87	0.87	82
Accuracy	0.85			247

Table 20. Classification report for hybrid LSTM+GRU on `trend60`

Class	Precision	Recall	F1-Score	Support
Down (−1)	0.91	0.83	0.87	70
Stable (0)	0.78	0.88	0.83	90
Up (1)	0.94	0.89	0.91	87
Accuracy	0.87			247

Analysis. The hybrid LSTM+GRU model exhibits exceptional performance across the three prediction horizons, achieving overall accuracies of 0.81, 0.85, and 0.87 for `trend20`, `trend40`, and `trend60`, respectively. For `trend20`, the model achieves an accuracy of 0.81. The Down (−1) class excels with an F1-score of 0.89, driven by high precision (0.91). On `trend40`, the accuracy increases to 0.85. The Down (−1) class performs exceptionally with an F1-score of 0.90 and a recall of 0.93. For `trend60`, the model attains its highest accuracy of 0.87. The Up (1) class stands out with an F1-score of 0.91 and precision of 0.94, indicating high confidence in upward trend predictions. The Down (−1) class maintains a strong F1-score of 0.87.

The hybrid LSTM+GRU model's superior performance likely stems from its four-layer architecture, combining LSTM's long-term memory with GRU's efficiency, allowing it to capture complex temporal patterns in the 16-feature input over a 60-day look-back.

5.7   Model Comparison and Discussion

Table 21 summarizes the overall accuracy of the five deep learning models evaluated in Sects. 5.2 to 5.6 across the three trend horizons on S&P 500 dataset. Tables 22 and 23 summarize the accuracy of the five deep learning models across the three trend horizons on Dow Jones and Nasdaq datasets, respectively.

Table 21. Accuracy comparison across models and trend horizons for S&P 500 dataset

Model	Trend20	Trend40	Trend60
Single-layer LSTM	0.79	0.75	0.81
Single-layer GRU	0.77	0.83	0.81
Stacked LSTM	0.81	0.83	0.86
Stacked GRU	0.79	0.83	0.85
Hybrid LSTM+GRU	0.81	0.85	0.87

Comparative Performance Analysis. The accuracy metrics, as presented in Tables 21, 22, and 23, reveal distinct patterns in model efficacy across the

Table 22. Accuracy comparison across models and trend horizons for Dow Jones dataset

Model	Trend20	Trend40	Trend60
Single-layer LSTM	0.74	0.78	0.83
Single-layer GRU	0.78	0.83	0.83
Stacked LSTM	0.77	0.82	0.85
Stacked GRU	0.78	0.85	0.87
Hybrid LSTM+GRU	0.79	0.85	0.89

Table 23. Accuracy comparison across models and trend horizons for Nasdaq dataset

Model	Trend20	Trend40	Trend60
Single-layer LSTM	0.56	0.66	0.88
Single-layer GRU	0.80	0.83	0.88
Stacked LSTM	0.77	0.67	0.90
Stacked GRU	0.77	0.83	0.91
Hybrid LSTM+GRU	0.80	0.86	0.92

three indexes and trend periods. For the S&P 500 dataset (Table 21), the hybrid LSTM+GRU model outperformed the other architectures, achieving accuracies of 0.81, 0.85, and 0.87 for the 20-, 40-, and 60-day horizons, respectively. The Stacked LSTM and Stacked GRU models followed closely, with accuracies ranging from 0.79 to 0.86, while the single-layer variants exhibited slightly lower performance, particularly at the 40-day horizon (e.g., single-layer LSTM at 0.75).

A similar trend emerged in the Dow Jones dataset (Table 22), where the hybrid LSTM+GRU model achieved the highest accuracies of 0.79, 0.85, and 0.89 across the three horizons. The Stacked GRU model matched the hybrid's 0.85 accuracy at the 40-day horizon and approached its 60-day performance with 0.87, suggesting competitive robustness among stacked architectures.

The Nasdaq dataset (Table 23) further underscored the hybrid LSTM+GRU model's superiority, with accuracies of 0.80, 0.86, and 0.92 for the 20-, 40-, and 60-day horizons, respectively. However, the Nasdaq results revealed greater variability among models, particularly at the 20-day horizon, where single-layer LSTM lagged significantly at 0.56 compared to the Hybrid's 0.80.

Across all datasets, the hybrid LSTM+GRU model demonstrated a consistent advantage, particularly at the 60-day horizons, suggesting its enhanced capacity to capture medium-term patterns. The stacked models generally outperformed their single-layer counterparts, highlighting the benefit of increased depth.

Strengths and Weaknesses. Each deep learning model showed distinct strengths and weaknesses in predicting medium-term trends across the S&P 500, Dow Jones, and Nasdaq datasets. The single-layer LSTM, though computationally efficient, struggled with short-term horizons (e.g., 0.56 on Nasdaq's 20-day trends) but improved over longer periods (e.g., 0.88 at 60 days). The single-layer GRU balanced efficiency and performance, often achieving competitive accuracies (e.g., 0.83 on Dow Jones at 40 days), though it lagged slightly behind stacked and hybrid models.

The Stacked LSTM and Stacked GRU leveraged deeper architectures for better accuracy (e.g., 0.91 for Stacked GRU on Nasdaq at 60 days), but risked overfitting, as seen in Stacked LSTM's 0.77 on Nasdaq's 20-day trends. The Hybrid LSTM+GRU model excelled across all datasets (e.g., 0.92 on Nasdaq at 60 days) by combining LSTM's memory with GRU's efficiency, though its higher computational cost is a drawback. Overall, simpler models suit specific cases, while the hybrid offers robust medium-term forecasting.

Implications and Insights. The Hybrid LSTM+GRU model's top performance across the S&P 500, Dow Jones, and Nasdaq datasets (e.g., 0.92 on Nasdaq at 60 days) highlights the value of combining LSTM's memory and GRU's efficiency for medium-term forecasting. This suggests hybrid architectures better handle stock market nonlinearity and volatility. Accuracy improved with longer horizons (e.g., 0.81 to 0.87 on S&P 500), indicating stable patterns emerge over time, while short-term noise challenges predictions, as seen in Nasdaq's 20-day results. These findings position the Hybrid model as a benchmark for reliable trend prediction, favoring it for high-accuracy needs, while simpler models suit less demanding cases. Variability across indexes suggests market-specific factors influence outcomes, meriting tailored approaches. Future work could integrate exogenous factors like macroeconomic data to boost resilience, especially for shorter horizons.

6   Conclusion

This study evaluated five deep learning models, single-layer LSTM, single-layer GRU, Stacked LSTM, Stacked GRU, and a Hybrid LSTM+GRU model, for predicting medium-term trends (20-, 40-, and 60-day horizons) of the S&P 500, Dow Jones Industrial Average, and Nasdaq indexes. Using a 10-year dataset of 2524 records from June 23, 2010, to June 30, 2020, with a 60-day look-back period and 16 input features, the models were trained and tested across three major U.S. equity indexes.

The Hybrid LSTM+GRU model consistently outperformed all others, achieving peak accuracies of 0.81, 0.85, and 0.87 on the S&P 500; 0.79, 0.85, and 0.89 on Dow Jones; and 0.80, 0.86, and 0.92 on Nasdaq for the 20-, 40-, and 60-day horizons, respectively. Its architecture, blending LSTM's long-term memory with GRU's efficiency, excelled in capturing medium-term patterns, with notable strength at longer horizons (e.g., 0.92 on Nasdaq at 60 days). Stacked models followed closely, benefiting from added depth (e.g., Stacked GRU at 0.91 on Nasdaq, 60 days), while single-layer models showed mixed results, with single-layer LSTM lagging at shorter horizons (e.g., 0.56 on Nasdaq, 20 days).

A key insight is the advantage of deeper architectures for longer-term trends, aligning the 60-day look-back with improved accuracy. However, variability across indexes, particularly Nasdaq's short-term fluctuations, suggests market-specific factors influence performance. The Hybrid model's robustness positions it as a benchmark for medium-term forecasting, though simpler models may suffice for less complex needs.

Future research could explore several avenues to enhance these findings. First, refining the feature set by incorporating additional technical indicators or reducing noise may improve prediction performance. Second, adjusting the trend thresholds (e.g., 0 to 2.5% for `trend20`, 0.5% to 4% for `trend40`, 1% to 5.5% for `trend60`) could better delineate stable periods. Third, experimenting with hyperparameter tuning (e.g., varying unit counts or look-back periods) or alternative architectures might further boost performance. Finally, extending the evaluation to other stock indices or timeframes could validate the generalizability of the hybrid model.

Acknowledgments. The research project was supported by the CSUSB College of Natural Sciences PATHS program during the summer of 2024. We would like to express our gratitude to the CSUSB High Performance Computing Center for providing the necessary computational resources.

References

1. Ta-lib - technical analysis library. https://ta-lib.org/. Accessed 10 Dec 2024
2. Yahoo finance - stock market live, quotes, business & finance news. https://finance.yahoo.com/. Accessed 10 Dec 2024
3. Bhandari, H.N., Rimal, B., Pokhrel, N.R., Rimal, R., Dahal, K.R., Khatri, R.K.: Predicting stock market index using LSTM. Mach. Learn. Appl. **9**, 100320 (2022)
4. Chen, J., Jing, H., Chang, Y., Liu, Q.: Gated recurrent unit based recurrent neural network for remaining useful life prediction of nonlinear deterioration process. Reliab. Eng. Syst. Saf. **185**, 372–382 (2019)
5. Gulli, A., Pal, S.: Deep Learning with Keras. Packt Publishing Ltd. (2017)
6. Gupta, U., Bhattacharjee, V., Bishnu, P.S.: Stocknet–GRU based stock index prediction. Expert Syst. Appl. **207**, 117986 (2022)
7. Liu, Y.W., Wang, Z.P., Zheng, B.Y.: Application of regularized GRU-LSTM model in stock price prediction. In: 2019 IEEE 5th International Conference on Computer and Communications (ICCC), pp. 1886–1890. IEEE (2019)
8. Medsker, L.R., Jain, L.: Recurrent neural networks. Des. Appl. **5**(64–67), 2 (2001)
9. Nabipour, M., Nayyeri, P., Jabani, H., Shahab, S., Mosavi, A.: Predicting stock market trends using machine learning and deep learning algorithms via continuous and binary data; a comparative analysis. IEEE Access **8**, 150199–150212 (2020)
10. Nelson, D.M.Q., Pereira, A.C.M., De Oliveira, R.A.: Stock market's price movement prediction with LSTM neural networks. In: 2017 International Joint Conference on Neural Networks (IJCNN), pp. 1419–1426. IEEE (2017)
11. Patra, G.R., Mohanty, M.N.: An LSTM-GRU based hybrid framework for secured stock price prediction. J. Stat. Manag. Syst. **25**(6), 1491–1499 (2022)

12. Salem, F.M., Salem, F.M.: Gated RNN: the gated recurrent unit (GRU) RNN. In: Recurrent Neural Networks: From Simple to Gated Architectures, pp. 85–100 (2022)
13. Sherstinsky, A.: Fundamentals of recurrent neural network (RNN) and long short-term memory (LSTM) network. Physica D **404**, 132306 (2020)
14. Trivedi, D.V., Patel, S.: An analysis of GRU-LSTM hybrid deep learning models for stock price prediction. Int. J. Sci. Res. Sci. Eng. Technol. **9**(3), 47–52 (2022)
15. Van Houdt, G., Mosquera, C., Nápoles, G.: A review on the long short-term memory model. Artif. Intell. Rev. **53**(8), 5929–5955 (2020). https://doi.org/10.1007/s10462-020-09838-1
16. Yu, Y., Si, X.S., Hu, C.H., Zhang, J.X.: A review of recurrent neural networks: LSTM cells and network architectures. Neural Comput. **31**(7), 1235–1270 (2019)

GA-CtabDiff: Graph-Augmented Diffusion Model for Mixed-Type Tabular Data Generation

Pengbo Gao, Feng Hu[✉], Jin Dai, Zuqiang Su, and Hong Yu

Chongqing Key Laboratory of Computational Intelligence, Chongqing University of Posts and Telecommunications, Chongqing 400065, People's Republic of China
2450233230@qq.com

Abstract. Diffusion models represent a novel approach to data generation. However, existing diffusion model-based methods for tabular data generation face challenges in adaptively learning feature relationships and enabling mutual learning between mixed features. This paper integrates graph neural networks to advance research on diffusion models. First, leveraging the graph topology of tabular features, we propose an adaptive edge weight computation method for feature extraction in diffusion models. Second, we analyze the limitations of independently training diffusion models for different feature types and introduce a collaborative evolution method for mixed-type tabular data generation by combining mutual interference between discrete and continuous diffusion models with pseudo-label fusion of generated features. Finally, we present a graph-augmented diffusion model-based algorithm for mixed-type tabular data generation. Experimental results demonstrate that, compared to existing algorithms in the literature, our method achieves superior performance in terms of AUC metrics, effectively reducing the distribution gap between generated and real data while enhancing the predictive performance of downstream models.

Keywords: Diffusion Model · Tabular Data · Tabular Data Generation · Graph Neural Network

1 Introduction

With the rising risks of data misuse and privacy breaches, data protection has become a focal point of societal concern. To effectively mitigate these risks, a series of data protection measures and regulations have been introduced, posing challenges to data-driven industries. Consequently, achieving knowledge discovery while complying with data privacy and government regulations has become a significant challenge. Additionally, the scarcity or difficulty in obtaining raw data in fields such as healthcare and industry limits algorithm training and model development. A primary solution to this problem is tabular data generation, where synthetic data not only statistically resembles the original data

© The Author(s), under exclusive license to Springer Nature Switzerland AG 2025
Q. Zhang et al. (Eds.): IJCRS 2025, LNAI 15710, pp. 405–420, 2025.
https://doi.org/10.1007/978-3-031-92741-6_29

but also demonstrates comparable utility in subsequent algorithmic research. Tabular data generation enables the creation of new data that simulates the original data, facilitating knowledge discovery while adhering to data privacy and regulatory constraints. Current tabular data generation methods primarily fall into three categories: (1) Variational Autoencoders (VAEs [3]), Map data to a latent space via an encoder and generate data from the latent space using a decoder. How-ever, they exhibit limited effectiveness in handling mixed-type data and categorical features. (2) Generative Adversarial Networks (GANs [5]), Approximate real data distributions through adversarial training. While capable of generating realistic samples, they suffer from complex training dynamics, mode collapse, gradient instability, and limited capacity for novel/diverse synthesis. (3)Diffusion Models [7], Diffusion models generate data by progressively adding noise and learning the reverse denoising process. They are capable of producing high-quality synthetic data.

Diffusion models, a class of probabilistic generative models, transform data distributions into simple distributions (e.g. Gaussian) by gradually adding noise and then generate high-quality data samples through the reverse denoising process. These models have achieved remarkable success in image generation, speech synthesis, and text generation. Notable examples include: Sora [1]: A video generation model by OpenAI, leveraging diffusion models for high-quality video synthesis. CoDi [2]: A multimodal generative model capable of handling and generating combinations of text, images, video, and audio. In the domain of tabular data generation, diffusion models have also garnered significant attention. Tabular data typically includes numerical and categorical features. Due to the powerful generative capabilities and ability to model complex data distributions, researchers have conducted targeted studies on continuous and categorical features. For numerical features, normalization is applied, and Gaussian diffusion models are used for modeling. For categorical features, one-hot encoding is used as input, and multinomial diffusion models are employed. Each categorical feature undergoes an independent forward diffusion process, with noise components sampled independently. The reverse diffusion process is modeled by a multi-layer neural network. However, tabular diffusion models face the following challenges. (1) Difficulty in Adaptively Learning Relationships: Tabular data often contains heterogeneous features with complex implicit relationships, such as nonlinear interactions, hierarchical dependencies, and conditional dependencies. Diffusion models struggle to fully capture these relationships, limiting their ability to understand deep feature interactions and generalize effectively. (2) Separate Training for Mixed Features: To achieve precise modeling of different feature types, continuous and categorical models are trained separately. This approach prevents mutual learning and co-evolution between discrete and continuous features, hindering the model's ability to capture the complex relationships within tabular data.

This paper proposes a novel tabular data synthesis model that integrates Gaussian diffusion for continuous features and multinomial diffusion for discrete features. Building upon this framework, we introduce a novel Graph Estimator

(GE) to construct feature relationship graphs, where nodes represent features and edges indicate relationships between features, with edge weights predicted by the GE module. The GE module learns to assign edge weights by estimating static feature relationships, thereby guiding feature interactions and generating higher-level features. Additionally, a co-evolution mechanism is introduced to enable mutual interaction between the two diffusion models, thereby improving joint modeling capabilities for deep feature semantics and combinatorial relationships. Key innovations include:

1. Graph-Augmented Diffusion Models: We propose a graph topology and adaptive edge weight method for tabular diffusion models, enabling the capture of implicit relationships between samples and features while preserving data distribution characteristics and sample correlations to enhance generated data diversity and authenticity.
2. Collaborative Diffusion with Pseudo-Label Fusion: We analyze the training differences between numerical and categorical diffusion models and introduce a collaborative diffusion method with pseudo-label fusion. By enabling mutual interference between diffusion models and leveraging pseudo-labels predicted from initial models, we enhance the model's ability to capture complex relationships between features and target distributions.

2   Related Work

Tabular data generation is an active research area in the field of data generation. Current tabular data generation methods primarily include VAE models, GAN models, and tabular diffusion models.

1. Variational Autoencoders (VAEs): VAEs [3] is a generative model that learns latent representations of input data through an encoder-decoder framework. The encoder outputs the mean and variance of latent variables, which are sampled via the reparameterization trick, and the decoder reconstructs the data. However, VAE has limitations: regularization may lead to blurry reconstructions with unclear details, and it may struggle to accurately model complex data distributions or capture intricate structures and patterns.
2. Generative Adversarial Networks (GANs): GANs, proposed by Ian Goodfellow et al. in 2014, are powerful deep learning models consisting of two key components: a generator and a discriminator. The generator aims to produce fake data that closely resembles real data, while the discriminator attempts to distinguish between real and fake data. Through adversarial training, the generator learns to produce increasingly realistic data. CTGAN [3] is a conditional generator for tabular data that integrates the PacGAN [4] structure and uses a combination of generator loss, WGAN loss, and gradient penalty (GP) to train the conditional GAN framework. Zhao et al. later developed CTAB-GAN [5], which effectively addresses data imbalance and distribution

issues. CTAB-GAN+ [6] extends this by modeling continuous and categorical variables, combining the strengths of existing techniques such as Wasserstein GAN with gradient penalty for enhanced training effectiveness, classifier supervision for improved utility in machine learning applications, and information loss and generator loss for optimizing data quality. Despite their success, GANs face challenges such as unstable training, mode collapse, and limited diversity in generated data.

3. Tabular Diffusion Models: Diffusion models are emerging as a leading generative modeling paradigm for various data types, including tabular data. As a mainstream model in computer vision, diffusion models have gained attention in speech, natural language processing, and time series domains. Recently, Kotelnikov et al. developed TabDDPM [7], demonstrating the effectiveness and advantages of diffusion models in tabular data modeling. TabDDPM can generate high-quality synthetic tabular data, addressing issues such as data scarcity and privacy protection. TabCSDI [8] is a tabular data generation method based on the Conditional Score Diffusion Model (CSDI). It unifies continuous and categorical variables into a Gaussian diffusion process using three preprocessing techniques: one-hot encoding, simulated bit encoding, and feature tokenization. This method excels in missing value imputation tasks. Despite their advancements, tabular diffusion models still face several limitations: (1) Insufficient Modeling of Implicit Feature Relationships: Tabular data often contains heterogeneous features, where different attributes may have distinct properties and implicit relationships. Diffusion models may fail to adequately capture these complex dependencies when modeling feature relationships. (2) Limitations in Mixed-Feature Modeling of Tabular Diffusion Models: Current methods use Gaussian diffusion for continuous features and multinomial diffusion for discrete features. This separation lacks coordination in preprocessing, noise addition, and training, hindering the model's ability to capture tabular data's intrinsic structure and relationships, and worsening label-related feature distribution fitting.

3   Proposed Method

In the overall architecture, the method first employs Graph Neural Networks (GNNs) to adaptively extract features from tabular data. A Graph Estimator (GE) is used to construct a Feature Relation Graph (FR-Graph), and a Graph Transformer is utilized to aggregate features layer by layer, extracting comprehensive semantic information. Subsequently, the extracted features are fed into a collaborative diffusion optimization module. Through mutual perturbation of discrete and continuous diffusion models, relevant data records are generated. Additionally, pseudo-label fusion optimization is incorporated to further enhance the model's ability to learn the relationships between labels and features, ultimately outputting the generated tabular data.

3.1   Adaptive Feature Extraction with Graph Neural Networks

Tabular data often contains heterogeneous features, where different features may have entirely different properties, and implicit relationships may exist between various attributes. Diffusion models may struggle to fully capture these complex dependencies when modeling relationships between features. Therefore, an automatic adaptive feature enhancement method is needed to optimize the diffusion model, helping it to extract implicit relationships between features. An ideal approach to handle such complex decision-making processes is to construct a graph with adaptive edge weights (Fig. 1). In this paper, inspired by the principles of T2G-Former [10], we propose a novel Graph Estimator (GE) in the aspect of table diffusion model data generation, which is used to organize independent tabular features into a Feature Relation Graph (FR-Graph). The GE automatically estimates relationships between tabular features and assigns edge weights to related features to construct the graph. Edge weights represent the strength of relationships based on specific feature value relationships, while the static graph topology describes the underlying knowledge that provides foundational constraints for feature interactions, such as the meaning of feature pairs. In this way, the FR-Graph organizes independent tabular features into a graph data structure, enabling feature interactions to occur in an orderly manner. Furthermore, a customized Transformer network is used for tabular learning, selectively performing feature interactions by stacking GE modules. This network conducts feature interactions under the guidance of the relation graph, effectively processing tabular data. In the Graph Transformer, each layer transforms input features into graph data using the FR-Graph and performs heterogeneous feature interactions in an orderly manner based on edge weights. This approach allows us to capture more nuanced interactions that might be mishandled by traditional grouping strategies.

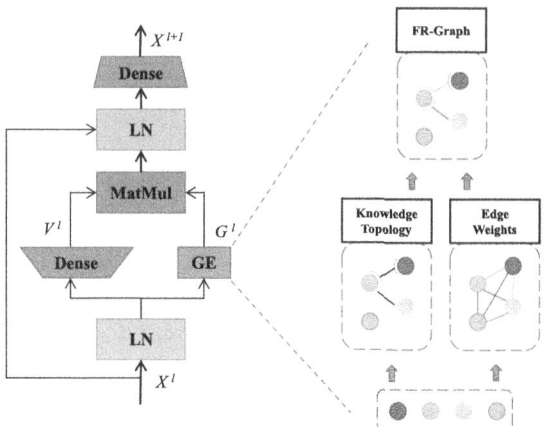

Fig. 1. Graph Neural Network Optimization Module.

3.1.1   Graph Estimation

Graph Estimation (GE) is used to automatically construct the Feature Relation Graph (FR-Graph), where tabular features are treated as nodes, and feature relationships are estimated as edges. The design of GE is inspired by Knowledge Graph Completion (KGC), with the core idea being to estimate the strength of relationships between two entities based on their semantic similarity. The basic form of measuring semantic similarity is:

$$f_r(h, t) = h^T M_r t \tag{1}$$

where $h, t \in R_n$ represents the encoded head and tail entity nodes, and the learnable matrix $M_r \in R_{n \times n}$ represents the relation r shipin the knowledge graph (KG). The difference lies only in the relationship embeddings and scoring functions.

1) Structural Components of FRGraph

To uncover the relationships between table features, we construct the FRGraph by treating table features as candidate graph nodes and predicting the edges between them. Edges are generated from two perspectives: adaptive edge weights represent data-specific information, while static edge topology represents foundational knowledge applicable to all data.

Adaptive Edge Weights: Given two table feature embedding vectors $X_i, x_i \in R_n (i, j \in \{1, 2, ..., N\})$ where N is the number of input features, we employ the following pairwise scoring function to evaluate their interaction likelihood:

$$G_w[i, j] = g_w(f_i^h, f_j^t) = f_i^{h^T} diag(r) f_j^t \tag{2}$$

$$f_i^h = W^h x_i, f_i^t = W^t x_i, \begin{cases} W^h \equiv W^t \, if \, symmetric, \\ W^h \neq W^t \, if \, asymmetric, \end{cases} \tag{3}$$

here, two learnable parameters $W^h, W^t \in R_{m \times n}$ represent the transformations for the head and tail features, and $diag(r) \in R_n \times n$ is a diagonal matrix parameterized by learnable relation vectors $r \in R_n$, which semantically represent feature interaction relationships. If the pairwise feature edge weights are symmetric (i.e. $G_w[i, j] = G_w[j, i]$), then W^h and W^t share parameters; in the asymmetric case (i.e. $G_w[i, j] = G_w[j, i]$), W^h and W^t are independent. For notational simplicity, all bias vectors are omitted. Consequently, the adaptive weight scores g_w for all feature pairs constitute a fully connected weighted relation graph G_w. When r is filled with scalar value $1, diag(r)$ becomes an entity matrix, and the edge weight scores degenerate into an attention score, thereby enabling the measurement of weighted feature similarity.

Static Knowledge Topology: Although we introduce soft edge weights for all feature pairs, it is also crucial to globally consider the underlying knowledge of the tabular data. Therefore, we use a series of column embeddings to represent the semantics of table features. The static relation topology scores can be computed as follows:

$$G_t[i,j] = g_t(e_i^h, e_j^t) = \frac{e_i^{h\,T} e_j^t}{\|e_i^h\|_2 \|e_i^t\|_2},$$
$$e_i^h = E^h[:,i], e_i^t = E^t[:,i] \tag{4}$$

here, $E \in \{E^h, E^t\}$ represents the learnable column embeddings for the head view or tail view $E = (e_1, e_2, ..., e_N) \in R_d \times N$, and d is the embedding dimension. Similarly, the relation topology score g_t has symmetric and asymmetric counterparts, where E_h and E_t share parameters in the case of symmetric relation topology (i.e. $G_t[i,j] \equiv G_t[j,i]$), but are parameter-independent in the asymmetric case (i.e. $G_t[i,j] \neq G_t[j,i]$). We employ $L2$ normalization in the g_t scoring function to transform the embeddings into a similar scale, thereby improving training stability. Based on the score in the equation, we generate the G_t static relation topology:

$$A = f_{top}(G_t) = [\sigma_1(G_t + b) > T] \tag{5}$$

here, σ_1 is an activation function parameterized by a learnable bias b, G_t is the adjacency matrix score composed of relation topology scores g_t, T is a constant threshold for signal clipping, and 1 denotes the indicator function. In this way, we obtain a global graph topology (adjacency matrix A) to constrain feature interactions, and this topology can be regarded as static knowledge for the entire task.

2) Relational Graph Assembly
When we obtain "soft" adaptive edge weights from the data view and "hard" static relation graph topology from the knowledge view, we combine them to generate the FRGraph, following the idea of "making decisions based on specific data and underlying knowledge." Specifically, we assemble these two components as follows:

$$G = \sigma_2(f_{nsi}(a) \odot G_w) \tag{6}$$

here, σ_2 is an activation function to limit the degree of each "feature node", and $\odot$ denotes the Hadamard product. The resulting relation graph G is a weighted graph based on adaptive feature matching and static knowledge topology. To help the FRGraph focus on learning meaningful interactions between different features, a "no self-interaction" function f_{nsi} is applied to explicitly exclude self-loops in G. We use the FRGraph to guide subsequent feature interactions.

3.1.2   Graph Transformer
We integrate GE into attention-based basic blocks and construct a Graph Transformer by stacking multiple blocks for selective tabular feature interaction. The Graph Transformer uses the estimated FRGraph to interact features and obtain higher-level features layer by layer. A cross-layer readout module sequentially transforms the feature space of each layer and selectively collects significant features for final prediction. By adding a shortcut path to retain information from previous layers and performing gated fusion across different feature levels, the model's performance is improved.

Basic Block: We construct a single block equipped with GE for selective feature interaction.Given input features $X_1 \in R_n \times N$ to the l-th layer, we obtain higher-level featuresas $X_1 + 1$ follows:

$$G^l = GE(X^l), V^l = W_v X^l,$$
$$H^l = G^l V^l + g(X^l), \qquad (7)$$
$$X^{l+1} = FFN(H^l) + g(H^l)$$

here,$W_v \in R_m \times n$ represents learnable parameters for feature transformation,V^l is the transformed input feature, and FFN denotes a feedforward network. Since self-interactions are excluded in G^l, a shortcut path g is added to preserve information from previous layers, which is implemented as a simple layer $Dropout$ in experiments.

Cross-Level Readout: A global readout node is designed to selectively collect significant features from each layer and obtain comprehensive semantics for final prediction. Specifically, we carefully fuse selected features at the current layer and combine them with lower-level features from the previous layer via a shortcut path. Given the current readout state $z^l \in R_n$, the selection process at the l-th layer is defined as:

$$\alpha_i^l = g_w(h^l, f_i^t) \cdot f_{top}(g_t(e^l, e_i^t)), h^l = W^h z^l \qquad (8)$$

$$r^l = \text{softmax}(\alpha^l)^T V^l + z^l \qquad (9)$$

$$z^{l+1} = FFN(r^l) + r^l \qquad (10)$$

here, α_i^l represents the weight of the i-th feature in the weight vector $\alpha^l \in R_N, e^l \in R_d$ is a learnable vector representing the semantics of the readout node at the i-th layer,f_i^t encodes features at each layer, and e_i^t denotes layer-wise column embeddings.V_1 is the transformed input feature. We advance FFN using the same z^l transformation to map the current readout to the feature space of the $(l+1)$-th layer for the next round of collection. The shortcut path is directly added without information removal. This collection process is repeated from input features to the highest-level features, encouraging interactions across feature levels.

Overall Architecture: Basic blocks are stacked in the Graph Transformer. Unless specified otherwise, we default to using 8 heads GE in each block in experiments. Predictions are made based on the readout state after processing the final layer L, as follows:

$$\hat{y} = FC(\text{Re} LU(LN(z^L))) \qquad (11)$$

here, LN and FC denote layer normalization and fully connected layers, respectively. Continuously optimizing the static graph topology A throughout the training phase may lead to performance instability for simpler tasks. Therefore, we freeze it after convergence for further training with a fixed topology.

3.2 Collaborative Diffusion with Pseudo-label Fusion

Tabular data often contains continuous and discrete features, which have entirely different properties. To accurately model different feature types, existing methods train discrete and continuous diffusion models separately, preventing mutual learning and co-evolution. This limits the model's ability to capture the intrinsic structure and relationships in tabular data. To address this issue, inspired by the codi [11] model, it is essential to incorporate mechanisms into the model design that enable discrete and continuous features to mutually learn and co-evolve, thereby better capturing and simulating the complex relationships in tabular data.

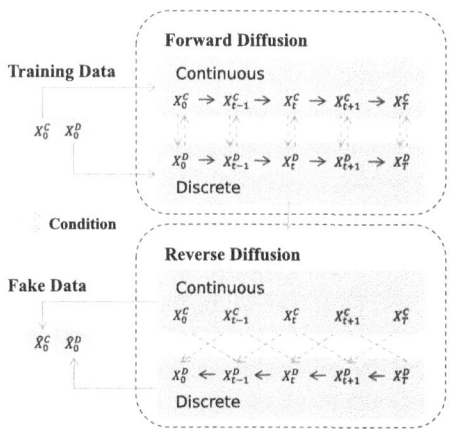

Fig. 2. Tabular Diffusion Model Co-Evolutionary Process Diagram.

3.2.1 Collaborative Diffusion Method

To enable the two diffusion models to synthesize tabular data, we make them read each other's conditions, as shown in Fig. 2. Both models perturb continuous and discrete variables simultaneously at each forward step. Specifically, the continuous (discrete) model reads the perturbed discrete (continuous) samples as conditions. For the continuous reverse process diffusion model, the model takes sample X_t^C, conditioned on continuous sample X_{t+1}^C and discrete sample X_{t+1}^D from its previous step. Given a sample X_0 composed of mixed-type variables, we generally assume X_0 contains N_C continuous columns $C = \{C_1, C_2, ..., C_{N_C}\}$ and N_D discrete columns $D = \{D_1, D_2, ..., D_{N_D}\}$, where $X_0 = (X_0^C, X_0^D)$. To synthesize samples from the space of each type, we train two diffusion models for the two variable types, which read conditions from each other, linking their diffusion and denoising processes.

To generate a correlated data pair using the two models, we input each other's outputs as conditions. Then, at each forward time step t, the pair (X_0^C, X_0^D) is perturbed simultaneously in each space. The perturbed data X_t^C in the continuous diffusion model will be the condition for X_t^D in the discrete diffusion model, and vice versa. The model parameters θ_C are updated based on the following equations:

$$L_{Diff_C}(\theta_C) := E_{t,X_0^C,\varepsilon}[\|\varepsilon - \varepsilon_{\theta_C}(X_t^C,t|X_t^D)\|^2] \tag{12}$$

$$L_{Diff_D}(\theta_D) = E_q[\underbrace{D_{KL}(Q(X_T^D|X_0^D)\|P(X_T^D)}_{L_T} - \underbrace{\log P_{\theta_D}(X_0^D|X_1^D,X_1^C)}_{L_0} +I] \tag{13}$$

$$I = \sum_{t=2}^{T} \underbrace{D_{KL}(Q(X_{t-1}^D|X_t^D,X_0^D)\|P_{\theta_D}(X_{t-1}^D|X_t^D,X_t^C))}_{L_{t-1}} \tag{14}$$

here, $L_{Diff_C}(\theta_C)$ and $L_{Diff_D}(\theta_D)$ are the loss functions for the continuous and discrete diffusion models, respectively. For the reverse process, generated samples $\hat{X}_0^C$ and $\hat{X}_0^D$ are synthesized stepwise from each noise space. The prior distributions of the two models are $P(X_T^C) = N(X_T^C;0,I)$ and , $P(X_T^{Di}) = C(X_T^{Di};0,1/K_i)$ where $\{K_i\}_{i=1}^{N_D}$ is the discrete column and $\{D_i\}_{i=1}^{N_D}$ is the number of categories. After sampling noise vectors, the two diffusion models transform the noise into fake samples while conditioning on the denoised samples from the previous time step. To go from X_{t+1}^C (respectively X_{t+1}^D) to X_t^C (respectively X_t^D), the continuous diffusion model (respectively discrete diffusion model) conditions on continuous sample X_{t+1}^C and discrete sample X_{t+1}^D, allowing the two models to collaborate in generating correlated data records $\hat{X}_0 = (\hat{X}_0^C, \hat{X}_0^D)$. The forward processes for the continuous and discrete diffusion models are defined as:

$$Q(X_t^C|X_0^C) = N(X_t^C; \sqrt{\bar{\alpha}_t}X_0^C, (1-\bar{\alpha}_t)\iota) \tag{15}$$

$$Q(X_t^{Di}|X_0^{Di}) = C(X_t^{Di}; \bar{\alpha}_t X_0^{Di} + (1-\bar{\alpha}_t)/K_i) \tag{16}$$

here, $1 \leq i \leq N_D$ $\alpha_t := 1-\beta_t$ and $\bar{\alpha}_t := \prod_{i=1}^{t} \alpha_i$. Additionally, the reverse process for the co-evolution conditional diffusion model is defined as:

$$P_{\theta_C}(X_{0:T}^C) := P(X_T^C) \prod_{t=1}^{T} P_{\theta_C}(X_{t-1}^C|X_t^C, X_t^D) \tag{17}$$

$$P_{\theta_D}(X_{0:T}^{Di}) := P(X_T^{Di}) \prod_{t=1}^{T} P_{\theta_D}(X_{t-1}^{Di}|X_t^{Di}, X_t^C) \tag{18}$$

3.2.2   Pseudo-label Fusion Optimization

In semi-supervised learning, pseudo-labels are labels predicted by the model itself and are used for training on unlabeled data. This method helps the model learn from unlabeled data, thereby improving its generalization ability. In the

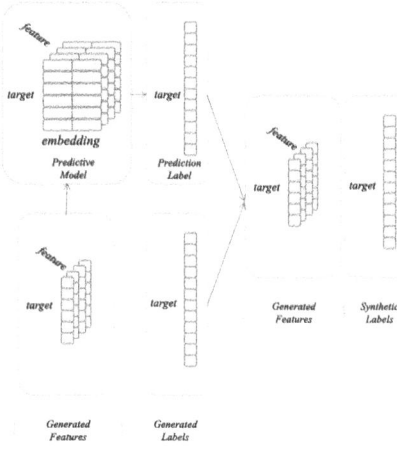

Fig. 3. Pseudo Label Optimization Process Diagram.

context of diffusion models, pseudo-labels can help the model learn richer data representations and reduce errors and biases introduced by the diffusion model's generated labels (Fig. 3). We train a latent diffusion label correction model that calibrates the labels generated by the diffusion model using pseudo-label categories, thereby improving the label distribution. Existing generated labels heavily rely on the quality of the diffusion model's outputs, which can lead to supervision errors and confirmation bias. To address these issues, we propose a latent diffusion label correction method for semi-supervised optimization. First, the model is trained on the original dataset, and pseudo-labels are predicted from the generated features. These predicted pseudo-labels are then fused with the generated labels to enhance the model's learning of the relationships between labels and features, thereby improving the distribution of generated labels. By combining pseudo-labels with the diffusion model's own generated labels, the model's performance on specific tasks can be further enhanced.

3.3 Overall Training and Sampling Algorithms

In the overall architecture, a Graph Neural Network (GNN) is first used to perform adaptive feature extraction on the tabular data. The Graph Estimator (GE) organizes the independent features in the table into a Feature Relation Graph (FR-Graph), where nodes represent features and edges represent relationships between features. The edge weights, predicted by the GE module, reflect the dynamic strength of the relationships between features. Subsequently, the Graph Transformer aggregates features layer by layer, with each layer using the FR-Graph to guide feature interactions. A cross-layer readout mechanism is employed to gradually extract comprehensive semantic information, generating higher-level feature representations (Fig. 4). Next, the features processed by the GNN are fed into the collaborative diffusion optimization module. In this

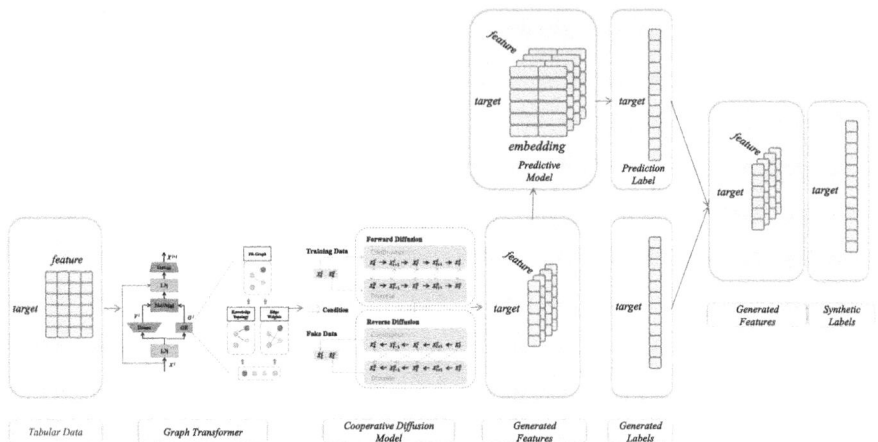

Fig. 4. Overall Algorithm Flowchart of GA-CtabDiff.

module, the discrete and continuous diffusion models generate correlated data records through mutual perturbation and collaboration. During the forward diffusion process, the continuous and discrete diffusion models read each other's perturbed samples as conditions. During the reverse diffusion process, they condition on the denoised samples from the previous time step, gradually generating samples of continuous and discrete features. Through this co-evolutionary approach, the model can fully capture the complex relationships between continuous and discrete features. Additionally, to further strengthen the model's learning of the relationships between labels and features, we introduce pseudo-label fusion optimization. During training, the model is first trained on a small amount of labeled data, and then pseudo-labels are predicted from the generated features. These predicted pseudo-labels are fused with the generated labels. This method not only helps the model learn from unlabeled data but also reduces errors and biases introduced by the diffusion model's generated labels, thereby enhancing the model's performance and generalization capabilities. Finally, the tabular data generated by the graph-enhanced collaborative diffusion optimization module not only resembles real data in statistical properties but also closely aligns with the real data distribution in terms of complex feature relationships.

4   Experimental Analysis

To systematically evaluate the performance of tabular generative models, we selected 15 real-world public datasets with varying scales, feature counts, and distributions, including: AB (Abalone), AD (Adult), BU (Buddy), CA (California Housing), CAR (Cardio), CH (Churn Modeling), DI (Diabetes), FB (Facebook Comm. Vol.), GE (Gesture Phase), HI (Higgs Small), HO (House 16H), IN (Insurance), KI (King), MI (MiniBooNE), and WI (Wilt), most of which were previously used in studies by Zhao, Z [5] and Gorishniy [9] et al. We primarily evaluated GA-CtabDiff and its comparative models, using machine learning (ML) efficiency as the core metric, measured by the performance of classification or regression models trained on synthetic data and evaluated on real test sets. In the experiments, we employed XgBoost (a leading GBDT implementation) to assess ML efficiency and optimized the hyperparameters of GA-CtabDiff and baseline models using the Optuna library(Akiba et al., 2019). The tuning process was based on the average ML efficiency scores from five different sampling seeds.

4.1 Qualitative Comparison

In most cases, GA-CtabDiff generates more realistic feature distributions compared to TVAE and CTABGAN+. This advantage is more pronounced in the following scenarios: (1) numerical features with uniform distributions, (2) categorical features with high cardinality, and (3) mixed-type features combining continuous and discrete distributions. We measured the Wasserstein distance (WD) for numerical features, the Jensen-Shannon (JS) divergence for categorical features, and the L2 distance between correlation matrices. The results are reported in Table 1 as average scores across all datasets (lower is better). Lower ranks indicate smaller WD, JS divergence, and L2 distance.

Table 1. Distance Metrics between Features (Wasserstein, Jensen-Shannon, L2)

	WD (Num.)	JS (Cat.)	L2 (Corr. matrix)
CTGAN	38.8676	0.7282	0.8448
TVAE	348.3719	nan	nan
CTABGAN	129.9185	0.7425	0.6803
CTABGAN+	14.4022	0.7398	0.8110
TabDDPM	10.5803	0.7425	0.6803
CoDi	36.8868	0.7389	0.8448
GA-CTabDiff	34.3051	0.7236	0.3970

4.2 Experimental Results

In this section, we compare the machine learning efficiency of GA-CtabDiff with other generative models. Synthetic datasets are sampled from each generative model, used to train classification/regression models, and then evaluated on real test sets. Classification performance is measured using AUC and F1 scores, while regression performance is evaluated using explained variance and MAE scores. We employ XgBoost for assessment, with hyperparameters derived from the report by (Gorishniy et al., 2021). This evaluation approach better demonstrates

Table 2. Performance Metrics (Classification AUC for the first 9 rows, Regression Explained Variance for the last 6 rows).

Data	CTGAN	TAVE	CTABGAN	CTABGAN+	TabDDPM	CoDi	GA-CtabDiff
AD	0.7242	0.7592	0.7536	0.7391	0.5845	0.8595	**0.9347**
CAR	0.7310	0.5021	0.7130	0.7308	0.7300	0.7559	**0.8142**
CH	0.6686	0.7965	0.6458	0.7090	0.6135	0.8015	**0.8978**
DI	0.7183	0.6504	0.7618	0.7574	0.7524	0.8864	**0.8897**
HI	0.6218	0.5285	0.6220	0.6718	0.6909	0.7617	**0.8796**
MI	0.6951	0.7193	0.8711	0.7005	0.9148	0.9486	**0.9886**
WI	0.8480	0.9262	0.6848	0.8236	0.8435	0.9651	**0.9858**
BU	0.9497	0.9651	0.9672	0.9710	0.9752	0.9738	**0.9869**
GE	0.7626	0.8218	0.8564	0.8440	0.8352	0.7309	**0.9508**
AB	0.4276	0.2287	0.4523	0.4777	0.5742	0.5833	**0.8183**
CA	0.5918	0.3530	0.6115	0.6435	0.7783	0.5750	**0.8303**
FB	0.4568	0.0713	0.4539	0.4700	**0.6923**	0.6374	0.1263
HO	0.3434	0.2016	0.4114	0.4643	**0.5652**	0.3376	0.5597
IN	0.7014	0.4804	0.7469	0.7713	0.7931	0.7958	**0.9148**
KI	0.7548	0.4821	0.8115	0.8301	0.4922	0.8011	**0.9189**

Table 3. Performance Metrics (Classification F1 for the first 9 rows, Regression MAE for the last 6 rows)

Data	CTGAN	TAVE	CTABGAN	CTABGAN+	TabDDPM	CoDi	GA-CtabDiff
AD	0.7551	0.6614	0.7712	0.7668	0.5861	0.6550	**0.7950**
CAR	0.7293	0.3356	0.7059	0.7298	0.7295	0.6990	**0.7353**
CH	0.6941	0.7062	0.6771	0.7356	0.6332	0.6178	**0.7518**
DI	0.6696	0.5112	0.7677	0.7640	0.7573	0.8424	**0.8666**
HI	0.6111	0.3655	0.6195	0.6687	0.6912	0.7257	**0.8031**
MI	0.7250	0.6018	0.8834	0.7289	0.9128	0.8951	**0.9162**
WI	0.5534	0.4862	0.7457	0.8624	0.8600	**0.8898**	0.8842
BU	0.8979	0.8982	0.8357	0.8890	0.8934	0.8703	**0.9334**
GE	0.5143	0.4932	0.3965	0.4058	0.4834	0.3696	**0.8938**
AB	0.5357	1.0361	0.4913	0.4881	0.3697	0.1552	**0.0961**
CA	0.5562	0.5782	0.4569	0.5238	0.3170	0.2442	**0.1384**
FB	0.2667	2.0154	0.1832	**0.1713**	0.8560	1.7576	1.3030
HO	0.6237	1.0529	0.3860	0.3796	0.3016	0.4026	**0.2500**
IN	0.2943	0.7393	0.2946	0.2695	0.2514	0.2654	**0.1574**
KI	0.1540	0.4376	0.2423	0.2278	0.8359	0.1788	**0.1073**

the practical value of synthetic data, as weak or suboptimal classifiers/regressors are rarely used in real-world scenarios. The first nine datasets are classification datasets (including BU and GE multi-class datasets), while the last six datasets are regression datasets (Tables 2 and 3).

4.3 Ablation Study

As shown in Table 4, we conducted ablation experiments to demonstrate the efficacy of GA-CtabDiff. We incrementally added our innovative methods, and across the 15 datasets, collaborative diffusion improved the metric scores, with the graph-enhanced diffusion model further enhancing the results. The experimental results indicate that the collaborative diffusion and graph-enhanced design choices are reasonable and can synthesize better samples. Method1 denotes collaborative diffusion optimizationMethod2 denotes adaptive Graph Neural Network.

Table 4. Ablation Study Results.

Data	Basemodel	Basemodel+Method1	Basemodel+Method1+Method2
AD	0.5845	0.9231	**0.9347**
CAR	0.7300	0.7997	**0.8142**
CH	0.6135	**0.9004**	0.8978
DI	0.7524	0.8480	**0.8897**
HI	0.6909	0.7548	**0.8796**
MI	0.9148	0.9805	**0.9886**
WI	0.8435	0.9493	**0.9858**
BU	0.9752	**0.9891**	0.9869
GE	0.8352	**0.9724**	0.9508
AB	0.5742	0.7129	**0.8183**
CA	0.7783	0.8288	**0.8303**
FB	0.6923	**0.7182**	0.1263
HO	0.5652	**0.5798**	0.5597
IN	0.7931	0.8983	**0.9148**
KI	0.4922	0.9134	**0.9189**

5 Conclusion

This paper proposes a graph-enhanced diffusion model for tabular data generation, which effectively addresses the limitations of existing approaches in adaptive feature learning and mixed-type feature modeling through the integration of graph neural networks with collaborative diffusion mechanisms. Experimental results demonstrate significant improvements in both synthetic data quality and downstream task performance, establishing a novel and effective solution for tabular data generation. However, diffusion models and graph neural networks still suffer from high computational costs and limitations in handling mixed-type features. Future work will focus on further optimizing computational efficiency and enhancing feature extraction and representation capabilities to more accurately capture complex feature relationships in tabular data.

Acknowledgements. This study was funded by the Natural Science Foundation Project of CQ CSTC (CSTB2023NSCQ-LZX0006), the National Key R&D Program of China (grant number 2021YFF0704103), the key project of the National Natural Science Foundation of China (62233018) and Chengdu Key Research and Development Plan Project (2023-YF11-00059-HZ).

References

1. Liu, Y., Zhang, K., Li, Y., et al.: Sora: a review on background, technology, limitations, and opportunities of large vision models. arXiv preprint arXiv:2402.17177 (2024)
2. Tang, Z., Yang, Z., Zhu, C., et al.: Any-to-any generation via composable diffusion. Adv. Neural Inf. Process. Syst. **36**, 16083–16099 (2023)
3. Xu, L., Skoularidou, M., Cuesta-Infante, A., et al.: Modeling tabular data using conditional GAN. Adv. Neural Inf. Process. Syst. **32** (2019)
4. Lin, Z., Khetan, A., Fanti, G., et al.: PacGAN: the power of two samples in generative adversarial networks. IEEE JSAIT **1**(1), 324–335 (2020)
5. Zhao, Z., et al.: CTAB-GAN: effective table data synthesizing. In: Asian Conference on Machine Learning. PMLR (2021)
6. Zhao, Z., et al.: CTAB-GAN+: enhancing tabular data synthesis. Front. Big Data **6**, 1296508 (2024)
7. Kotelnikov, A., et al.: TabDDPM: modelling tabular data with diffusion models. In: International Conference on Machine Learning. PMLR (2023)
8. Zheng, S., Charoenphakdee, N.: Diffusion models for missing value imputation in tabular data. arXiv preprint arXiv:2210.17128 (2022)
9. Gorishniy, Y., Rubachev, I., et al.: Revisiting deep learning models for tabular data. Adv. Neural Inf. Process. Syst. **34**, 18932–18943 (2021)
10. Yan, J., Chen, J., Wu, Y., et al.: T2G-former: organizing tabular features into relation graphs promotes heterogeneous feature interaction.. In: Proceedings of the 37th AAAI Conference on Artificial Intelligence, pp. 10720–10728 (2023)
11. Lee, C., Kim, J., Park, N.: CoDi: co-evolving contrastive diffusion models for mixed-type tabular synthesis. In: Proceedings of the 40th International Conference on Machine Learning, pp. 18940–18956. PMLR (2023)

Affinity Based Semantic Collaborative Hashing for Image Retrieval

Yi Li, Wentao Fan, Ziqi Meng, and Huaxiong Li$^{(\boxtimes)}$

Department of Control Science and Intelligence Engineering, Nanjing University,
Nanjing 210093, Jiangsu, People's Republic of China
{ylee,fanwentao0955,zqmeng}@smail.nju.edu.cn, huaxiongli@nju.edu.cn

Abstract. Hashing methods have captured increasing attention in image retrieval due to their computational efficiency and low storage cost. Most existing supervised hashing approaches focus on preserving the semantic similarity for discriminative hash codes learning. However, they may not discover intrinsic latent features embedded in high-dimensional feature space, and ignore the underlying affinity structures of data. In this paper, we proposed a novel Affinity based Semantic Collaborative Hashing (ASCH) method for image retrieval, which obtains the hash codes by the collaboration of latent discriminative features and semantic labels. Specifically, ASCH explores the data correlations by learning a self-representation based affinity matrix. To facilitate the semantic latent features learning, an asymmetric strategy is designed by factorizing an affinity matrix into the inner product of low-dimensional label matrix and latent feature matrix. Thus, the learned latent features not only explore the underlying low-rank data structure, but also encode the semantic class information. The hash codes are jointly learned in a unified framework by collaboratively projecting the latent features and labels. Based on the hash codes, a flexible regression model with adaptive marginalization is used to learn the hash functions for out-of-sample extension. Our proposed model can be efficiently solved with linear time complexity. We compare the ASCH with state-of-the-art methods across five benchmark datasets. The experimental results validate the feasibility and superiority of our proposed method.

Keywords: Image retrieval · hashing · data affinity · collaborative regression

1 Introduction

As the image data on the internet grows exponentially, achieving efficient and accurate large-scale image data retrieval has become a critical challenge. The high dimensionality and massive scale of image data introduce significant obstacles, necessitating advanced techniques to address these issues effectively. As an approximate nearest neighbor (ANN) search approach, hashing has recently

© The Author(s), under exclusive license to Springer Nature Switzerland AG 2025
Q. Zhang et al. (Eds.): IJCRS 2025, LNAI 15710, pp. 421–442, 2025.
https://doi.org/10.1007/978-3-031-92741-6_30

been applied to tasks such as image and video retrieval [7,21,34]. Hashing compresses high-dimensional features into compact binary codes while preserving the intrinsic relationships of the original data [23,40]. This approach facilitates fast retrieval by enabling efficient similarity comparisons through simple XOR operations, significantly reducing both storage requirements and computational costs.

In recent years, a variety of advanced hashing-based retrieval approaches have emerged [5,15,25,33]. Existing hashing methods can be broadly categorized into two branches: supervised hashing [16] and unsupervised ones [1,2,8,35]. When class labels are unavailable, unsupervised hashing methods generate hash codes and functions by exploiting and preserving the underlying structure (e.g., neighborhood relations) of data. For example, spectral hashing [1] regards the hash function learning as a spectral segmentation problem, relaxing the discrete constraints on hash codes and generating binary hash codes by computing the eigenvectors of the graph Laplacian.

In contrast, supervised hashing methods leverage the prior label information to guide the learning of hash codes and functions, resulting in better retrieval performance over unsupervised ones. Some methods treat hash codes learning as a classification task by establishing the connections between hamming space and label space [28,29,41]. While this strategy of utilizing supervised information is straightforward and effective, it may not fully explore the underlying relationships among the data. Some methods pre-define an $n \times n$ similarity matrix to indicate the semantic relations between data points, and then preserve it into hash codes [31,43]. For instance, supervised adaptive similarity matrix hashing [31] adaptively learns an $n \times n$ similarity matrix for semantic preserving, but it is difficult to adapt to large-scale image retrieval due to the high computational complexity. Instead of explicitly constructing a large similarity matrix, Li et al. [15] used the inner product of label matrix for similarity preserving. These methods mainly focus on leveraging the class information, neglecting intra-class variations. As a result, they provide a coarse similarity description that may fail to capture the nuanced differences within the same class, leading to sub-optimal results. Therefore, certain methods take into account both the high-level semantic similarity in the label space and the low-level sample relations in the feature space [13,14,33]. Lei et al. [14] proposed to learn fine-grained label distributions by leveraging lables and original features for more accurate similarity construction. Lan et al. [13] incorporated data manifold structure into similarity learning. However, these methods may not excavate the intrinsic latent features embedded in the original high-dimensional feature space and also neglect the latent affinity structures among samples.

To address above mentioned issues, we propose a novel hashing model for image retrieval, termed Affinity-Based Semantic Collaborative Hashing (ASCH), which jointly explores the data affinity structure and label information for discriminative hash learning. It learns a self-representation based affinity matrix and designs an asymmetric strategy to factorize the affinity matrix into the inner product of label matrix and latent feature matrix. The latent features not

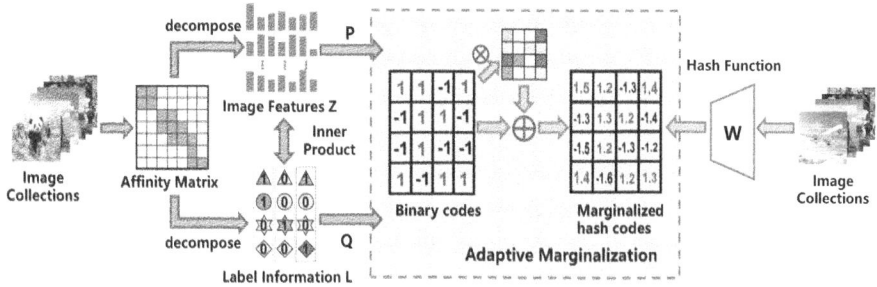

Fig. 1. The overall workflow of ASCH.

only explore the underlying low-rank data structure, but also encode the semantic class information. With the collaboration of latent features and semantic labels, the hash codes are obtained in a common Hamming space. Moreover, a discriminative regression model with adaptive margins is used to learn the hash functions for out-of-sample extension. Figure 1 illustrates the overall framework of ASCH. The main contributions of this paper are summarized as follows:

1) We propose a novel hashing model named ASCH for image retrieval. It jointly explores the data affinity structure and semantic labels to seek the latent intrinsic representation that bridges across the low-level image representation and high-level semantic labels for hash codes learning.
2) An asymmetric factorization strategy is integrated with self-representation seamlessly, which factorizes the data affinity matrix into the product of label matrix and latent representation that are used to learn binary codes. A flexible adaptive marginalized regression model is introduced for hash functions learning.
3) The proposed model can be efficiently solved with linear time complexity, allowing it scalable to large-scale data. Extensive experiments in several datasets demonstrate the effectiveness and superiority of our method.

The rest of this paper is organized as follows. In Sect. 2, some existing related works are reviewed. Section 3 presents the details of ASCH and the optimization scheme. Section 4 reports the experimental results and analysis. Section 5 concludes of this paper.

2   Related Work

In this section, some typical image hashing approaches are reviewed [1,2,6,8,9, 11,15,22,24,26,27,32,35–37].

Unsupervised approaches measure the similarities between training samples by mining the intrinsic structure and distribution of data, enabling hash learning without reliance on class labels [1,2,35]. To reduce time complexity, Spectral hashing [1], anchor graph hashing [2], and discrete graph hashing method [8] adopted the same strategy of constructing approximate adjacency graphs. Scalable graph hashing [37] introduces a bit-by-bit optimization strategy through feature transformation to achieve the optimal mapping for hash code. Unsupervised discrete hashing(UDH) [8] employs a balanced graph-based semantic loss to explore prior similarities and designs orthogonal consistency loss to enhance the independence of the hash codes. [24] leverages matrix decomposition to extract latent semantic information from original image features and generates binary codes by minimizing coding loss. The aforementioned unsupervised hashing methods enhance image retrieval accuracy to some degree. However, as data volume grows, these methods often face challenges in accurately capturing the intricate semantic structures embedded in the data.

Supervised hashing methods leverage the label information throughout the learning process. Under the guidance of high-level semantic information, such as category labels, supervised hashing can achieve higher retrieval accuracy with shorter binary codes. In order to fully utilize label information, the pairwise similarity matrix is constructed to capture the semantic correlations between training samples [15,31]. SSDH [27] aims to bridge the hamming space and the label space through a linear relationship, and employs non-relaxation optimization strategies to minimize quantization errors. COSDISH [9] extracts specific columns from similarity matrices in each iteration and then discretely learns binary encodings from semantic labels. RSLH [22] integrates mutual regression, semantic pair similarity, and relaxation strategies to improve short-length hashing through model reinforcement techniques. The aforementioned methods predominantly rely on label information, where the latent feature information embedded in original data is ignored, which may cause a loss of semantic information. To address this issue, certain approaches [31,32] consider both high-level semantic information and appropriately utilize low-level sample features. REPH [32] uses a linear autoencoder to reduce information loss during latent representation learning. Additionally, SASH [31] leverages the similarity matrix to model the relationships between features and labels, which extracts meaningful information from the label space while maintaining consistency between the feature space and the label space. However, these methods such as SASH [31] are time-consuming since the $n \times n$ similarity matrix consumes more training time to make similarity approximation. To address the issue of slow retrieval speed, some methods have been developed to optimize the efficiency of supervised hashing, such as FSDH [6], which maps class labels to corresponding hash code matrices and simplifies the solution process. FSSH [26] introduces pre-designed intermediate terms to enhance hash learning speed, while HCSDH [11] uses the Hadamard matrix to accelerate the training process. These approaches collectively improve both retrieval speed and accuracy to varying extents. Nevertheless, the accuracy of these methods is still not satisfactory. Therefore, in this paper, we aim to explore the data correlation while ensuring the speed of image hash retrieval.

3   The Proposed Method

In this section, the proposed ASCH is described in detail, including notations used in this paper, formulation of our method, alternative optimization algorithm and complexity analysis discussion.

3.1   Notations

Let $\mathbf{X} = [\mathbf{x}_1, \mathbf{x}_2, \mathbf{x}_3, ..., \mathbf{x}_n] \in \mathbb{R}^{d \times n}$ denote a training data of the image samples, where n denote the number of training samples, and d is the dimension of each image. $\mathbf{L} \in \mathbb{R}^{c \times n}$ is the ground-truth label matrix, where c is the number of classes. $\mathbf{L}_{ij} = 1$ if the j-th instance belongs to the i-th class, and 0 otherwise. $\mathbf{B} \in \mathbb{R}^{k \times n}$ represents the hash codes matrix, where k denotes the code length. Besides, some shorthand notations are explained. $Tr(\cdot)$ denotes the trace operator. $sgn(\cdot)$ is a sign function that returns -1 when the input is negative, otherwise 1. $\mathbf{I}$ denotes an identity matrix.

3.2   Model Formulation

The goal of supervised image hashing is to learn a set of hash functions to encode the high-dimensional visual features into low-dimensional discriminative binary hash codes with semantic information preserved. In this paper, we construct a latent subspace to discover the intrinsic low-dimensional features, and also bridge the original visual space and the target Hamming space. Under the collaboration of latent features and sample labels, the shared semantic-aware hash codes are obtained by dual projections for hash functions learning.

Affinity Based Discriminative Features Learning. To obtain discriminative hash codes by leveraging the label information, previous image hashing approaches usually define a pairwise similarity matrix to indicate whether any two samples are semantically similar or not, and then preserve it into hash codes. However, such a strategy mainly relies on the coarse-grained sample similarity in the label space, but ignores the fine-grained sample correlations in the visual space caused by intra-class variations. Although some methods consider the pairwise correlations among data, they generally use graph Laplacian technique for similarity preserving, which is time-consuming and unscalable to large-scale data. In this paper, we adopt self-representation to seek the latent affinity structure among data by minimizing the following objective function:

$$\min_{\mathbf{A}} \|\mathbf{X} - \mathbf{X}\mathbf{A}\|_F^2, \tag{1}$$

where $\mathbf{A} \in \mathbb{R}^{n \times n}$ is a latent subspace representation that explores the internal affinity relationship between samples. It is noted that $\mathbf{A}$ constructs an $n \times n$ affinity matrix, resulting in high computational complexity when dealing with large-scale datasets, substantially increasing the time required for hash retrieval processes. Traditional self-representation learning methods primarily focus on capturing the affinity structure within the original data, often overlooking the valuable semantic information embedded in the label matrix. This oversight can lead to a reduction in the discriminative power of the learned binary hash codes. To address these challenges, we propose an approach that employs asymmetric matrix decomposition. This technique not only circumvents the explicit computation of the affinity matrix $\mathbf{A}$, thereby enhancing computational efficiency, but also facilitates the integration of label information into the learning framework. By establishing a connection between the original data and the label information, this method ensures the preservation of semantic consistency, ultimately leading to more discriminative and efficient hash codes. We can rewrite Eq. (1) as follows:

$$\min_{\mathbf{Z}} \|\mathbf{X} - \mathbf{X}\mathbf{L}^T\mathbf{Z}\|_F^2, \tag{2}$$

where $\mathbf{L} \in \mathbb{R}^{c \times n}$ is the label matrix, and $\mathbf{Z} \in \mathbb{R}^{c \times n}$ is a latent discriminative feature matrix. By the above formulation, our latent discriminative feature matrix $\mathbf{Z}$ effectively integrates the feature information of the original data with the semantic information of the label matrix, enhancing the performance of subsequent hash code learning. The advantages of our asymmetric factorization strategy has three folds. (1) By calculating the low-dimensional representation $\mathbf{Z}$ instead of $\mathbf{A}$ (where $c \ll n$), the model can be efficiently solved with significantly lower computational complexity. (2) By enforcing $\mathbf{A} = \mathbf{L}^T\mathbf{Z}$, since $rank(\mathbf{L}^T\mathbf{Z}) \leq \min(rank(\mathbf{L}), rank(\mathbf{Z}))$, it encourages to seek the intrinsic low-rank affinity structure among samples, leading to a more robust and discriminative representation. (3) Both sample affinity structure and semantic class information are encoded into $\mathbf{Z}$, which helps to discover the latent intrinsic data representation for semantic-aware hash codes learning. Furthermore, by treating $\mathbf{D} = \mathbf{X}\mathbf{L}^T$ as a unified entity, $\mathbf{D} \in \mathbb{R}^{d \times c}$ can be interpreted as a semantic dictionary, in which each column represents a class prototype, encapsulating the latent features of a specific class. By coding the original features with this semantic dictionary, the obtained coefficients become more discriminative.

Semantic Collaborative Hashing. Based on the above latent discriminative features, we can obtain the hash codes by projecting the features into Hamming space. Considering the label matrix $\mathbf{L}$ contains the high-level semantic information, we adopt dual linear regression to project the latent features and semantic labels into a common subspace for hash codes learning.

$$\min_{\mathbf{B},\mathbf{P},\mathbf{Q}} \|\mathbf{B} - \mathbf{P}\mathbf{Z}\|_F^2 + \beta\|\mathbf{B} - \mathbf{Q}\mathbf{L}\|_F^2 + \lambda R(\mathbf{P},\mathbf{Q})$$
$$s.t.\, \mathbf{B} \in \{-1, +1\}^{k \times n}, \tag{3}$$

where β and λ are two trade-off parameters, $\mathbf{P} \in \mathbb{R}^{k \times c}$ and $\mathbf{Q} \in \mathbb{R}^{k \times c}$ are projection matrices, $\mathbf{B} \in \mathbb{R}^{k \times n}$ is the hash code matrix with code length k, and $R(\mathbf{P},\mathbf{Q}) = \|\mathbf{P}\|_F^2 + \|\mathbf{Q}\|_F^2$ is a regularization term to avoid over-fitting.

Overall Objective Function. By combining Eq. (2) and Eq. (3), we can obtain a unified framework for joint latent discriminative features learning and semantic collaborative hashing. However, real-world data often contain nonlinear structures that are overlooked by Eq. (1). To account for the nonlinear relationships in the data, it is common practice to map the data from a linear space to a nonlinear one, and kernel trick is one of the most popular techniques. In this paper, we use RBF kernel function ϕ to map the data $\mathbf{x}$ from the observed space to $\phi(\mathbf{x})$ in a reproducing kernel Hilbert space. The kernel function is defined as

$$\phi(\mathbf{x}) = \left[\exp(\frac{-\|\mathbf{x} - \mathbf{u}_1\|}{2\sigma^2}, ..., \exp(\frac{-\|\mathbf{x} - \mathbf{u}_m\|}{2\sigma^2}) \right]^T, \tag{4}$$

where $\{\mathbf{u}_i\}_{i=1}^m$ are m randomly selected anchors and σ is a bandwidth parameter defined as the mean distance between all data points and anchors. By using the kernel features, we derive our overall objective function of ASCH as follows:

$$\min_{\mathbf{Z},\mathbf{P},\mathbf{Q},\mathbf{B}} \Psi(\mathbf{Z},\mathbf{P},\mathbf{Q},\mathbf{B}) = \|\phi(\mathbf{X})(\mathbf{I} - \mathbf{L}^T\mathbf{Z})\|_F^2$$
$$+ \alpha\|\mathbf{B} - \mathbf{PZ}\|_F^2 + \beta\|\mathbf{B} - \mathbf{QL}\|_F^2 + \lambda R(\mathbf{P},\mathbf{Q}) \tag{5}$$
$$s.t. \, \mathbf{B} \in \{-1, +1\}^{k \times n}.$$

By solving the above model, we can obtain the optimal hash codes. In the following, we will present an efficient iterative algorithm to optimize it and derive a discriminative regression for an out-of-sample extension.

3.3 Optimization

We employ an alternative optimization strategy to solve the objective function iteratively [42]. Specifically, in each iteration, we decompose the overall optimization into multiple sub-problems, and solve each sub-problem w.r.t. one variable while keeping other variables fixed at each step.

Update Z: With other variables fixed, we can get the formulation w.r.t. $\mathbf{Z}$:

$$\min_{\mathbf{Z}} \|\phi(\mathbf{X})(\mathbf{I} - \mathbf{L}^T\mathbf{Z})\|_F^2 + \alpha\|\mathbf{B} - \mathbf{PZ}\|_F^2. \tag{6}$$

By setting its derivative over $\mathbf{Z}$ to 0, the above optimization problem can be converted to

$$\mathbf{LKL}^T\mathbf{Z} - \mathbf{LK} + \alpha\mathbf{P}^T\mathbf{PZ} - \alpha\mathbf{P}^T\mathbf{B} = \mathbf{0}, \tag{7}$$

where $\mathbf{K} = \phi(\mathbf{X})^T\phi(\mathbf{X})$. The optimal solution is given by

$$\mathbf{Z} = (\alpha\mathbf{P}^T\mathbf{P} + \mathbf{LKL}^T)^{-1}(\alpha\mathbf{P}^T\mathbf{B} + \mathbf{LK}). \tag{8}$$

P-Step: By fixing other variables, we can get the following problem w.r.t. $\mathbf{P}$:

$$\min_{\mathbf{P}} \alpha\|\mathbf{B} - \mathbf{PZ}\|_F^2 + \lambda\|\mathbf{P}\|_F^2. \tag{9}$$

Setting the derivative of the above equation w.r.t $\mathbf{P}$ to $\mathbf{0}$, the solution can be easily derived:

$$\alpha \mathbf{P}\mathbf{Z}\mathbf{Z}^T - \alpha \mathbf{B}\mathbf{Z}^T + \lambda \mathbf{P} = \mathbf{0}, \tag{10}$$

the result of this problem can be obtained as follows:

$$\mathbf{P} = (\alpha \mathbf{B}\mathbf{Z}^T)(\alpha \mathbf{Z}\mathbf{Z}^T + \lambda \mathbf{I})^{-1}. \tag{11}$$

Q-Step: The sub-problem of $\mathbf{Q}$ is

$$\min_{\mathbf{Q}} \beta \|\mathbf{B} - \mathbf{Q}\mathbf{L}\|_F^2 + \lambda \|\mathbf{Q}\|_F^2. \tag{12}$$

The optimal solution $\widehat{\mathbf{Q}}$ is

$$\mathbf{Q} = \beta \mathbf{B}\mathbf{L}^T(\beta \mathbf{L}\mathbf{L}^T + \lambda \mathbf{I})^{-1}. \tag{13}$$

B-Step: Fix other variables and update $\mathbf{B}$ by solving the following problems,

$$\min_{\mathbf{B}} \alpha \|\mathbf{B} - \mathbf{P}\mathbf{Z}\|_F^2 + \beta \|\mathbf{B} - \mathbf{Q}\mathbf{L}\|_F^2. \tag{14}$$

It can be equivalently converted to a trace norm maximization problem:

$$\max_{\mathbf{B}} Tr(\mathbf{U}\mathbf{B}^T), \\ s.t.\, \mathbf{B} \in \{-1, 1\}^{k \times n}. \tag{15}$$

where $\mathbf{U} = \alpha \mathbf{P}\mathbf{Z} + \beta \mathbf{Q}\mathbf{L}$. According to the definition of trace norm, it holds that $Tr(\mathbf{U}\mathbf{B}^T) = \sum_{ij} U_{ij}B_{ij}$. Therefore, the above problem can be transformed into:

$$\max_{B_{ij}} U_{ij}B_{ij}, \\ s.t.\, B_{ij} \in \{-1, 1\}^{k \times n} \tag{16}$$

for each B_{ij}. It is obvious that the maximum is obtained when $U_{ij}B_{ij} > 0$, i.e., $B_{ij} = sgn(U_{ij})$. Then we can get the optimal solution of the hash codes $\mathbf{B}$:

$$\mathbf{B} = sgn(\alpha \mathbf{P}\mathbf{Z} + \beta \mathbf{Q}\mathbf{L}). \tag{17}$$

3.4   Out-of-Sample Extension

After generating the optimal hash codes $\mathbf{B}$, ASCH needs to learn the hash functions for out-of-sample extension, which can project the data into binary codes. Some previous methods adopt the simple but effective least squares regression to learn hash functions directly, which can be written as follows:

$$\min_{\mathbf{W}} \|\mathbf{B} - \mathbf{W}\phi(\mathbf{X})\|_F^2 + \gamma \|\mathbf{W}\|_F^2, \tag{18}$$

where $\mathbf{W} \in \mathbb{R}^{k \times d}$ is the hash function to map the kernel features into hash codes. However, the binary codes are too rigid for effective regression, and the distance

Algorithm 1. ASCH Algorithm

Input: Data matrix $\mathbf{X}$, label matrix $\mathbf{L}$, hash code length k, and balance parameters
$\alpha, \beta, \lambda, \gamma$.
Output: Hash codes $\mathbf{B}$ and Hash functions $\mathbf{W}$.
 1: Map the original data into kernel space via Eq. (4);
 2: Random initialize $\mathbf{B}$, Initialize $\mathbf{P}$, $\mathbf{Q}$, $\mathbf{W}$, $\mathbf{M}$ to $\mathbf{0}$;
    % Hash Codes Learning
 3: **repeat**
 4:   Update $\mathbf{Z}$ via Eq. (8);
 5:   Update $\mathbf{P}$ via Eq. (11);
 6:   Update $\mathbf{Q}$ via Eq. (13);
 7:   Update $\mathbf{B}$ via Eq. (17);
 8: **until** Convergence;
    % Hash Functions Learning
 9: **repeat**
10:   Update $\mathbf{W}$ via Eq. (20);
11:   Update $\mathbf{M}$ via Eq. (22);
12: **until** Convergence;

---

between positive code "+1" and negative code "−1" is fixed (i.e., 2). To achieve a flexible regression, we introduce a positive margin matrix $\mathbf{M} \in \mathbb{R}^{k \times n}$ to enlarge the distances between positive code and negative code. Then, we develop Eq. (18) as follows:

$$\min_{\mathbf{W}, \mathbf{M}} \|\mathbf{B} + \mathbf{B} \odot \mathbf{M} - \mathbf{W}\phi(\mathbf{X})\|_F^2 + \gamma \|\mathbf{W}\|_F^2$$
$$s.t.\ \mathbf{M} \geq \mathbf{0}, \tag{19}$$

where $\odot$ denotes the element-wise product. Since $\mathbf{B} \in \{-1, +1\}^{k \times n}$, by introducing the $\mathbf{B} \odot \mathbf{M}$ term, the positive code becomes $+1 + m_{ij}$ and the negative code becomes $-1 - m_{ij}$, which helps to enhance the discrimination of hash functions. The above problem also can be efficiently solved by performing the alternate optimization scheme as follows:

Update W: Set the derivative of Eq. (19) w.r.t. $\mathbf{W}$ to zero, the optimal solution can be obtained by

$$\mathbf{W} = (\mathbf{B} + \mathbf{B} \odot \mathbf{M})\phi(\mathbf{X})^T (\mathbf{K} + \gamma \mathbf{I})^{-1}. \tag{20}$$

Update M: Remove the terms without $\mathbf{M}$, and the sub-problem is

$$\min_{\mathbf{M}} \|\mathbf{B} + \mathbf{B} \odot \mathbf{M} - \mathbf{W}\phi(\mathbf{X})\|_F^2\ s.t.\ \mathbf{M} \geq \mathbf{0}. \tag{21}$$

The optimal solution of Eq. (21) can be obtain by simple mathematical operations:

$$\mathbf{M} = \max\left(\mathbf{B} \odot (\mathbf{W}\phi(\mathbf{X}) - \mathbf{B}), 0\right). \tag{22}$$

After obtaining the hash function, for a new query sample $\mathbf{x}_q$, its hash codes can be generated by $\mathbf{b}_q = sgn(\mathbf{W}\phi(\mathbf{x}_q))$. The detailed procedure of ASCH is summarized in Algorithm 1.

3.5   Computational Complexity Analysis

As observed from Algorithm 1, the optimization is designed in a two-step fashion. Specifically, for hash code learning, the computational complexity primarily lies in four sub-problems, including $O(t(c^2k + cnd + c^2n + +c^3 + ckn))$ for $\mathbf{Z}$-step, $O(t(c^2k + c^2n + c^3 + ckn))$ for $\mathbf{P}$-step, $O(t(c^2k + c^2n + +c^3 + ckn))$ for $\mathbf{Q}$-step, and $O(t(knc))$ for $\mathbf{B}$-step, where t denotes the number of iterations. To obtain hash functions, obtaining $\mathbf{W}$ requires $O((k + kd + d^2)n + d^3 + d^2k)$, and updating $\mathbf{M}$ requires $O((k + kd)n)$. Given that $d, c, k \ll n$, the proposed ASCH is linear with respect to the size of the training samples. Consequently, ASCH is an efficient algorithm and can be applied to large-scale datasets.

4   Experiments

In this section, a series of experiments are performed to illustrate the effectiveness of the proposed approach on five large-scale datasets.

4.1   Datasets

Five widely-used benchmark datasets are employed to evaluate the image retrieval performance of ASCH, including MNIST [19], CIFAR-10 [11], Caltech-256 [44], Place205 [39] and NUS-WIDE [12]. **MNIST** comprises handwritten digits from various individuals, totaling 70,000 images, with 69000 for training and 1000 for testing. Each image is a 28×28 pixel representation. **CIFAR-10** consists of 60,000 samples, each a 32×32 pixel color image. The dataset is divided into 59000 training images and 1000 test images, spanning 10 categories labeled from 0 to 9. **Caltech-256** is an object recognition dataset comprising 29,780 real-world images of varying sizes, across 257 classes (256 object categories and one clutter category). Each class contains at least 80 images. **Place205** is a large-scale scene-centric dataset, which contains 104,100 samples covering 205 common scene categories. **NUS-WIDE** is a large-scale, multi-label web image dataset sourced from Flickr. Images are represented by 500-dimensional Bag-of-Words (BoW) features. This study uses 195,834 images tagged with 21 frequently occurring labels, represented by 500-dimensional BoW feature vectors. Table 1 lists the general statistics of the five datasets.

4.2   Baselines and Implementation Details

To illustrate the effectiveness of ASCH, we compare it with the state-of-the-art hashing approaches, including SDH (2015) [29], COSDISH (2016) [9], FSDH (2017) [6], FSSH (2018) [26], SCDH (2020) [4], POPSH (2021) [45], OLGH (2022) [44] and REPH (2024) [32]. The codes of all baselines are publicly available. Several widely-recognized metrics are employed to assess the performance, including mean Average Precision (mAP), top-50 precision (Precision@50), NDCG@100, and Precision-Recall (PR) curves [30,38]. The hyper-parameters of our method

Table 1. General statistics of five datasets used in experiments.

Statistics	MNIST	CIFAR-10	Caltech-256	Place205	NUS-WIDE
Total	70,000	60,000	29,780	104,100	195,834
Training	69,000	5,9000	28780	100,000	193,734
Query	1,000	1,000	1,000	4,100	2,100
Categories	10	10	256	205	21
Features	784	512	1024	128	500

are set as $\alpha = 1e - 4$, $\beta = 1e3$, $\lambda = 1e - 4$, and $\gamma = 1e - 3$. The dimension of kernel features is set to 1,500 [38]. The maximum iterative number t is set to 10. All the experiments are conducted on a personal computer using MATLAB R2023b with Win10 system, Intel i5-12500H CPU @2.50GHz, 64GB RAM.

4.3   Image Retrieval Results

Table 2 reports the mAP values of ASCH and all baselines on five datasets with code lengths ranging from 8 bits to 512 bits, and – indicates that the corresponding method is not applicable in this case due to its excessively long running time. We can obtain the following observations. (1) The proposed ASCH outperforms existing hashing methods across nearly all hash lengths. For example, with 64-bit hash codes, ASCH achieves improvements of 0.49%, 1.02%, 3.42%, 1.71%, and 0.20% over the second-best competitors on the five datasets, respectively. These results indicate that the affinity-based semantic collaborative hashing method effectively establishes the association between original feature information and label information, enhancing category differentiation and thereby improving overall retrieval performance across all datasets. (2) The performance of ASCH and most baselines improves progressively with increasing bit length, indicating that the representation with longer lengths can preserve more semantic information. Notably, some methods like OLGH experience performance degradation when the hash length exceeds 128 bits, showing its instability for hash learning. (3) For large-scale datasets (NUS-WIDE and Place205) or datasets with significant variations (Caltech256 and Place205), retrieval performance is particularly limited for short hash codes. For instance, methods like REPH with 8-bit codes do not perform optimally on some datasets due to substantial information loss. Since 8-bit binary code images contain less information, information loss is relatively high. However, ASCH remains stable due to the collaborative semantic learning framework, enabling superior performance even with relatively short hash codes.

Table 3 records the Precision@50 results of ASCH and all other compared methods on five datasets. Similar to the mAP results, ASCH outperforms nearly all baselines. On Caltech-256, when the length of hash codes increases from 16 bits to 128 bits, ASCH improves retrieval performance by 3.36%, 4.63%, 2.36%, and 2.42%, respectively. Retrieval performance is limited at 8 bits, especially for large-scale datasets(such as NUS-WIDE) and datasets with significant variations

Table 2. Comparison of mAP values on five datasets(The best results are highlighted in bold).

Dataset	Method	8bits	16bits	32bits	64bits	128bits	256bits	512bits
MNIST	SDH	0.7326	0.7483	0.8025	0.8190	0.8217	0.8171	–
	COSDISH	0.8973	0.9184	0.9265	0.9304	0.9341	0.9297	–
	FSDH	0.8988	0.9089	0.9165	0.9181	0.9228	0.9236	0.9228
	FSSH	0.9276	0.9303	0.9352	0.9506	0.9501	0.9536	0.9543
	SCDH	0.9042	0.9101	0.9209	0.9186	0.9250	0.9258	0.9266
	POPSH	0.9285	0.9439	0.9535	0.9557	0.9588	0.9591	0.9548
	OLGH	0.9322	0.9530	0.9525	0.9393	0.8676	0.7711	0.6646
	REPH	0.9546	0.9656	0.9685	0.9687	0.9706	0.9732	0.9733
	ASCH	**0.9620**	**0.9698**	**0.9754**	**0.9736**	**0.9733**	**0.9746**	**0.9775**
CIFAR-10	SDH	0.3703	0.4162	0.4578	0.4839	0.5012	0.5090	–
	COSDISH	0.5363	0.5735	0.6219	0.6391	0.6306	0.6371	–
	FSDH	0.5791	0.6160	0.6553	0.6686	0.6833	0.6901	0.6926
	FSSH	0.5900	0.6298	0.6810	0.6983	0.7089	0.7152	0.7152
	SCDH	0.6411	0.6805	**0.7279**	0.7259	0.7347	0.7360	0.7409
	POPSH	0.6097	0.6580	0.7171	0.6994	0.7234	0.7249	0.7285
	OLGH	0.6252	0.6503	0.5720	0.4771	0.3739	0.3129	0.2725
	REPH	0.6289	0.6993	0.7178	0.7273	0.7304	0.7347	0.7405
	ASCH	**0.6629**	**0.7041**	0.7259	**0.7375**	**0.7381**	**0.7441**	**0.7425**
Caltech-256	SDH	0.1560	0.2901	0.3705	0.4191	0.4531	0.4718	–
	COSDISH	0.1438	0.2575	0.4271	0.5949	0.7086	0.7414	–
	FSDH	0.2214	0.3020	0.2998	0.2983	0.2990	0.3006	0.3001
	FSSH	0.1407	0.4467	0.5598	0.6170	0.6258	0.6546	0.6689
	SCDH	0.2532	0.4380	0.5558	0.6293	0.6798	0.6959	0.7037
	POPSH	**0.3356**	0.5536	0.6553	0.7094	0.7559	0.7554	0.7662
	OLGH	0.2842	0.4961	0.5595	0.5141	0.4789	0.4527	0.4231
	REPH	0.2640	0.5081	0.6465	0.7231	0.7595	0.7873	0.7923
	ASCH	0.3122	**0.5907**	**0.7115**	**0.7573**	**0.7828**	**0.7938**	**0.8058**
Place205	SDH	0.0616	0.1457	0.2022	0.2491	0.2780	0.2950	–
	COSDISH	0.0824	0.2080	0.3553	0.4502	0.4913	0.5116	–
	FSDH	0.1906	0.2708	0.2716	0.2730	0.2719	0.2728	0.2719
	FSSH	0.0217	0.0508	0.1210	0.2246	0.2770	0.4586	0.4859
	SCDH	0.1936	0.3629	0.4522	0.5111	0.5393	0.5568	0.5613
	POPSH	0.2169	0.3682	0.4610	0.5158	0.5416	0.5564	0.5644
	OLGH	0.2070	0.3295	0.3100	0.2267	0.1886	0.1576	0.1356
	REPH	0.1701	0.3612	0.4682	0.5199	0.5432	0.5591	0.5646
	ASCH	**0.2246**	**0.4130**	**0.4993**	**0.5370**	**0.5573**	**0.5656**	**0.5719**
NUS-WIDE	SDH	0.4335	0.4383	0.4679	0.4739	0.4769	0.4801	–
	COSDISH	–	–	–	–	–	–	–
	FSDH	0.3398	0.3449	0.3580	0.3874	0.4139	0.3942	0.3549
	FSSH	0.4502	0.4274	0.5060	0.3825	0.4877	0.4467	0.5129
	SCDH	0.4825	0.4894	0.4951	0.4983	0.4968	0.5013	0.5021
	POPSH	0.3144	0.3144	0.3144	0.3144	0.3144	0.3144	0.3144
	OLGH	0.5319	0.5560	0.5413	0.5102	0.4776	0.4424	0.4100
	REPH	0.5721	0.6009	0.6266	0.6444	0.6458	0.6512	0.6545
	ASCH	**0.5774**	**0.6098**	**0.6309**	**0.6464**	**0.6539**	**0.6585**	**0.6614**

(such as Caltech256). While ASCH's performance is slightly lower than the top-performing method, it remains relatively strong.

Table 4 lists the NDCG@100 values of ASCH and all baselines on five datasets. ASCH still demonstrates superior performance across all comparisons

Table 3. Performance Comparison of Precision@50 values on five datasets(The best results are highlighted in bold).

Dataset	Method	8bits	16bits	32bits	64bits	128bits	256bits	512bits
MNIST	SDH	0.8076	0.8268	0.8701	0.8908	0.8997	0.9012	
	COSDISH	0.8813	0.9027	0.9071	0.9099	0.9147	0.9096	
	FSDH	0.8881	0.8962	0.9020	0.9036	0.9070	0.9064	0.9030
	FSSH	0.9164	0.9136	0.9192	0.9366	0.9354	0.9384	0.9395
	SCDH	0.8855	0.8913	0.8997	0.8972	0.9034	0.9020	0.9030
	POPSH	0.9205	0.9332	0.9419	0.9446	0.9473	0.9465	0.9471
	OLGH	0.9243	0.9442	0.9492	0.9535	0.9504	0.9506	0.9428
	REPH	0.9480	0.9575	0.9585	0.9589	**0.9690**	0.9640	0.9640
	ASCH	**0.9565**	**0.9634**	**0.9693**	**0.9667**	0.9665	**0.9674**	**0.9710**
CIFAR-10	SDH	0.4484	0.5116	0.5467	0.5571	0.5722	0.5755	
	COSDISH	0.4747	0.4985	0.5429	0.5506	0.5423	0.5495	
	FSDH	0.5227	0.5573	0.5932	0.5980	0.6107	0.6148	0.6153
	FSSH	0.5435	0.5701	0.6170	0.6260	0.6346	0.6404	0.6430
	SCDH	0.5963	0.6248	0.6630	0.6620	0.6684	0.6683	0.6744
	POPSH	0.5564	0.5990	0.6595	0.6305	0.6546	0.6533	0.6581
	OLGH	0.5583	0.6106	0.5770	0.5667	0.5263	0.4920	0.4651
	REPH	0.5728	0.6445	0.6600	0.6630	0.6636	0.6675	0.6726
	ASCH	**0.6124**	**0.6459**	**0.6666**	**0.6730**	**0.6703**	**0.6726**	**0.6825**
Caltech-256	SDH	0.1879	0.3924	0.4912	0.5310	0.5614	0.5804	
	COSDISH	0.1389	0.2416	0.4054	0.5683	0.6756	0.7032	
	FSDH	0.2161	0.2940	0.2929	0.2920	0.2920	0.2940	0.2920
	FSSH	0.1290	0.4285	0.5364	0.5915	0.5925	0.6142	0.6275
	SCDH	0.2380	0.4170	0.5315	0.5984	0.6422	0.6482	0.6570
	POPSH	**0.3203**	0.5332	0.6331	0.6789	0.7213	0.7171	0.7242
	OLGH	0.2658	0.4808	0.5804	0.5738	0.5795	0.5794	0.5701
	REPH	0.2454	0.4894	0.6254	0.6973	0.7239	0.7516	0.7556
	ASCH	0.2982	**0.5668**	**0.6794**	**0.7209**	**0.7481**	**0.7578**	**0.7698**
Place205	SDH	0.1019	0.2670	0.3440	0.3854	0.4016	0.4095	
	COSDISH	0.0740	0.1650	0.3038	0.3978	0.4268	0.4391	
	FSDH	0.1814	0.2622	0.2631	0.2624	0.2628	0.2624	0.2620
	FSSH	0.0187	0.0403	0.0997	0.1913	0.2345	0.3927	0.4094
	SCDH	0.1776	0.3386	0.4225	0.4655	0.4807	0.4905	0.4898
	POPSH	0.1979	0.3441	0.4283	0.4686	0.4831	0.4887	0.4925
	OLGH	0.1890	0.3189	0.3484	0.3296	0.3190	0.3058	0.2954
	REPH	0.1475	0.3389	0.4383	**0.4761**	0.4849	0.4927	0.4948
	ASCH	**0.2045**	**0.3706**	**0.4390**	0.4678	**0.4879**	**0.4944**	**0.4979**
NUS-WIDE	SDH	0.4940	0.4881	0.5070	0.5259	0.5358	0.5418	
	COSDISH	-	-	-	-	-	-	
	FSDH	0.2865	0.4010	0.4656	0.5124	0.3262	0.4408	0.4981
	FSSH	0.4607	0.4515	0.5905	0.4405	0.5376	0.4921	0.6203
	SCDH	0.4886	0.5531	0.5613	0.5475	0.5412	0.5726	0.5747
	POPSH	0.3635	0.3635	0.3635	0.3635	0.3635	0.3635	0.3635
	OLGH	0.5014	0.4879	0.4871	0.5130	0.5297	0.5324	0.5241
	REPH	**0.6493**	0.6462	0.6876	0.6982	0.7207	0.7339	0.7319
	ASCH	0.6301	**0.7077**	**0.7324**	**0.7466**	**0.7508**	**0.7535**	**0.7628**

on five datasets. Notably, it excels on large-scale datasets such as NUS-WIDE and Caltech256, affirming its capability in handling extensive datasets. Specifically, compared to competitors, ASCH achieves absolute improvements of 3.51%, 4.88%, 2.51%, and 2.41% for hash lengths ranging from 16 bits to 128 bits on Caltech256. This superior performance is attributed to the ability of ASCH to

Table 4. Performance Comparison of NDCG@100 values on five datasets(The best results are highlighted in bold).

Dataset	Method	8bits	16bits	32bits	64bits	128bits	256bits	512bits
MNIST	SDH	0.8097	0.8224	0.8691	0.8878	0.8969	0.8989	–
	COSDISH	0.8811	0.9027	0.9070	0.9098	0.9146	0.9096	–
	FSDH	0.8884	0.8964	0.9018	0.9036	0.9070	0.9064	0.9030
	FSSH	0.9164	0.9138	0.9194	0.9366	0.9355	0.9385	0.9395
	SCDH	0.8857	0.8910	0.8997	0.8969	0.9032	0.9020	0.9030
	POPSH	0.9201	0.9331	0.9419	0.9446	0.9470	0.9465	0.9471
	OLGH	0.9233	0.9435	0.9482	0.9525	0.9482	0.9466	0.9337
	REPH	0.9482	0.9576	0.9588	0.9589	**0.9690**	0.9640	0.9640
	ASCH	**0.9571**	**0.9635**	**0.9695**	**0.9669**	0.9665	**0.9674**	**0.9710**
CIFAR-10	SDH	0.4563	0.5123	0.5446	0.5579	0.5724	0.5768	–
	COSDISH	0.4759	0.4981	0.5430	0.5505	0.5419	0.5497	–
	FSDH	0.5212	0.5562	0.5936	0.5980	0.6110	0.6149	0.6154
	FSSH	0.5451	0.5707	0.6168	0.6259	0.6345	0.6404	0.6430
	SCDH	0.5958	0.6243	0.6630	0.6623	0.6684	0.6684	0.6744
	POPSH	0.5556	0.5971	0.6606	0.6308	0.6557	0.6540	0.6592
	OLGH	0.5596	0.6083	0.5754	0.5614	0.5151	0.4744	0.4390
	REPH	0.5723	0.6447	0.6593	0.6628	0.6635	0.6673	0.6725
	ASCH	**0.6122**	**0.6463**	**0.6666**	**0.6728**	**0.6705**	**0.6727**	**0.6825**
Caltech-256	SDH	0.1867	0.3454	0.4343	0.4758	0.5052	0.5215	–
	COSDISH	0.1418	0.2505	0.4153	0.5724	0.6769	0.7054	–
	FSDH	0.2173	0.2940	0.2930	0.2921	0.2920	0.2940	0.2922
	FSSH	0.1315	0.4302	0.5387	0.5930	0.5953	0.6167	0.6298
	SCDH	0.2388	0.4177	0.5335	0.6002	0.6442	0.6499	0.6588
	POPSH	**0.3239**	0.5354	0.6347	0.6807	0.7235	0.7188	0.7276
	OLGH	0.2694	0.4802	0.5609	0.5343	0.5209	0.5064	0.4878
	REPH	0.2487	0.4907	0.6270	0.6989	0.7269	0.7548	0.7584
	ASCH	0.2999	**0.5705**	**0.6835**	**0.7240**	**0.7510**	**0.7603**	**0.7722**
Place205	SDH	0.1045	0.2635	0.3408	0.3820	0.3985	0.4055	–
	COSDISH	0.0732	0.1767	0.3188	0.4001	0.4268	0.4391	–
	FSDH	0.1820	0.2622	0.2631	0.2624	0.2628	0.2624	0.2620
	FSSH	0.0190	0.0402	0.0995	0.1912	0.2345	0.3927	0.4095
	SCDH	0.1774	0.3386	0.4224	0.4655	0.4808	0.4905	0.4897
	POPSH	0.1988	0.3439	0.4282	0.4685	0.4831	0.4886	0.4925
	OLGH	0.1880	0.3161	0.3429	0.3145	0.2988	0.2816	0.2676
	REPH	0.1491	0.3387	0.4384	**0.4763**	0.4850	0.4928	0.4949
	ASCH	**0.2046**	**0.3708**	**0.4390**	0.4675	**0.4878**	**0.4943**	**0.4980**
NUS-WIDE	SDH	0.3708	0.3681	0.3864	0.3992	0.4085	0.4151	–
	COSDISH	–	–	–	–	–	–	–
	FSDH	0.1949	0.2967	0.3399	0.3692	0.2064	0.3237	0.4071
	FSSH	0.3413	0.3441	0.4190	0.3286	0.4103	0.3612	0.4342
	SCDH	0.3875	0.4258	0.4315	0.4318	0.4350	0.4295	0.4292
	POPSH	0.2773	0.2773	0.2773	0.2773	0.2773	0.2773	0.2773
	OLGH	0.3812	0.3824	0.3800	0.4004	0.4098	0.4104	0.3991
	REPH	0.4140	0.4497	0.5003	**0.5309**	0.5305	0.5434	0.5426
	ASCH	**0.4440**	**0.4898**	**0.5051**	0.5217	**0.5463**	**0.5530**	**0.5554**

establish a robust linkage between the original feature data and the label data, thereby generating more precise hash codes. Most NDCG@100 results exhibit a gradual increase with longer hash lengths, but certain methods show a decline in performance with increased code lengths.

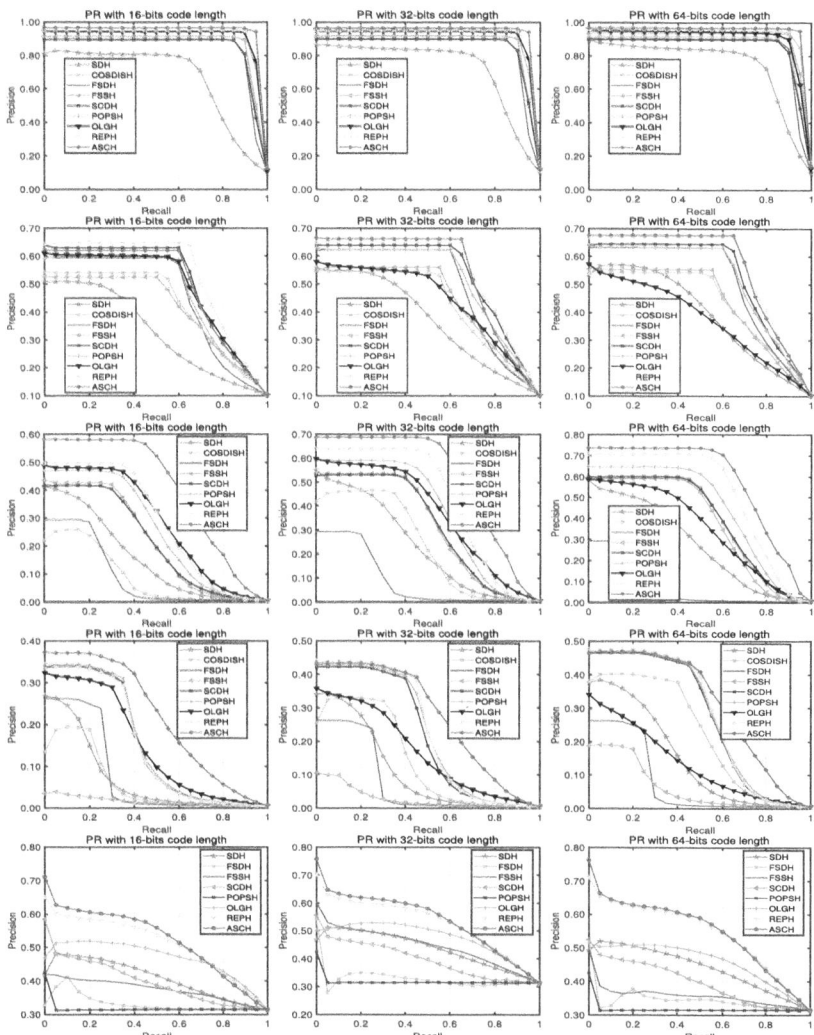

Fig. 2. PR curves on five datasets with 16, 32, and 64-bit hash codes.

The Precision-Recall (PR) curves for all compared methods across five datasets with varying hash lengths (16 bits, 32 bits, and 64 bits) are presented in Fig. 2. From these figures, ASCH overall outperforms other methods. This observation underscores the efficacy and superiority of ASCH in generating more discriminative hash codes. The PR results further validate the effectiveness of our collaborative semantic learning and affinity self-representation learning structure.

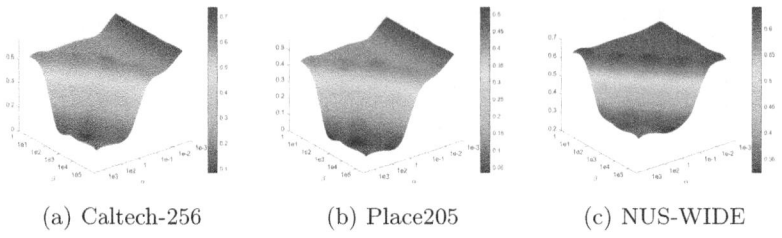

(a) Caltech-256   (b) Place205   (c) NUS-WIDE

Fig. 3. mAP values versus α and β on Caltech-256, Place205 and NUS-WIDE.

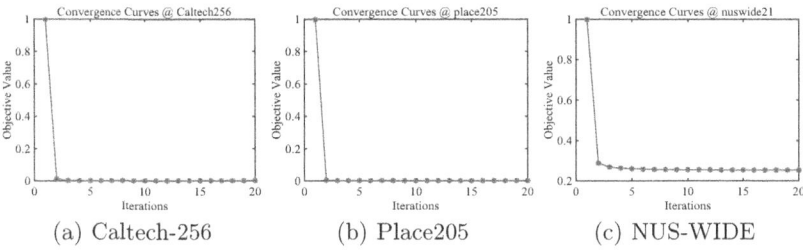

(a) Caltech-256   (b) Place205   (c) NUS-WIDE

Fig. 4. Convergence curves on Caltech-256, Place205 and NUS-WIDE.

4.4   Parameter Analysis

In this section, experiments have been conducted on Caltech-256, Place205 and NUS-WIDE to test the sensitivity of parameters, including α and β in hash codes learning stage. Firstly, we define a candidate set for each parameter:

* $\alpha : \{10^{-4}, 10^{-3}, 10^{-2}, 10^{-1}, 1, 10^2, 10^3\}$
* $\beta : \{1, 10^1, 10^2, 10^3, 10^4, 10^5\}$

In this test, two parameters α and β are varied within their respective candidate sets to observe the changes in retrieval performance, while the regularization parameters λ and γ are fixed at their optimal values of 10^{-4} and 10^{-1}, respectively. We set the hash code length to 32, and plot the mAP values in Fig. 3. Parameter α influences the quantization term of hash codes learning in latent semantic space $\mathbf{Z}$. We can observe that the mAP values drop significantly in all cases when α reaches 1. However, when α is located in a wide range of $[10^{-4}, 10^{-1}]$, the mAP results are stable. Parameter β affects the quantization term in the label information $\mathbf{L}$. It can be found that when β is less than 1e3, the retrieval results are suboptimal. Jointly considering five datasets, ASCH can yield good results when α is set to $[10^{-4}, 10^{-3}]$ and β is set to 10^3.

4.5   Ablation Study

In order to obtain a deep insight into the proposed ASCH method, ablation studies are conducted to evaluate the individual contributions of different components. We compare the ASCH model with three variants: ASCH-A, ASCH-C,

Table 5. Comparison of ablation results of ASCH on NUS-WIDE(The best results are highlighted in bold).

Methods	8bits	16bits	32bits	64bits	128bits	256bits	512bits
ASCH-A	0.5218	0.5531	0.5840	0.6063	0.6253	0.6348	0.6279
ASCH-C	0.5717	0.5936	0.6225	0.6318	0.6496	0.6533	0.6538
ASCH-M	0.5688	0.6061	0.6240	0.6359	0.6498	0.6531	0.6541
ASCH	**0.5774**	**0.6098**	**0.6309**	**0.6464**	**0.6539**	**0.6585**	**0.6614**

Table 6. Training time (in second) comparison of various approaches on Caltech-256 and NUS-WIDE.

Dataset	Method	8bits	16bits	32bits	64bits	128bits	256bits	512bits
Caltech-256	SDH	9.22	12.18	16.96	29.03	62.32	310.06	–
	COSDISH	0.96	1.75	4.25	15.44	59.14	423.85	–
	FSDH	3.00	2.97	3.15	3.42	4.01	5.20	7.38
	FSSH	1.81	1.89	1.98	2.19	2.49	3.08	6.50
	SCDH	7.74	7.90	8.35	9.01	10.22	35.16	167.08
	POPSH	1.44	1.46	1.58	1.78	2.15	3.14	5.32
	OLGH	2.31	2.40	2.41	2.62	2.99	3.43	4.80
	REPH	8.46	8.39	8.96	9.09	10.15	12.80	17.91
	ASCH	6.31	6.35	6.61	7.07	7.77	9.39	13.14
NUS-WIDE	SDH	54.61	60.04	81.56	168.70	480.32	1924.28	–
	COSDISH	–	–	–	–	–	–	–
	FSDH	18.28	19.03	19.90	21.53	24.35	31.31	43.68
	FSSH	17.73	18.06	18.48	20.32	22.33	27.25	34.73
	SCDH	32.24	33.13	43.68	76.19	193.30	570.43	2183.87
	POPSH	9.29	9.27	9.66	10.78	12.92	17.12	25.08
	OLGH	11.94	11.92	12.16	13.17	14.50	17.72	24.25
	REPH	48.98	49.90	51.93	55.63	62.90	77.60	106.90
	ASCH	27.05	27.57	28.52	30.78	36.28	45.58	62.12

and ASCH-M. In specific, ASCH-A only adopts the semantic labels for hash codes learning, i.e.,

$$\min_{\mathbf{Q},\mathbf{B}} \|\mathbf{B} - \mathbf{Q}\mathbf{L}\|_F^2 + \lambda \|\mathbf{Q}\|_F^2 \quad s.t.\, \mathbf{B} \in \{-1,1\}^{k \times n}. \tag{23}$$

ASCH-C discards the latent semantic representation learning, and directly uses collaborative regression of kernel features and labels for hash learning, i.e.,

$$\min_{\mathbf{P},\mathbf{Q},\mathbf{B}} \|\mathbf{B} - \mathbf{P}\phi(\mathbf{X})\|_F^2 + \alpha \|\mathbf{B} - \mathbf{Q}\mathbf{L}\|_F^2 + \lambda(\|\mathbf{P}\|_F^2 + \|\mathbf{Q}\|_F^2)$$
$$s.t.\, \mathbf{B} \in \{-1,1\}^{k \times n}. \tag{24}$$

ASCH-M evaluates the effectiveness of adaptive marginalized regression, and it adopts the least square regression for hash function learning, i.e.,

$$\min_{\mathbf{W}} \|\mathbf{B} - \mathbf{W}\phi(\mathbf{X})\|_F^2 + \gamma \|\mathbf{W}\|_F^2. \tag{25}$$

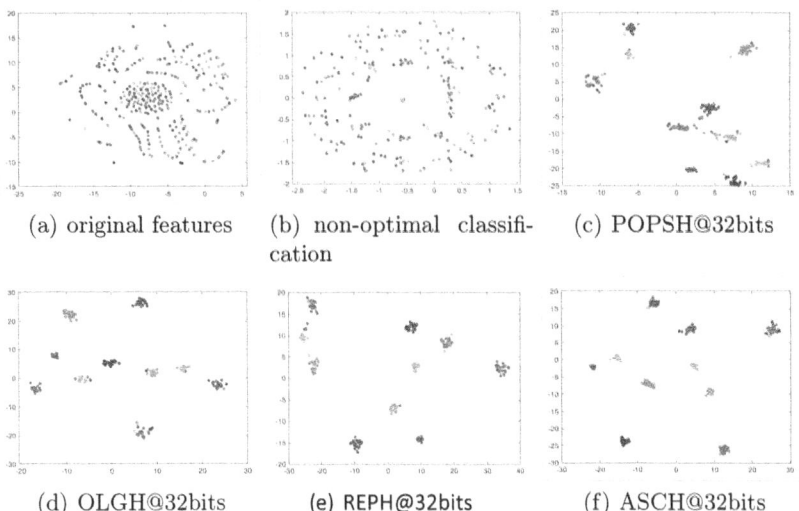

(a) original features   (b) non-optimal classifi-   (c) POPSH@32bits
                          cation

(d) OLGH@32bits   (e) REPH@32bits   (f) ASCH@32bits

Fig. 5. t-SNE visualization of original features, non-optimal classification, and the 32-bit codes of POPSH, OLGH, REPH, and ASCH on MNIST.

Table 5 shows the mAP values of ASCH and its variants on NUS-WIDE with different code lengths. ASCH yields the best retrieval results in all cases, verifying the effectiveness of self-representation learning, collaborative semantic similarity presentation and the marginalized regression. ASCH-A performs the worst, since it ignores the data features in hash codes learning. ASCH-C obtains better results than ASCH-A, but lower than ASCH, emphasizing the effectiveness of our proposed latent semantic features learning based on self-representation and asymmetric matrix factorization. Although ASCH-M achieves comparable results, it is still inferior to ASCH, indicating that the marginalized matrix contributes to more discriminative hash functions learning.

4.6   Convergence and Training Time

In this subsection, we investigate the convergence property of ASCH algorithm. It is clear that the value of the objective function $\Psi(\mathbf{Z}, \mathbf{P}, \mathbf{Q}, \mathbf{B})$ of Eq. (5) will decrease from iteration to iteration until it is stabilized, which ensures the theoretical feasibility of the subsequent convergence analysis:

$$
\begin{aligned}
\Psi(\mathbf{Z}^t, \mathbf{P}^t, \mathbf{Q}^t, \mathbf{B}^t) &\geq \Psi(\mathbf{Z}^{t+1}, \mathbf{P}^t, \mathbf{Q}^t, \mathbf{B}^t) \\
&\geq \Psi(\mathbf{Z}^{t+1}, \mathbf{P}^{t+1}, \mathbf{Q}^t, \mathbf{B}^t) \\
&\geq \Psi(\mathbf{Z}^{t+1}, \mathbf{P}^{t+1}, \mathbf{Q}^{t+1}, \mathbf{B}^t) \\
&\geq \Psi(\mathbf{Z}^{t+1}, \mathbf{P}^{t+1}, \mathbf{Q}^{t+1}, \mathbf{B}^{t+1}).
\end{aligned}
\tag{26}
$$

Then we plot the convergence curves conducted on the Caltech-256, Place205, and NUS-WIDE datasets with 32-bit codes. For better observation, the objective

Table 7. MAP Comparisons with Deep-Hashing Models on CIFAR-10.

Methods	12bits	24bits	32bits	64bits
DSDH	0.7400	0.7860	0.8010	0.8200
DPSH	0.6818	0.7204	0.7341	0.7464
DTQ	0.7311	0.7410	0.7645	0.7736
DPQ	0.7410	0.7528	0.7523	0.7525
DCRH	0.7353	0.7439	0.7625	0.7646
POPSH	0.7501	0.7998	0.8178	0.8271
ASCH	0.7169	0.7594	0.7663	0.7748

function values are normalized by dividing them by their maximum value. As illustrated in Fig. 4, the objective function value monotonically drops in the initial iteration and then stabilizes within five iterations, demonstrating the good convergence characteristic.

Table 6 exhibits the training time (seconds) of all approaches on CIFAR-10, Place205 and NUS-WIDE to further assess the efficiency of ASCH. Generally speaking, FSSH exhibits the fastest training speed due to its utilization of pre-designed intermediate terms to enhance the speed of hash learning. FSDH is efficient on small datasets, but its training time increases significantly with larger datasets. SDH and COSDISH are efficient with short codes, but their training time escalates rapidly as the length of hash code bits increases. COSDISH is not scalable to Caltech-256 with long codes and large NUS-WIDE dataset. Generally, our ASCH is efficient, and can achieve superior performance compared to other baselines with an acceptable training time.

4.7   Visualization

In this subsection, we perform visualization on the MNIST dataset to evaluate the capability of our model in capturing semantic features. We compare our model with POPSH, OLGH, and REPH, which also demonstrate strong performance in the aforementioned comparative experiments. The t-SNE tool is utilized to visualize the original features and the learned semantic features, as illustrated in Fig. 5. It is evident that the original sample features are difficult to distinguish. Compared to POPSH, OLGH, and REPH, ASCH yields more intuitive results with clear intra-class cohesion and inter-class separability, showing the effectiveness of our method.

4.8   Comparisons with Deep Hashing Methods

Deep hashing has been widely recognized for its superior performance in image retrieval. To further evaluate the effectiveness of the proposed method, we conduct experiments on the CIFAR-10 dataset. We compare the mean average precision (MAP) performance of six prominent deep hashing models: DSDH [17], DPSH [18], DTQ [20], DPQ [10], DCRH [3], and POPSH [45]. For a fair comparison, we use AlexNet as the backbone for all deep-hashing methods. For the models being compared, we either re-implement the released code or directly refer to the experimental results presented in the original papers. In our approach, we use the fine-tuned parameters of the AlexNet network from DPSH [18] to extract image features from the final fully connected layer, resulting in 4096-dimensional feature vectors, which are then input into our method. The MAP comparison results on the CIFAR-10 dataset are summarized in Table 7. As shown, our method generally demonstrates excellent performance when compared to the deep hashing models, thus confirming the effectiveness of the proposed learning framework.

5   Conclusion

In this paper, we present a novel hashing method called ASCH for image retrieval. ASCH learns a self-representation based affinity matrix to explore the data correlations. To facilitate the semantic latent features learning, an asymmetric strategy is designed by factorizing the affinity matrix into the inner product of low-dimensional label matrix and latent feature matrix. The learned latent features not only explore the underlying low-rank data correlations, but also encode the semantic class information. The hash codes are obtained by collaboratively projecting the latent features and labels. A flexible regression model is used to learn the hash functions for out-of-sample extension. Extensive experimental results on five datasets demonstrate the effectiveness of our proposed method.

Acknowledgements. This work was partially supported by the National Natural Science Foundation of China (Nos. 62176116, 62276136), and the Natural Science Foundation of the Jiangsu Higher Education Institutions of China (No. 20KJA520006).

References

1. Balasundaram, R., Sudha, G.F.: Retrieval performance analysis of multibiometric database using optimized multidimensional spectral hashing based indexing. J. King Saud Univ. Comput. Inf. Sci. **33**(1), 110–117 (2021)
2. Chen, R., Wang, H.: Supervised consensus anchor graph hashing for cross modal retrieval. IEEE Access **12**, 1805–1821 (2024)
3. Chen, Y., Lu, X.: Deep category-level and regularized hashing with global semantic similarity learning. IEEE Trans. Cybern. **51**(12), 6240–6252 (2020)
4. Chen, Y., Tian, Z., Zhang, H., Wang, J., Zhang, D.: Strongly constrained discrete hashing. IEEE Trans. Image Process. **29**, 3596–3611 (2020)
5. Fan, W., Zhang, C., Li, H., Jia, X., Wang, G.: Three-stage semisupervised cross-modal hashing with pairwise relations exploitation. IEEE Trans Neural Netw. Learn. Syst. **36**(1), 260–273 (2025)
6. Gui, J., Liu, T., Sun, Z., Tao, D., Tan, T.: Fast supervised discrete hashing. IEEE Trans. Pattern Anal. Mach. Intell. **40**(2), 490–496 (2017)

7. Huang, F., Zhang, L., Gao, X.: Domain adaptation preconceived hashing for unconstrained visual retrieval. IEEE Trans Neural Netw. Learn. Syst. **33**(10), 5641–5655 (2022)

8. Jin, S., Yao, H., Zhou, Q., Liu, Y., Huang, J., Hua, X.: Unsupervised discrete hashing with affinity similarity. IEEE Trans. Image Process. **30**, 6130–6141 (2021)

9. Kang, W.-C., Li, W.-J., Zhou, Z.-H.: Column sampling based discrete supervised hashing. In: Proceedings of the AAAI Conference on Artificial Intelligence, vol. 30 (2016)

10. Klein, B., Wolf, L.: End-to-end supervised product quantization for image search and retrieval. In: Proceedings of the IEEE/CVF Conference on CVPR, pp. 5041–5050 (2019)

11. Koutaki, G., Shirai, K., Ambai, M.: Hadamard coding for supervised discrete hashing. IEEE Trans. Image Process. **27**(11), 5378–5392 (2018)

12. Lai, Z., Chen, Y., Wu, J., Wong, W.K., Shen, F.: Jointly sparse hashing for image retrieval. IEEE Trans. Image Process. **27**(12), 6147–6158 (2018)

13. Lan, R., Tan, Y., Wang, X., Liu, Z., Luo, X.: Label guided discrete hashing for cross-modal retrieval. IEEE Trans. Intell. Transp. Syst. **23**(12), 25236–25248 (2022)

14. Lei, F., Zhang, C., Li, H., Gao, Y., Chen, C.: Label distribution guided hashing for cross-modal retrieval. ACM Trans. Knowl. Discov. Data **19**(1), 1–23 (2024)

15. Li, H., Zhang, C., Jia, X., Gao, Y., Chen, C.: Adaptive label correlation based asymmetric discrete hashing for cross-modal retrieval. IEEE Trans. Knowl. Data Eng. **35**(2), 1185–1199 (2021)

16. Li, L., Shu, Z., Yu, Z., Wu, X.-J.: Robust online hashing with label semantic enhancement for cross-modal retrieval. Pattern Recogn. **145**, 109972 (2024)

17. Li, Q., Sun, Z., He, R., Tan, T.: Deep supervised discrete hashing. Adv. Neural. Inf. Process. Syst. **27**(12), 5996–6009 (2018)

18. Li, W.-J., Wang, S., Kang, W.-C.: Feature learning based deep supervised hashing with pairwise labels, arXiv preprint arXiv:1511.03855

19. Li, X., Hu, D., Nie, F.: Large graph hashing with spectral rotation. In: Proceedings of the AAAI Conference on Artificial Intelligence, vol. 31, pp. 2203–2209 (2017)

20. Liu, B., Cao, Y., Long, M., Wang, J., Wang, J.: Deep triplet quantization. In: Proceedings of the 26th ACM International Conference on Multimedia, pp. 755–763 (2018)

21. Liu, H., Li, X., Zhang, S., Tian, Q.: Adaptive hashing with sparse matrix factorization. IEEE Trans Neural Netw. Learn. Syst. **31**(10), 4318–4329 (2019)

22. Liu, X., Nie, X., Dai, Q., Huang, Y., Lian, L., Yin, Y.: Reinforced short-length hashing. IEEE Trans. Circuits Syst. Video Technol. **31**(9), 3655–3668 (2020)

23. Liu, Y., Ji, S., Fu, Q., Chiu, D.K., Gong, M.: An efficient dual semantic preserving hashing for cross-modal retrieval. Neurocomput **492**, 264–277 (2022)

24. Lu, X., Zheng, X., Li, X.: Latent semantic minimal hashing for image retrieval. IEEE Trans. Image Process. **26**(1), 355–368 (2016)

25. Luo, K., Zhang, C., Li, H., Jia, X., Chen, C.: Adaptive marginalized semantic hashing for unpaired cross-modal retrieval. IEEE Trans. Multimed. **25**, 9082–9095 (2023)

26. Luo, X., Nie, L., He, X., Wu, Y., Chen, Z.-D., Xu, X.-S.: Fast scalable supervised hashing. In: The 41st International ACM SIGIR Conference on Research & Development in Information Retrieval, pp. 735–744 (2018)

27. Luo, X., Wu, Y., Xu, X.-S.: Scalable supervised discrete hashing for large-scale search. In: Proceedings of the 2018 WWW, pp. 1603–1612 (2018)

28. Qin, J., et al.: Discrete semantic matrix factorization hashing for cross-modal retrieval. In: 2020 25th ICPR, pp. 1550–1557. IEEE (2021)

29. Shen, F., Shen, C., Liu, W., Tao Shen, H.: Supervised discrete hashing. In: CVPR, pp. 37–45 (2015)
30. Shen, H.T., et al.: Exploiting subspace relation in semantic labels for cross-modal hashing. IEEE Trans. Knowl. Data Eng. **33**(10), 3351–3365 (2021)
31. Shi, Y., Nie, X., Liu, X., Zou, L., Yin, Y.: Supervised adaptive similarity matrix hashing. IEEE Trans. Image Process. **31**, 2755–2766 (2022)
32. Sun, Y., Dai, J., Ren, Z., Li, Q., Peng, D.: Relaxed energy preserving hashing for image retrieval. IEEE Trans. Intell. Transp. Syst. **25**(7), 7388–7400 (2024)
33. Sun, Y., Ren, Z., Hu, P., Peng, D., Wang, X.: Hierarchical consensus hashing for cross-modal retrieval. IEEE Trans. Multimed. **26**, 824–836 (2023)
34. Tian, D., Gong, C., Gong, M., Wei, Y., Feng, X.: Modeling cardinality in image hashing. IEEE Trans. Cybern. **53**(1), 114–123 (2023)
35. Wang, L., Yang, J., Zareapoor, M., Zheng, Z.: Cluster-wise unsupervised hashing for cross-modal similarity search. Pattern Recogn. **111**, 107732 (2021)
36. Wang, M., Zhou, W., Tian, Q., Li, H.: Neighborhood pyramid preserving hashing. IEEE Trans. Multimed. **22**(6), 1507–1518 (2019)
37. Wang, W., Zhang, H., Zhang, Z., Liu, L., Shao, L.: Sparse graph based self-supervised hashing for scalable image retrieval. Inf. Sci. **547**, 622–640 (2021)
38. Wang, Y., Luo, X., Nie, L., Song, J., Zhang, W., Xu, X.-S.: BATCH: a scalable asymmetric discrete cross-modal hashing. IEEE Trans. Knowl. Data Eng. **33**(11), 3507–3519 (2020)
39. Weng, Z., Zhu, Y.: Online hashing with bit selection for image retrieval. IEEE Trans. Multimed. **23**, 1868–1881 (2020)
40. Yao, T., Li, Y., Guan, G., Li, Y., Yan, L., Tian, Q.: Discrete robust matrix factorization hashing for large-scale cross-media retrieval. IEEE Trans. Knowl. Data Eng. **35**(2), 1391–1401 (2023)
41. Yao, T., et al.: Efficient supervised graph embedding hashing for large-scale cross-media retrieval. Pattern Recogn. **145**, 109934 (2024)
42. Zhang, C., Li, H., Chen, C., Qian, Y., Zhou, X.: Enhanced group sparse regularized nonconvex regression for face recognition. IEEE Trans. Pattern Anal. Mach. Intell. **44**(5), 2438–2452 (2020)
43. Zhang, C., Li, H., Gao, Y., Chen, C.: Weakly-supervised enhanced semantic-aware hashing for cross-modal retrieval. IEEE Trans. Knowl. Data Eng. **35**(6), 6475–6488 (2023)
44. Zhang, Z., Pun, C.-M.: Learning ordinal constraint binary codes for fast similarity search. Inform. Process. Manag. **59**(3), 102919 (2022)
45. Zhang, Z., Zhu, X., Lu, G., Zhang, Y.: Probability ordinal-preserving semantic hashing for large-scale image retrieval. ACM Trans. Knowl. Discov. Data **15**(3), 1–22 (2021)

Graph Neural Networks with Direct Reach to Labeled Nodes

Qing Teng[1], Qihang Guo[2(✉)], Xibei Yang[1], Keyu Liu[1], and Tianrui Li[3]

[1] School of Computer, Jiangsu University of Science and Technology,
Zhenjiang 212100, Jiangsu, China
221110701104@stu.just.edu.cn, {jsjxy_yxb,kyliu}@just.edu.cn
[2] School of Economics and Management, Jiangsu University of Science
and Technology, Zhenjiang 212100, Jiangsu, China
just_gqh@stu.just.edu.cn
[3] School of Computing and Artificial Intelligence, Southwest Jiaotong University,
Chengdu 611756, Chengdu, China
trli@swjtu.edu.cn

Abstract. Graph Neural Networks (GNNs) have become a leading technique for semi-supervised node classification. However, the distribution of labeled nodes in the graph is often uneven, resulting in only a small portion of unlabeled nodes being able to directly reach labeled nodes. This misalignment between labeled and unlabeled nodes significantly weakens the performance of GNNs. To address this issue, this paper introduces a plug-and-play GNN architecture called RLGNNs. The core idea of RLGNNs consists of two parts. First, by utilizing a graph transformer, we establish direct global connections between unlabeled and labeled nodes, enhancing the reachability of labeled nodes and improving the representations of unlabeled nodes. Second, we incorporate GNNs to capture the original structural information of the graph that was overlooked in the first part, allowing for a more comprehensive understanding of the graph's contextual information. Through a series of experiments, we demonstrate that RLGNNs achieve superior performance in semi-supervised node classification tasks compared to other state-of-the-art GNNs.

Keywords: Graph Neural Networks · Graph Transformers · Node Classification · Semi-supervised Learning

1 Introduction

Graphs, as a fundamental data structure, effectively capture the complex relationships and dependencies between entities in various domains, making them an essential representation for modeling intricate structures across multiple disciplines [1–3]. For example, in social networks, graphs are used to represent directed relationships between users [4,5], while in bio-informatics, they model

© The Author(s), under exclusive license to Springer Nature Switzerland AG 2025
Q. Zhang et al. (Eds.): IJCRS 2025, LNAI 15710, pp. 443–455, 2025.
https://doi.org/10.1007/978-3-031-92741-6_31

interactions between genes [6–9]. Graph Neural Networks (GNNs) [10,11], leveraging the message-passing mechanism, have emerged as the predominant approach for semi-supervised node classification tasks. Recent studies suggest that the success of GNNs is primarily attributed to how message passing reduces the distribution gap between labeled and unlabeled nodes (distribution alignment), enabling the model to make reliable inferences on unlabeled nodes [12,13].

However, recent studies have shown that low-degree nodes are often situated farther from labeled nodes, while high-degree nodes are relatively closer [14]. Due to the limited depth of typical GNNs models, low-degree nodes face challenges in passing messages to distant labeled nodes. As a result, in shallow GNNs architectures, labeled nodes often struggle to gather information from a wide range of unlabeled nodes, hindering effective distribution alignment. Consequently, trained GNNs tend to perform well on nodes near labeled nodes, but their classification performance is limited for nodes that were not encountered during training.

It is worth noting that various strategies have been explored to address this issue [15–17]. One common approach is to increase the number of stacked GNNs layers to facilitate long-range information propagation. However, this introduces new issues, such as over-smoothing, which makes node representations difficult to distinguish, and the potential loss of distant information due to excessive compression. Another strategy to improve labeled node accessibility involves designing a node generator to create auxiliary labeled nodes [18,19]. However, this approach alters the original spatial layout of the graph, and if the generated node features lack meaningful semantics, it complicates the model's understanding of the contribution of these nodes to the classification results, thereby reducing interpretability. Moreover, the success of this approach is highly contingent on the quality of the node generator's design. Therefore, developing a flexible and effective solution to optimize the accessibility of labeled nodes remains a critical challenge.

To address the over-smoothing and over-compression challenges associated with stacking GNN layers, while preserving the original spatial layout of nodes, we propose a novel solution in this paper. Specifically, we introduce a plug-and-play GNN architecture, termed RLGNNs. First, to facilitate long-range information propagation between nodes without the need for additional layers, we leverage the long-range interaction capabilities of graph transformers. By integrating graph transformers, we establish direct global associations between labeled and unlabeled nodes, thereby improving the accessibility of labeled nodes and enhancing the alignment of distributions across both labeled and unlabeled nodes. Second, to ensure that potentially valuable structural information from the original graph is retained, we incorporate a traditional GNN module. This module captures contextual information that may have been overlooked in the previous step, complementing the graph transformers and further improving the overall model performance. Through the synergistic combination of these two approaches, RLGNNs not only enhance labeled node accessibility but also preserve the integrity of the original graph structure, effectively mitigating the issues of over-smoothing and over-compression inherent in deep GNN models. Our main contributions are summarized as follows:

1. We propose a plug-and-play GNN architecture, termed RLGNNs, to address the over-smoothing and over-compression issues associated with stacking GNN layers, while preserving the original spatial layout of the nodes.
2. By incorporating graph transformers, direct global associations between unlabeled and labeled nodes are established, thereby enhancing the accessibility of labeled nodes without the need for stacking additional GNN layers.
3. Experiments results conducted on four graph datasets for semi-supervised node classification demonstrate that the proposed RLGNNs architecture is compatible with a variety of popular GNN models and outperforms other state-of-the-art GNN methods.

2   Related Work

2.1   GNNs for Semi-supervised Node Classification Tasks

Graph neural networks (GNNs) [1,6], particularly graph convolutional networks (GCNs) [10] and graph attention networks (GATs) [11], have shown significant potential in semi-supervised node classification, leveraging graph structural information to model complex relationships between nodes. However, a key challenge in this task is the uneven distribution of labeled nodes [12,15], which limits the access of many unlabeled nodes to labeled information. Although increasing model depth can aid long-range information propagation, it often results in over-smoothing, where node representations become indistinguishable, or over-compression, leading to the loss of distant information [20,21]. Expanding the receptive field can address some of these issues [15–17], but it tends to increase computational complexity and only indirectly improves labeled node accessibility. To address this, this paper proposes a solution that differs from previous studies by effectively mitigating the over-smoothing and over-compression issues caused by network layer stacking, without altering the original spatial layout of the nodes.

2.2   Graph Transformers

In recent years, the rapid advancement of transformer architectures, particularly graph transformers [22], has significantly impacted the study of graph-structured data, especially in node classification tasks. Utilizing their global attention mechanism, graph transformers excel at capturing long-range interactions between nodes, complementing the local structural learning of traditional GNNs [23,24]. This expansive receptive field allows graph transformers to efficiently extract valuable information from all nodes, enhancing the model's representation power [22,25]. However, most research has focused on expanding the receptive field through global attention, with limited attention to directly improving label accessibility. Nevertheless, the ability of graph transformers to capture long-range dependencies without altering node spatial layouts presents a promising solution to the over-smoothing and over-compression issues inherent in stacked GNN layers. This paper explores the application of graph transformers in designing a GNN architecture that directly enhances labeled node accessibility.

3   Graph Neural Networks with Direct Reach to Labeled Nodes

This section provides a detailed description of the RLGNNs, with its structure illustrated in Fig. 1. The framework consists of two primary modules: the labeled node direct access module and the graph structure learning module.

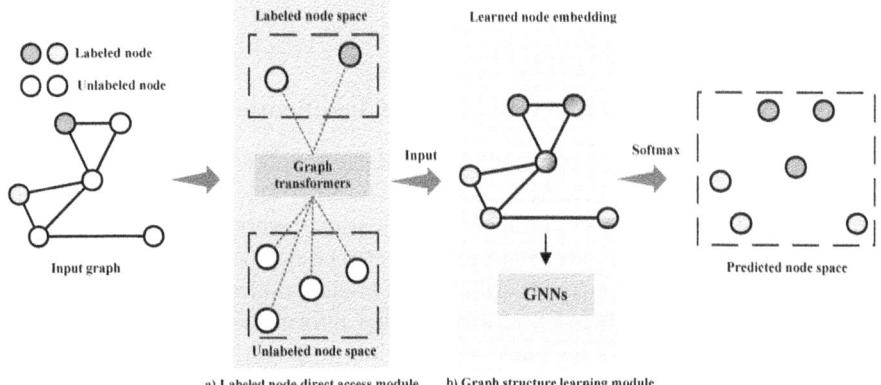

Fig. 1. The framework of RLGNNs

To illustrate the method proposed in this paper, the following basic definitions are provided: Given an undirected graph $\mathcal{G} = (\mathcal{V}, \mathcal{E})$, $\mathbf{A} \in \mathbb{R}^{n \times n}$ is the symmetric adjacency matrix of the n-th node in graph $\mathcal{G}$, representing the edge weights between two nodes. $\mathbf{X} \in \mathbb{R}^{n \times d}$ is the node feature matrix, where d is the dimensionality of the node features, meaning each node v_i in the graph has d-dimensional feature vector $x_i \in \mathbb{R}^d$.

3.1   Labeled Node Direct Access Module

This module leverages the multi-head global attention mechanism of graph transformers [22] to establish direct global associations between labeled and unlabeled nodes. Specifically, self-attention can be interpreted as a learnable, fully connected weighted graph, while multi-head attention facilitates the capture of diverse correlations between nodes. Given the labeled node embeddings $\mathbf{Z}^{L,(l)}$ and unlabeled node embeddings $\mathbf{Z}^{UL,(l)}$ at the l-th layer, we have:

$$\mathbf{Z}^{UL,(l+1)} = \text{COMBINE}(head_1, ..., head_B)\mathbf{W}^c \tag{1}$$

$$head_b(\mathbf{Z}^{L,(l)}, \mathbf{Z}^{UL,(l)}) = \text{ReLU}(\phi_q(\mathbf{Z}^{UL,(l)}\mathbf{W}_b^q)\phi_k(\mathbf{Z}^{L,(l)}\mathbf{W}_b^k)^T)\mathbf{Z}^{L,(l)}\mathbf{W}_b^v \quad (2)$$

$$\mathbf{Z}^{UL,(l+1)} = \text{COMBINE}(\mathbf{Z}^{UL,(l)}, \mathbf{Z}^{UL,(l+1)}) \quad (3)$$

where $head_b$ represents the information of the b-th head, and ϕ_q and ϕ_k denote two normalization functions. COMBINE() is the function used to merge different embeddings, and the concatenation method is employed for combining the embeddings. $\mathbf{W}_b^q, \mathbf{W}_b^k$ and $\mathbf{W}_b^v$ are trainable parameter matrices.

For convenience, the embedding learning process for unlabeled nodes is defined by Eqs. (1)–(3): $\mathbf{Z}^{UL,(l+1)} = f_{trans}(\mathbf{Z}^{L,(l)}, \mathbf{Z}^{UL,(l)})$. Similarly, the embedding learning process for labeled nodes can be defined as: $\mathbf{Z}^{L,(l+1)} = f_{trans}(\mathbf{Z}^{UL,(l)}, \mathbf{Z}^{L,(l)})$. Therefore, the embeddings of all nodes learned can be represented as: $\mathbf{Z}^{(l+1)} = [\mathbf{Z}^{UL,(l+1)}, \mathbf{Z}^{L,(l+1)}]$.

$$\begin{cases} \mathbf{Z}^{(l+1)} = \mathbf{Z}^{(l)} + \mathbf{Z}^{(l+1)} \\ \mathbf{Z}^{(l+1)} = \mathbf{W}_2 \, \text{ReLU}\left(\mathbf{W}_1 \, \text{LN}\left(\mathbf{Z}^{(l+1)}\right)\right) + \mathbf{Z}^{(l+1)} \end{cases} \quad (4)$$

where $\mathbf{W}_1$ and $\mathbf{W}_2$ represent two trainable parameter matrices, and $\text{LN}(\cdot)$ denotes the layer normalization operation.

To enhance the learning process of graph transformers and encourage more focused attention allocation between similar node pairs, this module incorporates a graph contrastive loss function. This loss function captures local graph information, particularly for node pairs directly connected by edges, thereby guiding the graph transformers to assign higher attention weights to these pairs during training.

Given the feature matrix $\mathbf{X}$, we first apply a multilayer perception (MLP) with learnable parameters $\mathbf{W}$ to generate a new embedding of $\mathbf{X}$.

$$\mathbf{Z}^{(0)} = \text{Dropout}(\sigma(\mathbf{X}\mathbf{W})) \quad (5)$$

Next, using the learned embeddings $\mathbf{Z}^{(0)}$ and the adjacency matrix $\mathbf{A}$, we define a graph contrastive loss to capture local information between nodes.

$$\mathcal{L}_{\text{gsi}} = \frac{1}{|\mathcal{V}_L|} \sum_{i \in \mathcal{V}_L} -\log \frac{\sum_{j \in \mathcal{V}_{UL}} \mathbb{1}_{[j \neq i]}^{\mathcal{G}} \exp(\text{sim}(\mathbf{z}_i^{(0)}, \mathbf{z}_j^{(0)})/\tau)}{\sum_{k \in \mathcal{V}_{UL}} \mathbb{1}_{[k \neq i]}^{\mathcal{G}} \exp(\text{sim}(\mathbf{z}_i^{(0)}, \mathbf{z}_k^{(0)})/\tau)} \quad (6)$$

where τ represents the temperature parameter, $\text{sim}(\cdot)$ denotes the similarity measure, and $\mathcal{V}_L$ represents the set of labeled nodes. Additionally, the matrix $I^{\mathcal{G}}$ is used to indicate the presence of edges in the graph $\mathcal{G}$.

3.2 Graph Structure Learning Module

This module leverages GNNs to capture the raw graph structure information that may be overlooked in the node embedding learning process of direct modules for labeled nodes. The embedding learning process for each layer of the GNN can be defined as:

$$z_{N(i)}^{(l+1)} = \text{AGGREGATE}\left(\left\{z_v^{(l)}, \forall v \in N(i)\right\}\right) \quad (7)$$

$$z_i^{(l+1)} = \text{COMBINE}\left(z_i^{(l)}, z_{N(i)}^{(l+1)}\right) \tag{8}$$

where AGGREGATE() represents the neighbor aggregation function, such as the symmetric normalized convolution aggregation operation using the adjacency matrix in GCN or the self-attention aggregation operation in GAT. COMBINE() represents the combination function, which merges the node's own information with the aggregated information.

For convenience, this paper defines the GNN learner as f_{GNNS}. The predicted node label matrix is obtained through the learning of multiple layers of GNNs: $\hat{Y}_{i,c} = f_{\text{GNNs}}(Z, A) = \{\hat{y}_{i,c}\} \in \mathbb{R}^{n \times C}$ where C is the number of labels, and Z represents the embedding learned by the direct module for the labeled nodes. Then, this paper adopts the cross-entropy loss $\mathcal{L}_L$ to train the GNNs for node classification.

$$\mathcal{L}_L = -\sum_{i \in V_L}\sum_{c=1}^{C} y_{i,c}\ln\hat{y}_{i,c} \tag{9}$$

where V_L is the set of labeled nodes, and $y_{i,c}$ denotes the true node label matrix.

3.3   Model Training and Algorithms

We adopt a global loss function $\mathcal{L}$ to train RLGNNs, which includes the graph contrastive loss function from Eq. (6) and the cross-entropy loss function from Eq. (9).

$$\mathcal{L} = \eta \cdot \mathcal{L}_{\text{gsi}} + (1 - \eta) \cdot \mathcal{L}_L \tag{10}$$

where η is a hyperparameter that controls the trade-off between the graph contrastive loss and the cross-entropy loss.

The specific RLGNNs model algorithm is shown in Algorithm 1.

Algorithm 1. RL-GNNs Algorithm

Input: V: node feature matrix X, adjacency matrix A, node label matrix Y, GNN model f_{GNNs}, number of training epochs T.
Output: predicted node labels $\hat{Y}$.
1: **for** $t = 1 : T$ **do**
2:     Obtain node classification embeddings Z via Eq. (1) - (4);;
3:     Obtain the graph similarity loss $\mathcal{L}_{gsi}$ via Eq. Eq. (5) - Eq. (6);
4:     Obtain the graph structure learning loss $\mathcal{L}$ via Eq. Eq. (7) to Eq. (8);
5:     Use GNN model f_{GNNs} to obtain the t-th node label prediction $\hat{Y}$;
6:     Obtain the total loss function $\mathcal{L}$ via Eq. Eq. (9);
7:     Update the model parameters using random gradient descent to minimize $\mathcal{L}$;
8: **end for**
9: Use the trained model to output the predicted node labels $\hat{Y}$.

---

4   Experiment

4.1   Datasets

Table 1 summarizes the detailed characteristics of the four datasets used in the experiments, where nodes represent publications and edges represent citation links.

Table 1. The description of datasets

ID	Datasets	Nodes	edges	features	classes	label rate
1	Cora	2708	5429	1433	7	5.2%
2	Citeseer	3327	4732	3703	6	3.6%
3	Pubmed	19717	44338	500	3	0.3%
4	Acm	3025	13128	1807	3	2.0%

4.2   Baselines

To validate the effectiveness of the framework, eight advanced GNNs models including **GCN** [10], **GAT** [11], **MIXHOP** [12], **GCNII** [17], **MOGCN** [16], **PA-GCN** [22], **GCN-PGBSF** [26], **MHMOGAT** [20] are used for the comparative study.

To facilitate the experimental analysis, the classic GNN models, GCN and GAT, are embedded within the RLGNN architecture, yielding two variants: RLGCN and RLGAT.

4.3   Parameter Settings

The model is optimized using the Adam optimizer. The experimental results are evaluated based on the classification accuracy metric. For consistency across all models, the following hyperparameters are used: 500 iterations, a learning rate of 0.003, a dropout rate of 0.5, and hyperparameter values chosen from the set $\{0.00, 0.10, 0.20, ..., 0.90\}$, with performance evaluated at the optimal setting. For comparison methods, hyperparameters are set according to the default values specified in the respective papers or code implementations. To ensure robustness, each experiment is repeated 50 times, and the average result is reported, as this provides a more reliable measure of the model's stability.

4.4   Analysis of Architecture and Label Rate

In this study, two novel models, RLGCN and RLGAN, are developed by integrating RLGNNs with GCN and GAT, respectively. These models are then compared against baseline models across varying label rates. The label rates for the Cora

Table 2. Classification accuracy of models with different label rates

Datasets	Label rate	GCN	RLGCN	improvement	GAT	RLGAT	improvement
Cora	0.05%	43.82	47.32	7.99%	44.55	50.11	9.43%
	0.10%	62.31	73.52	17.99%	66.82	77.84	16.49%
	5.20%	81.33	83.93	3.20%	83.26	84.92	2.00%
Citeseer	0.05%	43.62	55.80	27.92%	58.41	67.72	15.94%
	0.10%	55.35	65.70	18.70%	61.16	70.60	15.43%
	3.60%	70.35	73.65	4.69%	72.51	74.11	2.21%
Pubmed	0.08%	57.52	68.50	19.09%	58.60	65.79	12.26%
	0.15%	64.31	73.20	13.82%	70.70	76.47	8.17%
	0.30%	79.00	81.02	2.51%	79.13	81.05	2.43%
ACM	0.50%	65.63	70.12	6.84%	64.91	69.35	6.84%
	1.00%	79.67	85.23	6.98%	80.00	84.58	5.73%
	2.00%	87.82	91.11	3.75%	87.45	90.53	3.52%

and Citeseer datasets range from 0.05% to 5.20% and 0.05% to 3.60%, respectively, while the label rates for the Pubmed and ACM datasets vary from 0.08% to 0.30% and 0.50% to 2.00%, respectively. The average classification accuracy is computed over 1000 test nodes, and the results are presented in Table 3. The findings indicate that RLGNNs consistently outperform the baseline models, GCN and GAT, in terms of classification accuracy across all datasets and label rates. Moreover, RLGNNs exhibit particularly strong performance under low-label-rate conditions, demonstrating its remarkable adaptability and robustness in sparse-label settings (Table 2).

4.5 Classification Performance Comparison

Table 3. Node classification accuracy

Datasets	Cora	Citeseer	Pubmed	ACM
GCN	81.33	70.35	79.00	87.82
GAT	83.26	72.51	79.13	87.45
GCNII	**85.45**	73.41	80.32	90.85
MIXHOP	81.92	71.32	80.71	88.05
MOGCN	82.43	72.47	79.22	90.15
PA-GCN	83.64	70.42	79.33	90.92
GCN+PGBSF	82.48	71.96	80.67	90.13
MHMOGAT	82.45	70.66	79.68	89.15
RLGCN	83.93	73.65	80.92	**91.11**
RLGAT	84.92	**74.11**	**81.05**	90.53

The semi-supervised classification results of RLGNNs, obtained under optimal hyperparameters and benchmark models, are summarized in Table 3. Overall, RLGNNs consistently outperform other baseline models in terms of classification accuracy across all datasets. Notably, RLGNNs exhibit a substantial performance advantage across various datasets, demonstrating its robustness.

Moreover, when compared to other models addressing the label reachability problem, such as GCNII, MIXHOP, and MHMOGAT, RLGNNs consistently outperform these models across most datasets, with particularly pronounced improvements observed on the Citeseer dataset. These findings substantiate the effectiveness of the proposed RLGNNs architecture.

4.6 T-SNE Visualization

Figures 2 and 3 illustrate the t-SNE visualization results for 1000 test samples from the Cora and Citeseer datasets. By projecting the high-dimensional data into a two-dimensional space, we can visually compare the classification performance of the two models.

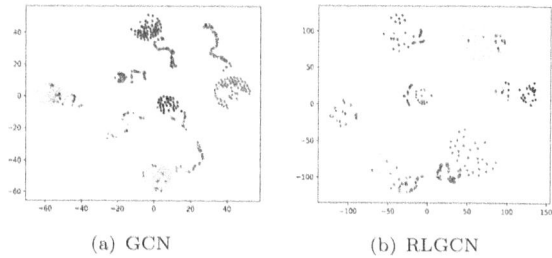

(a) GCN   (b) RLGCN

Fig. 2. The T-SNE visualization results of the Cora dataset

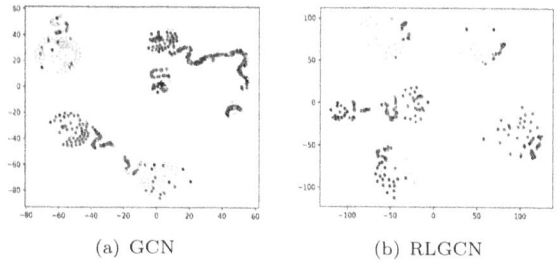

(a) GCN   (b) RLGCN

Fig. 3. The T-SNE visualization results of the Citeseer dataset

As observed in Figs. 2 and 3, the RLGCN model consistently outperforms GCN in terms of classification performance across both Cora and Citeseer datasets. Specifically, RLGCN demonstrates stronger intra-class node cohesion and more distinct class boundaries. This suggests that RLGCN is more effective at capturing global dependencies between labeled and unlabeled nodes, leading to enhanced classification performance on complex graph data.

4.7   Ablation Study

This subsection provides an analysis of the performance of each component within RLGNNs. For simplicity, we focus on GCN, which serves as the baseline model, alongside RLGCN, a variant of GCN that omits the graph contrastive loss. To facilitate this analysis, we design two model variants:

- GCN: Removing the entire label direct access module and only learns the graph structure information.
- w/o$\mathcal{L}_{gsi}$: Discarding the local information learning in the label-direct module, i.e., without the inclusion of the graph contrastive loss $\mathcal{L}_{gsi}$.

Table 4. Classification accuracy of RLGNNs variants

Datasets	Cora	Citeseer	Pubmed	ACM
GCN	81.33	70.35	79.00	87.82
w/o$\mathcal{L}_{gsi}$	81.67	70.82	78.74	87.56
RLGCN	83.93	73.65	80.92	91.11

The classification accuracy of the respective variants is presented in Table 4. After removing the graph contrastive loss $\mathcal{L}_{gsi}$, the model's performance improved by approximately 0.42% on the Cora dataset and by about 0.67% on the Citeseer dataset. However, it decreased by around 0.33% on the Pubmed dataset and by about 0.30% on the ACM dataset. This indicates that contrastive loss $\mathcal{L}_{gsi}$ has a positive effect on performance in some datasets but may introduce some noise in others.

The RLGCN with the labeled node direct access module demonstrated significant improvements across all datasets. Specifically, on the Cora dataset, performance improved by approximately 3.20%. These improvements indicate that the label propagation module effectively enhances the model's ability to utilize node label information. However, if the labeled node direct access module does not incorporate local information learning, its performance may fall short of the GCN baseline. This highlights the critical role of local information in the learning process, enabling the model to more accurately capture relationships between nodes. Additionally, the graph contrastive loss $\mathcal{L}_{gsi}$ guides graph transformers to focus attention on similar node pairs, preventing attention dispersion and thereby improving the label reachability of similarly labeled nodes. Overall, the introduction of the labeled node direct access module and graph contrastive loss significantly enhances the performance of RLGNNs, but the learning of local information and the design of loss functions require careful adjustment based on specific datasets.

4.8   Hyperparameters Sensitivity Analysis

Hyperparameters control the weights of the two loss functions in RLGNNs. To simplify the analysis, in this section, we take RLGCN as an example and observe the impact of the change in the hyperparameter on model performance by plotting the curve of the model's classification accuracy (ACC) as a function of the hyperparameter value η, as shown in Fig. 4. Across all datasets, the classification accuracy peaks when the hyperparameter is within the range of 0.4, 0.6, and the accuracy decreases significantly when it exceeds the optimal value. This indicates that selecting appropriate hyperparameters is crucial for enhancing the classification performance of the model, while excessively high hyperparameters may cause the model to pay too little attention to supervised information, leading to a decrease in the model's generalization ability.

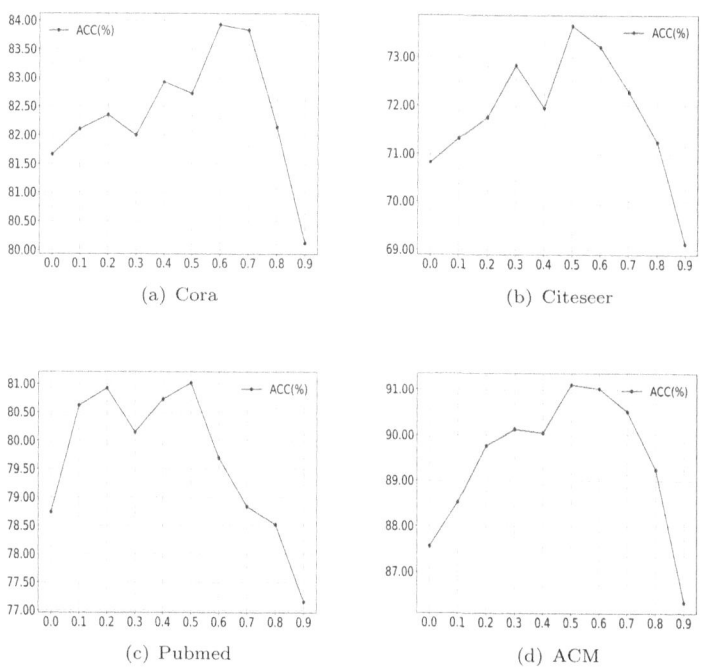

Fig. 4. Classification results of RLGCN under different hyperparameter values

5   Conclusion

This paper presents RLGNNs, a novel graph neural network architecture that combines graph transformers and traditional GNN modules to enhance the accessibility of labeled nodes by improving global dependencies. RLGNNs mitigate over-smoothing and over-compression issues without requiring deeper layers or auxiliary nodes. Experimental results demonstrate superior performance on benchmark datasets compared to state-of-the-art GNNs. The architecture is also highly compatible with existing GNN models, maintaining the original node layout while improving performance.

In future work, we will focus on further optimizing RLGNNs by exploring two primary directions: first, reducing the complexity of the direct connection module for labeled nodes to enhance model efficiency; second, investigating potential architecture integration and extensions, examining the compatibility of RLGNNs with other GNN models, and expanding its application to domains such as social networks and bio-informatics.

References

1. Wang, K., An, J., Zhou, M., Shi, Z., Shi, X., Kang, Q.: Minority-weighted graph neural network for imbalanced node classification in social networks of internet of people. IEEE Internet Things J. **10**, 330–340 (2022)
2. Zhong, Y., et al.: Learning motif-based graphs for drug-drug interaction prediction via local-global self-attention. Nat. Mach. Intell. **6**, 1094–1105 (2024)
3. Guo, Q., Yang, X., Li, M., Qian, Y.: Collaborative graph neural networks for augmented graphs: a local-to-global perspective. Pattern Recogn. **158**, 111020 (2025)
4. Wang, Y., Yang, X., Sun, Q., Qian, Y., Guo, Q.: Purity skeleton dynamic hypergraph neural network. Neurocomputing **610**, 128539 (2024)
5. Bao, Y., Liu, J., Shen, Q., Cao, Y., Ding, W., Shi, Q.: PKET-GCN: prior knowledge enhanced time-varying graph convolution network for traffic flow prediction. Inf. Sci. **634**, 359–381 (2023)
6. Zhu, X., Li, C., Guo, J., Dietze, S.: CNIM-GCN: consensus neighbor interaction-based multi-channel graph convolutional networks. Expert Syst. Appl. **226**, 120178 (2023)
7. Peng, D., Zhang, Y.: MA-GCN: a memory augmented graph convolutional network for traffic prediction. Eng. Appl. Artif. Intell. **121**, 106046 (2023)
8. Wang, Y., Liu, Y., Zhang, J.: MGCN: medical relation extraction based on GCN. Comput. Inform. **42**, 411–435 (2023)
9. Zheng, S., Qiu, L., Lan, F.: TSO-GCN: a graph convolutional network approach for real-time and generalizable truss structural optimization. Appl. Soft Comput. **134**, 110015 (2023)
10. Kipf, T.N., Welling, M.: Semi-supervised classification with graph convolutional networks. arXiv preprint arXiv:1609.02907 (2016)
11. Veličković, P., Cucurull, G., Casanova, A., Romero, A., Lio, P., Bengio, Y.: Graph attention networks. arXiv preprint arXiv:1710.10903 (2017)
12. Abu-El-Haija, S., et al.: MixHop: higher-order graph convolutional architectures via sparsified neighborhood mixing. In: Proceedings of the 36th International Conference on Machine Learning, pp. 21–29 (2019)

13. Zhong, L., Chen, Z., Wu, Z., Du, S., Chen, Z., Wang, S.: Learnable graph convolutional network with semisupervised graph information bottleneck. IEEE Trans. Neural Netw. Learn. Syst. **36**, 433–446 (2023)
14. Lu, W., Guan, Z., Zhao, W., Yang, Y., Jin, L.: NodeMixup: tackling under-reaching for graph neural networks. In: Proceedings of the AAAI Conference on Artificial Intelligence, vol. 38, pp. 14175–14183 (2024)
15. He, L., Bai, L., Yang, X., Du, H., Liang, J.: High-order graph attention network. Inf. Sci. **630**, 222–234 (2023)
16. Wang, J., Liang, J., Cui, J., Liang, J.: Semi-supervised learning with mixed-order graph convolutional networks. Inf. Sci. **573**, 171–181 (2021)
17. Chen, M., Wei, Z., Huang, Z., Ding, B., Li, Y.: Simple and deep graph convolutional networks. In: Proceedings of the 37th International Conference on Machine Learning, pp. 1725–1735 (2020)
18. Xia, S., Zheng, S., Wang, G., Gao, X., Wang, B.: Granular ball sampling for noisy label classification or imbalanced classification. IEEE Trans. Neural Netw. Learn. Syst. **34**, 2144–2155 (2021)
19. Sang, B., Chen, H., Yang, L., Li, T., Xu, W.: Incremental feature selection using a conditional entropy based on fuzzy dominance neighborhood rough sets. IEEE Trans. Fuzzy Syst. **30**(6), 1683–1697 (2021)
20. Ben, J., Sun, Q., Liu, K., Yang, X., Zhang, F.: Multi-head multi-order graph attention networks. Appl. Intell., 1–16 (2024)
21. Sang, B., Chen, H., Yang, L., Wan, J., Li, T., Xu, W.: Feature selection considering multiple correlations based on soft fuzzy dominance rough sets for monotonic classification. IEEE Trans. Fuzzy Syst. **30**(12), 5181–5195 (2022)
22. Zhu, Y., Feng, L., Deng, Z., Chen, Y., Amor, R., Witbrock, M.: Robust node classification on graph data with graph and label noise. In: The AAAI Conference on Artificial Intelligence, vol. 38, pp. 17220–17227 (2024)
23. Xing, Y., Wang, X., Li, Y., Huang, H., Shi, C.: Less is more: on the over-globalizing problem in graph transformers. arXiv preprint arXiv:2405.01102 (2024)
24. Sang, B., Xu, W., Chen, H., Li, T.: Active antinoise fuzzy dominance rough feature selection using adaptive k-nearest neighbors. IEEE Trans. Fuzzy Syst. **31**(11), 3944–3958 (2023)
25. Sang, B., Yang, L., Xu, W., Chen, H., Li, T., Li, W.: VCOS: multi-scale information fusion to feature selection using fuzzy rough combination entropy. Inf. Fusion **117**, 102901 (2025)
26. Cong, H., Sun, Q., Yang, X., Liu, K., Qian, Y.: Enhancing graph convolutional networks with progressive granular ball sampling fusion: a novel approach to efficient and accurate GCN training. Inf. Sci., 120831 (2024)

Recognizing Plastic Bottles on Water Surface Based on Flipping Transformation

Jiangli Duan(iD), Xiufeng Cao(iD), Huilin Jiang(iD), Xin Hu$^{(\boxtimes)}$(iD), Mei Liao(iD), and Jun Luo(iD)

College of Big Data and Intelligent Engineering, Yangtze Normal University, Chongqing 408100, China
{duanjl,huxin,luojun}@yznu.edu.cn,
{222154101401,222154101413,202106661328}@stu.yznu.edu.cn

Abstract. Plastic bottles on the water surface are important factors causing water source pollution. The low efficiency of traditional manual salvage makes the automated fishing technology of plastic bottles on the water surface extremely urgent. However, the recognition methods of plastic bottles on the water surface generally have the problem of dataset limitations. The limited datasets lead to insufficient extraction of plastic bottle feature information, limited generalization ability of the model, and thus reduced recognition efficiency. Therefore, to address this problem, this paper proposes a data augmentation method based on flip transformation. The 1200 images with plastic bottles in the original dataset are flipped horizontally and vertically, and then 1200 horizontally-flipped images and 1200 vertically – flipped images are generated, which increased the number of images from 1200 to 3600. Experimental results show that after the dataset is processed by flip transformation, the performance of both the Faster R-CNN (Regions with Convolutional Neural Network) and YOLOv5 algorithms has been significantly improved. In the multi-object and single-object recognition tasks of the Faster R-CNN algorithm, the result on the comprehensive dataset with 3600 images perform better than that on the original dataset. Moreover, in the single-object recognition process of the YOLOv5 algorithm, the result on the comprehensive dataset has obvious advantages, with an accuracy reaching 100%. In conclusion, flip transformation can effectively optimize the performance of object recognition algorithms and significantly improve algorithm's recognition ability caused by limited datasets.

Keywords: Target recognition · Fliping transformation · Faster R–CNN · YOLOv5

1 Introduction

In the current context that the global ecological environment is facing severe challenges, the problem of water pollution has become increasingly prominent

Q. Zhang et al. (Eds.): IJCRS 2025, LNAI 15710, pp. 456–467, 2025.
https://doi.org/10.1007/978-3-031-92741-6_32

and has become one of the key factors restricting sustainable development. As a common component of water surface garbage, a large number of plastic bottles float on the water surfaces of rivers and lakes, which has already become one of the main sources of water pollution. They not only seriously affect the aesthetics of water bodies, but also cause incalculable damage to the aquatic ecosystem. At present, the work of cleaning plastic bottles floating on the water surface still mainly relies on the traditional manual fishing method. This method is not only inefficient and labor-intensive, but also has certain safety risks. The emergence of water surface cleaning robots provides an effective way to liberate labor force and improve the cleaning efficiency. In recent years, remarkable progress has been made in the research and development of water surface cleaning robots [1,2]. The primary prerequisite for achieving intelligent water surface garbage fishing is to accurately complete the recognition of target garbage.

According to the technical focuses and task types, the relevant models can be roughly divided into object recognition models [3–5], image classification models [6–8], and instance segmentation models [9–11]. 1) Object recognition models focus on the accuracy of localization and classification. Xie et al. [3] improved the neck network and head network of the YOLOv model, replaced the loss function with WIOUv3, and improved the color recognition accuracy of waste plastic bottles. Zeng et al. [4] proposed a method for plastic bottle recognition and localization based on YOLOv3, which utilized shallow enhanced features for convolution operations and combined them with the K-means clustering algorithm to optimize the anchor boxes. Zhang et al. [5] proposed a real-time water surface target recognition method based on Faster R-CNN to address the problem that water surface recognition is affected by sunlight and waves, and optimized the scale and aspect ratio of the anchor points. 2) Image classification models emphasize the generalization ability of the model and the classification accuracy. Chen et al. [6] designed four recognition models for plastic floating objects on lake surfaces based on VGG (Visual Geometry Group), and used multi-channel convolution to extract texture features. Li et al. [7] proposed fine-tuned the AlexNet model using the gradient descent method and processed the image illumination to improve the recognition accuracy of small-sample water surface floating objects. Wei [8] used the VGG network to calculate the gram matrix to distinguish floating objects. 3) Instance segmentation models focus on the segmentation accuracy and discrimination ability. Liu et al. [9] proposed a recognition method for water surface floating objects based on Mask R-CNN, which enhanced the generalization ability in natural environments. Tong et al. [10] improved the Mask R-CNN network model, introduced the attention mechanism SENet module into the FPN network, and improved the recognition and segmentation accuracy. Su et al. [11] utilized the Mask R-CNN model to detect and recognize water surface obstacles and improved the monocular ranging method.

Although the above-mentioned research achievements are fruitful, they all have a common problem, that is, the datasets are too limited. The limited datasets lead to insufficient extraction of plastic bottle feature information, limited generalization ability of the model, and thus reduced recognition efficiency. In the field of machine learning, the scale of the training dataset has a crucial impact on the performance of the algorithm. It is difficult for a small-scale dataset to enable the algorithm to learn enough feature information.

To effectively address this issue, this paper proposes flipping transformation to expand the number of images. The image dataset is expanded by flipping the original images horizontally and vertically to obtain mirrored images, which enables the model to learn more abundant features. Meanwhile, this paper applies the Faster R-CNN algorithm and the YOLOv5 algorithm respectively to train and recognize the expanded samples. The experimental results show that after the adoption of flip transformation, the experimental results are significantly improved compared with those without it, and the overall performance is more optimized. This is a new idea for the accurate recognition of plastic bottles on the water surface, which can improve the intelligent development of water surface cleaning robots.

Fig. 1. Sample images in plastic bottle dataset

2   Plastic Bottles on the Water Surface

All images of plastic bottles on the water surface are sourced from the public dataset of the PaddlePaddle Community (https://aistudio.baidu.com/datasetdetail/247554), This dataset encompasses images of plastic bottles on the water surface in various forms under different environmental conditions. All the images are stored in JPG format, and their resolutions are either 1280×720 or 1280×640. Some images of the dataset are shown in Fig. 1. According to the states of the plastic bottles on the water surface, these images can be classified into images of cylindrical plastic bottles, square plastic bottles, extruded plastic bottles, plastic bottles along the boundary, plastic bottles under strong light, plastic bottles under weak light, plastic bottles in the distant view, plastic bottles in the close view, plastic bottles floating on the water surface, plastic bottles half-immersed in the water surface, plastic bottles blocking each other, and plastic bottles blocked by external objects, which corresponds to subgraphs in Fig. 1 respectively.

3   Recognition Method of Plastic Bottles

3.1   Flipping Transformationn

When using the machine learning algorithm to extract small target features, the small proportion of small targets in the image will lead to the lack of features, and then the feature information is extremely likely to be significantly reduced or even disappear during the extraction process. This seriously affects the target recognition performance, resulting in a decrease in recognition accuracy and an increase in the missed recognition rate. To improve the model's ability to recognize small targets, the flipping transformation is proposed. Flipping the dataset vertically or horizontally can expand the scale of the dataset, enabling the model to learn better feature representations on diverse data. Data from different angles can enhance the model's ability to recognize small targets, reduce the model's sensitivity to the target direction, and enable it to more stably recognize small targets in the face of various actual scenarios, thereby improving the overall recognition performance.

Flipping Horizontally. Let the width of the original target area be width. For any point (x, y) in the original image, after flipping horizontally, the new coordinates (x', y') are calculated as follows:

$$x' = \text{width} - x, \tag{1}$$
$$y' = y. \tag{2}$$

Example 1. Assuming that the width of the image is width $= 1280$, and a pixel point $(x, y) = (360, 120)$. According to the above formula, $x' = \text{width} - x = 1280 - 360 = 920, \quad y' = y = 120$. Therefore, the coordinates of this pixel point after flipping horizontally are $(x', y') = (920, 120)$.

Flipping Vertically. Let the height of the original target area be height. For any point (x, y) in the original image, after flipping vertically, the new coordinates (x', y') of this point are calculated as follows:

$$y' = \text{height} - y, \tag{3}$$
$$x' = x. \tag{4}$$

Example 2. Assuming that the height of the image is height $= 720$, and a pixel point $(x, y) = (360, 120)$. According to the above formula, $x' = x = 360$,

$y' = \text{height} - y = 720 - 120 = 600$. Therefore, the coordinates of this pixel point after flipping vertically are $(x', y') = (360, 600)$.

An original image, its horizontally flipped image, and its vertically flipped image correspond to the subgraphs in Fig. 2.

(a) original image   (b) horizontally flipped image   (c) vertically flipped image

Fig. 2. Images before and after flipping transformation

3.2 Faster R-CNN Algorithm

In the field of object recognition, Faster R-CNN, as a typical two-stage object recognition algorithm, holds a pivotal position. Although the early R-CNN and Fast R-CNN [12] achieved certain performance, their efficiency was not satisfactory. Faster R-CNN innovatively replaced the traditional selective search algorithm with the Region Proposal Network (RPN), optimized the model architecture, significantly reduced the computational amount, and remarkably improved the object recognition efficiency, enabling a qualitative leap in the overall performance. It mainly consists of five key modules: the feature extraction network, the RPN [13], ROI (Regions of Interest) Pooling [14], the fully connected (FC) layer, and the classification and regression module. The feature extraction network extracts key features from the input image, providing an information basis for subsequent recognition; the RPN efficiently generates multi-scale anchor boxes on the feature map to locate the target regions; ROI Pooling adjusts candidate regions of different sizes to a fixed size to adapt to the fully-connected layer; the fully connected (FC) layer deeply processes the features of the candidate regions and learns the feature representations; the classification and regression module screens and accurately classifies the target candidate regions to determine the target category and location information.

The execution process, as shown in Fig. 3, of the Faster R-CNN algorithm is rigorous and efficient. 1) The original image is input into the VGG-16 (visual geometry group-16), a classic deep convolutional neural network for feature extraction, generating a shared feature map containing rich semantic information [15], which provides a data basis for subsequent recognition. 2) The feature map enters the RPN layer and the ROI Pooling layer. In the RPN layer, a 3×3 sliding window, with a step size and padding set to 1, performs convolution and activation operations on the feature map, generating anchor boxes with various aspect ratios point-by-point. The RPN network calculates the foreground probability, background probability, and offset of each anchor box to locate the target.

Subsequently, the feature map is mapped back to the original image proportionally, and the Non-Maximum Suppression (NMS) algorithm is used to remove the candidate boxes with high overlap and low confidence. The optimized candidate regions are then passed to the ROI Pooling layer. The ROI Pooling layer combines the feature map with the candidate regions and generates a fixed-size feature map by cropping and scaling. Finally, the FC layer further extracts features, and the classification and regression module classify the targets based on these features and outputs the class confidence and position coordinates.

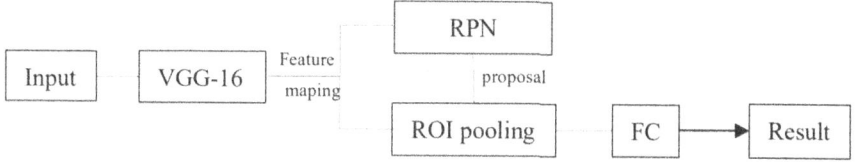

Fig. 3. Workflow of Faster R-CNN

3.3   YOLOv5 Algorithm

In the field of object recognition, the YOLOv5 algorithm [16], as a leading one-stage object recognition algorithm, occupies an important position with its high efficiency and accuracy. One-stage algorithms directly predict the categories and positions of objects from the input image, avoiding the complex processes of two-stage algorithms and significantly improving the recognition speed. Compared with the previous generation, YOLOv5 adopts the Path Aggregation Network (PANet) [17] in the neck part, effectively fusing feature information at different levels and enhancing the perception of objects at different scales. The YOLOv5 network structure consists of four parts: the input terminal, the backbone network (Backbone), the neck network (Neck), and the recognition head (Head), as shown in Fig. 4. The input terminal pre-processes the input data to make it more suitable for network training and recognition; the backbone network extracts image features through multiple convolutional layers; the neck network, as a key innovation point, fuses multi-scale features, optimizes the recognition performance, and improves the recognition ability for objects of different sizes; the recognition head processes the feature maps fused by the neck network, and through calculation and judgment, outputs the final object recognition results.

The input terminal pre-processes the images using mosaic data augmentation and adaptive anchor box calculation [18]. The former stitches four different images together to increase the image diversity and enhance the model's generalization ability. The latter automatically adjusts the size and proportion of the anchor boxes according to the training data, laying a more accurate foundation for recognition. The pre-processed images are entered the backbone network, and features are extracted through multiple convolutional layers. The cross-stage partial network (CSPNet) module [19] fuses feature across stages, reducing the computational amount while extracting deeper-level features and enhancing the network's feature representation ability. Subsequently, in the neck network, the features are fused through the Feature Pyramid Network (FPN) [20] and the Path Aggregation Network (PAN) [21]. The FPN extracts multi-scale features, and the PAN further integrates and optimizes them to generate a more representative feature map, enhancing the recognition ability for objects of different scales. Finally, the recognition head analyzes and judges it using multiple recognition layers of different scales. Each recognition layer is responsible for objects within a specific scale range and outputs the object category and position information.

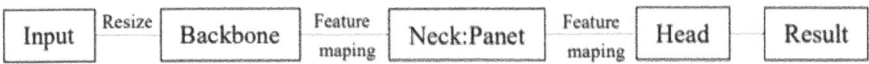

Fig. 4. Workflow of YOLOv5

4   Experiments

4.1   Experimental Dataset

The original dataset of this experiment is sourced from the open dataset on the Paddle community (https://aistudio.baidu.com/datasetdetail/247554), which contains 1200 images. To expand the dataset, the flip-transformation method is adopted to perform horizontal and vertical flip operations on the 1200 images respectively, generating 1200 new images each. Then, "data1", "data2", "data3", and "data4" refer to the original dataset, the union set of the original dataset and the horizontal flipped dataset, the union set of the original dataset and the vertical flipped dataset, and the union set of the original dataset, the horizontal flipped dataset, and the vertical flipped dataset, respectively. And the numbers of images in these datasets are 1200, 2400, 2400, and 3600, respectively.

The models were trained on the NVIDIA GeForce GTX 1050 Ti hardware environment equipped with 4 GB of video memory. For the Faster R-CNN model, the initial learning rate is 0.0001, which increases and decreases periodically with the training process, with a minimum value of 0.000001. The Batch Size is 4 during frozen layer training and adjusted to 2 during fully thawed training. The total training epochs are 150, and the Intersection over Union (IoU) threshold is 0.5. For the YOLOv5 model, the initial learning rate is 0.01 and exhibits linear decay, the Batch Size is 16, the IoU threshold is 0.5, and the total number of iterations is 300.

4.2   Performance Evaluation

P, R, and F1 represent precision, recall, and the harmonic mean of precision and recall, respectively, and are selected as the experimental indicators. The calculation formulas for these three indicators are as follows:

$$P = \frac{A \cap B}{A} \tag{5}$$

$$R = \frac{A \cap B}{B} \tag{6}$$

$$F1 = \frac{2 \times P \times R}{P + R} \tag{7}$$

Where A represents the set of plastic bottles recognized by the algorithm, and B represents the set of plastic bottles annotated in the dataset.

4.3   Multi-object Recognition

Multi-object recognition means that the algorithms R-CNN and YOLOv5 need to recognize all plastic bottles in the images, and the test result are shown in Fig. 5, Fig. 6, and Fig. 7. Among them, the Score_Threshold is a preset score value in the object recognition process. When the score of an object is greater than Score_Threshold, the object will be determined as a valid object.

The precision P of the algorithms Faster R-CNN and YOLOv5 varies with the change of Score_Threshold, as shown in Fig. 5. As the Score_Threshold gradually decreases, the algorithms return more recognition objects, so the precision of two algorithms show a downward trend. For the Faster R-CNN algorithm, when the data4 is selected as the test dataset, its precision is significantly higher than that of other datasets. For the YOLOv5 algorithm, although the precision P on data4 is not the best, it is still significantly better than the precision P on data1. Overall, no matter how Score_Threshold changes, the precision P on data4 is significantly better than that on data1, which means that flipping transformation are very effective.

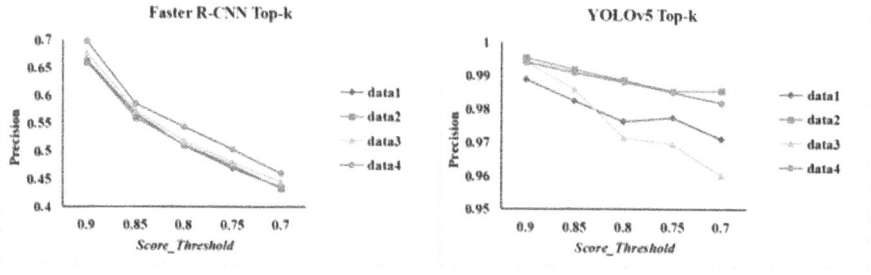

Fig. 5. The precision P for multi-object recognition

The recall R of the algorithms Faster R-CNN and YOLOv5 varies with the change of Score_Threshold, as shown in Fig. 6. As Score_Thresholdgradually decreases, the algorithms return more positive object, so the recall R of two algorithms show an upward trend. For the Faster R-CNN algorithm, when the data1 is selected as the test dataset, its recall R is significantly higher than that of other datasets. For the YOLOv5 algorithm, the recall R of data2 is significantly higher than that of other datasets.

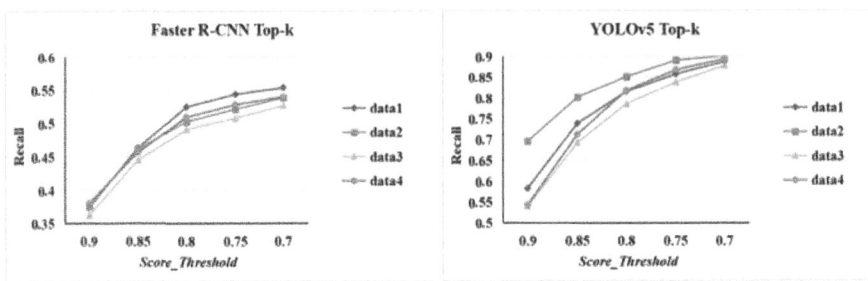

Fig. 6. The recall R for multi-object recognition

As the harmonic mean of precision and recall, the $F1$ score can comprehensively reflect the performance of the algorithms. The $F1$ score of the algorithms Faster R-CNN and YOLOv5 varies with the change of Score_Threshold, as shown in Fig. 7. As the Score_Threshold gradually decreases, the $F1$ score of the algorithm Faster R-CNN shows an upward trend followed by a downward trend, and the $F1$ score of the algorithm YOLOv5 shows an upward trend. For the Faster R-CNN algorithm, when the Score_Threshold = 0.8, the $F1$ score of each dataset reach their peaks, indicating that the precision and recall of the algorithm Faster R-CNN are in the optimal balance at this point. For the YOLOv5 algorithm, the $F1$ score over data2 are higher than that over other dataset.

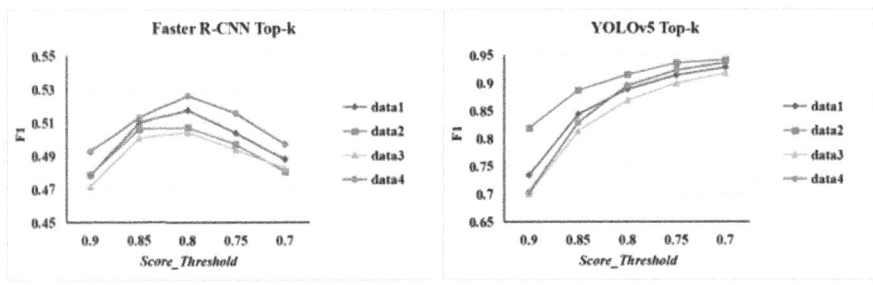

Fig. 7. The $F1$ scores for multi-object recognition

4.4   Single-Object Recognition

Single-object recognition means that the algorithms R-CNN and YOLOv5 only need to recognize one plastic bottle in each image, and the test result are shown in Fig. 8. The purpose of this paper is to assist the water surface cleaning robot in achieving efficient recognition of plastic bottles on the water surface. In practical applications, even if there are multiple plastic bottles on the water surface simultaneously, the robot can only salvage them one by one. Based on this, this paper conducts experiments for single-object recognition, and the accuracy rate is selected as the evaluation index because the algorithms generate only one result on each image. From Fig. 8, no matter how Score_Threshold changes, the precision P on data4 is significantly better than that on data1, which means that flipping transformation are very effective. Especially, for the YOLOv5 algorithm, the precision P on data4 reaches 100%.

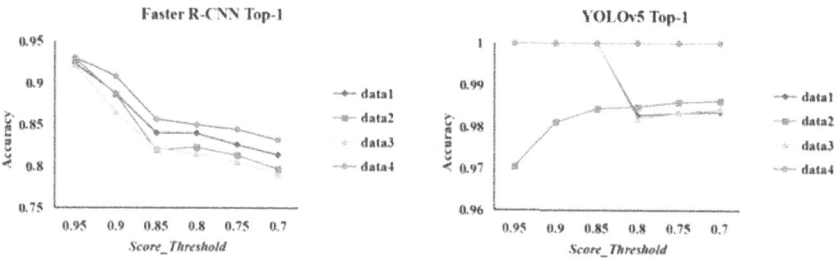

Fig. 8. The accuracy for single-object recognition

4.5 Recognition Time

Testing times of the algorithms Faster R-CNN and YOLOv5 are shown in Table 1. From Table 1, regardless of the longest, shortest, or average time, the testing time for Faster R-CNN is longer than that of YOLOv5, because the algorithm Faster R-CNN uses the region proposal network (RPN) to generate candidate regions, and its image processing process is more complex, thus needing more time.

Table 1. Recognition times of Faster R - CNN and YOLOv5

	Longest time	Shortest time	Average time
Faster R - CNN	1.8170	1.6482	1.6798
YOLOv5	0.0951	0.0281	0.0337

5 Conclusion

Plastic bottles on the water surface are important factors causing water source pollution. They not only seriously affect the aesthetics of water bodies, but also cause incalculable damage to the aquatic ecosystem. The low efficiency of traditional manual salvage makes the automated fishing technology of plastic bottles on the water surface extremely urgent. The primary prerequisite for achieving intelligent water surface garbage fishing is to accurately complete the recognition of target garbage. However, the recognition methods of plastic bottles on the water surface generally have the problem of dataset limitations. The limited datasets lead to insufficient extraction of plastic bottle feature information, limited generalization ability of the model, and thus reduced recognition efficiency.

Therefore, to address this problem, this paper proposes a data augmentation method based on flipping transformation. Flipping the dataset vertically or horizontally can expand the scale of the dataset, enabling the model to learn better feature representations on diverse data.

To verify the effectiveness of flipping transformation, this paper organized multi-object and single-object recognition experiments based on the algorithms Faster R-CNN and YOLOv5. Regardless of the multi-object recognition or the single-object recognition, the precision P on dataset after flipping transformation is significantly better than that on dataset before flipping transformation, which means that flipping transformation are very effective. Moreover, for the single-object recognition, the accuracy of the YOLOv5 algorithm on the dataset after flipping transformation reaches 100%.

Acknowledgments. This work is supported by the Chongqing Social Science Planning Project (2024BS034), the National Natural Science Foundation of China (62106024), the China Postdoctoral Science Foundation (2022M711458), the Special Funding for Postdoctoral Research Projects in Chongqing (2023CQBSHTB3118), the Chongqing Language Research Project (yyk22105), the project of Chongqing Education Commission of China (KJZD-K202401405, KJQN202401 432, KJQN202201410, KJQN202301415, KJQN202201413, KJZD-M202201401, KJQN202301416, KJQN20 2303123), the Fuling District Research Project (FLKJ 2024BAG5134), and the College Student Innovation and Entrepreneurship Training Program (202410647005, 202310647004).

Disclosure of Interests. The authors have no competing interests to declare that are relevant to the content of this article.

References

1. Tang, W., Gao, H., Liu, S.: Design and implementation of small waters intelligent garbage cleaning robot system based on Raspberry pi. Sci. Technol. Eng. **19**(34), 239–247 (2019)
2. Zhu, J., Yang, Y., Cheng, Y.: SMURF: a fully autonomous water surface cleaning robot with a novel coverage path method. J. Mar. Sci. Eng. **10**(11), 1620–1620 (2022)
3. Xie, S., Wu, H., Mao, W., Chu, X., Yang, X.: Research on efficient color recognition method for waste plastic bottles based on deep learning. Plast. Sci. Technol. **52**(11), 140–146 (2024)
4. Zeng, W., Yin, S., Zhang, F.: Research on recognition and location algorithm of waste plastic bottle based on computer vision. Electron. Meas. Technol. **44**(23), 12–17 (2021)
5. Zhang, L., Zhang, Y., Zhang, Z., Shen, J., Wang, H.: Real-time water surface object detection based on improved faster R-CNN. Sensors **19**(16), 3523–3523 (2019)
6. Chen, Y.: Recognition of plastic products on lake based on VGG. Plast. Sci. Technol. **48**(08), 77–80 (2020)
7. Li, N., Wang, Y., Xu, S., Shi, L.: Recognition of floating objects on water surface with small sample based on AlexNet. Comput. Appl. Softw. **36**(02), 245–251 (2019)

8. Wei, L.: Research on surface floating object monitoring based on binocular vision. Dalian Maritime University (2019)

9. Liu, W., Wang, Y., Jiang, S., Ma, T., Xiao, W.: Research on recognition method of floating objects on water surface based on Mask R-CNN. Yangtze River **52**(11), 226–233 (2021)

10. Tong, J., Yu, Y., Han, S.: Water surface object detection algorithm based on improved Mask R-CNN network. J. Qilu Univ. Technol. **37**(05), 12–18 (2023)

11. Su, P., Zhu, X.: Research on water surface target recognition and ranging based on monocular vision. Comput. Technol. Dev. **31**(02), 80–84 (2021)

12. Shilpa, R., Deepika, G., Sandeep, K.: Object detection and recognition using contour based edge detection and fast R-CNN. Multimedia Tools Appl. **81**(29), 42183–42207 (2022)

13. Ren, S., He, K., Girshick, R., Sun, J.: Faster R-CNN: towards real-time object detection with region proposal networks. IEEE Trans. Pattern Anal. Mach. Intel. **39**(6), 1137–1149 (2015)

14. Tran Duy, L., Arai, M.: Multi-scale subnetwork for rol pooling for instance segmentation. Int. J. Comput. Theory Eng. **10**(6), 207–211 (2018)

15. Girma, T., et al.: Advanced image preprocessing and integrated modeling for UAV plant image classification. Drones **8**(11), 645–645 (2024)

16. Chen, R., Liu, Z., Ou, W., Zhang, K.: Small target detection algorithm based on improved YOLOv5. Electronics **13**(21), 4158 (2024)

17. Wang, W., Li, Y., Zhang, Y., Han, P., Liu, S.: MPANet-YOLOv5: multi-path aggregation network for complex sea object detection. J. Hunan Univ. (Nat. Sci.) **49**(10), 69–76 (2022)

18. Wang, K., Ding, Q.: Remote sensing images small object detection algorithm with adaptive fusion and hybrid anchor detector. J. Electron. Inf. Technol. **46**(07), 2942–2951 (2024)

19. Senussi, M., Kang, H.: Occlusion removal in light-field images using CSPDarknet53 and bidirectional feature pyramid network: a multi-scale fusion-based approach. Appl. Sci. **14**(20), 9332 (2024)

20. Han, B., He, L., Ke, J., Tang, C., Gao, X.: Weighted parallel decoupled feature pyramid network for object detection. Neurocomputing **593**, 127809 (2024)

21. Wang, C., Xu, C., Wang, C., Tao, D.: Perceptual adversarial networks for image-to-image transformation. IEEE Trans. Image Process. **27**(8), 4066–4079 (2018)

A Multi-level Tree Mapping for Hyper-parameter Assigning in Network Representation Learning

Hao Li[1], Shun Fu[1], and Huajian Xie[2($\boxtimes$)]

[1] School of Artificial Intelligence and Big Data, Chongqing Industry Polytechnic College, Chongqing 401120, China
[2] Chongqing Academy of Education Science, Chongqing 400015, China
huajian28@gmail.com

Abstract. Network data is ubiquitous in real-life applications, such as social media, advertising recommendation and intelligent decision making. Network representation learning techniques map elements in the network onto a low-dimensional vector space, and have found widespread application in downstream tasks such as node classification. In recent years, node2vec methods in the field of network representation learning have been used to generate low-dimensional representation vectors of network nodes by collecting network structural features through random walk and using the word2vec language model. However, the two hyperparameters controlling breadth-first search and depth-first search in node2vec are given globally and uniformly, which makes the random walker insufficiently adaptive. In the network, different nodes have different structural characteristics, and the more specific hyperparameters need to be set according to the structural characteristics of the nodes. The multi-level tree structure mapping method proposed in this paper first maps nodes to different levels by calculating the importance of nodes. Different depth and breadth search intensities are adopted for network nodes at different levels, allowing the random walker to adopt different search strategies at different nodes. Thus the algorithm captures the structural features more effectively. Through experiments on realistic network data, it can be seen that the proposed multi-level tree mapping (MTMap) method can effectively improve the performance of the node2vec method.

Keywords: network representation learning · random walk · data mining · machine learning

1 Introduction

Network representation learning simplifies the complexity of traditional adjacency matrix network representations by mapping the nodes in the network data to a low-dimensional vector space. At the same time, the structural characteristics of the original network are preserved in the low-dimensional representation vectors. For example, nodes that have a community arrangement in the original network will have their low-dimensional representation vectors distributed in a

Q. Zhang et al. (Eds.): IJCRS 2025, LNAI 15710, pp. 468–478, 2025.
https://doi.org/10.1007/978-3-031-92741-6_33

cluster in the vector space. When nodes are represented as low-dimensional vectors, subsequent machine learning based network analysis methods become convenient. Typical applications include: node classification, link prediction, community detection, visualisation, etc.

The core task of network representation learning techniques is extracting features in the network. In this paper, we are primarily concerned with structural features in the network, i.e. features of the topology formed by the connections between nodes. Among the network representation learning methods that are focusing on structural feature extracting, there are three main categories of methods: matrix decomposition-based methods [8], random walk-based methods [6] and graph neural network-based methods [9].

In the random walk-based approaches, the node2vec [6] framework learns low-dimensional representations for nodes in a graph by optimizing a neighborhood preserving objective. The objective is flexible, and the algorithm accommodates for various definitions of network neighborhoods by simulating biased random walks. Specifically, it provides a way of balancing the exploration-exploitation trade-off that in turn leads to representations obeying a spectrum of equivalences from homophily to structural equivalence. As Figure 1 show, node2vec framework defines two hyperparameter: p and q. After the walker transitioning to node (v) from node (t), the return hyperparameter, p and the inout hyperparameter, q control the probability of a walk staying inward revisiting nodes (t), staying close to the preceding nodes (x_1), or moving outward further away (x_2, x_3). In detail, If the value of p is set to be less than 1 and the value of q is greater than 1, the random walker prefers to explore a wider range of nodes. This paper hence uses the term BFS (breadth-first search) for this strategy. Conversely, if the p-value is greater than 1 and the q-value is less than 1, then the random walker prefers to explore deeper nodes, hence the term DFS (depth-first search) is in use. The node2vec framework introduces a parameter setting strategy for adjusting the random walks, and this method influences the structural features that the random walker attends to.

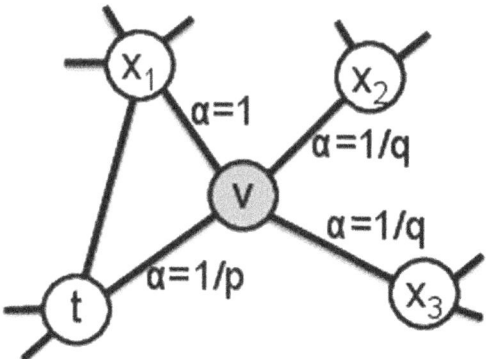

Fig. 1. The strategy of random walk in node2vec framework.

However, different nodes in the network are not in the same structural position and each node has different structural characteristics. If the BFS is applied equally on each node, nodes at the centre of the community or connected by more nodes will be sampled and even over-sampled, while the remote nodes that has less connections will tend to be under-sampled. If the DFS is applied equally on each node, nodes at the centre of the community or connected to more communities (high importance) will be under-sampled and the remote nodes will receive more random walker visits, which may result in high importance nodes being under-sampled and the remote nodes being over-sampled. As shown in Fig. 2, node 4 is located at the intersection of multiple communities, and a breadth-first search strategy is appropriate for node 4 when performing random walks on this subgraph structure. However, for node 3, as it is located on a unique path, a depth-first search strategy is appropriate; otherwise, the random walker will backtrack, resulting in under-sampling.

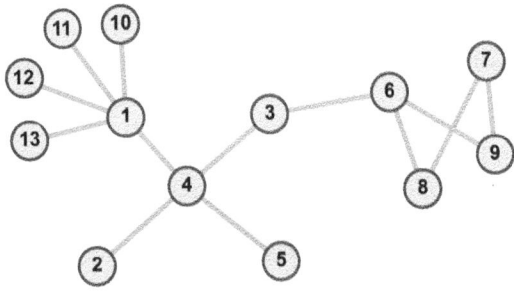

Fig. 2. Nodes with different local structural should adopt personalized random walk hyperparameters.

From this case, we can see that the uniform p, q setting is used by node2vec. This is not an optimal search strategy in some cases. We argue that the random walker should be personalised to specify the transition probability according to the structural characteristics of the nodes, thus optimising the sampling corpus of the nodes and ultimately improving the performance of network representation learning.

In this paper, the original graph nodes are evaluated according to their structural importance and the pq-value assignment problem is personalised by mapping the graph nodes onto a tree structure. By mapping the tree structure to different levels, the pq-values of the nodes are thus assigned individually according to their importance, rather than the equally pq setting strategy in the node2vec framework. This unequally assigning pq strategy allows the random walker to better balance the DFS and BFS, and hence the visiting resource allocation. In particular, such a unequally allocation strategy makes sense in terms of saving the number of walks, and walk length of the random walker. Experiments on the real datasets show that the unequally pq allocation strategy in this paper achieves improvement over a global uniform allocation strategy.

In order to address the limitations of global hyperparameter settings in node2vec, this paper proposes a novel multi-level tree mapping (MTMap) method that assigns personalized p, q values to nodes based on their structural importance. This individualized assignment enables a more balanced exploration of breadth-first search (BFS) and depth-first search (DFS) strategies, leading to improved performance in network representation learning. The rest of this paper is structured as follows. In the Sect. 2, the related work is briefly reviewed. The innovative approach and technical details are described in Sect. 3. In Sect. 4, we empirically evaluate our approach and the most closely related methods on classification tasks over nodes on various real-world data sets. The conclusion which contains an outlook on possible future directions is allocated in Sect. 5.

2   Related Works

Network representation learning, also known as network embedding, obtains the low-dimensional vector by either applying a general dimensionality reduction or by methods that are specifically tailored towards network-specific properties. A wide variety has been proposed in the survey proposed by Goyal [5]. In the early 2000s, the representative works such as IsoMap [17], Locally Linear Embedding (LLE) [13] and Laplacian Eigenmap [2] calculate the similarity between pairwise data points to construct an affinity graph and then represent the affinity graph into a new space having much lower dimensionality. With the development of language model and neural networks, the research efforts have shifted to scalable algorithms. The representative models like DeepWalk [12], LINE [15], SDNE [18], GraRep [4] embed the network structural proximities into latent, low-dimensional space while the models like [19,20], and [21] attach the rich content and side information on attributes. Recently, some variety of random walk based graph embedding are published such as HALK [14], NRL-RWCE [7]. NRL-RWCE [7] estimates three random walk parameters: the walk length, the number of walks, and the sliding window size. HALK [14] represents the graph features at different levels of detail by removing a fraction of the least frequent nodes from the original walks at different levels. These work fine-tune the parameters of the model so that the strategy of the model can adapt to the specific needs of network analysis. Network representation leaning shows remarkable performance for various applications such as node classification, link prediction, community detection, visualization, etc. In this paper, from the perspective of random walk strategy optimization, the original graph structure is processed in layers. Compared with the unified random walk strategy, hierarchical strategy and personalized transfer can better balance the resources of the walker and reduce the problems of over sampling and under sampling.

3   Methods and Models

3.1   Definitions

In this paper, we use the following notation throughout:

- $\mathcal{G}$: Graph
- $\mathcal{V}$: Set of nodes, where each element v_i represents a node
- $\mathcal{E}$: Set of edges, which may be weighted or unweighted
- $\mathbf{A}$: Adjacency matrix, where a_{ij} indicates the connection status from v_i to v_j
- $f : \mathcal{V} \to \mathbb{R}^d$: Mapping function that maps each node to a d-dimensional vector
- Other symbols (e.g., pr_i, y_i) will be explicitly defined in their respective sections

Network and Graph. A graph $\mathcal{G}$ is defined as a pair of sets $(\mathcal{V}, \mathcal{E})$, where $\mathcal{V}$ denotes the set of nodes and $\mathcal{E}$ the set of edges. Each element $v_i \in \mathcal{V}$ represents a node, and each edge $e_{ij} \in \mathcal{E}$ may be associated with a weight through an explicit partial weight function $w : \mathcal{E} \to \mathbb{R}^+$.

The adjacency matrix $\mathbf{A} = (a_{ij})_{|\mathcal{V}| \times |\mathcal{V}|}$ is defined by:

$$a_{ij} = \begin{cases} 1, & \text{if there exists an edge from } v_i \text{ to } v_j, \\ 0, & \text{otherwise.} \end{cases} \tag{1}$$

Network Representation Learning (NRL). Given a network $\mathcal{G} = (\mathcal{V}, \mathcal{E})$. The goal of *Network representation learning* is to learn a mapping function $f : v \mapsto \mathbf{z} \in \mathbb{R}^d$ where $\mathbf{z}$ is the learned representation vector, and $d \ll |V|$ is the dimension of $\mathbf{z}$. The mapping function f preserves the original network information, such that the distribution of vectors in the hidden vector space can reflect the structural proximity between nodes in the original graph.

3.2   Multi-level Tree Mapping Model

In this section, we describe a model that maps nodes from a given set $\mathcal{V}$ onto a tree structure by generating a spanning tree. The procedure involves two main parts: constructing a directed graph from the original node set and determining node hierarchy via a PageRank-based importance measure. The overall process is detailed as follows:

1. **Initialization:**
   - Let $\mathcal{V}$ be the set of all nodes.
   - Define the directed graph $\mathcal{G}_1 = (\mathcal{V}_1, \mathcal{E}_1)$ such that $\mathcal{V}_1 = \mathcal{V}$ and the edge set $\mathcal{E}_1$ is initially empty.
2. **Compute Node Importance:**
   - For each node $v_i \in \mathcal{V}_1$, calculate an importance value pr_i using the PageRank algorithm [11].

3. **Parent Node Assignment:** For every node $v_i \in \mathcal{V}_1$, perform the following:
   - Identify a node $v_k \in \mathcal{V}_1$ that satisfies both:
   (a) $pr_k > pr_i$, meaning v_k is more important than v_i;
   (b) The distance (evaluated using a chosen metric such as the Jaccard distance or SimRank distance) between v_i and v_k is minimized relative to other nodes with higher PageRank values.
   - Assign v_k as the parent of v_i and add the directed edge e_{ik} to $\mathcal{E}_1$.
4. **Root Node Identification:**
   - The node with the highest PageRank value is not assigned any parent. This node is designated as the root node, denoted by v_{root}.
   - Consequently, the total number of edges satisfies $|\mathcal{E}_1| = |\mathcal{V}| - 1$.
5. **Determination of Node Levels:**
   - For each node v_i, the layer number y_i is defined as the length of the shortest path from v_i to the root node v_{root} within the tree $\mathcal{G}_1$.

In addition, the PageRank values are computed via either an iterative method (akin to the power iteration method [1]) or an algebraic approach. In the iterative approach, the computation proceeds as follows:

1. **Initialization:** At time $t = 0$, assign an equal probability to each node:

$$PR(v_i; t = 0) = \frac{1}{|\mathcal{V}|}. \tag{2}$$

2. **Iterative Update:** For each subsequent iteration ($t \geq 0$), update the PageRank of node v_i according to:

$$PR(v_i; t + 1) = \frac{1 - e}{|\mathcal{V}|} + e \sum_{v_j \in M(v_i)} \frac{PR(v_j; t)}{L(v_j)}, \tag{3}$$

where e is the damping factor, $M(v_i)$ is the set of nodes that connected with v_i, and $L(v_j)$ is the degree of v_j.

3.3   Assign the p,q Value for Node v_i

In this subsection, we discuss how to implement the p, q assigning. Assume the layer of v_i is y_i, $y_0 < y_i < y_{max}$, $y_0 = 0$. Define the middle layer number $y_{med} = \text{floor}(0.5 * (y_{max}))$. The floor function of x is defined to be the greatest integer that is less than or equal to x. Define a parameter d_{max} for the strength of DFS and BFS, $d_{max} > 1, d_{max} \in \mathbb{Z}$, where the $\mathbb{Z}$ is the set of integers. We borrow the Sigmoid function for p, q calculation:

For the $y_i < y_{med}$, node v_i has larger importance, apply the BFS with equation below:

$$q = 1, \text{ and } \frac{1}{p} = \begin{cases} \frac{d_{max}}{1 + e^{(y_{med} - y_i)}} & y_i < y_{med} \\ 1 & otherwise \end{cases} \tag{4}$$

For the $y_i > y_{med}$, node v_i has less importance, apply the DFS with equation below:

$$p = 1, \text{ and } \frac{1}{q} = \begin{cases} \frac{d_{max}}{1+e^{(y_i - y_{med})}} & y_i > y_{med} \\ 1 & otherwise \end{cases} \tag{5}$$

With the methods described in Subsect. 3.3 and equation for p, q assignment, the MTMap model can obtain the p, q value for node v_i. Note the value of d_{max} is pre-defined by users. A typical value selection, such as 2, 4, 5 is preferred.

4   Experiment

In this section, we'll examine the effects of personalized p, q-value assignments on three actual social network data sets. Through statistics on visits that the nodes received and evaluation on network analysis task such as node classification, it's observed that the personalized assignment proposed by this paper had an obvious positive effect on the balance of random walk sampling resources.

4.1   Data Sets

Les Misérable. Data Set [10]. The Les Misérable is a French historical novel by Victor Hugo, first published in 1862, that is considered one of the greatest novels of the 19th century. This small scalre network contains 77 nodes corresponding to characters of the novel, and 254 vertices connecting two characters whenever they appear in the same chapter. This data set is used for statistic on the changing of visits received by nodes if p, q globally changes.

Cora. Data Set [3]. The Cora data set consists of 2,708 scientific publications of 7 classes. The graph has 5,429 links that indicate citation relations between documents. Each document has text attributes that are expressed by a binary-valued vector of 1,433 dimensions. We only use its linkages and labels.

blogCatalog. data set [16]. The blogCatalog data set is a social network. Nodes and edges represent the bloggers and the friendship between them respectively. It contains 10312 nodes, 333983 undirected edges, 39 labels.

4.2   Experimental Settings

For investigating the visits numbers each vertex received under different setting of p, q-value, the number of walk is set to 100 and length of walk is set to 50. This is done on the Les Misérable network. For investigating the F1-score of node classification on the different strategy of assigning p, q-value, the dimension of vectors is set to 64, the number of walk start on random node is set to 5, and the walk length is 10. This setting makes the random walk visiting resources relatively insufficient, which can make the significance of personalized p, q-value assigning highlighted.

4.3   Variation of Received Visits

In this section, we investigate the change of the random walk visits obtained by nodes under the different settings, i.e. the breadth-first search (BFS) and the depth-first search (BFS). We use the LesMisérable dataset to initially explore the relationship between the importance of nodes (defined in Sect. 3.2) and the random walk visits that the nodes in different graph structure can obtain.

To investigate how DFS and BFS affect node sampling, five nodes of degree 11 are picked out for statistics. With different pq-value settings, these nodes can obtain a variety of random walk visits. Table 1 shows the statistical results of the visits number obtained by nodes.

Table 1. The visits that nodes can obtain on different p, q-value settings.

Name of node	Degree	$p = q = 1$ (General)	$p = 1, q = 0.5$ (DFS)	$p = 1, q = 2$ (BFS)
Cosette	11	19153	17432	20926
Combeferre	11	17624	16333	18407
MmeThenardier	11	9663	9717	10001
Feuilly	11	9908	9078	10423
Eponine	11	5324	6374	4726

From Table 1, it can be observed that the node 'Cosette' can initially obtain the most visits, i.e. 19153 visits obtained on $p = 1, q = 1$ settings. Changing the setting from DFS to BFS makes the visits it obtained increase (see the columns on the right side). Nevertheless, for the node 'Eponine' which has less importance, the changing from DFS to BFS makes the visits it obtained decrease.

Compare the number of visits in column 3 and column 4, the number of visits obtained by 'Cosette' node decrease while the number of visits obtained by 'Eponine' node obtains increases. The nodes of 'Combeferre', 'MmeThenardier' and 'Feuilly' exhibit the same property with 'Cosette', except for 'Eponine'. If we look into the original graph structure, the node 'Eponine' has the less importance value in the 5 nodes listed in Table 1. This means the node with relatively less important can get more random walk visits in DFS, while the more important nodes get more random walk visits in BFS.

4.4   Optimal p, q-Value for Node Classification

On the Cora dataset, we first adopt the global p, q-value setting and obtain the low-dimensional representation vectors for the nodes by running node2vec algorithm [6]. Based on low-dimensional vector obtained, the SVM model is used to conduct the node classification experiments. Table 2 shows the node classification precision scores obtained by the model under different global pq-value Settings.

Table 2. The precision obtained on different sampling strategy on Cora dataset.

Strategy	Micro-F1	Macro-F1
$p = 1, q = 1$, General:	0.2551	0.1996
$p = 1, q = 2$, BFS1:	0.2526	0.1994
$p = 1, q = 4$, BFS2:	0.2493	0.1990
$p = 0.25, q = 4$, BFS3:	0.2405	0.1876
$p = 1, q = 0.5$, DFS1:	0.2622	0.1997
$p = 1, q = 0.25$,DFS2:	0.2592	0.2035
$p = 1, q = 0.1$, DFS3:	0.2516	0.1929
$p = 4, q = 0.25$, DFS4:	0.2457	0.1860
p, q assigned by MTMap	**0.2847**	**0.2344**
p, q inverse assign	0.2385	0.1855

Table 3. The precision obtained on different sampling strategy on BlogCatalog dataset.

Strategy	Micro-F1	Macro-F1
$p = 1, q = 1$, General:	0.2394	0.0716
$p = 1, q = 2$, BFS1:	0.2433	0.0745
$p = 1, q = 4$, BFS2:	0.2469	0.0781
$p = 1, q = 0.5$, DFS1:	0.2282	0.0645
$p = 1, q = 0.25$,DFS2:	0.2226	0.0622
$p = 4, q = 0.25$, DFS3:	0.2240	0.0616
p, q assigned by MTMap	**0.2494**	**0.0794**
p, q inverse assign	0.2421	0.0733

It can be seen that, in Table 2, on the Cora data set, using the global p, q-value setting and the global DFS/BFS sampling strategy, the low-dimensional vector representation obtained cannot achieve more than 26% microF1 score on the node classification task. However, using the method presented in this paper, i.e. the personalized assignment of p, q-value, can achieve a performance improvement of about 15%. In order to illustrate the effectiveness of MTMap personalized assignment of p, q, we reversely assign the p and q values, that is, we apply BFS strategy for nodes that should be applied by DFS, and DFS for nodes that should be configured with BFS's p, q setting. The results are recorded in the row named as 'p, q inverse assign' in Table 2. It can be seen that the precision obtained by this reverse operation is significantly lower, and even lower than the result of global p, q-value assignment.

Table 3 shows the precision of node classification on BlogCatalog dataset. The improvement of precision obtained by the p, q assigned by MTMap is not obvious on the BlogCatalog dataset. This can have reason that the structure of graph in BlogCatalog dataset has not a relative equably distribution of importance. For a dataset with a large number of nodes in a decentralized community that is not important for global graph connection, the strategy of adjusting p and q may not be as effective as just using global p, q Settings.

5   Conclusions

In this paper, we proposed the MTMap method to optimize the assignment of p and q values in node2vec by leveraging a multi-level tree mapping based on PageRank-calculated node importance. The experimental results demonstrate that our personalized hyperparameter assignment significantly improves node classification performance compared to the global uniform setting, achieving approximately 15% improvement in micro-F1 score on the Cora dataset.

The key contribution of our work is demonstrating that adaptive, structure-aware hyperparameter assignment can substantially outperform global uniform settings. This is particularly effective in networks where nodes exhibit diverse structural characteristics and importance levels.

While the current method has shown promising results, further research is needed to explore the deeper relationship between graph structure and random walk strategies, which may lead to even more effective hyperparameter allocation schemes. Future work could focus on developing more sophisticated importance metrics and exploring how the proposed approach scales to larger, more complex networks.

References

1. Arasu, A., Novak, J., Tomkins, A., Tomlin, J.: PageRank computation and the structure of the web: experiments and algorithms. In: Proceedings of the Eleventh International World Wide Web Conference, Poster Track, pp. 107–117 (2002)
2. Belkin, M., Niyogi, P.: Laplacian eigenmaps and spectral techniques for embedding and clustering. In: Advances in Neural Information Processing Systems, pp. 585–591 (2002)
3. Cabanes, C., et al.: The CORA dataset: validation and diagnostics of in-situ ocean temperature and salinity measurements. Ocean Sci. **9**(1) (2013)
4. Cao, S., Lu, W., Xu, Q.: GraRep: learning graph representations with global structural information. In: Proceedings of the 24th ACM International on Conference on Information and Knowledge management, pp. 891–900 (2015)
5. Goyal, P., Ferrara, E.: Graph embedding techniques, applications, and performance: A survey. Knowl.-Based Syst. **151**, 78–94 (2018)
6. Grover, A., Leskovec, J.: node2vec: scalable feature learning for networks. In: Proceedings of the 22nd ACM SIGKDD International Conference on Knowledge Discovery and Data Mining, San Francisco, CA, USA, 13–17 August 2016, pp. 855–864. ACM (2016)
7. Guo, K., Wang, Q., Lin, J., Wu, L., Guo, W., Chao, K.-M.: Network representation learning based on community-aware and adaptive random walk for overlapping community detection. Appl. Intell., 1–19 (2022)
8. Hamilton, W.L.: Graph representation learning. Synth. Lect. Artif. Intell. Mach. Learn. **14**(3), 1–159 (2020)
9. Kipf, T.N., Welling, M.: Semi-supervised classification with graph convolutional networks. arXiv preprint arXiv:1609.02907 (2016)
10. Mardewi, T., Larashandy, H.R., Indriyani, F., Dewi, N.: The leadership of Jean Valjean in Les Miserable movie directed by Tom Hooper. Engl. Educ. J. Engl. Teach. Res. **6**(1), 87–100 (2021)

11. Page, L., Brin, S., Motwani, R., Winograd, T.: The pagerank citation ranking: bringing order to the web. Technical report, Stanford InfoLab (1999)
12. Perozzi, B., Al-Rfou, R., Skiena, S.: DeepWalk: online learning of social representations. In: Proceedings of the 20th ACM SIGKDD International Conference on Knowledge Discovery and Data Mining, pp. 701–710. ACM (2014)
13. Roweis, S.T., Saul, L.K.: Nonlinear dimensionality reduction by locally linear embedding. Science **290**(5500), 2323–2326 (2000)
14. Schlötterer, J., Wehking, M., Rizi, F.S., Granitzer, M.: Investigating extensions to random walk based graph embedding. In: 2019 IEEE International Conference on Cognitive Computing (ICCC), pp. 81–89. IEEE (2019)
15. Tang, J., Qu, M., Wang, M., Zhang, M., Yan, J., Mei, Q.: LINE: large-scale information network embedding. In: Proceedings of the 24th International Conference on World Wide Web, pp. 1067–1077 (2015)
16. Tang, L., Liu, H.: Leveraging social media networks for classification. Data Min. Knowl. Disc. **23**(3), 447–478 (2011)
17. Tenenbaum, J.B., De Silva, V., Langford, J.C.: A global geometric framework for nonlinear dimensionality reduction. Science **290**(5500), 2319–2323 (2000)
18. Wang, D., Cui, P., Zhu, W.: Structural deep network embedding. In: Proceedings of the 22nd ACM SIGKDD International Conference on Knowledge Discovery and Data Mining, pp. 1225–1234 (2016)
19. Yang, C., Liu, Z., Zhao, D., Sun, M., Chang, E.: Network representation learning with rich text information. In: Twenty-Fourth International Joint Conference on Artificial Intelligence, pp. 2111–2117 (2015)
20. Zhang, D., Yin, J., Zhu, X., Zhang, C.: Collective classification via discriminative matrix factorization on sparsely labeled networks. In: Proceedings of the 25th ACM International on Conference on Information and Knowledge Management, pp. 1563–1572 (2016)
21. Zhu, S., Yu, K., Chi, Y., Gong, Y.: Combining content and link for classification using matrix factorization. In: Proceedings of the 30th Annual International ACM SIGIR Conference on Research and Development in Information Retrieval, pp. 487–494 (2007)

Author Index

Q. Zhang et al. (Eds.): IJCRS 2025, LNAI 15710, pp. 479–482, 2025.
https://doi.org/10.1007/978-3-031-92741-6

The manufacturer's authorised representative in the EU is Springer
Nature Customer Service Centre GmbH, Europaplatz 3, 69115 Heidelberg,
Germany. If you have any concerns regarding our products, please
contact ProductSafety@springernature.com

Printed and bound by CPI Group (UK) Ltd, Croydon, CR0 4YY
28/04/2026
02098515-0009